HANDBOOK of COSMETIC SCIENCE and TECHNOLOGY

1st Edition

HANDBOOK of COSMETIC SCIENCE and TECHNOLOGY

1st Edition

John Knowlton, CChem, MRSC
Steven Pearce, BSc, CChem, MRSC

ELSEVIER
ADVANCED
TECHNOLOGY

ISBN 1 85617 197 3

Other books in this series include:
Hydraulic Handbook
Seals and Sealing Handbook
Handbook of Hose, Pipes, Couplings and Fittings
Handbook of Power Cylinders, Valves and Controls
Pneumatic Handbook
Pumping Manual
Pump User's Handbook
Submersible Pumps and their Applications
Centrifugal Pumps
Handbook of Valves, Piping and Pipelines
Handbook of Fluid Flowmetering
Handbook of Noise and Vibration Control
Handbook of Mechanical Power Drives
Industrial Fasteners Handbook
Geotextiles and Geomembranes Manual

Published by

Elsevier Advanced Technology
Mayfield House, 256 Banbury Road, Oxford OX2 7DH, UK
Tel 010 44 (0) 865-512242
Fax 010 44 (0) 865-310981

Printed in Great Britain by Professional Book Supplies, Abingdon, Oxon

Preface

For many years we have both felt the urgent need for a book in the Cosmetics and Toiletries Industry, which provides a comprehensive overview of all aspects of the science associated with it. These feelings have been endorsed by the many students of cosmetic science, with whom we have been associated throughout our period of extensive involvement with education in the industry.

In this FIRST EDITION of the HANDBOOK OF COSMETIC SCIENCE AND TECHNOLOGY we hope we have satisfied that need, providing a publication which touches on all aspects of the science and technology involved in the wide field of cosmetics and toiletries. Although this book is aimed principally at new students of cosmetic science, it is also written for people who work in other disciplines. Buyers, salespeople, engineers, packaging technologists and marketeers, will find this volume provides a useful introduction to the technology associated with their business. The book is also designed as a valuable teaching aid to academics and institutions involved in the teaching of cosmetic science up to degree level.

The inspiration for this work has come from our strong belief that education and training is the foundation for the future growth and prosperity of cosmetic science. If, in this book, we have managed to lay a cornerstone of that foundation, then our efforts will have been truly rewarded.

J. L. Knowlton & S. E. M. Pearce, August 1993

Acknowledgements

The authors gratefully acknowledge the significant contributions made to this book by the following people:

Mr Philip Alexander (Consultant)
Mr Keith Capper (Fragrance Oils International)
Dr Alan Collings (Product Safety Assessment Ltd)
Mr Kenneth Daykin (Consultant)
Dr Jean Ann Graham (Consultant)
Mr Bundusiri Jayasekara (Microbiology Lab)
Mr Patrick Love (The Boots Company PLC)
Mr Len McNair (Consultant)
Dr Anthony Morton (The Robert McBride Group Ltd)
Ms Debra Redbourne (CTPA Ltd)
Mr David Williams (Consultant)

Acknowledgements are also due to all lecturers who were involved with teaching on the Diploma Course in Cosmetic Science, run by the Society of Cosmetic Scientists, during the period 1984 to 1989.

The following organisations also granted permission for use of their material:

Aerosol Research & Development Ltd
Givaudan-Roure Ltd
Soap, Perfumery and Cosmetics magazine

Thanks are also due to Ms Dee Seaward and Ms Sarah Bullen for their assistance in the preparation of texts.

Finally, the preparation of this book would not have been possible, were it not for the unerring patience and understanding of Dawn and Elizabeth.

Contents

SECTION 1 – Raw Materials

Rheology of Cosmetics Systems 1
Thickeners & Gums 11
Oils, Fats and Waxes 21

SECTION 2 – Safety and Legislation

Product Safety 35
Legislation 52

SECTION 3 – Surface Chemistry

Surface Chemistry 67
Emulsions 95

SECTION 4 – Decorative Cosmetics

Decorative Cosmetics 121

SECTION 5 – Personal Care

Skin and Skin Products 167
Hair and Hair Products 201
Bath Products 233
The Teeth and Toothpaste 244

SECTION 6 – Packaging

Aerosols 263
Packaging 286

SECTION 7 – Perfumery

Perfumery 317

SECTION 8 – Manufacturing

Production 365
Quality Control and Assurance 397
Stability Testing 435
Industrial Microbiology, Hygiene and Preservation 440

SECTION 9 – Marketing

Product Evaluation 461
Consumer Research 480
Psychology 493

SECTION 10 – Glossary of Terms
Glossary of Terms 505

SECTION 11 – Reference Section
Reference Section 523

Advertisers' Buyer's Guide
Advertisers by Product Category 545
Alphabetical Index of Advertisers 563

Editorial Index 569

SECTION 1

Raw Materials

RHEOLOGY OF COSMETICS SYSTEMS
GUMS & THICKENERS
OILS, FATS AND WAXES

$$\tau = \eta D$$

where

τ = shear stress
D = shear rate

In order to derive the units for dynamic viscosity, it is convenient to consider the case of liquid flow between two parallel plates, as shown in Figure 1.

FIGURE 1

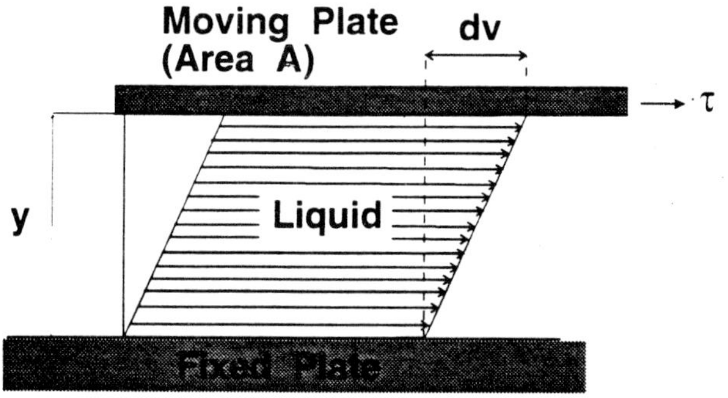

For a section of the liquid subjected to a horizontal force, F, on a plate of area, A,

$$\tau = \frac{F}{A} \qquad \frac{\text{Newtons}}{(\text{metre})^2} = \text{Nm}^{-2} \text{ (Pascals)}$$

and

$$D = \frac{dv}{dy} = \frac{\text{metre/sec}}{\text{metre}} = s^{-1}$$

Therefore the units of dynamic viscosity, η, are,

$$\eta = \frac{\tau}{D} = \text{Nm}^{-2}\text{s} \qquad \text{(Pascal seconds)}$$

Although the units of dynamic viscosity are properly quoted in Pascal seconds (Pa.s), it is more common in the field of cosmetic products to quote viscosity in centipoises (cP), where 1 centipoise (cP) is equal to 1 milliPascal second (mPa.s). The kinematic viscosity of a fluid, v, is related to the dynamic viscosity as follows.

$$v = \frac{\eta}{\sigma}$$

where

σ = density

The units of kinematic viscosity are m^2s^{-1}.

Rheological flow curves

One of the best ways of describing the rheological behaviour of liquids is to examine their rheological flow curves, normally plotted as shear stress (τ) on the y-axis against shear rate (D) on the x-axis. Alternatively, a plot of dynamic viscosity (η) against shear rate (D) may be made, although this does not result in the classical types of flow curve normally used when discussing rheological behaviour. There are four main types of rheological behaviour important to the cosmetic chemist, Newtonian, dilatent, pseudoplastic and pseudoplastic with yield point (sometimes referred to as plastic behaviour). The shear stress:shear rate and dynamic viscosity:shear rate flow curves for each of these flow behaviours is shown in Figure 2.

FIGURE 2

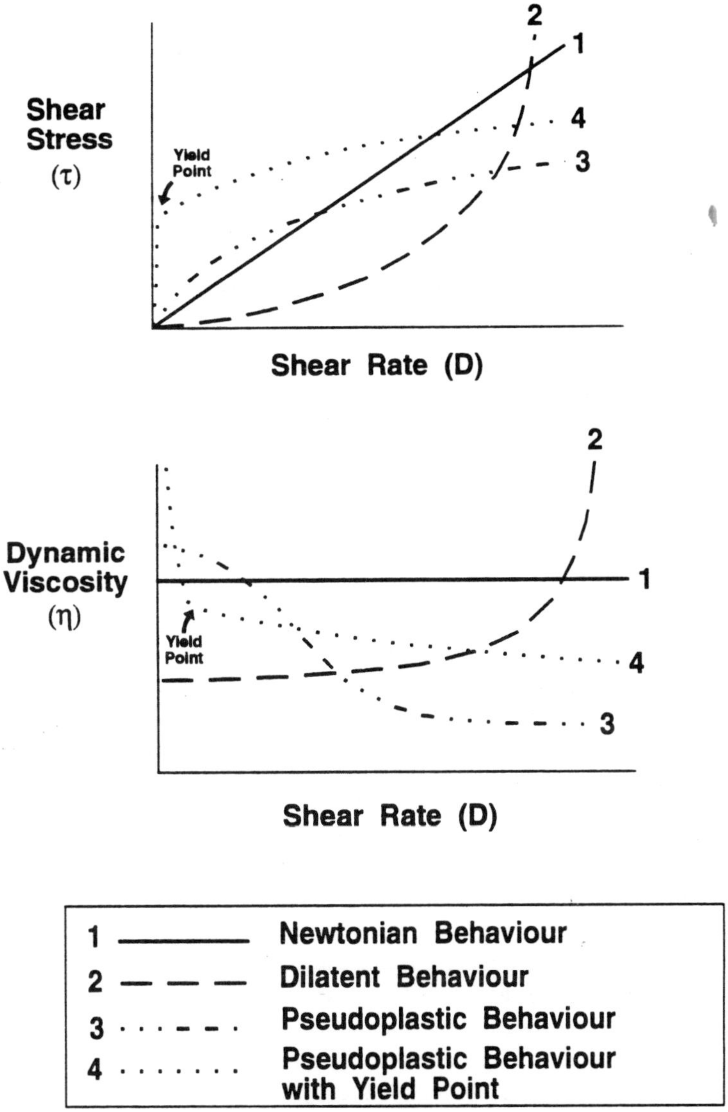

Each type of rheological flow exhibits certain characteristics, as discussed below.

2.1. *Newtonian behaviour*

Newtonian flow is only exhibited by ideal, or "Newtonian" liquids, and describes a linear increase in shear stress with increasing shear rate. It is important to note that, for this type of liquid flow, the value of the dynamic viscosity (η) is independent of shear rate. A Newtonian flow curve follows a mathematically straight-line relationship, as illustrated in Figure 3.

FIGURE 3

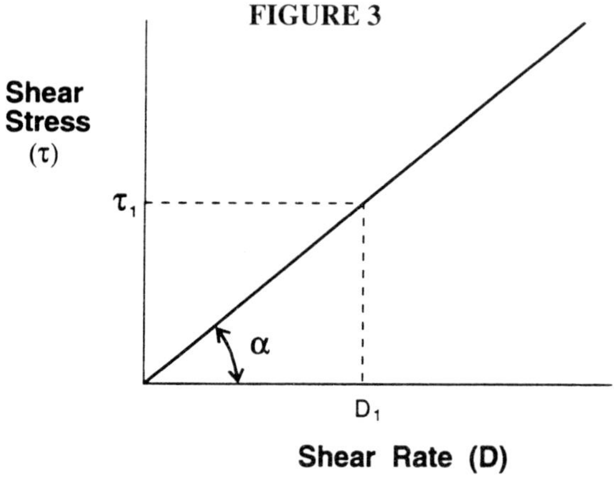

The dynamic viscosity can therefore be calculated from the following equation.

$$\text{Dynamic viscosity } (\eta) = \tan\alpha = \frac{\tau_1}{D_1}$$

Examples of liquids which exhibit Newtonian flow characteristics are water and mineral oil.

2.2. *Dilatent behaviour*

Dilatent behaviour is used to describe a system in which the shear stress increases markedly as the shear rate is increased. In terms of apparent viscosity, the liquid thickens upon the application of shear and therefore exhibits what is sometimes known as "shear-thickening" behaviour. Dilatent behaviour is infrequently seen in cosmetic products, except in the case of systems containing a high solids content, in suspension. At low shear rates the solid particles move easily around each other but, as the shear rate is increased, the particles become less mobile and may even "lock together".

2.3. *Pseudoplastic behaviour*

Pseudoplasticity is used to describe a system in which the shear stress drops as the shear rate is increased. Such behaviour is consistent with a drop in apparent viscosity with an increase in shear, often referred to as "shear-thinning". Pseudoplastic behaviour, which is exhibited by a large number of cosmetic emulsions, is associated with an increase in the orientation of the molecules or other orientable species in the system, as the shear rate is increased. Frequently, the apparent loss in viscosity is totally reversible when the shear rate is reduced, although complete recovery of viscosity may be time dependent. This latter case is known as thixotropic behaviour and is discussed in more detail later in this chapter.

2.4. *Pseudoplastic behaviour with yield point*

Pseudoplastic behaviour with a yield point, sometimes referred to as plastic behaviour, is similar to pseudoplastic behaviour, except that the system resists flow until a certain value of externally applied force is reached, after which the rheological behaviour is pseudoplastic in nature. The point at which pseudoplastic flow takes place is called the yield point, as shown in Figure 2. The existence of a yield point is due to the fact that "network" forces in the system are so great, that they confer a "solid" nature on the liquid and provide infinitely high viscosity. Only when the outside forces exceed a certain point, known as the yield value, does the network collapse and flow take place. Pseudoplastic liquids of this type often regain their viscosity very quickly after the removal of shear. Examples of cosmetic products which exhibit this type of rheology are lipstick masses and toothpastes. Also many cosmetic gels exhibit pseudoplastic behaviour with a yield point, the gel state being equivalent to a plastic state.

3. *Thixotropy*

Thixotropy itself is not a type of rheological flow but rather a feature exhibited by some pseudoplastic systems. Whilst pseudoplastic liquids lose their apparent viscosity on the application of shear, many regain it instantaneously when the shear is removed. Sometimes, however, the complete recovery of product viscosity is a function of time and, in such cases, the system is said to be thixotropic. Shear stress:shear rate and viscosity:shear rate curves for pseudoplastic systems exhibiting thixotropy are illustrated in Figure 4.

FIGURE 4

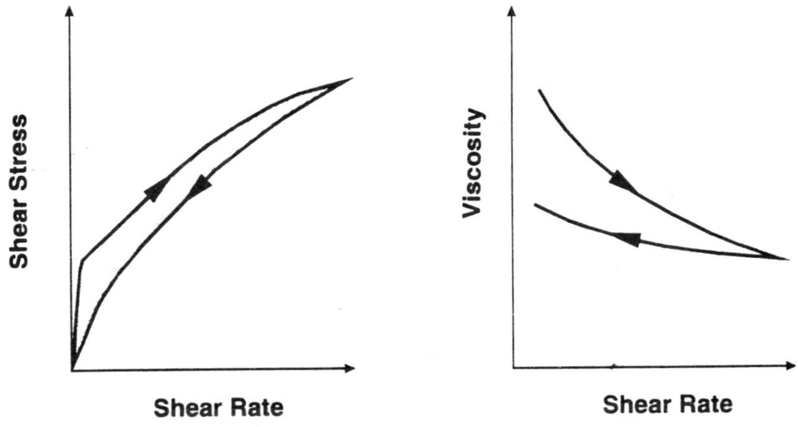

In both cases, the curves representing the application of shear, and the subsequent removal of the shear, are illustrated. The dependency of the recovery of viscosity, with time, is best illustrated by examination of the viscosity:time curve, as illustrated in Figure 5.

FIGURE 5

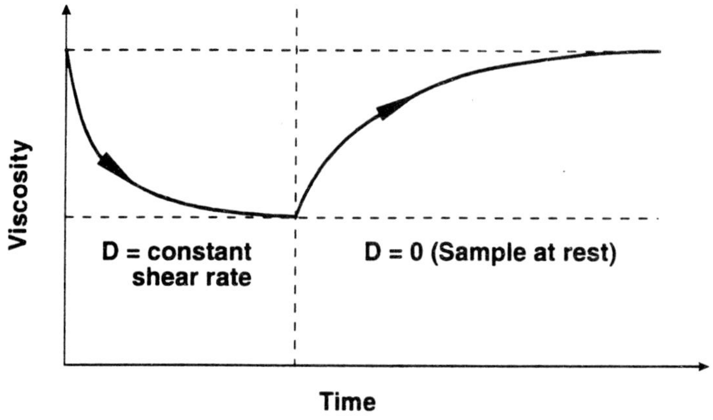

The measurement of rheological characteristics

From a practical viewpoint, the most frequently measured rheological characteristic is viscosity. The measurement of viscosity in isolation will not give any indication of rheological behaviour, as it merely involves measuring the shear stress of a system at a single predetermined shear rate. The obvious advantage of measuring viscosity is the relative ease and rapidity with which it can be done, making it an ideal choice for simple quality control applications. In order to understand the rheology of the system, many measurements of shear stress, under varying magnitudes of shear rate, must be made and the former plotted as a function of the latter. Such a plot results in the classical shear stress:shear rate curves described earlier.

1. *Measurement of viscosity*

Unless viscosity measurements are determined carefully, with due consideration given to the type of apparatus used and its relevance to the product being tested, the result obtained can be useless and, at worst, misleading. Before making any viscosity measurement the following precautions should be observed.

- the test sample should be homogeneous in all respects and should not contain any entrapped air or foreign matter
- all viscosity measurements should be carried out at constant temperature, under thermostatically controlled conditions
- the applied shear must lead only to laminar, rather than turbulent, flow in the system and the sample under test must be uniform throughout. Turbulent flow may cause large errors in the measurement of viscosity
- all viscosity readings should be taken under steady state conditions

inflexibility and the lack of a thermostatic control for the sample. The latter problem is normally overcome by subjecting the sample to external thermostatic control, using a water bath at either 20°C or 25°C, before measurements are made.

2.2. *Coaxial viscometers*

There are two main types of coaxial viscometer, the Searle viscometer and the Couette viscometer, both of which operate on a similar principal. Both viscometers are illustrated, diagrammatically, in Figure 7.

FIGURE 7

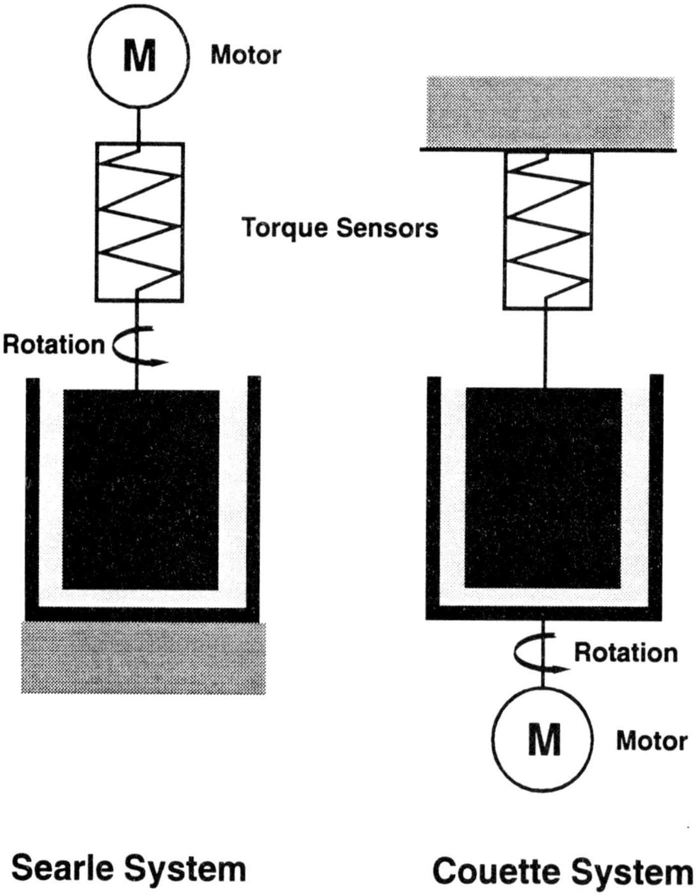

In the Searle viscometer, the sample is put into an accurately machined cup and a solid cylinder, connected to a motor and torque sensor, is lowered into the sample in a configuration which is concentric to the cup. The cylinder can then be rotated, at various speeds, in the sample, each speed representing a different shear rate. The shear stress is measured, as a function of viscous drag, by the torque sensor. By continuously accelerating the cylinder, monitoring the change in shear stress throughout, a full rheological profile of the sample under test can be determined. The Searle system is good for measuring the viscosities of both Newtonian and non-Newtonian materials although the viscosity must be in excess of 2-3 mPa.s for accurate measurements. One particular problem that may be encountered when using the Searle system is the generation of heat with very viscous liquids.

The Couette viscometer works on an identical principle, except that in this case the cup, and not the cylinder, is rotated. Once more shear stress is measured as a function of viscous drag by the torque sensor mounted to the fixed cylinder. Couette viscometry is excellent for small samples with very low viscosities and for viscosity measurements using low shear rates. This method is not suitable for the measurement of highly viscous samples or for determinations of shear stress at high shear rates.

Both the Searle and Couette systems are expensive, although they are relatively sophisticated in operation.

2.3. *Cone and plate viscometers*

In many ways, cone and plate viscometers are very similar to coaxial systems and operate using similar principles. Diagrammatic representation of two different configurations of the cone and plate viscometer are shown in Figure 8.

FIGURE 8

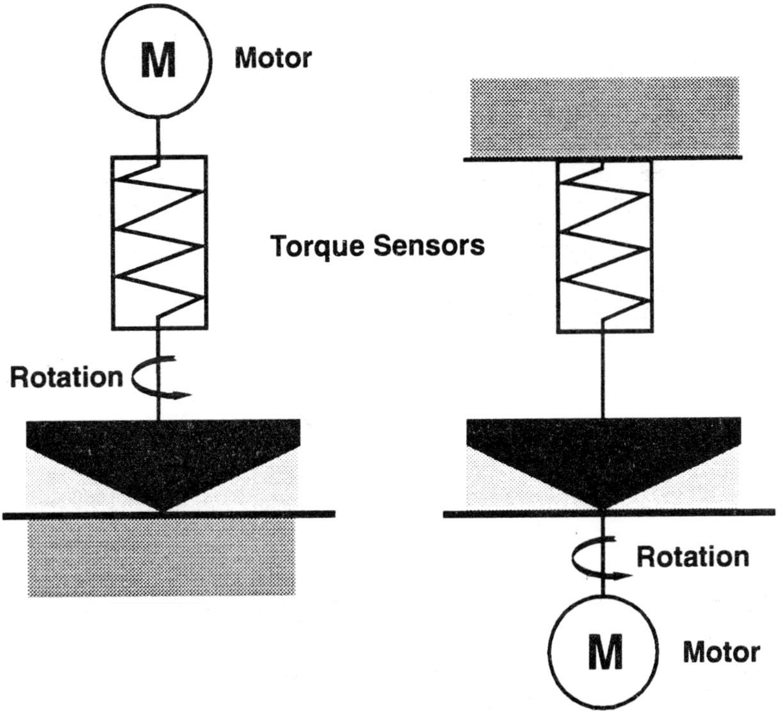

Cone and plate viscometers are both accurate and flexible although, like coaxial viscometers, they are expensive to buy. They are excellent for small sample sizes (<1 g) and where high throughput is required, because of the relative ease with which they can be cleaned. Cone and plate viscometers are particularly suitable for measurements involving high shear, such as in the rheological examination of highly viscous materials. This methodology is not suitable for materials which contain particulate matter and will accentuate any inhomogeneity in the test sample. Finally, cone and plate viscometry is more sensitive to errors caused by inherent elasticity in the sample, than the other forms of viscometry discussed earlier.

2.4. *Capillary and orifice viscometers*

Capillary viscometers operate, as their name suggests, by quantifying the time taken for a fixed quantity of the test sample to flow through a capillary orifice of known dimension. Liquid flow through the orifice may be assisted, using a piston mechanism which applies a known force to the liquid mass, or may occur simply under the force of gravity. The use of a capillary orifice prevents the acceleration of liquid flow during the measurement. Capillary viscometers are excellent for determining the absolute viscosity of low-medium viscosity Newtonian liquids.

Orifice viscometers operate under a similar principle, in this case the liquid flowing through a small orifice of known dimensions. Whilst this type of viscometer is very cheap to purchase and simple to operate, results obtained bear little relevance to absolute viscosity measurements, as the liquid is continuously accelerating during the measurement process.

2.5. *Falling ball viscometers*

This type of viscometer is of very limited use but provides accurate measurements for transparent liquids exhibiting Newtonian behaviour. Viscosity measurements are determined by measuring the time required for a ball of known mass to fall, under gravity, through a column of the liquid under test.

THICKENERS AND GUMS

Introduction

The primary purpose of a cosmetic thickener is to enhance the feel or appearance of a product by increasing its viscosity. There are many different types of cosmetic thickener available and choice will depend upon the type of product to be thickened, be it a detergent, emulsion or suspension and the type of rheology required in the finished product.

Thickeners are also added to cosmetic products for a variety of reasons, in addition to the primary function described above. These include:

- to produce or promote the viscosity of a product, so enhancing its sensory properties
- to modify the rheology of a product, thus producing the desired flow characteristics when shear is applied
- to increase the stability of inorganic/organic pearling agents, principally in detergent-based systems
- to modify the foaming characteristics of detergent-based systems
- to increase the stability of cosmetic emulsions by increasing the viscosity of the continuous phase, thereby reducing the likelihood of phase separation
- to increase the stability of cosmetic suspensions by increasing the viscosity of the continuous medium, thereby reducing the chance of settling

Selection of a thickener for any cosmetic product must be made carefully and a knowledge of the product/thickener compatibility is essential. Raw materials which may act as effective thickeners in certain cosmetic systems can produce severe thinning and instability in others. An exhaustive review of all cosmetic thickeners currently available is beyond the scope of this chapter but some of the more commonly used classes are described below.

Electrolyte thickeners

Electrolytes, principally chloride salts of sodium, potassium and magnesium, are not external thickeners in their own right, but can be added to detergent systems to produce an overall thickening effect. The increase in viscosity observed when electrolytes are added to detergent products is due to modification of the micelle structure in the system, both in terms of micelle shape and size and the distribution of ionic charge. The increase in viscosity is caused by the interaction of the modified micelles. The extent to which viscosity is increased is affected by a number of factors including the type and ratio of components in the surfactant system, electrolyte level, pH and the ionic strength of the electrolyte used. In a simple detergent system based on sodium lauryl ether (2) sulphate, the addition of low levels of an electrolyte, for example sodium chloride, will produce a marked viscosity increase, particularly in the presence of an alkanolamide detergent. As the electrolyte level is increased, viscosity will continue to rise until a point is reached where no further viscosity increase is observed. Addition of further electrolyte, even in small amounts, will produce a viscosity decrease in the system, accompanied by destabilisation. This process is illustrated graphically in Figure 1.

FIGURE 1

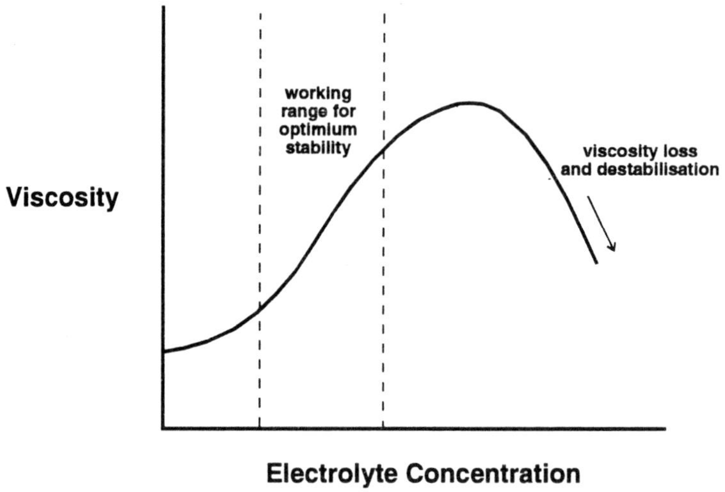

The bell-shaped curve observed is characteristic of electrolyte addition to surfactant systems. The destabilisation and thinning observed beyond the point of maximum viscosity is caused by destruction of the micelle structure by the high level of electrolyte present. This effect is often referred to as "salting out". Whilst it is difficult to specify the quantities of electrolyte than can be used to effectively thicken a detergent system, typical levels incorporated are in the order of 1%-2%. Electrolytes should be added in the form of an aqueous solution to maximise even distribution throughout the detergent system and prevent adverse effects due to localised electrolyte concentration.

Nonionic thickeners

Historically, one of the prime ways of thickening detergent systems is to include an alkanolamide. Typically both monoalkanolamides and dialkanolamides are used, the latter being a more effective thickener for any given alkyl chain length. Probably the most commonly employed alkanolamide is coconut diethanolamide, $CH_3(CH_2)_{12/14}CON(CH_2CH_2OH)_2$, although lauric and myristic derivatives are also used. Coconut diethanolamide is readily obtainable and, being a liquid, can be easily incorporated into a detergent system to give a cost-effective method of thickening. The viscosity increase observed when alkanolamides are added is due to interaction with the anionic detergent micelle. For this reason, the presence of an alkanolamide also significantly enhances the viscosity increase obtained by electrolyte addition and, in practice, a combination of alkanolamide and electrolyte is often used.

In addition to their viscosity-building characteristics, alkanolamides also enhance the foaming properties of primary surfactant systems, producing a more copious thicker, tighter lather, with a higher degree of stability. Over recent years, the alkanolamides have received some criticism, over their potential to form toxic nitrosamines when incorporated into a cosmetic product. The potential for nitrosamine formation is derived from the presence of free amines, principally diethanolamine, often found as impurities in commercially available alkanolamides. At the time of writing, legislative guidelines are being established for maximum permissible levels of free amines in alkanolamides used for cosmetic products.

Some types of surfactant system, for example those based on acyl sarcosinates, are particularly difficult to thicken and will not respond to the addition of either electrolytes or alkanolamides. These methods are also less effective in surfactant systems with low active detergent content, such as frequent-use or baby shampoos. In these cases, other types of nonionic surfactant are often used to enhance product viscosity. Although ethoxylated alcohols have been successfully used for this purpose in some surfactant systems, perhaps the most commonly used class of nonionic thickener is the polyethylene glycol diesters. Perhaps the most frequently used material in this class is PEG-6000 distearate, a high molecular weight polymer which produces a marked viscosity increase when dispersed in detergent systems. Viscosity increase is achieved principally through the dissolution of high molecular weight polymer chains and is relatively independent of the surfactant system itself. Another advantage of thickeners of this type is that they are stable to hydrolysis and produce better viscosity stability under the influence of adverse conditions of pH and temperature. Care should be exercised when using this material however, as excess levels of addition can produce undesirable sensory characteristics.

Recently more sophisticated nonionic thickeners have been developed, amongst these PEG-55 propylene glycol oleate. This material is available in both solid and pre-dispersed forms and can be incorporated into surfactant systems at low levels to give effective thickening.

Cellulosic thickeners

Cellulosic thickeners, primarily cellulose ethers, are one of the most widely used category in all types of cosmetic products. There are many different types of cellulosic thickeners available, choice depending upon the type of product to be thickened, method of addition, cost and the rheological characteristics that are required. All cellulosic thickeners are manufactured from the precursor cellulose, itself an effective thickener in water-based systems. The structure of cellulose is illustrated in Figure 2.

FIGURE 2

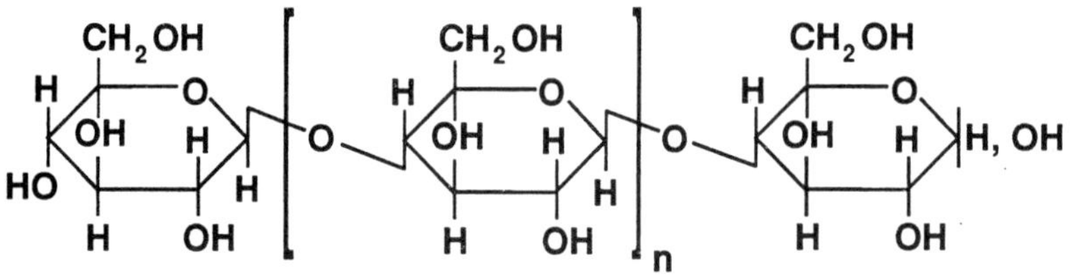

Cellulose is a natural carbohydrate containing a basic repeating structure of anhydroglucose units. Each anhydroglucose unit has three hydroxyl groups which can react in a variety of ways to produce different cellulose derivatives. It is the presence of these three hydroxyl groups that leads to the possibility of producing cellulose derivatives, principally the cellulose ethers. When forming a cellulose ether, either one, two or all three of the hydroxyl groups may be substituted to give a variety of products, exhibiting different properties. The number of hydroxyl groups substituted dictates the degree of substitution (DS). Furthermore, in the case of reaction with ethylene oxide for example, more than one mole of ethylene oxide may react with a hydroxyl group. The number of moles of ethylene oxide reacting with the anhydroglucose hydroxyl groups gives rise to the second means of characterising cellulose ethers, moles of substitution (MS).

When using cellulosic thickeners, one of the most important considerations is their incorporation into the cosmetic product. The most commonly used materials, the cellulosic ethers, are hydrophilic in nature and straightforward addition of the powdered material into an aqueous system will cause lumping and incomplete hydration. In order to obtain effective dispersion of these materials it is essential to wet-out the individual powder particles, before allowing them to hydrate. This can normally be achieved in one of two ways. The first method is to pre-disperse the powder in a non-aqueous material to achieve wetting. This pre-dispersion can then be added to the water phase, whereupon dissolution and thickening can take place. The second method for dispersing cellulosic ethers relies upon their inverse solubility/temperature characteristics. The powder particles are dispersed in hot water, allowing the particle surfaces to wet-out and the dispersion subsequently cooled to allow dissolution and thickening to occur. Some of the more modern grades of cellulose ethers are available in surface-treated powdered form. The surface treatment, typically glyoxal, acts as a solubility retarding agent and allows time for the powder to wet-out fully, before dissolution occurs. These materials, normally referred to as cold water dispersible grades, are much easier to handle in a manufacturing environment and should be used wherever possible.

The range of application for the cellulose ethers is enormous, although their hydrophilic nature means that their use is largely restricted to aqueous systems. Generally, cellulose ethers are effective in the thickening and stabilisation of emulsions, forming protective colloids, the thickening and stabilisation of powder/liquid suspensions such as toothpastes and the thickening of liquid detergent products such as shampoos and bath foams.

One disadvantage of cellulosic thickeners is that they are particularly prone to microbial attack, as they provide an ideal substrate for microbial growth. In the case of microbial attack product spoilage will rapidly occur, accompanied by complete loss of viscosity due to the enzymatic breakdown of the cellulosic gum structure. In addition to this, long term storage of products thickened with cellulosic ethers, particularly at higher temperatures, sometimes results in partial hydrolysis of the polymer accompanied by a slight decrease in product viscosity.

1. *Cellulose gum*

Purified cellulose can be incorporated as a cosmetic thickener, although its use is somewhat less common nowadays due to the variety of other cellulosic thickeners available. Cellulose gum forms structured solutions in aqueous products which are often thixotropic in nature. Its rheological characteristics makes it suitable as a suspension aid in products such as anti-dandruff shampoos. Cellulose gum can also be used to stabilise cosmetic emulsions and its high water-binding capacity helps reduce the occurrence of synerisis.

2. *Hydroxyethylcellulose*

Hydroxyethylcellulose is formed by the reaction of cellulose with ethylene oxide, in the presence of sodium hydroxide. The ethylene oxide reacts with one or more of the hydroxyl groups on the anhydroglucose units to produce the ether derivative. The higher the degree of substitution, the more soluble the product is in water. Hydroxyethylcellulose is also readily available in surface-treated grades which are dispersible in cold water.

Hydroxyethylcellulose is a highly efficient thickener and emulsion stabiliser, the solutions of which exhibit pseudoplastic, or shear-thinning, behaviour. It is compatible with a wide variety of cosmetic materials including anionic and cationic surfactants, electrolytes and nonionic materials. In detergent systems it provides a slight enhancement of lather, providing a creamier, more lubricious foam. Hydroxyethylcellulose is stable over a wide range of pH values and is commercially available in a variety of molecular weights, providing viscosity enhancements from approximately 50cps to in excess of 10000cps, for a 1% aqueous solution.

Hydroxyethylcellulose is commonly used as a thickener in hair conditioners, clear hair rinses, surfactant-based cleansing products, shaving products and decorative cosmetics.

3. *Hydroxypropylcellulose*

Hydroxypropylcellulose is not as hydrophilic as hydroxyethylcellulose and is less commonly used in cosmetic detergent-based products. In common with other cellulose ethers it is insoluble in hot water and soluble in cold water and the method of use is similar to that for hydroxyethylcellulose described above. Hydroxypropylcellulose has excellent organosolubility and for this reason can be used to formulate clear, viscous, alcohol-based gels. It also exhibits good surface activity and film-forming capability, making it particularly suitable for use in styling gels where it acts as a co-fixative and film plasticiser.

Hydroxypropylcellulose is also incorporated into shaving products where its surface activity and foam stabilising properties enable the production of a lubricious, stable foam on the face.

4. *Methylcellulose*

Methylcellulose and its derivatives, hydroxypropyl methycellulose and hydroxybutyl methylcellulose, exhibit the peculiar property of having a thermal gelation point. As such, these materials are water-soluble up to a certain temperature, specific for each type, whereupon gelation occurs as a result of the inverse solubility characteristics. This gelation phenomenon is completely reversible and cooling of the gel will result in the production of a cellulosic solution.

In use, methyl cellulose and its derivatives are normally incorporated into the water-phase, which has been heated to a temperature in excess of the gelation point. Under these conditions the methylcelluloses are insoluble and wetting-out of the powder can take place. As the water-phase is cooled, a gel will form at the gelation point for the particular material under consideration. Continued cooling below the gelation point produces a fully dispersed, homogeneous solution of the methylcellulose or methylcellulose derivative.

For methylcellulose itself, the gelation temperature is between 50-55°C, the actual point of gelation being dependent upon the concentration and molecular weight of the material. Generally, the higher the concentration and the higher the molecular weight, then the lower the gelation temperature. The gel structure of methylcellulose is fairly rigid.

Methylcellulose is still used for thickening some types of cosmetic product, although the use of its derivatives, particularly hydroxypropyl methylcellulose, is more common.

5. *Hydroxypropyl Methylcellulose*

Like methylcellulose, hydroxypropyl methylcellulose exhibits a thermal gelation point, in this case between 58-90°C, depending upon molecular weight, degree of substitution and moles of substitution. Similarly, the gel structures formed when gelation occurs range in texture from semi-firm to mushy. Gelation point is also affected by factors such as concentration and the presence of electrolytes or surfactants in the aqueous phase.

Hydroxypropyl methylcellulose is widely used in many different types of cosmetic product. In lotions and creams it provides viscosity and colloid stabilisation and functions synergistically with Carbomer resins. The use of hydroxypropyl methylcellulose in emulsion products can also modify the skin feel after rub-in.

Hydroxypropyl methyl cellulose is particularly useful in the formulation of shampoos. Not only does it provide a means of building viscosity and stabilising pearl systems but its relatively high surface activity also enables it to act as a foam stabiliser. For this reason, shampoos formulated with hydroxypropyl methylcellulose have a particularly stable, creamy, lubricious lather. Stabilisation of the foam takes place as a result of a micro-gel structure being formed between the individual lather bubbles, improving drainage time and increasing foam stability. Shampoo systems formulated with hydroxypropyl methylcellulose also demonstrate good high temperature stability.

In practice, hydroxypropyl methylcellulose is used to thicken and stabilise shampoos, liquid soaps, shower gels and a wide variety of emulsion-based skin care products.

6. *Sodium carboxymethylcellulose*

Sodium carboxymethylcellulose is made by the reaction of sodium chloroacetate with cellulose under alkaline conditions, and, unlike other cellulose ethers, it is anionically charged rather than nonionic. As such, sodium carboxymethylcellulose does not exhibit the high temperature inverse solubility characteristics of other cellulose ethers and is soluble in both hot and cold water.

Perhaps the most frequent use for sodium carboxymethylcellulose is as a gelling agent in toothpaste, variation in the degree of substitution on the cellulose chain providing the required variation in gel structure. Being anionic, sodium carboxymethyl cellulose is somewhat less tolerant to electrolyte level and for this reason it is not used in the formulation of detergent-based toiletry products.

Acrylic acid based thickeners

A class of thickeners that enjoy wide use in the cosmetic and toiletries industries are those based on polymers of acrylic acid crosslinked with allyl ethers of pentaerythritol or sucrose. These materials are commonly recognised by their assigned CTFA name, Carbomers.

Carbomer resins are commonly available in a variety of molecular weights, the higher the molecular weight the more pronounced the viscosity building characteristics. All grades come in the form of light, fluffy white powders with a slightly acidic odour and protective face masks should be used when handling them in significant quantities.

Being highly acidic in nature, the Carbomers exhibit limited solubility in water, a characteristic which allows the powder to be readily wetted-out when they are dispersed. Meticulous dispersion techniques are required if the full thickening potential of the material is to be realised. In order to effect solubilisation and gel formation, the Carbomer dispersion must be neutralised with an alkali or base, commonly sodium hydroxide or triethanolamine. Neutralisation to form the acid salt results in ionisation of the polymer, accompanied by rapid swelling of the polymer network to form a gel, or mucilage as it is more correctly called. Because gel formation relies on neutralisation of the acid polymer, effective gelation and viscosity enhancement can only be achieved between pH values of 4 and 10, maximum viscosity build occurring at approximately pH 6 to pH 8, depending upon the grade.

The most commonly used grades of Carbomer in cosmetics and toiletries are Carbomer 941, Carbomer 934 and Carbomer 940, in order of increasing viscosity building potential. More recently, low residual solvent equivalents have become available under the names of Carbomer 981, Carbomer 984 and Carbomer 980 respectively.

Carbomers have a distinct advantage over the cellulose ethers, in that they are much less prone to microbial contamination, being synthetically made and acidic in manufactured form. They do however have several disadvantages which limits their application. Carbomers are incompatible with cationic surfactants and show a significant reduction in viscosity building potential in the presence of electrolytes. For this reason, their use in the stabilisation of detergent-based products is very limited.

The most frequent application for Carbomers is in the production of clear, stable, cosmetic gels or in the viscosity enhancement and stabilisation of nonionic emulsion products. Carbomer 940/980 and Carbomer 934/984 are commonly used in the stabilisation of higher viscosity cosmetic creams, where usage levels range from about 0.1% (w/w) up to 1.0% (w/w), depending upon the effect required. Carbomer 941/981 is commonly used in lower viscosity cosmetic lotions at similar addition levels.

Fumed silica

Fumed silica, more properly referred to as amorphous silicon dioxide, is a synthetically produced inorganic thickener with a wide variety of industrial thickening applications but only of limited use in toiletry preparations. Its thickening and stabilising properties arise as a result of the surface Si-OH groups which, when brought into close proximity with each other, undergo hydrogen bonding causing the formation of an agglomerated silica network. The characteristics of the network are dramatically affected by the polarity of the system in which it is formed. Generally, non-polar systems require low levels of fumed silica to provide the thickening function, whilst polar systems demand much higher addition levels to achieve the same effect.

Products thickened with fumed silica are characteristically thixotropic in nature and particularly resistant to temperature variation. Being of synthetic origin, it is also less susceptible to microbial contamination than natural organic thickeners. Fumed silica is insoluble in water and, in the field of cosmetics and toiletries, is most commonly used in dry-powder aerosol antiperspirants and toothpastes.

In dry powder antiperspirants, fumed silica is added at low levels as a suspending and stabilising agent for the antiperspirant active, normally aluminium chlorhydrate powder. Compatibility with toothpaste ingredients is generally excellent and fumed silica is frequently used, not only as a gelling agent, but also as an abrasive in clear-gel toothpaste products. Its high refractive index value, approximately 1.45, assures that clarity in this type of product is maintained.

Natural thickeners and gums

In addition to the synthetic thickeners and stabilisers discussed already there are a large number of naturally occurring thickeners and gums that still enjoy wide use in cosmetic and toiletry products. An exhaustive review of these is beyond the scope of this text but some of the more important materials are briefly reviewed below:

1. *Bentonite*

Bentonite is a naturally occurring, hydrated aluminium silicate clay, which can be dispersed in aqueous media to produce a thixotropic gel. Clays of this type have a lattice structure which, when wetted, swells to produce a gel with what is often referred to as "a house of cards" structure. These hydrated, structured clays are generally capable of ion exchange with the surrounding environment and thickening can often be increased by adding controlled amounts of electrolyte.

Generally, clays are used to impart structure to detergent and emulsion-based liquid products, without greatly increasing the apparent viscosity. Bentonite is sometimes used in cosmetic emulsions to modify the rheology of the product and act as a colloid stabiliser, thus reducing the chances of phase separation. In common with other clays, it is also highly effective for reducing the oily feel of cosmetic emulsions. Historically, perhaps the most common use for bentonite clays is as a stabiliser in antidandruff shampoo formulations based on suspensions of zinc pyrithione.

A modified bentonite, commonly referred to under the CTFA name Quaternium-18 Bentonite, is used as a thickener and suspending agent in dry powder aerosol antiperspirant products.

2. *Carrageenan*

Carrageenan is a water-extract of the gigartinaceae or solieriaceae families of red seaweed and, chemically, is composed of a mixture of the linear sulphated polysaccharides of D-galactose and 3,6-anhydro-D-galactose. Its use in cosmetic and toiletry products is not widespread, although it does act as an effective thickener and gelation agent.

Carrageenan is supplied in the form of a free-flowing powder and is readily dispersed in aqueous media. It is compatible with both anionic and nonionic surfactants, although cationic surfactants destroy the gel structure, leading to re-precipitation of the carrageenan. Being of natural origin, carrageenan is also very susceptible to microbial attack and effective preservation is essential in any finished product in which it is used.

Carrageenan can be usefully incorporated in skin creams and lotions at levels of up to 1.0%, in toothpastes at levels of up to 1.2% and in aqueous gel products at levels of up to 0.8%.

3. *Guar gum and guar gum derivatives*

Guar gum is a natural hydrocolloid and the reserve polysaccharide of the endosperms of the seed of the Guar plant, Cyamopsis Tetragonolobus. Chemically, guar gum is a galactomannan with a very specific macromolecular configuration. The macromolecular backbone forms a long chain, consisting of ß-1,4-glycosidic bound mannose units, which statistically have a α-1,6-galactose unit forming a side branch on every second mannose unit.

Like most polysaccharides, guar gum consists of two or three free hydroxyl groups in each mannose/galactose unit, which can be substituted in a similar way to those in cellulose, as described earlier. It is therefore possible to modify the macromolecules in a variety of different ways and so adjust the viscosity and rheological properties of the resultant solutions, to suit the particular requirements of a given application.

Commercially available guar gums are free-flowing powders, readily soluble in both hot and cold water. In practice, they are introduced into the aqueous phase under conditions of vigorous agitation and subsequently stirred until full hydration is achieved. The resultant gum solution has good structural stability and exhibits a thixotropic rheological profile. Perhaps the most common use of guar gum in personal care products is as a thickener/stabiliser in toothpastes.

Of more interest to the cosmetic formulator are the modified guars, in particular the quaternised guar referred to by the CTFA name of guar hydroxypropyltrimonium chloride. This material is available in a variety of grades, differing in molecular weight and cationic charge density. Guar hydroxypropyltrimonium chloride exhibits the viscosity building characteristics of guar gum and, in addition, is substantive to hair and skin. The level of substantivity is related to the charge density on the quaternised gum. Guar hydroxypropyltrimonium chloride is added to the aqueous phase with rapid stirring and allowed to fully hydrate to produce a viscous solution. Many grades of guar hydroxypropyltrimonium chloride have an alkaline pH in solution form and the hydration/gelation process can be accelerated by reducing the pH to a value around 6.0. In cases where satisfactory dispersion of the gum is difficult to achieve, pre- dispersion in a non-aqueous solvent may assist.

Substantivity to both skin and hair gives guar hydroxypropyltrimonium chloride the ability to confer conditioning properties to both skin and hair care products. Cosmetic emulsions, stabilised with low levels of this material, often demonstrate superior feel, providing additional lubricity to the skin surface. Effective conditioning shampoos can be formulated using levels of 0.5% or less, the gum being deposited on to the hair providing excellent lubricity and wet-comb characteristics.

Quaternised guar gum derivatives can also be used to good effect in hair conditioners, bath and shower products, liquid soaps and other personal care products.

4. *Veegum*

Veegum is a clay, magnesium aluminium silicate, refined from naturally occurring mineral deposits and is normally supplied in flake form. Although insoluble, veegum swells in the presence of water, forming a viscous stabilising matrix which can act as a colloid stabiliser for both emulsions and suspension-based products.

The structure of the colloidal dispersions is very similar to that of bentonite, with the "house of cards" characteristics. As with other clays, thickening takes place by penetration of water into the veegum lattice, the development of the colloidal structure being highly dependent upon the efficiency of dispersion.

Veegum can be used in a variety of applications, including skin care creams and lotions, liquid soaps, shampoos and conditioners. It is also used to stabilise and modify the rheological properties of toothpaste.

5. *Xanthan gum*

Xanthan gum is a heteropolysaccharide produced by biopolymerisation from the bacteria *Xanthomonas campestris*. Commercially supplied in powdered form, this material is soluble in cold water, producing highly viscous stable solutions with a wide range of application in cosmetics and toiletry products.

In practice, the gum is added to water using vigorous agitation and, for best results, should be added in advance of any other ingredients. Although natural in origin, xanthan gum is somewhat resistant to enzymatic degradation and is stable over a wide pH range. A particularly unusual feature of xanthan gum solutions is that they exhibit pseudoplastic, or shear-thinning, behaviour.

Xanthan gum can be used as a viscosity builder and pearl stabiliser in shampoo systems, where it is normally incorporated at levels of between 0.1% and 0.3%. It can also be used as a thickener and colloid stabiliser in emulsion products, where it is added at levels of up to 0.5% to control product rheology and consistency.

OILS, FATS AND WAXES

Introduction

Oils, fats and waxes are terms that are frequently confused with each other. The term "oil" conveys a well understood physical condition and is generally applied to the materials comprised in three classes, fatty oils of vegetable and animal origin, mineral oils derived from petroleum and essential oils. The latter find extensive use in perfumery and the development of fragrance materials. These are covered, in detail, in the chapter on perfumery.

In the case of animal or vegetable oils, if the material is liquid at normal climatic temperatures, it is commonly spoken of as an oil. If the material is in solid form, it is normally referred to as a fat. Scientifically, there is no clear distinction between such oils and fats and the two terms are often used indiscriminately. For example, coconut oil is a solid fat in the normal climatic temperatures of Northern Europe but is a liquid oil in its countries of origin.

Fatty oils chiefly consist of glycerides, compounds derived from the combination of glycerol and fatty acids. The most frequently encountered composition for a fatty oil is a triglyceride, in which one molecule of glycerol has combined with three molecules of fatty acids, with the elimination of three molecules of water. A typical triglyceride contains approximately 90% fatty acid and 10% glycerol. Since the fatty acids constitute the greater part of the triglyceride molecule, and also represent the reactive portion, their properties are largely determined by the properties of the component fatty acids. The three fatty acid groups in the triglyceride are usually different and randomly distributed, with the fatty acid chain lengths commonly ranging from C_8–C_{18}. In order to illustrate this, typical fatty acid distributions for coconut oil and tallow, which is derived from an animal source, are shown in Table 1.

Apart from the varying number of carbon atoms in the fatty chain, the fatty acids may be saturated or unsaturated. Saturated refers to the fact that all the carbon atoms in the chain are bonded to four other atoms, whereas unsaturated is used when one or more pairs of carbon atoms have a double bond between them. Generally, the melting point of the fatty oil increases with an increase in molecular weight of the fatty acid group, or a decrease in

TABLE 1

Fatty acid	*% in coconut oil*	*% in tallow*
Caprylic acid [C_8]	3 to 15	<1
Capric acid [C_{10}]	3 to 15	<1
Lauric acid [C_{12}]	41 to 56	<1
Myristic acid [C_{14}]	13 to 23	3 to 6
Palmitic acid [C_{16}]	4 to 12	25 to 37
Stearic acid [C_{18}]	1 to 5	15 to 30
Oleic acid [$C_{18=1}$]	3 to 12	30 to 50
Linoleic acid [$C_{18=2}$]	1 to 4	approx. 5

its unsaturation. Examples of fatty acids, with varying degrees of unsaturation, are shown in Figure 1.

FIGURE 1

$CH_3(CH_2)_{10}COOH$ Lauric acid	saturated fatty acid
$CH_3(CH_2)_7CH{=}CH(CH_2)_7COOH$ Oleic acid	mono-unsaturated fatty acid
$CH_3(CH_2)_4CH{=}CHCH_2CH{=}CH(CH_2)_7COOH$ Linoleic acid	di-unsaturated fatty acid
$CH_3CH_2CH{=}CHCH_2CH{=}CHCH_2CH{=}CH(CH_2)_7COOH$ Linolenic acid	tri-unsaturated fatty acid

In addition to triglycerides, fatty oils may also contain varying low levels of monoglycerides, diglycerides, free fatty acids and unsaponifiable matter. Unsaponifiable matter usually consists of sterols, fatty alcohols, hydrocarbons and colouring matters.

Oils and fats

1. *Vegetable oils and animal fats*

Vegetable oils are obtained from various nuts, kernels and seeds by pressure extraction and/or solvent extraction. Nowadays it is common to remove the bulk of the oil, approximately 90%, using hydraulic pressure extraction and the remaining 10% by solvent extraction. The solvent used is normally either light petroleum based or, in some cases, chlorinated. The former is normally preferred, as it produces a much purer oil with better colour. Common examples of vegetable-based oils are coconut oil, palm kernel oil, groundnut oil, olive oil and cotton seed oil.

Animal fats are also composed of triglycerides, with the same types of fatty acids found in vegetable oils being present, but in a different distribution pattern. The most important animal fat for the cosmetics and toiletries industry is tallow, which comes predominantly from beef but also from mutton. This material is obtained from the animal carcasses by rendering in large vats of water, heated by steam. The temperature is kept below 100°C and the fat is skimmed off from the top of the water in the vat. Although the best quality tallow is produced in this way, poorer quality tallows can be obtained in the absence of water, at higher temperatures and sometimes with the assistance of mechanical pressure.

In order to reach the standard of purity and stability required for use in cosmetics and toiletries, vegetable oils and animal fats normally require further treatment by refining, bleaching, hydrogenation and, where appropriate, the addition of suitable anti-oxidants. Refining may be chemical or physical, or a combination of the two methods in the case of poorer quality oils. The refining process is designed to remove undesirable impurities such as free fatty acids, particularly those with shorter chain lengths, and carbonyl compounds such as aldehydes and ketones. A filtration step is also included to remove any remaining insoluble matter such as proteinaceous material. The aim is to achieve better clarity, acceptable odour and to produce a material with greater oxidative stability.

Chemical refining is carried out with sodium hydroxide, which converts the free fatty acids into soap, followed by water washing to remove the soap containing the free fatty acids. Physical refining uses a technique of combined vacuum and steam distillation, which removes the fatty acids and other undesirable volatile materials. Nowadays, the physical refining process is preferred because it results in a higher grade of distilled fatty acids, the resulting oil exhibiting a superior odour and being produced with better yields.

It is very often necessary to improve the colour of the extracted oil by bleaching or decolourising. This is normally carried out by treating the crude oil with a diatomaceous earth and activated charcoal, to remove the unwanted colour. In the chemical refining process, the bleaching step is carried out after the treatment with sodium hydroxide, whereas in the case of physical refinement, the bleaching step is carried out before the steam distillation process.

The oxidative stability of the oil is improved by hydrogenation to reduce the number of double bonds present. This is carried out by reacting the oil with hydrogen, under pressure, at a temperature of approximately 150-200°C, in the presence of a nickel catalyst. The necessary degree of hydrogenation depends upon the required specification of the finished oil.

Anti-oxidants may also be added to some oils, to protect against oxidation and the development of rancidity, which can take place during storage. The anti-oxidants function by scavenging the free radicals that are formed during the oxidation process. A typical example of a commonly used anti-oxidant is butylated hydroxy toluene (BHT).

2. *Mineral oils*

Mineral oils are derived from petroleum and the two most important materials are white oil and petroleum jelly. The refining process includes distillation, solvent extraction, crystallisation, alkali neutralisation and bleaching.

White oil, more commonly referred to as light liquid paraffin, is a mixture of aliphatic straight-chained or branched-chained hydrocarbons (C_{16}–C_{22}), together with cyclic and polycyclic hydrocarbons, some of which may include unsaturated aromatic rings. Petroleum jelly is a purified mixture of semi-solid hydrocarbons. It is obtained by chilling petroleum oil, in a suitable solvent, to crystallize the solid wax impurities which are then filtered off. Evaporation of the solvent from the separated mass yields petroleum jelly. White petroleum jelly is produced by treating the crude material with a suitable bleaching process.

Waxes

Waxes are obtained from a number of different sources and they all tend to be rather complicated mixtures of components with carbon chain lengths ranging from C_{18}–C_{32}. They may be classified into animal, vegetable and mineral types.

1. *Animal waxes*

The most commonly used animal wax in cosmetics and toiletries is lanolin, which is obtained from the wool of sheep by treatment with alkaline washes. It is chiefly composed of esters of C_{18}–C_{26} alcohols and various fatty acids, plus sterols (mainly cholesterol) and terpene alcohols. The fatty acids are typically aliphatic, hydroxy acids and branched chain fatty acids. In contrast to the triglycerides there is no glycerol present and therefore the term “wax” is a correct description. Until recent years, another commonly used animal derived wax was spermaceti, obtained from the head oil of the sperm whale. It consists chiefly of cetyl palmitate with significant quantities of cetyl alcohol and other esters of fatty acids and higher alcohols. This material is less commonly encountered nowadays because of the public concern for the use animal derived products and the protection of whales as an endangered species.

2. *Insect waxes*

The most commonly used insect wax is beeswax, which is secreted by the honey-bee to make the honeycomb. It is obtained by washing the honeycomb, to removed the residual honey, and boiling it with water, when the crude wax comes to the surface where it is skimmed off. White beeswax is obtained by bleaching the crude material. Beeswax is a complicated mixture of hydrocarbons, esters formed from high molecular weight alcohols/acids and free fatty acids.

3. *Vegetable Waxes*

A typical example of a vegetable-derived wax is carnauba, which is obtained from the carnauba palm tree, indigenous to Brazil. The leaves are dried and the wax beaten off as a fine powder which may then be decolourised by bleaching. It consists of high molecular weight esters formed from primary alcohols (C_{28}, C_{30}, C_{32}) and fatty acids, plus some free acids and hydrocarbons.

4. *Mineral waxes*

Two common examples of mineral derived waxes are paraffin wax, which is a crystalline wax and petroleum wax, which is a microcrystalline material. They are both derived from petroleum by the further refining and vacuum distillation of the lubricating oil fraction.

Both of these waxes chiefly consist of paraffins with carbon chain lengths of between C_{20} and C_{30}, cycloparaffins and aromatic hydrocarbons.

Derivatives from oils, fats and waxes

Animal and vegetable oils and fats are used to produce a whole range of derivatives, or oleochemicals, which find extensive application in the formulation of cosmetics and toiletries. These include fatty acids, fatty alcohols, esters, glycerol and many different types of surfactants and emulsifiers. An exhaustive description of the many derivatives available is beyond the scope of this text but Figure 2 illustrates many of the classes of materials that can be obtained and indicates the method of preparation in each case.

FIGURE 2

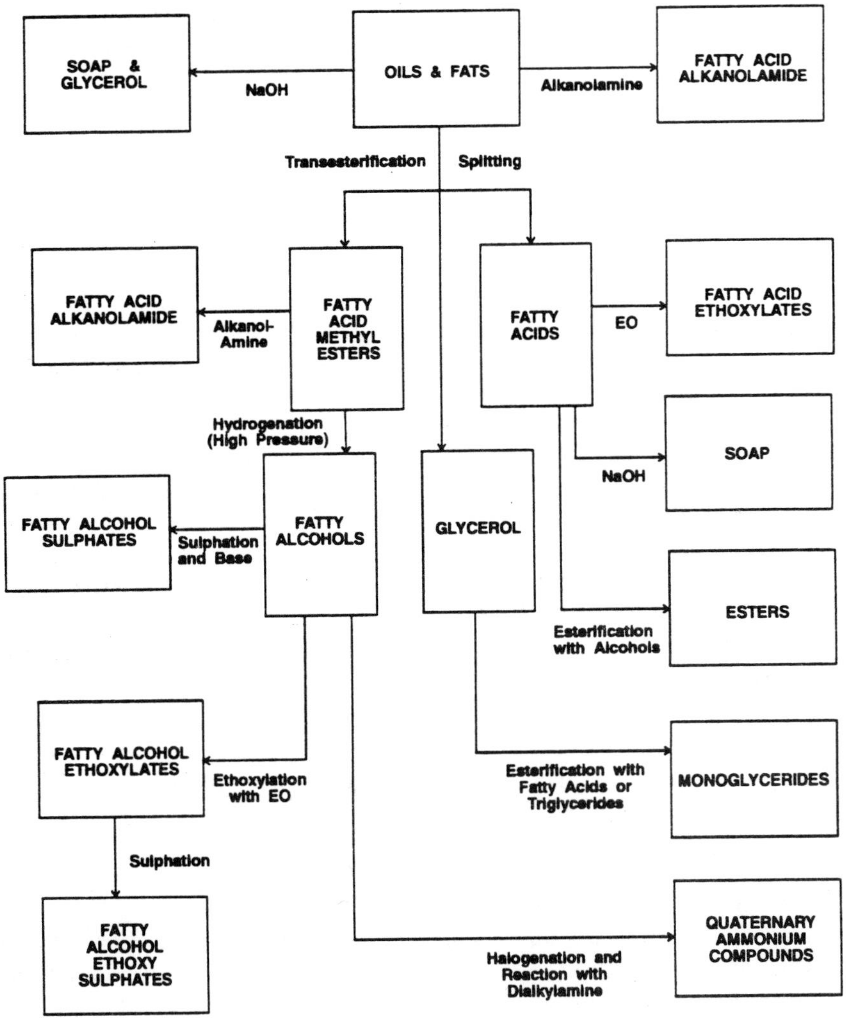

The testing of oils, fats and waxes

In view of the variety and inherent variability of the sources from which oils and fats are obtained, it is important to characterise these materials using a series of specially designed test procedures. Some of the physical and chemical test methods more frequently applied to the raw materials used in the manufacture of cosmetics and toiletries are discussed below. Many of them are used for the examination of waxes and mineral products as well as oils and fats, although the conditions of the tests may be slightly modified.

1. *Physical methods*

Before any testing of these materials can be undertaken, it is vital to recognise the importance of obtaining a representative sample. Frequently the analyst will have little control over the sample obtained from bulk storage or a delivery but there are some well-defined standardised methods that can be used. Sampling procedures are not discussed here but they are obtainable from organisations such as the British Standards Institute. Having obtained a representative sample, the following physical test methods may be applied.

1.1. *Colour and saponified colour*

Colour is obviously important as an indication of the quality of the material and a poor coloured raw material may adversely effect the colour of the finished product made from it. Colour measurements are frequently made using a Lovibond Tintometer, which compares the colour of the raw material, under a standard light source, against a series of standard coloured glasses. The coloured glasses are in sets of red, yellow and blue, each one in three series of numbers, 0.1–0.9, 1.0–9.0 and >10. Each series is additive, the colour of one glass of 3.0 units being equal to two or more glasses of the same series, for example 2.0 units and 1.0 unit, superimposed. If glasses of equal value of each series are combined and viewed against a white background, the colour transmitted appears neutral or grey. This method offers simplicity but does require a degree of experience on the part of the operator and all users must first be tested for colour blindness.

As the name implies, the saponified colour value is the colour of the material after saponification, with potassium hydroxide, under standard conditions. The resulting colour is frequently darker than the colour of the original material and is of particular importance in the manufacture of soap or any other circumstances in which the oil, fat or fatty acid may be saponified, wholly or partially, in the finished product.

1.2. *Specific gravity and refractive index*

Both of these properties are important in determining the consistency of raw materials and are normally determined using proprietary laboratory equipment.

1.3. *Melting point*

The recommended melting point test for fats and fatty acids is the slip point test, in which a open-ended capillary tube is dipped into the sample, so that a column of material about 10 mm long rises into it. The tube is then kept at about 15°C for a period of at least 16 hours, before being attached to a thermometer with a rubber band and suspended in a beaker of water. The water is then gently heated at a rate of approximately 1°C min^{-1}, until the column of sample begins to rise. This temperature is recorded as the slip point and may be used as a quality acceptance test as a guide to hardness of the material.

Other tests may be used for mineral products and waxes. For example, the melting point of paraffin wax, typically 50°C–57°C, is normally determined as a solidifying point, using the same technique as for the titre, described below. The melting point of petroleum jelly, typically 38°C–56°C, is determined as a drop point.

The melting point of these materials is affected by properties such as the degree of unsaturation and the ratios of the cis- and trans-isomers present.

1.4. *Titre*

Titre is generally determined on fatty acids, rather than triglycerides. When the melted fatty acids are allowed to solidify under carefully controlled conditions, the temperature at first drops and then, after crystallisation has begun, the temperature rises a little due to the latent heat liberated. This is illustrated, graphically, in Figure 3.

FIGURE 3

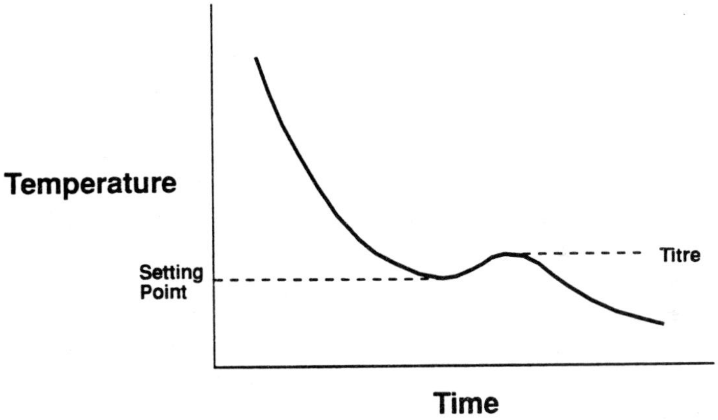

The highest temperature reached is referred to as the titre or solidifying point and is of value for characterising the fat and giving some idea of its hardness. This test is particularly important for the evaluation of fatty acids subsequently used in the manufacture of soap.

1.5. *Moisture content*

Moisture content is determined by centrifuging a melted sample of the material in a special centrifuge tube which tapers at the bottom and is calibrated in 0.1 ml graduations. The volume of moisture present in a known quantity of the melted sample can be assessed directly.

2. *Chemical methods*

There are many different types of chemical procedure that may be used to characterise oils, fats and waxes. The most frequently used of these are discussed below.

2.1. *Acid value*

Acid value is defined as the number of milligrams of potassium hydroxide required to neutralise 1 gram of the sample under test and is a measure of the total free fatty acid content of the material. Acid value will vary over a considerable range, depending on the quality of the material being tested. Nowadays, more attention is paid to the type of free fatty acids present and several different methods are available for their identification. The presence of short chain fatty acids often indicates that oxidation of the sample has already taken place.

The test procedure requires boiling the sample in neutral alcohol, or a mixture of alcohol and xylene in the case of waxes, and titrating the mixture with a standard solution of potassium hydroxide using phenolphthalein as a visible indicator of the end-point. The end-point can also be determined potentiometrically, if preferred. The acid value of the sample is then calculated from the following relationship.

$$\text{Acid Value} = \frac{(t - b) \times 56.1 \times M}{w}$$

where

t = sample titre value at the end-point

b = blank titre value (solvent with no test sample)

M = molarity of the potassium hydroxide solution

w = weight of sample used

2.2. *Saponification value and saponification equivalent*

Saponification value is defined as the number of milligrams of potassium hydroxide required to convert 1 gram of the sample into soap. The saponification value is used to characterise certain oils such as coconut oil and tallow, where typically specified values are 250–265 and 193–200, respectively. Mineral oil has a saponification value of 0. Saponification value is not sufficiently reliable to characterise some types of oil. For example, palm kernel oil has a typical saponification value of 240–255 which overlaps with the values for coconut oil. Clearly then, the saponification value of an unknown sample could not be used to determine if it was palm kernel oil, as opposed to coconut oil. When determining the saponification value, it is important that the triglyceride sample is reasonably pure for the value to have any meaning. The presence of free fatty acid impurities will increase the saponification value obtained, whilst the presence of any unsaponifiable matter will decrease it.

The saponification value of a material is determined by boiling it for about 30 minutes, under reflux, with an excess of alcoholic potassium hydroxide and then determining the excess alkali present by titration with a standard acid solution. Waxes must be boiled with a mixture of alcohol and xylene for longer periods, in order to determine their saponification value. The saponification value of the test sample is then determined from the following relationship.

$$\text{Saponification value} = \frac{(b - t) \times 56.1 \times M}{w}$$

where

t = sample titre value at the end-point

b = blank titre value (solvent with no test sample)

M = molarity of the potassium hydroxide solution

w = weight of sample used

The saponification equivalent is the number of grams of the sample saponified by 1 mole (56.1 g) of potassium hydroxide.

2.3. *Iodine value*

The iodine value is defined as the number of grams of iodine absorbed by 100 g of the sample, under standard conditions, and is a measure of the degree of unsaturation of the test material. The Wijs iodine determination is most commonly employed, in which the sample is dissolved in a mixture of acetic acid and carbon tetrachloride and a known volume of Wijs solution, iodine monochloride, added. The mixture is allowed to stand in the dark for between 1 and 2 hours, after which a potassium iodide solution is added. The liberated iodine, equivalent to the quantity of excess iodine monochloride, is titrated with standard sodium thiosulphate solution, using starch as an indicator. A blank solution determination is also carried out. The iodine value is then calculated from the following relationship.

$$\text{Iodine value} = \frac{126.9 \times M \times (b - t) \times 100}{1000 \times w}$$

where

t = sample titre value at the end-point
b = blank titre value
M = molarity of the sodium thiosulphate solution
w = weight of sample used

2.4. *Hydroxyl value*

The hydroxyl value is defined as the number of milligrams of potassium hydroxide required to neutralise the amount of acetic acid capable of combining, by acetylation, with 1 gram of the sample. It is a measure of the number of free hydroxyl groups in the sample. The hydroxyl value may be used to assist in characterising a given material but it is particularly important for the analysis of derivatives, such as ethoxylated fatty alcohols and polyglycols. It provides an indication of the degree of ethoxylation or polymerisation and whether or not the glycol has been esterified. The hydroxyl value may also be used as a guide to the amount of esterification present.

The hydroxyl value is determined by firstly acetylating the sample under test, with a measured quantity of acetic anhydride in pyridine. The excess anhydride is then hydrolysed by boiling with water to produce acetic acid, which is then titrated with an alcoholic potassium hydroxide solution. Two blanks or reference tests are carried out at the same time, one omitting the sample, to obtain a reagent blank, and the other omitting the acetic anhydride, to obtain the acid value. The hydroxyl value can then be calculated from the following relationship.

$$\text{Hydroxyl value} = \frac{56.1 \times M \times (b - t)}{w} + AV$$

where

t = sample titre value at the end-point

b = blank titre value

M = molarity of the potassium hydroxide solution

AV = acid value of the sample

w = weight of sample used

2.5. *Peroxide value*

The presence of peroxide oxygen in animal and vegetable fats is an indication that auto-oxidation has occurred. The peroxide value indicates, up to a point, the extent of oxidation that has taken place in a material and the onset of rancidity. It is expressed as milli-equivalents of peroxide oxygen per kilogram of fat. The peroxide is determined by treating the sample, dissolved in a mixture of acetic acid and chloroform, with a solution of potassium iodide and titrating the liberated iodine with a standard solution of sodium thiosulphate, using starch as indicator.

When considered in isolation, the peroxide value alone is not always a reliable indication of the extent of oxidation that may have occurred. This is because the peroxide value is only a measure of the extent of oxidation during the oxidation process itself. When the oxidation process nears completion, the peroxide value of the sample will start to drop. This is illustrated graphically in Figure 4.

The peroxide value should always therefore be considered in conjunction with other properties, such as odour and oxidative stability.

2.6. *Unsaponifiable matter*

The unsaponifiable matter present in most natural oils and fats is a measure of the material that does not react with alkali, to form soap. Typical unsaponifiable matter includes mineral oils, waxes, higher alcohols and sterols. The level of unsaponifiable matter is determined by refluxing the sample with alcoholic potassium hydroxide to saponify it and then extracting the matter which has not been saponified with diethyl ether.

2.7. *Fatty acid distribution*

The fatty acids present in any triglyceride or commercially available fatty acid can be

FIGURE 4

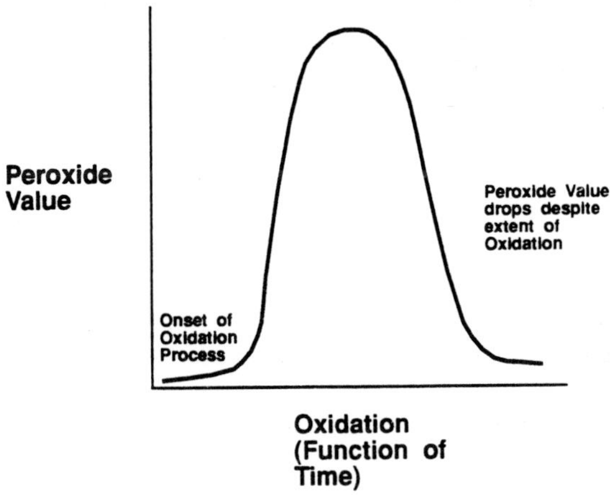

identified and quantified by means of gas-liquid chromatography. When using this technique, it is first necessary to convert the fatty acids into volatile derivatives, such as their methyl esters. The methyl esters are then injected on to the column of the chromatograph and separated from each other by passing the mixture through the stationary phase in the column, using a stream of inert gas. The fatty acids present can be determined by examination of the resultant gas-liquid chromatogram, a typical example of which is illustrated in Figure 5.

FIGURE 5

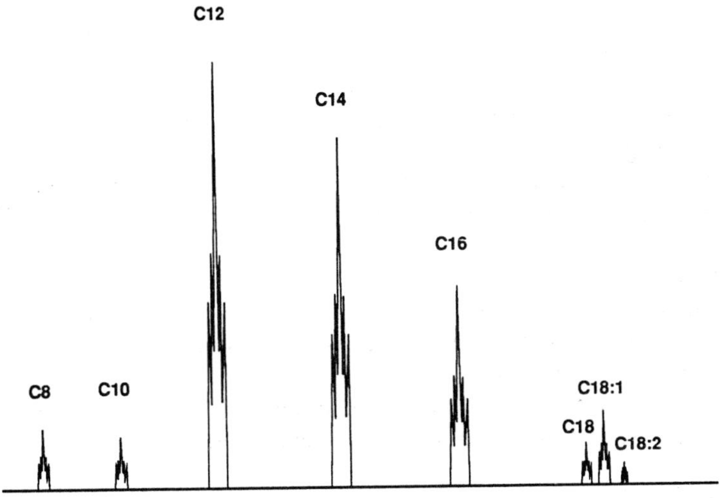

Typical Gas-Liquid Chromatogram for a Fatty Acid Mixture

The peaks representing the fatty acids are identified by injecting methyl esters of known composition and comparing retention times. Normally the percentage of each acid present is calculated by measuring the area under the peak and comparing it to the total area of all the peaks in the chromatogram. Alternatively, an internal standard may be used.

References sources

1. K. A. Williams "Oils, Fats and Fatty Foods".
2. R. G. Harry "Principles of Practice of Modern Cosmetics".
   Vol. 1 Modern Cosmeticology
   Vol. 2 Cosmetic Materials
3. British and United States Pharmacopoeia.
4. British Standards 684: 1982 "Methods of Analysis of Fats and Fatty Oils".
5. IUPAC 'Standard Methods for the Analysis of Oils, Fats and Derivatives' 6th Edition.
6 CTPA "Standards of Purity – Specifications".
7. Peter Tooley "Fats, Oils and Waxes".

COSMETICS AND TOILETRIES IS A FAST MOVING BUSINESS—ONE IN WHICH IT IS ESSENTIAL TO KEEP IN TOUCH WITH THE LATEST DEVELOPMENTS.

THAT'S WHERE COSMETICS & TOILETRIES MANUFACTURERS & SUPPLIERS COMES IN. A BI-MONTHLY MAGAZINE FOR ALL THOSE INVOLVED IN THE MANUFACTURE AND SUPPLY OF COSMETICS & TOILETRIES AND FRAGRANCES, IT KEEPS YOU UP TO DATE WITH EVERYTHING FROM THE LATEST INGREDIENTS AND PROCESS TECHNIQUES, TO PRODUCT LAUNCHES, PACKAGING, MARKETING INITIATIVES, EC LEGISLATION—IN FACT ANYTHING THAT IMPACTS ON THE RUNNING OF YOUR BUSINESS.

THE YEAR'S MOST EXCITING DEVELOPMENTS IN COSMETICS AND TOILETRIES . . .

WHETHER DIGGING DEEP INTO RAW MATERIALS OR LIFTING THE LID ON PACKAGING, IT IS DETAILED , ANALYTICAL—AND ABOVE ALL JOB-USEFUL. REGULAR FEATURES INCLUDE:

- **PRODUCT LAUNCHES AND MARKETING INITIATIVES**
- **MARKET STATISTICS AND BUSINESS TRENDS**
- **RAW MATERIAL DEVELOPMENTS**
- **INNOVATIONS IN LABELLING AND PACKAGING**
- **DEVELOPMENTS IN PROCESS TECHNOLOGY**
- **ENVIRONMENTAL MATTERS AND LEGISLATIVE CHANGES**
- **PROFILES OF INDUSTRY PERSONALITIES**

ALL IN READABLE STYLE ACCESSIBLE TO TECHNICAL AND NON-TECHNICAL ALIKE.

WHATEVER YOUR ROLE IN COSMETICS AND TOILETRIES—AS A CHEMIST, A BUSINESS MANAGER, MARKETING DIRECTOR OR PRODUCTION CONTROLLER, YOU'LL FIND CTMS ESSENTIAL READING.

SO MAKE SURE YOU DON'T MISS OUT ON THE YEAR'S MOST EXCITING DEVELOPMENTS. USE THE FORM TO APPLY FOR YOUR COPY TODAY.

I would like to take out a subscription to CTMS magazine for one year (6 copies): £42.00 (UK)/$110.00 (Overseas).

☐ Please invoice me **Please tick as applicable*

☐ I enclose a sterling cheque/international money order payable to Morgan-Grampian plc

☐ Please debit my credit card ☐ AMEX ☐ Master Card ☐ VISA

Number.. Expiry date ....

Name..

Company..

Position...

Address...

.... Tel.......................................

Signature.. Date...................................

Return to: CTMS, 30 Calderwood Street, Woolwich, London SE18 6QH, UK

CTMS/8081

SECTION 2

Safety and Legislation

PRODUCT SAFETY
LEGISLATION

PRODUCT SAFETY

Introduction

It is essential when marketing a cosmetic product, that the formulation sold does not harm the user under conditions of use or foreseeable misuse. The safety record of the cosmetics and toiletries industry is good, but this is not by chance. It is the result of careful evaluation of the safety (in use) of selected ingredients, together with the art of skilled formulation that ensures safety. In the context of this statement it is important to remember that safety does not just happen, it has to be made to happen.

The climate for assessing product safety has radically changed in recent years. In the past, development scientists and toxicologists were free to assess the safety of fully formulated products by using experimental animals as models of possible human response. Such approaches are now strongly discouraged. The 6th amendment of the EEC Directive on Cosmetics states that, after 1 January 1998 the use of animals to test ingredients or combinations of ingredients shall not be permitted, provided alternative validated in-vitro tests are available. Despite this regulation, the Directive still demands that cosmetic products must not cause damage to human health when applied under normal or foreseeable conditions of use, taking into account, in particular, the product's presentation, its labelling and any instructions for its use.

With these facts in mind, this chapter offers the reader a new approach to the safety testing of cosmetics and toiletries products, where product safety is fully integrated with development, ensuring that safety is in-built from the start. Unsafe raw materials should be avoided and full use should be made of safety test data currently available.

The product safety evaluation programme is a staged process and involves the following steps:

- Obtain awareness of legislative requirements.
- Review the product development brief.
- Evaluate the safety of individual ingredients.
- Determine the effects of combining ingredients into a formulation.
- Assess the intrinsic toxic properties of the formulation.

- Assess the consequences of consumer exposure to the product.
- Assess the risks to consumers using the product.
- Assess the risks to those manufacturing the product.
- Design the use of experimental studies to assess hazard.
- Monitor and assess customer complaints for products being sold in the market place

The legislative control of ingredients

In many countries legislation has been enacted, which strictly controls ingredients used in the manufacture of cosmetics. The type of legislation and nature of control varies from one country to another.

In Japan, the industry is controlled by listing all permitted ingredients with their physical and chemical specifications, together with the specific types of cosmetic products in which they may be used. In some cases, the maximum permitted concentrations are given and certain ingredients are restricted for use in specific types of products. Only ingredients that have been given governmental approval may be used in cosmetic products.

In the EEC, a different system is used to control cosmetic ingredients. Control is effected by the EEC Cosmetics Directive, which includes a series of Annexes, as follows:

Annex I	An illustrative list of cosmetic products by category.
Annex II	Substances which must not be used in cosmetic products.
Annex III (Part 1) Fully permitted	Substances which are controlled for use in cosmetic products. The controls restrict the type of product where a substance may be used and often sets an upper concentration limit.
Annex III (Part 2)	Provisionally permitted substances which are controlled for use in cosmetic products.
Annex IV (Part 1)	Colours fully permitted for use in cosmetic products.
Annex IV (Part 2)	Colours provisionally permitted for use in cosmetic products.
Annex V	Substances which are excluded from certain products.
Annex VI	Permitted preservatives with any concentration or limits, depending upon product type.
Annex VI	Permitted sunscreen agents with any concentration or limits, depending upon product type.

The EEC Cosmetics Directive sets limits for heavy metals and specific impurities in certain ingredients, but otherwise no strict specifications are issued.

By December 1994 the EEC are committed to compile an inventory of ingredients employed in cosmetic products. This will not be an approved list but simply a listing of those chemicals that may be, or have been, used in cosmetic products.

The control system used in the United States is different again, with the publishing of the Cosmetics Ingredients Dictionary (CID) which provides information on the specifications of certain ingredients, particularly colours. The presence of a substance named in the dictionary does not automatically mean that the substance is permitted to be used in cosmetics, or that the inclusion of such substances will ensure safe products. The Food and Drug Administration (FDA) exerts strict control over some cosmetic products and can put a prohibition order on the use of ingredients which may cause risk to health. The FDA also monitors any adverse reactions to cosmetics.

Another area of control is legislation on the use of new substances. In the EEC, all new substances have to be registered under the European Inventory of Existing Chemical Substances (EINECS). This requires the provision of data on the physical/chemical properties of the substance, together with information on its toxic and ecological properties.

It is important for development scientists to be aware of the legislative control of ingredients, thereby avoiding any frustration which may arise from the development of a highly performant product, which is subsequently found to contain an ingredient which is prohibited or restricted for use in cosmetic products.

The product development brief

Product development should not proceed until a clearly documented and specifically designed product development brief has been received. This document should include the following information:

- The generic type of the proposed product (eg shampoo, hand cream, eye shadow).
- The identification of similar competitive products already in the market place, which may be cited as a development action standard.
- The target cost of the formulation.
- Geographically, where it is proposed that the product be sold.
- The demographics of the target market.
- If the product concept or type is novel, how much and how often would the product be used.
- Any "not animal tested" claims to be made (see Appendix I).
- The colour, texture and fragrance for the product.
- The proposed weight/volume for the unit of sale.

Raw material selection

Where possible, the development scientist should use raw materials with a good history of safety. Frequently, these materials will have already been incorporated into similar types of products, at equivalent concentrations, and will not have caused any apparent adverse effects in health among those using such products.

Such raw materials can be normally obtained in a variety of grades, as determined by their specifications. This is particularly true for cosmetic ingredients. It is important therefore that any raw material used in the formulation of a new product has a fully documented and defined specification before use. Every ingredient must comply with any specifications laid down within the regulatory requirements pertaining to the end use for which it is intended that the product be sold. The presence of impurities or by-products in the raw material also needs to be considered. These may have a major influence on the compatibility of the ingredient with other substances. In some instances, such impurities can be markedly toxic, as in the case of dioxane in alcohol ether sulphates or nitrosamines in secondary amines. Some grades of talc may contain small amounts of free silica or asbestos. It is much more common however to find that most impurities are inert substances, such as inorganic salts.

Some raw materials, particularly some types of detergent, can be supplied as high active pastes or as dilute solutions. The paste may have a sufficiently high concentration so as not to require preservation, but the dilute solution may need an added preservative. Other materials may require the addition of substances to ensure that their physical or chemical stability is not compromised and stabilising substances such as antioxidants may be incorporated.

Obtaining specifications for natural raw materials can be especially difficult, particularly in view of the fact that the same trivial plant name can relate to several different plant species. For example, "cedar wood" may be derived from at least 8 different sorts of trees belonging to several species. With some natural ingredients currently being offered for use in cosmetics, it is impossible to identify the actual plant species from which it came, since the product may be derived from an ill-defined and potentially highly variable mixture (eg meadow flowers, spring flowers, orchids).

Different parts of a plant may also be used; for example the root, the bark, the leaves, the flower and the fruit, leading to further variability in raw material specification. The chemical composition of these extracts from differing parts of a plant can therefore vary enormously thus exhibiting markedly different toxicological properties. For example, the beans of the castor oil plant contain ricin, a highly toxic chemical, whereas castor oil itself is free from this contaminant. The time of harvest can also influence the chemical composition of any extract and, in ragweed, the concentration of allergen is enhanced 10 fold at plant maturity, when it flowers. Simply drying the plant can alter its chemical composition and, potentially, its toxicological profile. The parts of the plant may be extracted by pressing, distillation or solvent extraction. Both distillation and extraction may be carried out in stages, each giving a different composition and quality of product. Where an extract of a natural is being considered, it should be noted that the solvent used for the extraction process may be of questionable quality, either being potentially toxic in its own right or containing undesirable contaminants.

Plants may also be contaminated with pesticides, and the extracts thereof may have a toxicological profile which is affected by the presence of such materials. The harvested crop may be contaminated with other plant species such as lichens, and extracts may also be contaminated with metals from the vessels used in the extraction process. Many natural products require bleaching and this process can, once more, affect the final composition of the plant extract. Lastly, plant extracts can be cut back, or diluted, with synthetic forms of natural plant constituents and the toxicological profiles of such substances must also be borne in mind.

When reviewing the specifications of mineral raw materials, the full mineralogical identification should be established. This should include details on the source of the mineral and how it was mined, together with information on how variable the composition may be from within the same source. Naturally occurring minerals can be contaminated with a wide range of substances, including heavy metals, asbestos or crystalline silica. The mineral can also be contaminated during processing.

Finally, the microbiological status of the raw material should be assessed. This should comply with the guidelines issued by various industry organisations, such as the Cosmetics, Toiletries and Perfumes Association (CTPA). Many raw materials are preserved to maintain their microbiological status in storage and transfered to the customer.

In manufacture, good hygiene is essential. Should hygiene be inadequate, manufacturers may be tempted to add excessive amounts of preservatives to kill the microbes and to stop any further growth, should subsequent reinoculation occur. The raw material can then contain excessive amounts of preservative, which may affect its toxicological profile, therefore compromising the safety of the formulated product in which that material is subsequently used.

Where raw material specifications are being obtained, for the purposes of assessing their toxicological profile, the following information is required.

Plant-Derived Raw Materials

- The botanical name of the plant species from which it is derived.
- At what period during the plant's growth cycle it was harvested.
- Which part of the plant was extracted.
- The method used for extraction.
- The identity of any preservatives, additives or contaminants present.
- The main chemical composition of the extract.
- The physical properties of the extract.
- The microbiological quality of the extract.

Mineral-Derived Raw Materials

- The mineralogical name.
- The source from which the mineral was obtained.
- The particle size distribution of the mineral particles.
- The microbiological quality of the mineral.

Synthetic Raw Materials

- The main chemical composition.
- Details of any impurities present.
- The physical properties of the material.
- Details of any preservatives, additives or diluents present.
- The microbiological quality of the material.

Each raw material, irrespective of type or source, must also comply with the appropriate legislation.

When obtaining information of this type, it is important to note that sufficient information is needed to enable the design of quality controls, which ensure that all subsequent batches of the raw material are the same as that initially evaluated.

Collection of raw material toxicological data

Information is needed on the intrinsic toxicological properties of raw materials, as supplied, for setting safe handling procedures in the workplace and assisting in the safety assessment of the final formulation.

For the reasons outlined earlier, raw materials of natural origin need careful assessment. Plants contain a variety of complex chemical substances, some inert, but many with a wide range of interesting biological activities. Some can be used in beneficial ways whilst others may, under certain conditions, cause serious injury and even death. Of course, natural products have been used for many thousands of years and even the ancient civilisations were not completely unaware of the possible serious physiological side-effects. Even so, the safety margins which are in operation today are much higher for these types of material than they have ever been.

Contact dermatitis reports give many examples where a natural substance has been responsible for an adverse skin reaction. Classic, and well known, reactions are to substances such as poison ivy or stinging nettles. There are many examples of people showing a moderate to severe skin reaction to simple plants such as daisies, tulips, primrose, chrysanthemums and certain woods like olive and cedar. Even vegetables can produce adverse skin reactions in a small number of people.

When collecting toxicological data on natural raw materials, it is important to note that much of the original information on the use of these substances has been lost - particularly data relating to dose response. Whilst a substance may be reported as being good for a particular skin condition for example, vital information on the toxicology and dose response of that material must also be obtained if the product safety is to be accurately assessed.

When assessing the toxicological profile of a mineral-derived raw material, it is important to take into consideration the particle size distribution of the substance, as this can have a large influence on the toxic properties, especially if inhaled in the manufacturing environment.

In seeking information on the toxicological properties of any raw material to be used in a cosmetic formulation, the toxicologist should draw upon the following sources of reference:

- Suppliers literature
- Dermatological books and journals
- Recently published scientific literature

The review of raw material toxicology is best accomplished by the use of a comprehensive literature search. This is only possible if the raw material can be described in an unequivocal way, such as chemical name(s). Perhaps the most specific way of identifying a raw material is through the use of it's Chemical Abstracts Service (CAS) number. Information should be collected on the raw material itself and any data that exists on adverse effects following incorporation of the raw material into formulations. The collection and assessment of data should focus very clearly on information that has been properly obtained through sound and well-established scientifically validated techniques.

The result of the toxicological search should provide the toxicologist with the potential toxic properties of the raw material, along with information on the dose response and pharmacological characteristics. Information should be available about the conditions under which the raw materials will cause eye or skin irritation, sensitisation and/or toxicity, following ingestion, skin absorption or inhalation. Particular attention should be paid to raw materials with reported allergenic properties. Whilst this does not automatically exclude the use of a raw material it is likely that it may only be used in "rinse-off" products, at concentrations unlikely to cause sensitisation.

In conducting the search for raw material toxicological data, company records should be reviewed for any reported adverse reactions to products which contain the same raw material in a similar product type. This review may help to identify the maximum "safe" concentration for certain raw materials in specific product types and identify any suspect ingredients.

It is the combination of the intrinsic toxic properties of an ingredient and the concentration at which it is incorporated, along with the overall composition of the formulation and its mode of use, that will be used to determine whether a particular raw material may be suspect, from a product safety point of view, for the use intended.

Certain synthetic and natural substances, when used at specific concentrations, have the potential to cause a high incidence of adverse reactions among people exposed to them and these substances must be avoided at all costs. Perhaps one of the most obvious categories of raw material, about which such concerns abound, are preservatives. Correct use of preservatives ensures that cosmetic products will not become contaminated with micro-organisms, leading to spoilage or compromise of product safety, under conditions of normal use. In view of the fact that many preservatives are highly toxic or allergenic above certain in-use concentrations, it is essential that the recommended safe addition levels are not exceeded, if significant and sometimes serious product safety problems are to be avoided.

In a review of the biological and toxicological properties of raw materials, both benefits and potential hazards associated with their use will be found. These should be documented alongside the dose and method of application necessary to produce the reported effects. Such information will serve as a guide for development scientists who may wish to exploit the benefits of incorporating certain raw materials into a formulation, while minimising the potential hazards.

Assessing the toxic properties of a formulation

When a new product is being formulated, it is rare that all raw materials used will be novel and the majority of new developments are related, in some way, to existing technology held within an organisation. Development scientists should be guided towards the use of raw materials with a good history of safety, and those materials should be incorporated at concentrations which have been shown not to cause any problems in previously formulated products of a similar type. Where a new raw material is being incorporated into a formulation, care needs to be taken to ensure that the concentrations proposed do not markedly increase the potential irritancy or toxicity.

When assessing the toxic properties of a new product, the aim is to document the intrinsic ability of the formulation to cause injury, and identify those conditions under which such injury is likely to be caused. In carrying out this intrinsic toxicological assessment, it is necessary to take many factors into account. The toxic properties of the product must be assessed in the form in which the end-users will come into contact with it.

The chemical species present in a formulation are related to the raw materials used in its manufacture, although it is very unlikely that they will be identical. This is because there will be many different interactions between the raw materials present in a product. For this reason, it is very important not only to assess the toxicological profiles of all raw materials used in the formulation, but also of the interaction products present.

The effects of interaction on product safety can be clearly demonstrated by a simple acid-base interaction, where a base and an acid may be added separately to the formulation, where they react to form the salt insitu. This particular interaction often provides an advantage in the context of product safety. This is because the free unionised acid or base is usually more biologically available that the ionic form and therefore tends to be more lipid soluble and irritant to the skin and eyes. In view of its higher lipid solubility, it will also exhibit a more pronounced ability to be absorbed across the skin's barrier function.

Particular attention is required where a formulation contains a weak acid or base. If the pKa value is close to the pH of the formulation, then a small change in the pH of the product either during manufacture or subsequent storage, can radically effect its safety profile.

This similar observation can be made for formulations containing triethanolamine, which is widely used in creams and lotions. Triethanolamine has a pKa value of 7.76 and the effect of formulation pH on the degree of ionisation is illustrated in Table 1.

TABLE 1

pH	*% Ionised*	*% Unionised*
8.6	13	87
8.4	19	81
8.2	27	73
8.0	37	63
7.8	48	52
7.76 (pKa)	50	50
7.6	59	41
7.4	70	30
7.2	78	22
7.0	85	15

The Effect of pH on the Degree of Ionisation of Triethanolamine

It can be seen that just a small rise in pH can dramatically increase the proportion of free triethanolamine base in the formulation. As this base is particularly irritant to the eyes, a small rise in pH, around the pKa value, will therefore increase the irritancy profile of any product in which triethanolamine is incorporated. This could have serious consequences if the product is designed for use on the face, resulting in the possibility of eye irritation among users.

Soaps are also particularly sensitive to the effects of pH. In the Comite European D'Agents de Surface et Intermediaires Organiques (CESIO) classification of surfactants, as required under the Dangerous Substances Directive, some potassium soaps are classified as "corrosive" and this is probably due, at least in part, to excess amounts of base present. Where soaps are incorporated into cosmetic formulations, it is important that the pH is controlled to minimise the adverse effects of excessively high pH values.

Surfactants also warrant a special mention, particularly in view of their enormous application in a wide variety of cosmetic and toiletry products. These are complex molecules with, by definition, both hydrophilic and hydrophobic character. They can interact with themselves and other surfactants. In very dilute solutions, surfactants are truly soluble but as their concentration is increased and the molecules reach their limit of solubility, clusters of molecules, more commonly known as micelles, start to form. The presence of these micelles is vital in the formulation of cosmetics and toiletries and significantly affects the irritancy profile of the finished product.

When considering potential skin irritancy, it is the level of free surfactant in the formulation that has the major influence. As the concentration of surfacant increases, the irritancy potential increases, until the point is reached when micelles are formed, the so-called critical micelle concentration (CMC). Beyond the CMC, a plot of irritancy against surfactant concentration changes shows little increase in irritancy, despite marked increase in concentration .

Mixtures of different surfactants interact and form miscelles at much lower concentrations than single surfactant systems. As a result, it is possible to reduce the potential skin irritancy of a formulation by intentionally forming mixed miscelles, thus reducing the concentration of free surfactant in solution. Such micelles are often larger in size and, in many cases, this contributes to a lower irritancy profile.

It is also possible for one component of a formulated product to enhance the bioavailability, and therefore the potential toxicity, of another component in the formulation. A classic example of this is the enhanced skin absorption of materials, in the presence of a solvent incorporated in a formulation. Often the solvent will carry other substances through the skin, leading to enhanced skin penetration and irritancy.

After reviewing the intrinsic toxic properties of the raw materials present in the formulation, and any interactions thereof, an assessment can be made of the toxicological profile of the product, in the form in which the end-user will come into contact with it. Following this assessment, it is necessary to determine whether the product, when used as intended, is likely to cause an unacceptable incidence of adverse reactions. In real terms, it is the consumer who sets the safety standards for all consumer products and most companies set limits for the type and incidence of consumer complaint that would warrant the withdrawal of a product. This assessment of risk can be assisted by carrying out testing either *in vitro* or by the use of human volunteers in controlled studies.

Assessment of consumer exposure to the product

In addition to the collection of toxicological data, the route and extent of consumer exposure to the product should be assessed. Many questions need to be posed during this stage of product safety assessment. For example, how much of the product will be applied to the body, to which parts and how often? Is the product intended for "leave-on" or "rinse-off" use and will the product be used on adults, children or babies? If the product is volatile, or in aerosol form, will it be inhaled by the user and, if so, to what extent? All of these questions, which relate to the "dose" of the product experienced by the user, will inevitably affect any decision made about the product's safety. This information will also allow the toxicologist to further identify the type of toxicological information needed to assess the final product.

Whether a consumer will react adversely to a product will depend on a number of factors. These include the intrinsic toxic properties of the formulation, dose response, and the extent and frequency of exposure.

When a product is designed to be used in spray or powder form, an estimate should be made of the amount of product that the consumer may inhale during normal use. With particulate matter, for example aerosols or a talc dusting powder, the extent of exposure to particles smaller than 35 microns (the limit for particles likely to be drawn into the nose) and smaller than 5 microns (the limit for particles likely to enter the deep into the lungs) must be assessed. In assessing these specific particle size exposures, it should be remembered that small particles emitted from an aerosol can may lose any volatile solvent present in a very short space of time. Thus, the actual particle which is inhaled is much smaller than that which left the aerosol can at the time of spraying. Therefore, it is the inhalation toxicity of these particles, and any vapour from the volatile phase, that needs to be assessed for adverse effects on human health.

When a water-based emulsion cream is applied to the skin, the water is rapidly lost by evaporation, leaving a film of the non-volatile components in the formulation on the skin. Knowledge of the amount of product applied and the area covered provides information on the amounts of the various ingredients in the formulation that are deposited on the skin. In turn, this enables the amount of material likely to be absorbed through the skin to be estimated.

When a shampoo, hair dye or conditioner is applied to the hair, a finite amount is diluted with water, rubbed into the scalp and then rinsed off after a period of time. The concentration of product in contact with the skin, the area of contact and the contact time can therefore be estimated. The concentration and composition of material that may run into the eyes during use should also be considered.

Dependent upon the route, frequency and extent of exposure, the body is able to tolerate exposure to products with a wide range of skin irritancy properties. The greater the potential to cause skin irritancy, the shorter must be the contact time, or extent and frequency of any exposure. The irritancy potential of a shampoo is considerably greater than that of a skin cream but the shampoo is in contact with the skin for a very short time and then washed off, whereas the skin cream can be applied several times a day to the skin and left on.

The use of *in vitro* toxicological tests to assess the safety of formulated products

Many *in vitro* tests are being developed to replace the use of whole animals, in order to assess the toxic properties of substances and products. These tests have to be validated by comparing the results obtained from a number of test products, studied both in animals (*in vivo*) and *in vitro*. Whilst there may be good correlation between the results from the two types of test for certain chemical types, regulators are currently demanding that before any *in vitro* test protocol becomes acceptable from a regulatory point of view, it must be validated and proven for all chemical types and preparations. *In vitro* tests are usually designed to assess specific aspects of toxicity, such as eye irritancy or skin irritancy.

When a particular test has limited validation, (that is when there is a good correlation between the results from the *in vitro* test and animal studies for a group of closely related chemical substances) there is little point using that test if the chemical composition of the test product bears no resemblance to the chemicals used in the validation process. Where the compositions of validation chemicals and the test product are similar, such a test may be used.

The relevance of any result in a test procedure may be enhanced by the use of appropriate controls. For example, if there is an established product(s) of the same type and chemical composition as the test product on the market, these should be used as controls in any *in vitro* testing. The study can then be used to investigate whether or not the new formulation is likely to have a more severe effect on health than the existing product. If the existing product is deemed to exhibit an acceptable product safety profile in the marketplace and the test product demonstrates the same or less potential to cause injury, then it is likely that the new product will be acceptable within the specific context of the tests undertaken.

The evaluation of safety using human volunteer studies

Fully formulated products may be assessed for potential irritancy and allergy by using human volunteers and numerous test protocols are available for assessing the potential for human skin irritancy in different types of products. Such tests normally involve a minimum of 25 people, with patches upon which the test substance is placed being applied to the skin for periods up to 24 hours. Up to 6 patches can be applied to each volunteer simultaneously, so many comparisons are possible.

Not everyone exposed to a cosmetic product will respond in the same way. The normal population is a heterogeneous collection of people, with a wide range of sensitivity to substances applied to the skin. It is not uncommon within a group of 100 people, after applying the same concentration of a test product to the skin, for some to show no reaction, some to show a moderate to severe reaction and some to show a slight reaction. This variation in response can best be measured if a marketplace control product, of the same type as the development formulation, is used in any study. This control product may be one which is manufactured by the same organisation carrying out the product safety assessment, or it may be a competitive product which is designed to be used by the consumer in the same way as the one being developed. Assuming that the control product is well established, with a good record of safety in the marketplace, then by comparison a high degree of confidence in the safety of the new development. When using such a control, both test and control products can be evaluated in a multiple-patch, human volunteer, skin irritancy test, enabling the assessor to make a direct comparison, side-by-side, of the potential irritancy observed in the case of each volunteer.

All such studies, using human volunteers, must be carried out under the direction of an ethical committee, in accordance with good clinical practice. Before a human evaluation study is initiated, it is necessary to present evidence that under the conditions of the test, the volunteers are unlikely to suffer anything more than a slight skin irritation and that none of the ingredients has caused allergy in man when incorporated at levels similar to those in the development product. Usually, human tests are restricted to assessment of skin irritancy.

When designing test protocols for assessing comparative skin irritancy, it is frequently necessary to elicit a minor skin response in the volunteers. A negative response to both test and control substances is of limited value, since no comparison of potential irritancy is possible. Equally, deliberate elicitation of a severe skin response is unacceptable, both from a moral and ethical standpoint.

Where there is little knowledge on the potential irritancy profile of a formulation type, the concentration applied to the skin, and the exposure time used, should be investigated in a pilot study. In such studies, investigations always commence using a dilute solution of the test product and limited contact time in a small number (2 or 3) of volunteers. Both the product concentration and contact time can then be gradually increased, in stages, until the minimum required response is obtained.

A new development product can also be evaluated for safety in a limited user study. This involves asking a limited number of volunteers to use the new product and monitor any adverse effects on the skin, or irritant reaction in the eyes. Simultaneously, an effective performance assessment of the new product can be made. If all the indications point to the new formulation having an acceptable level of safety, a test launch can be planned. An essential part of this test launch is to monitor and assess all adverse reports concerning the safety of the product. If the test launch endorses the results obtained in the limited volunteer study, then a high degree of confidence that a full product launch would elicit no abnormal responses in the marketplace should exist.

Even after a full product launch, it is imperative that all customer complaints involving any adverse reaction, no matter how slight, are monitored. Accumulation of this data will supply the organisation with a valuable database of information that will provide a measure of consumer acceptance of the product. The number of complaints received for any recently launched product usually follows a specific pattern. A relatively high number of complaints are normally received immediately following a product launch and this number drops and stabilises to an acceptable level after a relatively short period of time. Obviously, the number of complaints can also be affected by any usage patterns for particular product types, for example, seasonal patterns in the case of sunscreens. By monitoring the number of adverse reactions to a product, early warning signs should be obtained if the incidence of customer complaints starts to exceed expected levels. Should the incidence of adverse reaction exceed the levels normally associated with the type of product concerned, then serious consideration should be given to a reformulation exercise or, in the case of a widespread adverse reaction, it may be necessary to withdraw the product from the marketplace.

The product dossier

The 6th Amendment to the EEC Cosmetics Directive requires that every manufacturer of cosmetic and toiletry products should keep a dossier on each product either manufactured within, or imported into, the EEC. This dossier should contain the following information.

- The physical and chemical specifications of the raw materials used in the product.
- Documentation regarding quality control on the raw materials in the product.
- Microbiological profile of the raw materials in the product.
- Toxicity profile of the raw materials in the product.
- Ecotoxicity of the raw materials in the product.
- The formulation of product.
- Details of the manufacturing process used to make the product.
- Quality control of the finished product.
- Microbiological challenge testing results for the finished product.
- Data on the safety of the final formulation.
- Results of human patch tests on the finished product.
- Data arising from the investigation of consumer complaints.
- Toxicological evaluation of the final formulation.
- Details of the qualified person responsible for the assessment of the safety of the finished product.

References

1. Council Directive 93/35/EEC, amending for the 6th time Directive 76/68/EEC.
2. Council Directive 92/32/EEC, amending for the seventh time Directive 67/548/EEC.
3. Calnan, C.D. – Dermatitis due to cedar wood pencils. Trans St John's Hosp. Derm. Soc. 58:43, 1972
4. Fisher, A. A. – Contact Dermatitis, 3rd Ed. Lea & Febiger, Philadelphia, 1985
5. van Ketal, W.G. – Allergic contact dermatitis from a new pesticide. Cont. Derm. 1:297,1975.
6. Mitchell, J. C. – Allergy to Frullania. Arch. Dermatol. 104:46, 1969.
7. CTPA Guidelines on the microbiological control of cosmetics.
8. Rook, A. – Plant dermatitis in general practice. Practicioner 188:627,1962
9. Mijnssen, G.A.W.V. – Pathogenesis and causative agent of "tulip finger". Br, J, Dermatol. 81:737, 1969.
10. Fregert, S., Hjorth, N., and Schutz, K.H. – Patch testing with synthetic primin in persons sensitive to Primula obconica. Arch. Dermatol. 98:144, 1968.
11. Agrup, G., Fregert, S., Hjorth, N., et at. – Routine patch testing with ether extract of Primula obcnica. Br J. Dermatol. 80:497, 1968.
12 Hausen, B.M., and Schultz, K.S. – Experimental studies on the identification of Chrysanthemum allergens. Cont. Derm. 4:244,1975.
13. Hausen, B.M., and Rothenborg, H.W. – Allergenic contact dermatitis caused by olive wood jewellery. Arch. Dermatol. 117:732, 1981
14. Amdur, M. O., Doull J., Klaasen C. D. – Casarett and Doull'Toxicology 4th Edition, Pergamon Press, 1991

15. Kern, A. B. – Contact dermatitis from cinnamon. Arch. Dermatol. 87:599, 1960
16. CESIO Classification and labelling of Surfactants 12th October 1990.
17. Directive 67/548/EEC on the approximation of the laws, regulations and administrative provisions relating to the classification, packaging and labelling of dangerous substances.

APPENDIX I

"NOT ANIMAL TESTED"

In light of the ethical and moral issues which surround the argument for and against animal testing, some marketers of cosmetics and toiletries use a "Not Animal Tested" statement as a feature supporting the products that they manufacture.

Many raw materials that were once incorporated into cosmetic and toiletry formulations are no longer used, since animal tests carried out in the past have shown them to be potentially hazardous to man. For example, there are incidences where potentially carcinogenic dyes have been withdrawn from use and materials have been severely restricted due to their potential to cause allergy. Thus, the choice of raw materials used to manufacture a cosmetic product is ultimately determined by the results obtained from the testing of that raw material, on animals, at some time in the past.

Some raw materials used in formulations may be purchased via an agent, who has the option to source a raw material from several different suppliers. One supplier may have tested the basic raw material in animals, to establish its toxic properties, so that its safety in use could be assured. If an organisation was making claims of the type "Not Tested on Animals" for the raw materials used in their products, then purchase of that raw material from that supplier would not be possible. However other suppliers may be able to supply the same raw material, using the original animal test data as an assurance of safety, although they have not actually tested the raw material on animals themselves. It is important to recognise that if an alternative source of a raw material is sought, it may have been manufactured to a different specification to that originally tested on animals and the results of the animal tests may not be directly applicable to the material from the new source.

Frequently, governmental and non-governmental agencies such as the Environmental Protection Agency (EPA), the Food and Drug Administration (FDA), the World Health Organisation (WHO) and the Food and Agriculture Organisation (FAO), initiate animal tests on chemical raw materials of general interest to the community. In addition, specific legislation in certain countries, including the EEC and the USA, demands that chemicals are tested on animals before they can be sold in the marketplace. It should be remembered that many of the raw materials used in the manufacture of finished products are complex mixtures of chemicals. The raw material, as supplied, may not have been tested on animals but some of the individual chemicals present in it may have been.

All new chemical substances manufactured within, or imported into the EEC, since the initiation of the European Inventory of Existing Commercial Substances (EINECS) in 1982, are subject to a formal notification scheme involving full documentation of the substance's physical and chemical properties. New substances registered must be accompanied by safety data obtained from a number of prescribed animal tests, to allow the classification of any hazards and risks associated with it.

These legal requirements make certain types of claim, both for the product and the raw materials used within them, very difficult to support. The level of support which can be provided for the claim depends upon the phraseology used in the claim itself.

Phrases such as "Raw materials not tested on animals" and "Raw materials not tested

Appendix 1 continued

for cosmetic purposes" are very difficult to support, as it is almost inevitable that at some time the safety of that raw material will have been assessed using animal experimentation.

Conversely, phrases such as "Only raw materials with a long history of safe use are used", or "Neither the retailer or the manufacturer of this product has initiated any animal tests on the raw materials used in its manufacture" are fairly easily supported.

The safety testing of finished cosmetics and toiletry products is an entirely different issue and, using some of the techniques reviewed in this chapter, there is no reason why any manufacturer should need to test a finished product on animals, to adequately assess its safety.

This observation is reflected in some of the typical product claims made in today's marketplace, where phrases such as "This product as sold has not been tested on animals" or "The safety of this product is assured by the careful choice of raw materials with a long history of safe use, coupled with professional product safety assessment prior to marketing" are easily supportable.

LEGISLATION

Introduction

The intention of this chapter is to give an introduction to the EEC Cosmetics Directive but reference will also be made to requirements in the USA and Japan. It must be emphasised that legislation is constantly changing and even in the time interval between writing this chapter and publication of this book at least two amendments to the EEC Cosmetics Directive are expected, one of which will radically change what is required. Readers requiring more than an introduction, or who have specific queries, should refer directly to the current legislation at the time of reading.

The structure of the EEC Cosmetics Directive

The EEC Cosmetics Directive is composed of two parts; the Articles and the Annexes. The Articles set down definitions, criteria, labelling requirements, obligations of Member States and details of how amendments can be made. The Annexes are lists of prohibited and controlled ingredients, with the exception of Annex I, which is an illustrative list by category of cosmetic products.

1. *Cosmetic Directive Articles*

Some of the requirements of the Articles are listed below:

1.1. *Article 1*

Article 1 includes the definition of a "cosmetic product", defined as *"any substance or preparation intended for placing in contact with the various external parts of the human body (epidermis, hair system, nails, lips and external genital organs) or with the teeth and the mucous membranes of the oral cavity with a view exclusively or principally to cleaning them, perfuming them or protecting them in order to keep them in good condition, change their appearance or correct body odours"*.

1.2. *Article 2*

Article 2 stipulates the important and perhaps obvious requirement, that cosmetic products are safe. The exact wording of this statement is "*Cosmetic products put on the market within the Community must not be liable to cause damage to human health when they are applied under normal conditions of use*".

1.3. *Article 3*

This Article is primarily concerned with the requirement that Member States only market products that comply with the EEC Cosmetics Directive and its amendments.

1.4. *Article 4*

Article 4 prohibits the marketing of cosmetic products that do not comply with Annexes II and Part 1 of Annexes III, IV, VI and VII (Part 1 refers to fully permitted ingredients and will be discussed with later). It also has a useful clause which allows the presence of traces of the substances listed in Annex II, provided that "*such presence is technically unavoidable in good manufacturing practice and that it conforms with Article II*".

1.5. *Article 5*

Article 5 allows the marketing of cosmetic products that contain ingredients listed in Part 2 (provisionally permitted), providing they comply with the concentration conditions laid down and are within the dates given for provisional listing.

1.6. *Article 6*

Article 6 lists the mandatory labelling requirements for the container and packaging which should be indelible, easily legible and visible. These requirements are:

1.6.1 The name and address of the registered office of the manufacturer or organisation responsible for marketing the cosmetic product. The country of origin may be required by Member States, if goods are manufactured outside the European Community

1.6.2 The nominal content at time of packaging, in either weight or volume. There may be exceptions to this requirement, for example as in the case of free samples and packages containing less than 5 g or 5 ml

1.6.3 The date of minimum durability, if less than 30 months. A product would be considered durable if it fulfils its designed function and, more importantly, is still safe at the time of use (conformance with Article 2, above)

1.6.4 The conditions of use and warning labels, such as those given in Annexes III, IV, VI and VII

1.6.5 The batch or lot number, or other means of reference for identifying the goods

1.7. *Article 7*

Article 7 does not allow member states to "*refuse, prohibit or restrict the marketing of any cosmetic product*", which complies with the Directive. It also allows Member States to require that some of the mandatory labelling is given in their own national or official language(s).

1.8. *Article 8*

Article 8 refers to *"the methods of analysis necessary for checking the composition of cosmetic products"*. At the time of writing only six Commission Directives on Methods of Analysis have been published, each of these containing more than one methodology. If an official body, for example a local trading standards officer in the UK, wanted to check that the concentration of formaldehyde does not exceed 0.05% (which would then require the warning label "contains formaldehyde"), the official method would be used. This does not necessarily mean that companies can only use official methods, if they exist, they can use their own preferred methods. However, if challenged, they would have to prove that their method gives the same results as the official method. Article 8 also mentions the development of *"criteria of microbiological and chemical purity of cosmetic products and methods for checking compliance with those criteria"*. As yet this has only been discussed, but industry realises that official criteria and methods could be introduced, possibly even in the sixth amendment which will be referred to later in this chapter.

The remainder of Article 8 and Article 8a refers to the Scientific Committee on Cosmetology and prior national approval, both of which will be discussed later in this chapter.

1.9. *Articles 9 to 15*

Articles 9 to 15 concern the Committee of Adaptation to Technical Progress, its procedures and the obligations of member states. This involves such issues as what should be done if a product represents a hazard to health and other such issues.

2. *Cosmetic Directive Annexes*

A summary of the Annexes is given below:

2.1. *Annex I*

Annex I is an *"illustrative list by category of cosmetic products"*. This does not mean that if a product is not listed it is not a cosmetic product but it would need the scrutiny of a legal practitioner to decided whether it could be considered a cosmetic under the definition in Article 1.

2.2. *Annex II*

Annex II is a *"list of substances which must not form part of the composition of cosmetic products"*. At the time of writing there are almost 400 prohibited ingredients. Many of those included in the list originated from poisons lists but now substances are added if their safety has been reviewed and they are considered unsafe for use in cosmetics. Traces of prohibited substances are allowed as per Article 4 described earlier.

2.3. *Annex III*

Annex III is a *"list of substances which cosmetic products must not contain, except subject to the restrictions and conditions laid down"*. Generally speaking, it is a list of active ingredients used in oral hygiene products, nail care products, depilatories, anti-perspirants and hair care products (including some hair dyes), which do not belong in the positive list of colours, preservatives or UV filters under Annexes IV, VI or VII. For example, selenium disulphide is permitted in anti-dandruff shampoos at a maximum concentration of 1% and must carry the warning labels "contains selenium disulphide" and "avoid contact with eyes or damaged skin". No other limitations or requirements are stipulated. In other words, the use of selenium disulphide is only acceptable in this one product type, providing it is used as described. Any other use of selenium disulphide in cosmetic products would be illegal.

2.4. *Annex IV*

Annex IV is a positive list of colours but does not include colouring agents intended to be used solely to colour the hair. It is often referred to as a positive list because if a colour is not on this list it is not allowed. Likewise, for the opposite reason, Annex II is sometimes called a negative list. Each colour is listed by its Colour Index (CI) number, which is an internationally recognised code, in numerical order. The uses of colours permitted are divided into four fields of application as follows:

1. colouring agents allowed in all cosmetic products
2. colouring agents allowed in all cosmetic products except those intended to be applied in the vicinity of the eyes, in particular eye make-up and eye make-up remover
3. colouring agents allowed exclusively in cosmetic products not intended to come into contact with mucous membranes
4. colouring agents allowed exclusively in cosmetic products intended to come into contact only briefly with the skin

The field of applications allowed depends on the safety data available for each colour. If the colour is permitted in field of application 2 (e.g. a lipstick), then it would also be permitted in fields of application 3 and 4 (e.g. a hand cream and shampoo respectively).

The lakes and salts of the colours are equally permitted, providing they are not prohibited under Annex II. For example CI 45410:1 is permitted, as are its aluminium, barium, calcium and zirconium lakes, as well as its sodium salt, CI 45410.

Some of the colours are accompanied by the letter "E" and, in these instances, they must also fulfil the purity criteria laid down in the EEC Directive of 1962 and its amendments concerning colouring matters in foodstuffs. For colours without an "E" there are, as yet, no official specifications for purity criteria.

2.5. *Annex V*

Annex V was formed when the Cosmetics Directive was originally prepared because there were a group of substances for which Member States could not agree the classification of, in the limited time available prior to adoption in 1976. The ingredients in Annex V are excluded from the Directive and Member States can regulate them as they wish. Over the years each of the ingredients in Annex V have been reviewed and either regulated elsewhere, usually in Annex III, or forbidden. At the time of writing, only strontium and its compounds remain in Annex V.

2.6. *Annex VI*

Annex VI is a positive list of preservatives. Preservatives according to the Directive are *"substances which may be added to cosmetic products for the primary purpose of inhibiting the development of micro-organisms in such products"*. Some substances, such as some alcohols and essential oils, have anti-microbial properties but they are added to a product for their primary function, such as solvent or olfactory properties, and therefore would not be listed in Annex VI.

2.7. *Annex VII*

Annex VII is a positive list of ultraviolet (UV) filters. The Directive defines UV filters as *"substances which, contained in cosmetic sunscreen products, are specifically intended to filter certain UV-rays in order to protect the skin from certain harmful effects of these rays"*. Annex VII does not apply to UV filters added to a product for the purpose of product protection, although it would be preferable to use those that are considered safe.

Annexes III, IV, VI and VII are in two parts. Part 1 lists fully permitted substances and Part 2 provisionally permitted substances. Part 2 also lists the dates up to which use of each substance is permitted.

Amendments to the EEC Cosmetic Directive

The Directive was first published in 1976 and, at the time of writing, there have been five amending Council Directives and fourteen amending Commission Directives. The flow diagram in Figure 1 illustrates the amendment procedure.

It is a working group under the chairmanship of the Commission, with Member States, COLIPA and consumers, that makes the recommendations for modifications to the Articles or Annexes. COLIPA (Comite de Liaison des Associations Europennes de L'industrie de la Parfumerie, des Produits Cosmetiques et de Toilette) is the European trade association based in Brussels. Each EEC country has a trade association for companies that either manufacture cosmetic, toiletry and perfumery products, or have a vested interest in this industry, such as raw material and packaging suppliers. Each trade association has its own subcommittees, made up of experts on specific issues, such as sunscreens, ingredients, colours and hair dyes. Representatives from the trade associations then meet at COLIPA to discuss relevant issues and ensure that their members' interests are taken into consideration. COLIPA representatives can then attend Commission meetings and express the united opinion of EEC industry.

The meetings at which amendments are discussed are those of the Ad Hoc Working Party (AHWP), which is chaired by the Commission and attended by representatives from the Member States, COLIPA and consumers. A vote is taken on proposed amendments to the Annexes at Committee of Adaptation to Technical Progress (CATP) meetings and the approved amendments are then published as a Commission Directive in the official journal. Recommendations to change the Articles (and sometimes the Annexes as well) go through the more complicated council procedure, with input from the Economic and Social Committee and the European Parliament before the final amendment is published in a Council Directive.

FIGURE 1

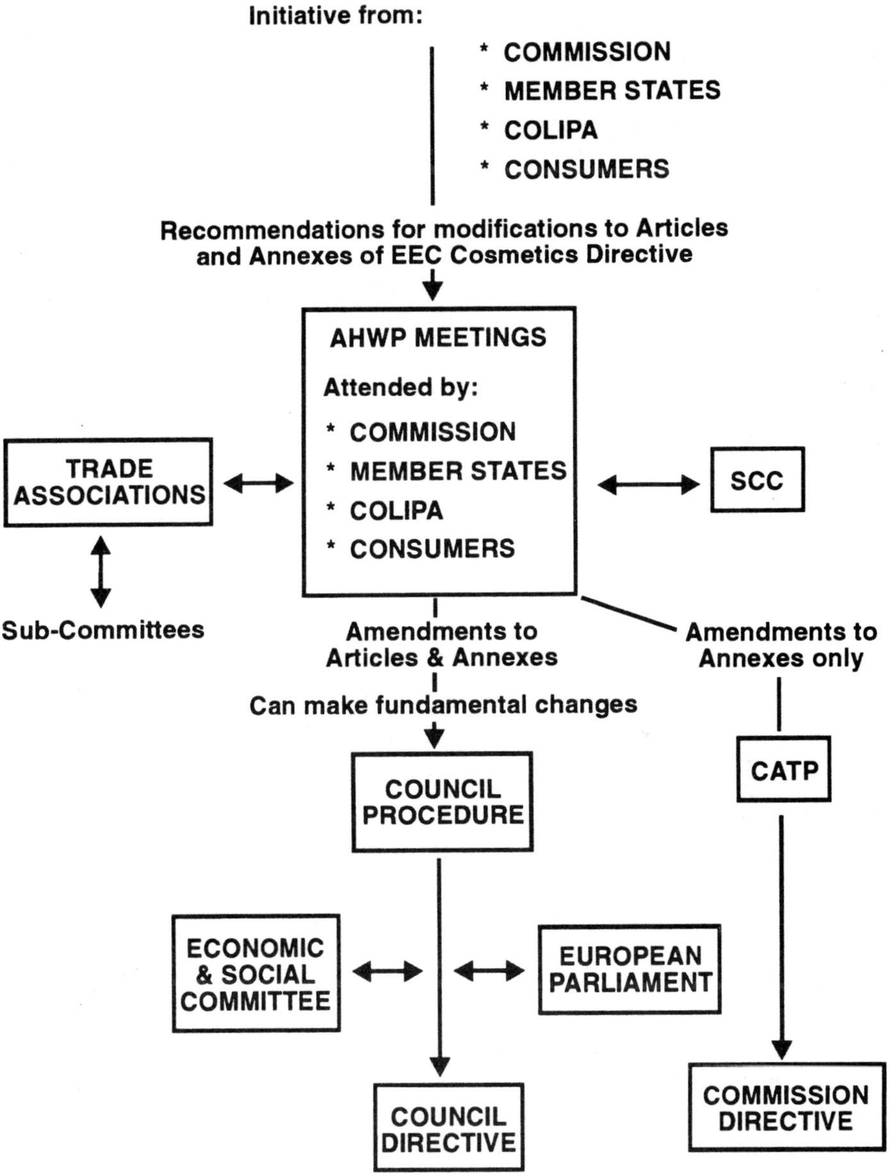

AHWP - Ad Hoc Working Party
SCC - Scientific Committee on Cosmetology
CATP - Committee for Adaptation to Technical Progress

The Scientific Committee on Cosmetology (SCC)

The Commission has an independent body of experts, such as dermatologists and toxicologists from Member States, to review the safety data in submissions for provisionally permitted ingredients and other ingredients for which safety is being questioned. This body is called the Scientific Committee on Cosmetology (SCC).

If a manufacturer wishes to market a colour, preservative or UV filter for cosmetic use, and it is not already listed in the appropriate Annex, then it would have to prepare a submission of general and safety data. The SCC has published guidelines giving details on what information should be included in these submissions. The SCC reviews any submissions sent to the Commission and issue their opinion, which is either that there is sufficient data to consider the substance safe for use in cosmetics, or that the substance appears to be safe but more data is required to confirm this. Alternatively, the SCC may believe that the substance is unsafe, in which case it would be proposed for inclusion in Annex II at an Ad Hoc Working Party meeting. A new substance is usually added to the provisional list and is allocated a date by when industry must provide further information for the SCC to review.

The Ad Hoc Working Party usually meet prior to the "permitted until" dates, to discuss, among other issues, the future of provisionally permitted substances. If data for an ingredient is still being prepared or reviewed, then Member States may allow more time in the next Commission Directive by prolonging the "permitted until" date. Data is provided until the SCC are satisfied with its safety, whereupon the ingredient is discussed at the Ad Hoc Working Party and Committee of Adaptation to Technical Progress meetings. Should the data be acceptable, the ingredient can be transferred to Part 1. If, however, industry did not provide the data requested by the SCC then when the "permitted until" date expires the substance would be removed from the positive list and the amendment would include dates by when industry must comply and stop using it.

This procedure of having a substance included on a positive list can be slow, often two to three years, and an alternative quicker route of getting a substance approved in one Member State (not the whole of the EEC) is prior national approval (PNA), which is detailed in Article 8a of the Directive.

The implementation of EEC Cosmetic Directive and its amendments into national legislation

So far, only the EEC Cosmetics Directive, which is directed to the member states, has been discussed. The Commission gives the Member States a period of time to implement the Council and Commission Directives, in their own national legislation. In the UK the Council and Commission Directives are published in Statutory Instruments titled "The Cosmetic Products (Safety) Regulations", which are available from Her Majesty's Stationery Office.

One of the main reasons that the Cosmetic Directive was originally published, was to harmonise the cosmetic legislation in each of the Member States and remove all barriers to trade, therefore making it possible to market the same product anywhere in the EEC.

Unfortunately, many Member States have interpreted the Directive quite differently and have also added extra requirements. For example, France requires the preparation of a dossier; Greece, Portugal, Spain and Italy require product notification. In France and Portugal formulation details have to be sent to poison centres and any promotive ingredients must be labelled. These are just examples that demonstrate some of the differences in the cosmetic legislation of EEC countries.

The sixth amendment

About ten years after the Directive was introduced, the Commission conducted an independent review of its successes and inefficiencies. At the same time industry, via COLIPA, also made their suggestions to the Commission on how the Directive could be improved. Many of the conclusions of these reviews are to be included in the "Sixth Amendment" (The Sixth Council Amendment to the Cosmetics Directive), which is expected to be published in 1993 and will introduce the greatest changes to EEC cosmetic legislation since the Directive was first published.

A proposal for the Sixth Amendment was published in the official journal in February 1991. A summary of the main points at the time of writing is given below but it is important to remember that, at the time of reading, some changes may have taken place.

Definition

Although a modification to the definition, which would have changed the status of some products previously considered to be borderline products, was proposed, it is now thought the definition will remain unchanged.

Inventory

The Commission wished Member States to prepare an inventory of all ingredients used in cosmetic products. This would include synthetic and natural substances but exclude perfumes and aromatic compounds.

Ingredient labelling

The Commission wants to introduce ingredient labelling to provide consumers with better information on the components of the products that they use. It is essential to have a common labelling nomenclature that is acceptable to all Member States, despite the multitude of official languages. It is envisaged that the ingredient labelling system will be based on CTFA (Cosmetics, Toiletries and Perfumes Association) nomenclature, the system that has been adopted in the USA for many years. There will, however, have to be modifications to this to accommodate the different languages and various other difficult areas such as colours, alcohol and natural ingredients.

Information for poison centres

A scheme will be introduced to ensure that poison centres have sufficient information to act promptly in the event of an accident.

Dossiers

Comprehensive documents giving qualitative and quantitative formulation details, physico-chemical and microbiological specifications of raw materials and the finished product, method of manufacture and safety data will have to be compiled for each product. Dossiers will have to be *"readily available to the competent authorities of the Member State concerned at the place of manufacture or, in the case of importation from a non-member country, at the place of initial importation into Community territory".*

Animal testing

This issue will most probably be addressed. Due to consumer and political concerns further moves to reduce the number of animals used are expected.

Other legislation effecting cosmetics

So far, only legislation specific to cosmetics has been discussed but it is essential to remember that other EEC and national legislation can effect products marketed in the EEC. Typical examples are discussed below:

EINECS

This is an acronym for European Inventory of Existing Chemical Substances and all substances used in the European market, within a set time period, should be listed in EINECS.

Council Directive of 15 January, 1981

This is the approximation of the laws of the Member States relating to the ranges of nominal quantities and nominal capacities permitted for certain pre-packaged products (80/232/EEC). It is often referred to as the Standards Ranges Directive. It is not obligatory for Member States to implement this Directive but some of the Member States do enforce it.

Packaging Directives

EEC pre-packaging Directives 75/106/EEC(O.J.no.L42,15.2.1975,p1) and 76/211/EEC (O.J.no.L46,17.2.1976,p.1) introduce a system for the quality control for pre-packed goods sold by weight and volume and is often referred to as the "average weight system".

Alcohol legislation, which differs in each Member State, aerosol legislation, health and safety legislation, legislation on consumer protection/rights and advertising restrictions are all further examples for consideration when examining the legal status of any cosmetic or toiletry product.

Cosmetic legislation in the USA

The main legislation controlling the composition and labelling of cosmetics in the USA is the Federal Food, Drug and Cosmetic Act (FDC Act) and the Fair Packaging and Labelling Act (FPLA). One of the first things to consider when exporting to the USA, is whether the product in question is classified as a "cosmetic". The definition of a cosmetic is different in the USA to the EEC and products such as sunscreens and antiperspirants are classified as drugs and must therefore comply with the drug regulations. If a cosmetic claim is made, then the product would have to comply with the cosmetic legislation as well.

With the exception of a handful of ingredients that are prohibited and a positive list of colours, the USA does not have official positive and negative lists like the EEC. The list of colours is in two parts, those that are subject to certification and those that are not. In the USA there is also an independent body called the Cosmetic Ingredient Review Expert Panel (CIR). This body gathers and reviews ingredient safety data in not too dissimilar a manner to the Scientific Committee on Cosmetology in the EEC. To date, the CIR have published nearly 200 final reports on over 300 cosmetic ingredients and industry, on a self-regulatory basis, follows their recommendations as to whether an ingredient is safe and, if so, what the maximum concentrations and product types are it can be used in.

The labelling requirements in the USA are very detailed and include the need for full ingredient labelling on the outer container. The nomenclature used by the CTFA in their Cosmetic Ingredient Dictionary is the preferred source. Finally, it should be noted that there is a voluntary reporting programme where manufacturers and importers send details of their manufacturing sites, formulations and adverse reactions to the Food and Drug Administration (FDA).

Cosmetic legislation in Japan

In Japan, cosmetics are regulated by the Pharmaceutical Affairs Law (No. 78 May 1983) and many products that are considered cosmetics in the EEC, such as hair dyes, permanent waves and bath products, are classified as quasi-drugs.

Japan has a detailed licensing requirement. For each manufacturing site and importer's office, a licence must be obtained from the Minister of Health and Welfare (MHW). At one time each product also required a licence but this law has now been amended and a system developed whereby a licence is only required per category. This is referred to as the Comprehensive Cosmetic Licensing System by Category (CLS). The MHW have so far developed over 30 product categories. In each category there is a list of ingredients permitted with maximum concentrations and reference to specification standards. For each ingredient used it is necessary to check that it complies with the specifications of the Japanese Standards of Cosmetic Ingredients (JSCI). Only if a product and its ingredients comply with a CLS and the standard specifications can a licence be given per category, otherwise a separate licence application must be made. Quasi-drugs are treated in a similar way to drugs and require approval and a licence.

Apart from the ingredients permitted per product category, Japan also has a list of prohibited ingredients and a list of permitted colours, as well as a list of ingredients that must be labelled.

Trade associations

Interpreting legislation can be difficult and further information can be obtained from the trade association in any particular country. The addresses for the UK, American and Japanese trade associations are given below:

UK
Cosmetic Toiletry and Perfumery Association (CTPA)
35 Dover Street
London
W1X 3RA
England

USA
Cosmetic, Toiletry & Fragrance Association (CTFA)
Suite 300
1101 17th Street NW
Washington DC 20036
USA

Japan
Japan Cosmetic Industry Association (JCIA)
Fourth Floor
Hatsumei Building
9-14, Toranomon 2-chome
Minato-Ku
Tokyo
Japan

Cosmetic legislation reference sources

In the UK the following Cosmetic Products (Safety) Regulations should be used.

- The Cosmetic Products (Safety) Regulations 1989, SI 2233 – (ISBN O O 11 098233 9)
- The Cosmetic Products Safety (Amendment) Regulations 1990 SI 1812 – (ISBN 0 11 004812 1)
- The Cosmetic Products Safety (Amendment) Regulations 1991 SI 447 – (ISBN 0 11 013447 8)

These documents can be purchased from Her Majesty's Stationery Office at 49 High Holborn, London WC1V 6HB (mail order telephone: 071-873-9090).

In the USA, the Cosmetic Regulations are entitled "Federal Food, Drug and Cosmetic Act", as amended, and related laws (updated yearly) and "Code of Federal Regulations", title 21, parts 1–99 and 700–799 (contains final regulations implementing the Act and is updated yearly). These can be obtained from the Superintendent of Documents, US Government Printing Office, Washington DC 20402–9325, USA.

Trade associations often prepare useful publications, the best example of this being the "CTFA International Cosmetic Ingredient Dictionary, 4th Edition". The "CTFA International Resource Manual" is an extremely informative resource and has been used as a reference in the sections summarising American and Japanese legislation. The "CTFA

International Color Handbook" is another invaluable publication from the American Trade Association.

In Japan, the "Japanese Standards of Cosmetic Ingredients" (Second Edition and Supplement), the "Comprehensive Licensing Standards (CLS) of Cosmetics by Category", Parts I–VI, and the "Principles of Cosmetic Licensing in Japan", (Second Edition) may be referred to. These publications, and all relevant cosmetic legislation, can be obtained from Yakuji Nippo Limited, 1 Kanda Izumicho, Chiyoda-Ku, Tokyo 101, Japan. The JCIA also publish the "Japanese Cosmetic Ingredient Dictionary" Vol 1–12.

SECTION 3

Surface Chemistry

SURFACE CHEMISTRY
EMULSIONS

SURFACE CHEMISTRY

Principles of the colloidal state

1. *Colloidal systems*

Colloidal systems fall midway between true solutions (e.g. sugar or salt in aqueous solution) and suspensions (e.g. proteins or polysaccharides in aqueous solution) and show some of the properties of each. For example, the particles in a colloidal system are normally only visible under certain conditions, using special techniques such as light scattering. Unlike suspensions, colloidal systems do not always separate over time and the colloidal particles can only be filtered out of the system using special filtration membranes with very small pore sizes.

The characteristic property of a colloidal system is particle size, which is normally in the region of 1 micrometre or less. Ideal colloidal systems are composed of two phases, the disperse phase and the continuous phase. Normally, the continuous phase is in excess but as the quantity of disperse phase approaches that of the continuous phase, phase inversion can occur and the disperse phase is then in excess. Most colloidal theory is based on ideal two-phase systems, whereas in practice this is rarely true. Real colloidal systems tend to be multi-component in nature, with small amounts of other components present having quite significant effects on colloidal stability. For example, a change of surfactant in an oil-in-water emulsion can change the properties of that emulsion, such that phase inversion occurs. A second important characteristic of colloid systems is related to the presence of very small particles which create a very large interfacial area between phases. Interactions at these surfaces changes properties such as rheology of the colloid system.

The most important types of colloid system are *sols* (solids dispersed in liquids), *emulsions* (liquids dispersed in liquids) and *foams* (gases dispersed in liquids).

2. *Stabilisation of colloid systems*

There are two phenomena that are associated with the stabilisation of colloids and either one, or both, may be used in a particular colloid system. The first is to induce the presence of repulsive forces between colloid particles, ensuring that they remain dispersed in the continuous phase. This can be achieved in following ways:

1. Subject the dispersed phase aggregates to high shear conditions in the presence of the continuous phase, in order to break down the aggregates to colloidal size. The high shear also results in frictional forces, which may cause particles to acquire a charge resulting in mutual repulsion
2. Adsorption. The dispersed phase particles physically adsorb surrounding ions and become charged, resulting in mutual repulsion
3. Inducing repulsive charges onto colloid particles by changing specific physical parameters such as pH

The second method of colloid stabilisation is to induce interaction of the colloid particles with the continuous phase itself, such that they become solvated. In the case of a water continuous phase, this phenomenon is referred to as the hydration state. This can be achieved through the use of a surfactant, which bridges the interface between the dispersed and continuous phases.

In order to prepare a stable colloid, one of the above methods must be used. In practice, it is much easier to induce colloid stability using the second method, rather than the first. A schematic representation of colloid preparation is shown in Figure 1.

FIGURE 1

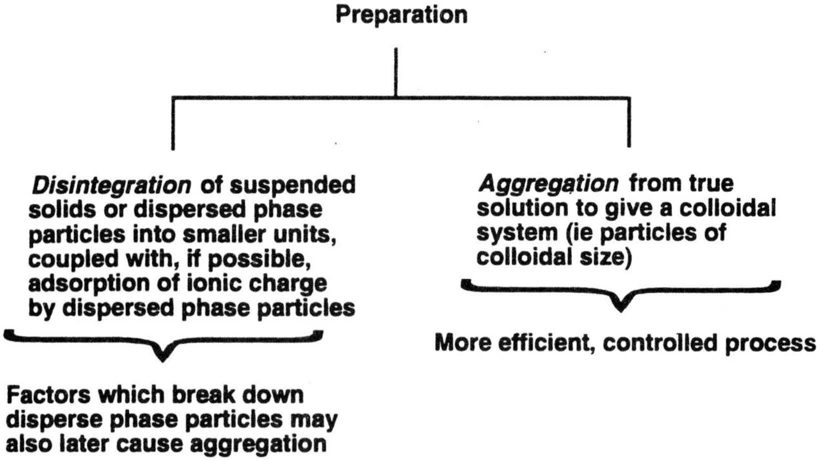

3. *Maintaining a stable colloid system*

Many other factors are associated with colloid stability. These are listed below:

1. Forces of mutual repulsion or solvation between the two phases
2. A reduction in size of the dispersed phase particle. The smaller the size of the dispersed phase particle, the more chance there is of the colloid being stable
3. The density difference between the dispersed phase and the continuous phase.

When the phases are of similar density, maximum colloid stability is observed. An example of this method is the addition of solvents or essential oils to the lipid phase of a lipid/water emulsion. The addition of these materials will increase the density of the lipid phase, thereby making the colloid more stable

4. The viscosity of the continuous phase. A higher viscosity continuous phase impairs movement of the dispersed phase particles and therefore impairs or inhibits re-aggregation. The viscosity of the continuous phase can be increased by the addition of stabilisers which normally produce gels if added in excess. Typical stabilisers are proteins, gums and polysaccharides
5. The presence of surfactants, which will reduce incompatibility between continuous and dispersed phases

Mathematically, points 2,3 and 4 above can be expressed by Stokes Law, the equation for which is shown below:

$$V = \frac{2a^2(\rho_1 - \rho_2)g}{9\eta}$$

where

V	=	rate of rise or fall of particles in a continuous phase
a	=	radius of dispersed phase particle
ρ_1	=	density of the dispersed phase material
ρ_2	=	density of the continuous phase
g	=	gravitational force
η	=	viscosity of continuous phase

Although an approximation of colloid stability can be predicted using Stokes Law, it makes several assumptions which, in real systems, are often not obeyed. Stokes Law is only valid if colloid particles are spherical which, in reality, is rarely the case. It also makes no provision for dispersed phase interparticle forces and takes no account of the fact that colloid destabilisation can cause a change in disperse phase particle size.

Intermolecular and interfacial forces

Intermolecular and interfacial forces are associated with both repulsion between dispersed phase particles and solvation of the dispersed phase particles in the continuous phase. Although both phenomena often occur in a colloid system, one type is always largely predominant. The predominant force determines the classification of the colloid system, which may be either lyophilic or lyophobic in type.

1. *Stability of lyophilic colloid systems*

Lyophilic colloids rely largely on the forces of solvation for their stability and are generally prepared from macromolecules (e.g. starch, protein) or by using dispersed phase material that is able to absorb macromolecular materials at its surface. Thus, in a lyophilic colloid, the dispersed phase has an affinity for the continuous phase.

Lyophilic colloids are often quite stable and will not easily coagulate if electrolyte is added. The primary factors that promote stability are the size of the dispersed phase particle, the viscosity of the continuous phase and the presence of surfactants. However, other forces also exist which promote stability in this type of system. These are the forces of attraction between the dispersed phase particles that are generally referred to as *Van der Waals forces*. The factors responsible for the existence of Van der Waals forces are listed below:

1 The presence of molecules containing permanent dipoles, which orientate themselves so that mutual attraction results
2. The presence of dipolar molecules, which induce dipoles in other molecules, thereby promoting attraction
3. The presence of non-polar molecules, which exhibit changing polarisation due to their electron distribution

Apart from being associated with the stability of lyophilic colloid systems, Van der Waals forces are also implicated in the miscibility of both polar and non-polar solvents.

Whilst lyophilic colloids are not readily destabilised by the addition of electrolytes, high concentrations of electrolyte or polar materials will cause, or contribute to, colloid destabilisation. This method of inducing colloid instability is referred to as *salting out*, an example being the precipitation of a proteinaceous lyophilic colloid when another polar material, such as ethyl alcohol, or high quantities of electrolyte are added. Salting out occurs because of competition between the added polar compound or electrolyte and the original continuous phase. For example, in the case of the proteinaceous colloid described above, the alcohol or electrolyte competes with the dispersed phase material for the water in the continuous phase, thus destabilising the colloid system.

Lyophilic systems are also destabilised by reducing the affinity of the dispersed phase for the continuous phase. This can be done by changing the pH of the system, which modifies the dipolar charge, thereby leading to destabilisation. Proteinaceous lyophilic colloids are particularly prone to destabilisation at the *isoelectric* point, which is the pH at which the *zwitterion* occurs. This destabilisation is due to mutual attraction between the zwitterions as illustrated in Figure 2.

FIGURE 2

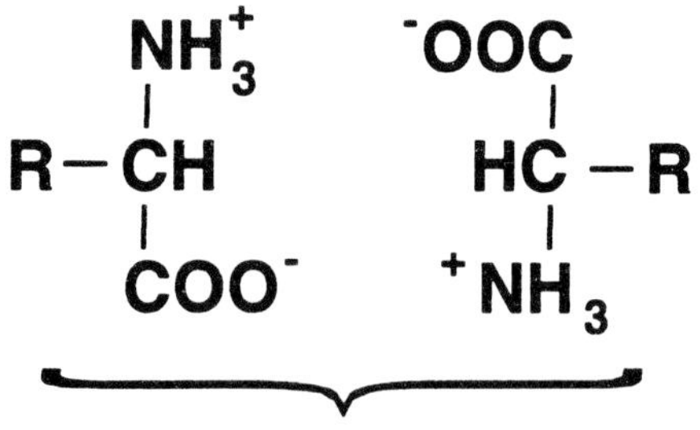

Aggregation

At the isoelectric point, even small quantities of electrolyte will destabilise the colloid system. Once a lyophilic colloid system has been destabilised, redispersion is very difficult to achieve.

2. *Stability of lyophobic colloid systems*

The forces concerned with stability of lyophobic colloid systems are those of repulsion and attraction.

Lyophobic colloid stability is a finely balanced phenomenon in which the dispersed phase particles carry a net electrostatic charge (unlike lyophilic systems in which charge is induced). The forces contributing to the stability of lyophobic systems are similar to those forces acting between molecules or ions. In making this analogy the forces of attraction and repulsion acting between two molecules or ions can be considered, thus:

FIGURE 3

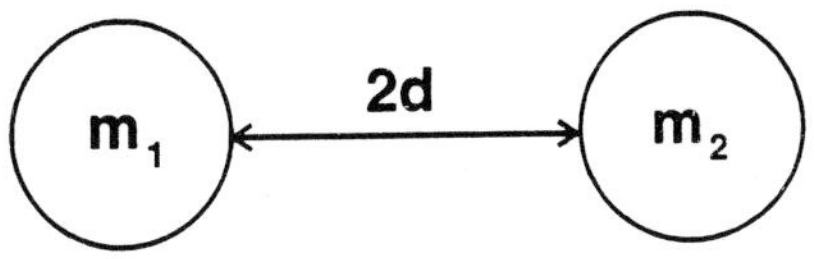

(Where m_1 and m_2 are two adjacent molecules/ions and 2d is the distance between them)

If 2d is large, the net force acting is attraction. As 2d is reduced the force of attraction rises until a point is reached, where the attractive force is replaced by a repulsive force. If 2d becomes very small then a second force of attraction, due to the interpenetrating electron clouds of m_1 and m_2, appears. Graphically, these forces can be represented as shown in Figure 4.

FIGURE 4

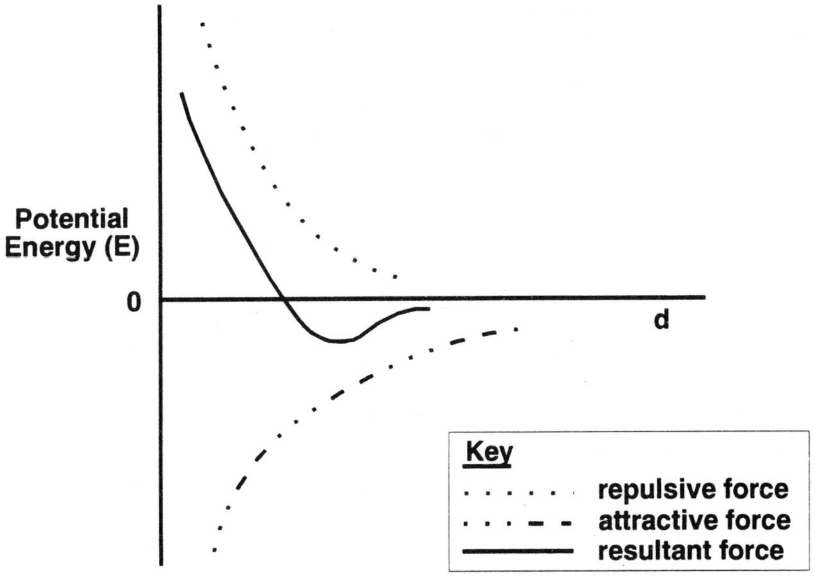

In practice, d is related to the concentration of the colloidal system. Therefore, in a low concentration colloid system d is large, whereas in a high concentration colloid system d is small. To optimise colloidal stability it is necessary to optimise the concentration, such that the repulsive forces described above have maximum effect.

3. *Quantification of repulsive/attractive forces*

Theories quantifying the attractive and repulsive forces in lyophobic colloid systems were proposed by two groups of workers, Deryagin/Landau and Verwey/Overbeek in the 1940s and 1950s. These theories have been combined to form the DLVO (Deryagin, Landau, Verway and Overbeek) theory, which quantifies these forces using classical electrostatic theory.

The forces of attraction are given by the following equation:

$$V_A = \frac{-A}{48\pi d^2}$$

where

V_A	=	potential energy of attraction between colloid particles
A	=	Hamaker-De Boer constant
d	=	half the distance between colloid particles

The forces of attraction are small in magnitude but operate over a large range.

The forces of repulsion are given by the equation:

$$V_R = \frac{64nkT\phi^2 e^{-2Kd}}{K}$$

where

V_R	=	potential energy of repulsion between colloid particles
k	=	Boltzmann constant
T	=	absolute temperature
n	=	number of ions in bulk solution, per unit volume
ϕ	=	constant (a function of particle charge and surface potential)
K	=	reciprocal Debye-Huckle length (a function of the thickness of the electrical double layer)
d	=	half the distance between the colloid particles

The forces of repulsion are large in magnitude but operate over a small range.

Depending on the net forces in the colloid system, attraction or repulsion can occur. If the net potential energy (the sum of V_A and V_R) is negative, a net force of attraction exists between colloid particles. Conversely, if the net potential energy is positive, a net force of repulsion between colloid particles is observed.

The DLVO theory was originally applied to discreet molecules or ions. When the DLVO theory is applied to colloidal systems, in which colloidal particles are "aggregates" of ions or molecules, each particle in the aggregate has its own forces of attraction and repulsion. Because each aggregate consists of many particles, the forces of attraction and repulsion in the colloid system are magnified and consequently operate over much larger ranges. The stability of the colloid system therefore involves much larger inter-particulate distances than the corresponding molecular/ionic system and is therefore a function of the concentration of the colloid system.

Another factor which also affects the repulsive forces in a colloid system is the existence of the electrical double layer, sometimes referred to as the Helmholtz double layer. The presence of an electrical double layer, around the disperse phase particle, extends the range over which forces of repulsion operate. An illustration of the electrical double layer is shown in Figure 5.

FIGURE 5

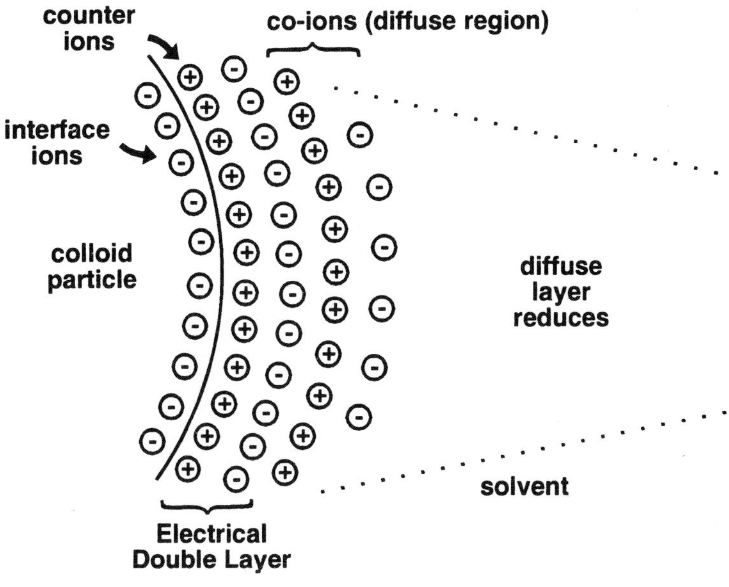

The electrical double layer provides a shielding effect on the repulsive forces. The reciprocal thickness of the double layer, designated the symbol K earlier, appears in the equation describing the repulsive forces between particles. These forces are reduced as the value of K is increased. In practice, K can be increased by adding further co-ions to the colloid system and therefore the net effect of adding electrolyte to the colloid is to reduce the repulsive forces, thereby reducing colloid stability. It is for this reason that lyophobic colloid systems are destabilised by very small quantities of electrolyte.

The type and concentration of electrolyte required to destabilise a lyophobic colloid system is related to the valency of the added ion, providing the assumption that the electrical charge of the added ion is opposite to that responsible for stabilising the colloid, is met.

Interfacial forces and their application

Interfaces involving a liquid component, that is liquid–gas, liquid–liquid and liquid–solid interfaces, are some of the most important in surface chemistry. If one of these phases consists of very small particles, and is dispersed in the other, a colloidal system will exist. The interface which exists between a liquid and a gas is more correctly called a "surface" and the work required to extend this surface is referred to as the surface energy or *surface tension*.

1. *Surface tension*

The surface tension of a liquid may defined as "the force acting in a liquid surface along any unit length" or, more accurately, "the work required to increase the surface by unit amount". Surface tension is relatively easy to measure and is affected by a number of factors including temperature and the presence of solutes in the liquid. The surface tension of a liquid also provides some indication of its behaviour in the presence of other liquids or solids.

Surface tension arises through an imbalance of Van der Waals forces at the liquid surface. This is represented diagrammatically in Figure 6.

FIGURE 6

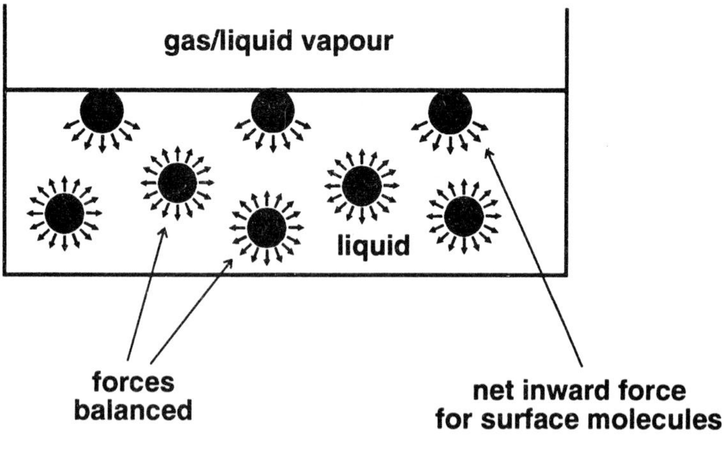

Within the body of the liquid, Van Der Waals forces are balanced but at the liquid surface there is a net Van der Waals attractive force back into the body of the liquid. If the surface molecules obtain sufficient kinetic energy to overcome these Van der Waals forces (through heating, for example) they may be released as the vapour phase. Thus, at the surface of a liquid, there exists an attractive force, not balanced by the gas/vapour phase, and therefore surface molecules are under tension. This tension, normally referred to as surface tension, can be used to describe behaviour at the other interfaces.

In a system consisting of two miscible liquids, a solution is formed. This has no interface and therefore no interfacial forces exist. However, the result of mixing two miscible liquids is a modification of the surface tension value of the total system, which may in itself affect other systems.

Both liquid/gas and miscible liquid/liquid systems are unimportant when considering the science of surface chemistry.

2. *Adsorption*

When considering other types of interface, such as those found in gas–solid, liquid–solid or immiscible liquid–liquid systems, a phenomenon known as adsorption occurs at the interface. Adsorption occurs when two phases, not completely miscible with each other, come into contact, and the particle concentrations at that interface differ from the concentrations in the adjacent bulk phases. The phase which adsorbs is the *adsorbent* and the phase which produces a surface excess is called the *adsorbate*, as illustrated in Figure 7.

FIGURE 7

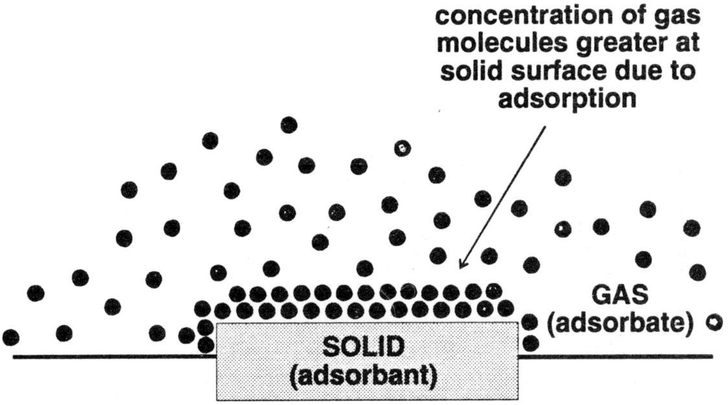

Adsorption is involved with many other aspects of surface chemical behaviour, including adhesion and cohesion, spreading, wetting and emulsification. Adsorption is also used in many industrial applications including decolourisation, chromatography, and emulsification.

3. *Adsorption measurements*

Techniques for the measurement of adsorption are based upon determining the extent of accumulation of the adsorbate at an interface. The measurement of adsorption at gas–solid and liquid–solid interfaces is relatively easy and direct methods can be used. Measurement of adsorption at liquid-liquid interfaces is much more difficult and indirect methods, using theoretical models, have to be used. The measurement of adsorption at each interface type will now be considered.

3.1. *Gas-solid interfaces*

The degree of adsorption at gas–solid (and gas–liquid) interfaces is relatively easy to measure. The adsorption profile is illustrated in Figure 8, below.

FIGURE 8

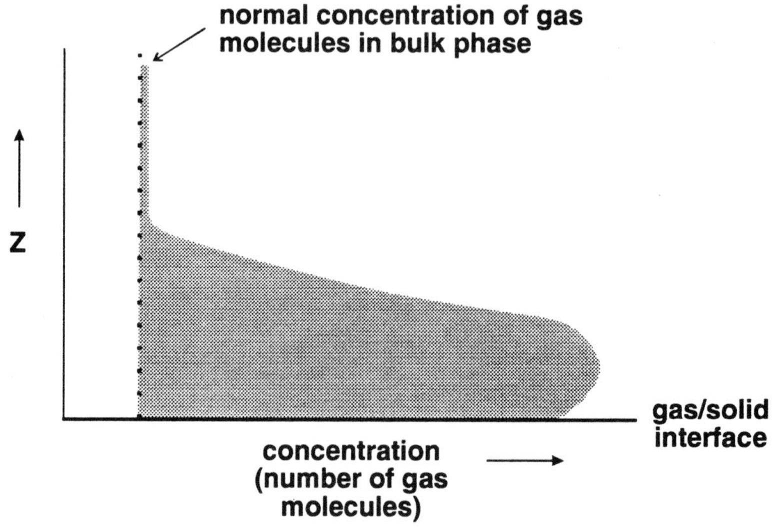

(Z = distance in a direction normal to the solid surface)

The adsorption profile shows the concentration of adsorbate in the region of the adsorbent and a high concentration of gas near the solid (or liquid), which falls off with distance, is observed. In addition to adsorption at the interface, absorption also occurs. For this reason, the profile described in Figure 8 is sometimes referred to as the sorption profile, which takes into account both the adsorption and absorption effects. The extent of adsorption can be illustrated in two ways, either by use of the adsorption profile, or by examining the *adsorption isotherms*. The latter actually enable quantification of the extent of adsorption at the interface. An adsorption isotherm is a graph which relates concentration of *adsorbent* to the extent of adsorption, at constant temperature. There are three types of isotherm normally found in the case of gas–solid adsorption systems, Langmuir, Freundlich and Braunauer, Emmett and Toller (BET). Because the isotherms relate to a constant temperature, a change in temperature results in a change in the shape of the isotherm. Adsorption is an exothermic process and therefore the extent of adsorption is increased at lower temperatures, as described by Le Chateliers principle.

Adsorption isotherms show the extent of adsorption which, for gas–solid systems, is a function of the pressure of the gas at the interface. A study of the isotherms and their shape indicates the mechanism of adsorption occurring and this method differs for the three types of isotherm described above.

3.1.1. *Langmuir isotherms*

In this case, the amount of gas adsorbed initially rises with pressure but then reaches a maximum, remaining constant despite a further pressure increase. An illustration of this isotherm type is shown in Figure 9.

FIGURE 9

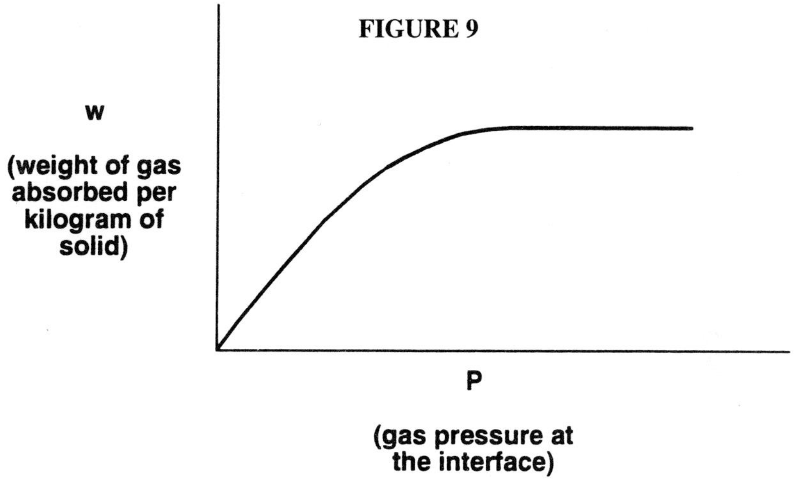

This type of isotherm indicates the formation of a monolayer of gas molecules on the solid surface, with no subsequent secondary or multi-layer formation. Langmuir isotherms frequently indicate the presence of chemisorption, in which adsorption actually involves the formation of chemical bonds. Because chemical bonds are actually involved in the adsorption, no more gas can be adsorbed on to the solid, thus prohibiting multi-layer formation. Interestingly, if an isotherm plot is made for increasing pressure to a constant value for the amount of gas adsorbed, and the gas pressure then reduced to cause desorption, the resultant isotherm frequently exhibits a hysteresis loop, as depicted in Figure 10.

FIGURE 10

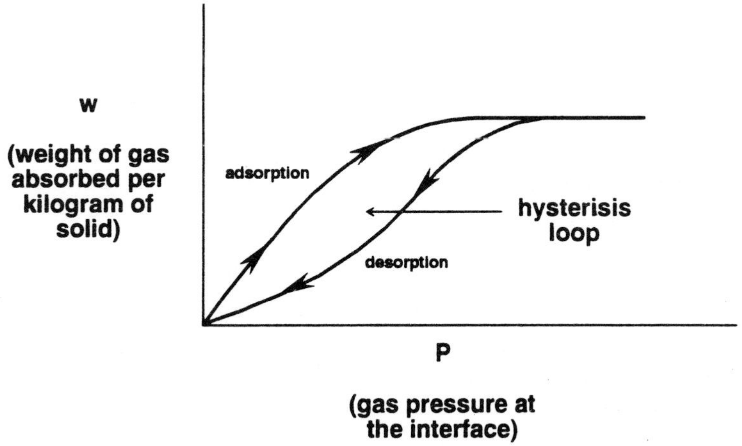

Hysteresis occurs because the desorbed gas is chemically different from the adsorbed gas. An example of this type of adsorption, which involves a chemical reaction, is the adsorption of oxygen onto carbon surfaces, which desorbs as a mixture of carbon monoxide and carbon dioxide.

A practical application of this type of adsorption is in the determination of surface areas of solids. The point at which the isotherm reaches a plateau indicates the formation of a complete monolayer on the solid, from which its surface area can be calculated.

3.1.2. *Freundlich isotherms*

Here a gradual increase in the amount of gas adsorbed with increasing pressure is observed, indicating multi-layer formation. This type of isotherm reflects true physical adsorption, which is dependent upon Van der Waals forces of attraction. There is rarely any chemisorption occurring with this type of adsorption. An illustration of this isotherm type is shown in Figure 11.

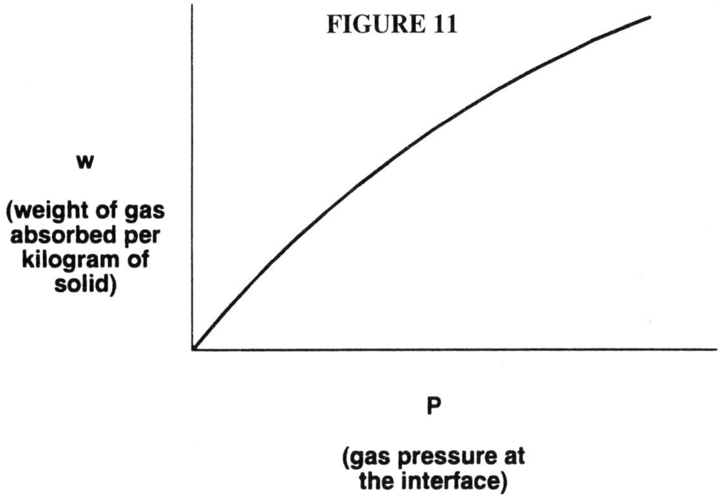

This type of adsorption normally occurs on solids which make good catalysts, due to the absence of chemical reaction. The adsorption of inert gases on solids generally exhibit Freundlich type isotherms.

3.1.3. *Braunaeur, Emmett and Toller (BET) isotherms*

This isotherm, characterised by a sigmoid curve, is normally exhibited by heterogeneous solids and reflects a mixture of adsorption mechanisms. An illustration of this isotherm type is shown in Figure 12.

From the point O to point A on the diagram, the formation of a monolayer by chemisorption, is observed. Between points A and B, little further adsorption occurs with increasing pressure until point B is reached, when an increase in adsorption occurs once more. This increase, which is due to condensation of gas in the pores or capillaries of the solid surface, is a result of absorption or occlusion, rather than adsorption. This type of adsorption isotherm is normally exhibited by very porous solid surfaces.

FIGURE 12

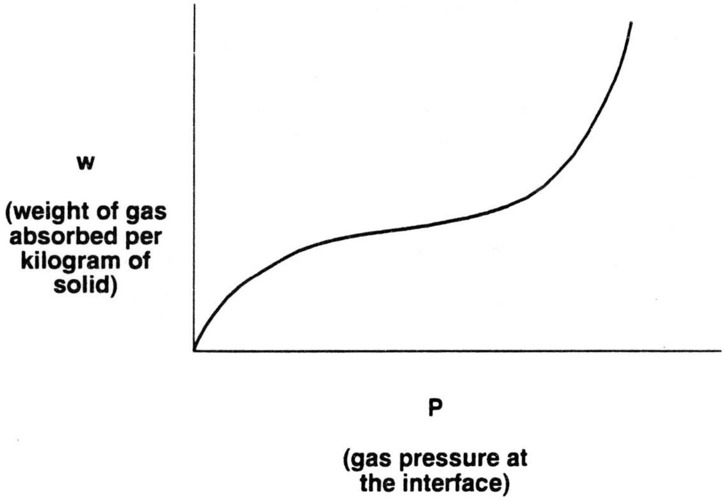

3.2. *Liquid-solid interfaces*

Adsorption at liquid–solid interfaces provides the basis for two very important industrial applications, *wetting* and *adsorption from solution*. At liquid–solid interfaces, the degree of adsorption is given by the adsorption profile illustrated in Figure 13.

FIGURE 13

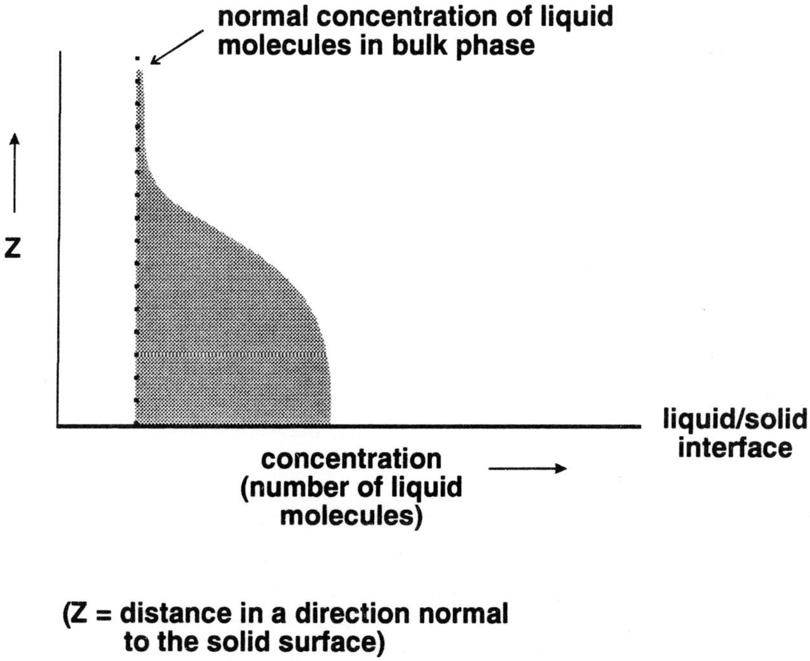

Comparison of this profile with the gas–solid adsorption profile indicates a lower extent of adsorption at the surface but a less rapid fall-off with distance from the surface, z. As in the case of gas–solid interfaces, quantification of adsorption can be studied using adsorption isotherms. Adsorption isotherms at liquid–solid interfaces, however, do not give as much information as gas–solid adsorption isotherms, due to the fact that liquid-solid isotherms cannot differentiate between adsorption and absorption. The processes of adsorption from solution and wetting will now be discussed in more detail.

3.2.1. *Adsorption from solution*

In the case of liquid–solid interfaces, more information on the extent of adsorption is obtained by examining adsorption from solution, where the extent of adsorption indicates the degree of competition between the solute and the solvent, for the solid. Both Langmuir and Freundlich isotherm types may apply in this case, as shown in Figure 14.

FIGURE 14

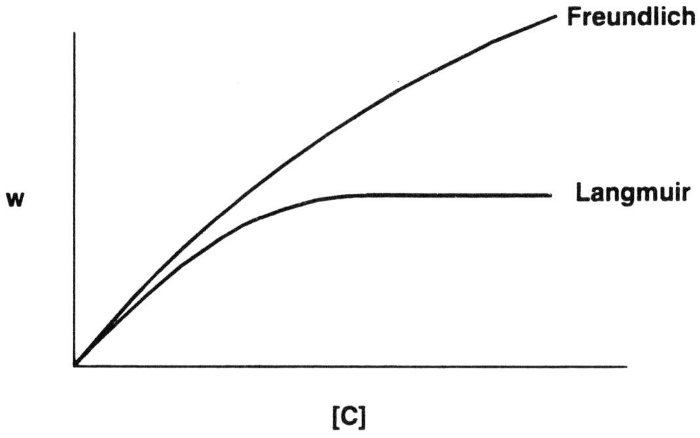

where

W = weight of solute absorbed by a mass of solid, m.
[C] = concentration of solute in the solution.

The occurrence of a Langmuir isotherm indicates a mechanism of chemisorption, where no further adsorption occurs despite an increase in the solute concentration, [C], beyond a certain point. The occurrence of a Freundlich isotherm indicates multi-layer adsorption. A practical industrial application of this type of adsorption is in decolourisation, where the Freundlich type of isotherm is most desirable.

3.2.2. *Wetting*

Wetting is a very important industrial application of adsorption and involves the displacement of one "fluid" from a surface by another. In the case of wetting, one of these "fluids" is normally air and the extent of adsorption of a liquid, on to a solid surface, is described.

The degree of wetting will determine whether or not a solid substance can be dispersed in a liquid. The extent of wetting can be increased by the addition of surface-active compounds (surfactants), often referred to as wetting agents, to the system. The degree of wetting or displacement is governed by the interfacial or surface tension and the contact angle. The contact angle is the angle that exists between the liquid and the vapour (in most cases air) interfaces and the solid, and by convention is taken in a direction through the denser fluid phase. Figure 15 illustrates the situation occurring at a solid surface, for both "wetting" and "non-wetting" liquids.

FIGURE 15

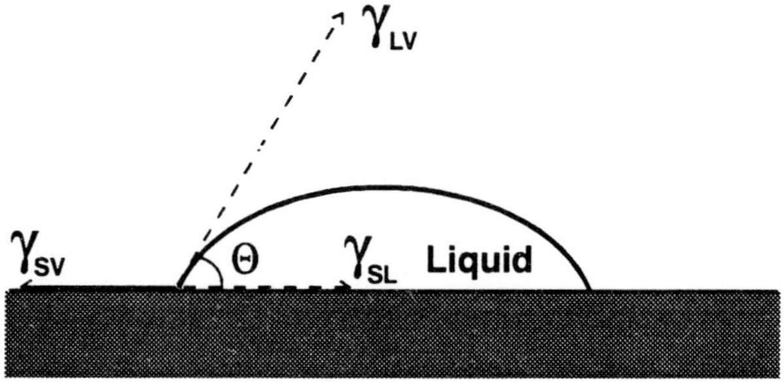

15a) "Wetting Liquid"

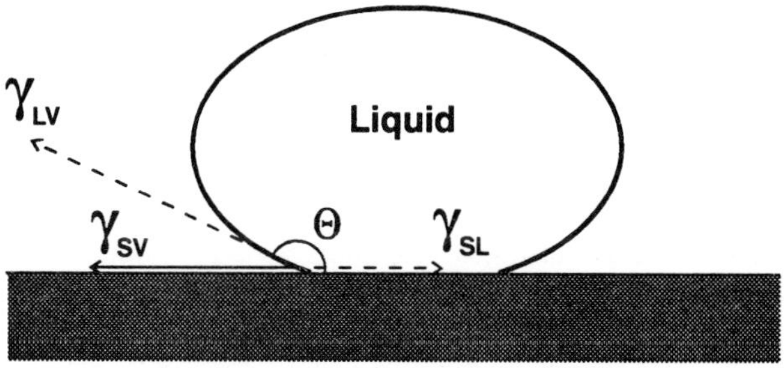

15b) "Non-Wetting" Liquid

where

θ = contact angle
γ_{SV} = solid–vapour interfacial tension
γ_{SL} = solid–liquid interfacial tension
γ_{LV} = liquid–vapour interfacial tension (surface tension)

At equilibrium, the forces acting parallel to the solid surface are balanced, giving the following relationship described by Young's equation:

$$\gamma_{SV} = \gamma_{SL} + \gamma_{LV} \cos \theta$$

It is important to note that this relationship is only valid for a sessile (stationary) droplet. The wetting ability of the liquid can be predicted from the contact angle, which itself depends upon the values of interfacial tension between the three phases. In most cases, the "vapour" is air and the interfacial tension is therefore the surface tension. Complete wetting or non-wetting, which would require values for θ of either 0° or 180°, never occurs.

In practice, the contact angle θ is normally measured by the tilting plate method, the simplified apparatus for which is illustrated in Figure 16.

FIGURE 16

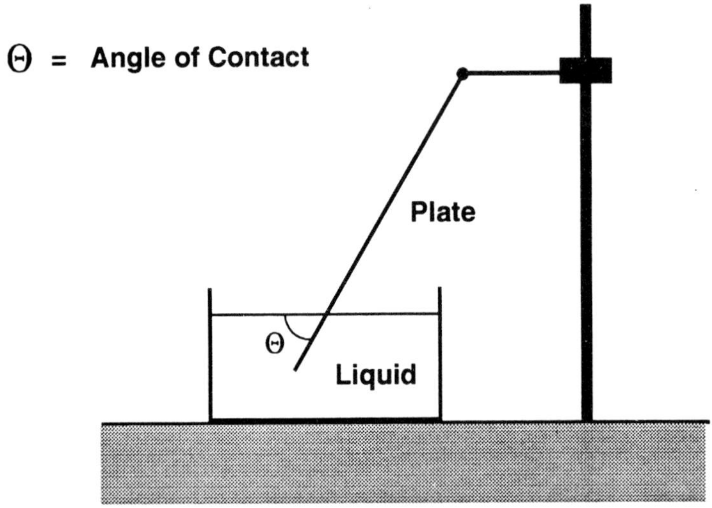

The angle of the plate is adjusted so that the liquid surface remains perfectly flat right up to the solid plate surface. If impurities are present in the liquid, a reduction in surface tension normally occurs, thus causing an erroneous result to be obtained. When using this method, the plate should be lowered slightly and then raised again, to ensure that measurement between dry and "wetted" plate surfaces is not being made.

3.2.3. *The Measurement of Surface Tension*

There are many practical methods available for determining the surface tension of liquids. The most common of these are listed below.

3.2.3.1. *Capillary rise method*

This method is not suitable for determining liquid/liquid interfacial tension. When a capillary tube is immersed in a liquid, the surface tension will cause the liquid to rise within the tube, as illustrated in Figure 17.

FIGURE 17

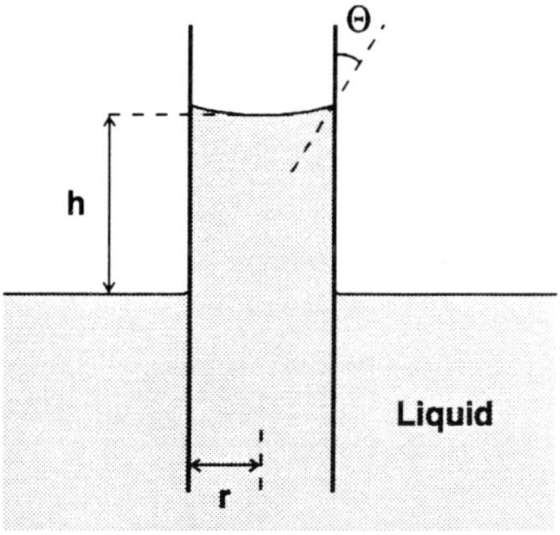

Surface tension, γ, is the force per unit length acting in the surface of the liquid; in the capillary tube, the length of liquid in contact with the glass is $2\pi r$, where r is the radius of the capillary tube. The upward force acting on the liquid, $2\pi r\gamma$, supports a column of liquid of height, h, and volume of πr^2h. The mass of the liquid, m, is given by πr^2hD, where D is the density of the liquid. The weight of the column is therefore given by the expression, πr^2hDg, where g is the force due to gravity. Therefore,

$$2\pi r\gamma = \pi r^2hDg$$

and

$$\gamma = \frac{rhDg}{2}$$

It is worth noting that if the capillary tube has a large bore, or is dirty, the upward force acting on the liquid will be a function of the angle of contact, θ (i.e. $2\pi r\gamma\cos\theta$ is equal to the weight of liquid). For aqueous solutions in clean, capillary tubes, $\theta = 0$ and $\cos\theta = 1$.

3.2.3.2. *Drop volume method*

This method is suitable for pure liquids, or for a measurement of interfacial tension in liquid mixtures. In this method, pendant drops of a known volume or mass are allowed to detach slowly from the tip of a capillary tube, by means of a peristaltic pump. Droplet size and rate of formation of can be monitored using a photocell drop detector.

The force, F, acting downwards at the point of detachment is given by:

$$F = mg = VDg$$

where

V = volume of the drop
D = density of liquid

Surface tension acts at the tip of the capillary tube, within the tube diameter, as a force resisting the downward force of the drop. Therefore:

$$F = 2\pi r\gamma$$

At the point of detachment

$$VDg = 2\pi r\gamma$$

therefore

$$\gamma = \frac{VDg}{2\pi r}$$

In the final equation a correction factor, f, is necessary. This factor is a function of $r/\sqrt[3]{V}$ and is determined by calibration with liquids of known surface tension. The corrected expression is given by:

$$\gamma = \frac{fVDg}{2\pi rf}$$

3.2.3.3. *Wilhemy plate method*

This method uses a thin mica plate or clean glass microscope slide, which is suspended from an analytical balance, and the force required to pull the plate out of the interface is measured. A typical apparatus is illustrated in Figure 18.

FIGURE 18

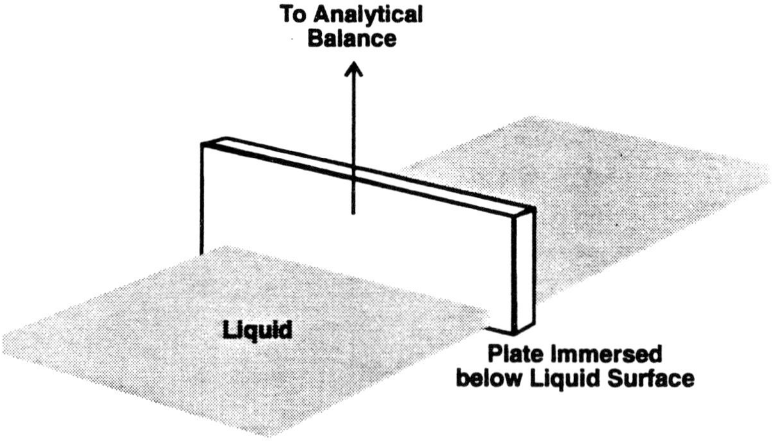

In experimental work, the plate is usually rectangular, so that its perimeter, l, is given by 2(length + thickness). Assuming a zero contact angle, the force required to detach the plate from the interface is given by:

$$\text{Force} = 2\,(\text{length x thickness})\gamma$$

At the point of detachment, the weight required to counter-balance this force is m, and the downward force is mg. Thus:

$$mg = 2(\text{length x thickness})\gamma$$

therefore

$$\gamma = \frac{mg}{2(\text{length x thickness})}$$

3.3. *Liquid-liquid interfaces*

Liquid–liquid interfaces also play a very important role in the practical application of surface chemistry. Three chosen aspects of liquid–liquid interface science are discussed below, *adsorption at liquid interfaces*, *spreading* and *emulsions*.

3.3.1. *Adsorption at liquid interfaces and the Gibbs' Adsorption Isotherm*

When two immiscible or partly miscible liquids are in contact, an interface is formed which has an interfacial tension approximately equal to the difference in surface tension of the two pure liquids. This reduction (or increase) in surface tension can be likened to the adsorption of solute molecules at the surface of a liquid to form a solution but in this case the "solute" is another liquid. Thus, at the liquid interface, one of the liquids will have a surface excess of molecules and is said to be adsorbed. This adsorption can be represented by the adsorption profile, illustrated in Figure 19.

FIGURE 19

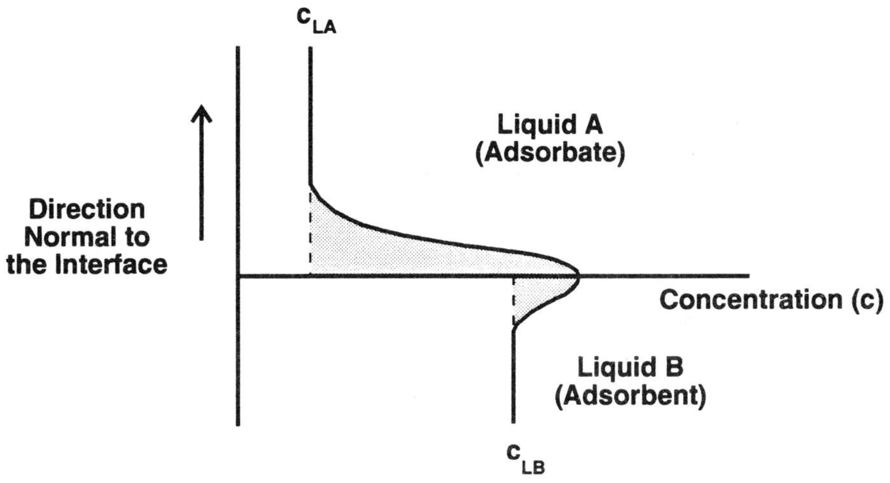

Where

c_{LA} = **concentration of Liquid A**

c_{LB} = **concentration of Liquid B**

In this case, there is a greater number of molecules at the interface of liquid A than liquid B and liquid A is therefore the adsorbate. Liquid B is the adsorbent. Adsorption at solid–gas and liquid–solid interfaces can be measured accurately but the adsorption, or surface excess Γ, at a liquid interface cannot be readily determined in the same way. The Gibbs adsorption isotherm enables the extent of adsorption at a liquid surface to be estimated from surface tension data. It is convenient to regard the interface between two liquids as a mathematical plane (Gibbs dividing surface or δ plane) in order to derive the isotherm equation. In practice, however, there is an interfacial region of varying composition with an appreciable volume in molecular dimensions.

Consider the following,

FIGURE 20

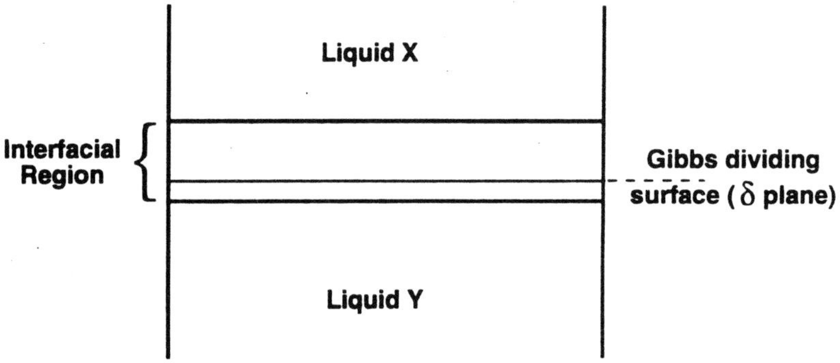

^ A Y A

where

Γ = surface excess concentrate
A = area of the interface
n_X = amount (mol) of liquid X in the interface, in excess of that in the bulk phase
n_Y = amount (mol) liquid Y in the interface, in excess of that in the bulk phase

The change in surface tension ($-d_\gamma$) due to the formation of the interface is given by:

$$-d_\gamma = \frac{n_X}{A} d\mu_X + \frac{n_Y}{A} d\mu_Y$$

or

$$-d_\gamma = \Gamma_X d\mu_X + \Gamma_Y d\mu_Y$$

where μ = chemical potential

The position of the δ plane is established such that the surface excess of liquid Y is zero. Therefore,

$$-d_\gamma = \Gamma_X d\mu_X$$

Chemical potentials are related to activity and the following relationship holds true:

$$d\mu_X = RT\, d\log_e a_X$$

where

R = universal gas constant
T = temperature
a_X = activity of liquid X

Substituting for $d\mu_X$ and rearranging gives,

$$\Gamma_X = \frac{-1}{RT} \frac{d_\gamma}{d\log_e a_X} = \frac{-1}{2.303\, RT} \frac{d_\gamma}{d\log_{10} a_X}$$

For ideal solutions, activity is equal to concentration and the above equation translates to the following relationship, known as the Gibbs' Adsorption Isotherm:

$$\Gamma_X = \frac{-1}{2.303\, RT} \frac{d_\gamma}{d\log_{10} c_X}$$

where

c_X = concentration of liquid X

3.3.2. *Spreading*

Spreading describes the behaviour of one liquid, in small quantities, spreading upon the surface of a second bulk liquid with which it is in contact, both liquids being immiscible. Such a system can give rise to either true spreading of the first liquid on the bulk liquid surface or the formation of "liquid lenses" of the first liquid at the surface of the bulk. Spreading will occur if the bulk liquid is "wetted" by the added liquid, whereas "liquid lenses" will be formed if it is not.

Factors affecting spreading are related to the relative surface tensions of the two liquids concerned, which in turn are dependent on the forces of cohesion and adhesion in the system. The cohesive force for a single pure liquid corresponds to the work required to pull apart a volume of liquid of unit cross-sectional area, as illustrated in Figure 21.

FIGURE 21

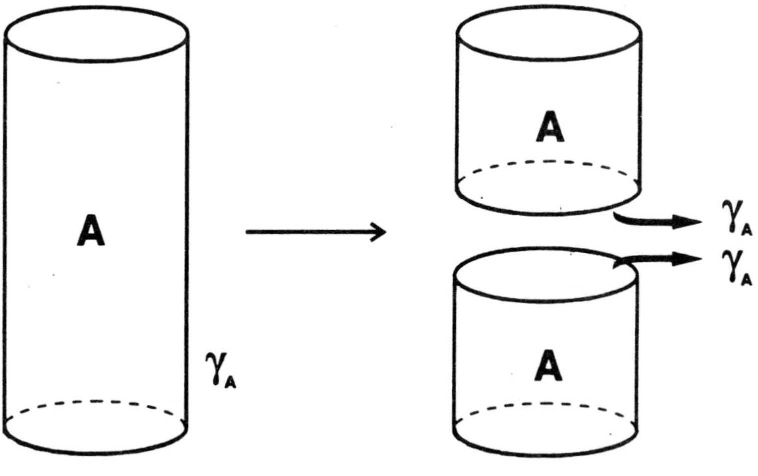

In this case the following relationship is true:

$$W_{coh} = \gamma_A + \gamma_A = 2\gamma_A$$

where

W_{coh} = work of cohesion (measured in mNm^{-1})

The work of cohesion is the work required to create two new liquid surfaces and surface energies. If spreading occurs, the cohesive force is not very strong.

The adhesive force is equal to the work required to separate unit cross-sectional area of two immiscible liquids at an interface, to form two separate liquid–vapour interfaces. This is illustrated in Figure 22.

FIGURE 22

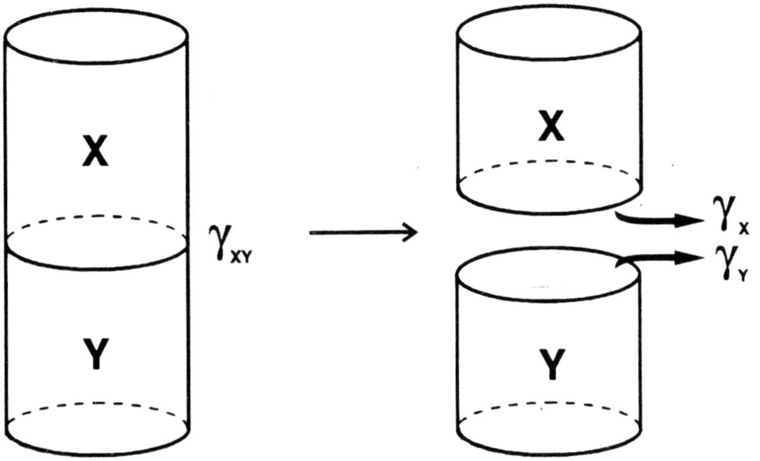

The work of adhesion is given by the following relationship, known as the *Dupre equation*:

$$W_{adh} = \gamma_X + \gamma_Y - \gamma_{XY}$$

where

W_{adh} = work of adhesion
γ_{XY} = interfacial tension between liquids X and Y
γ_X = surface tension of liquid X
γ_Y = surface tension of liquid Y

Spreading will occur if the force of adhesion is greater than the force of cohesion. Figure 23 represents the contact of a small amount of liquid Y with a bulk liquid X, liquids X and Y being immiscible.

FIGURE 23

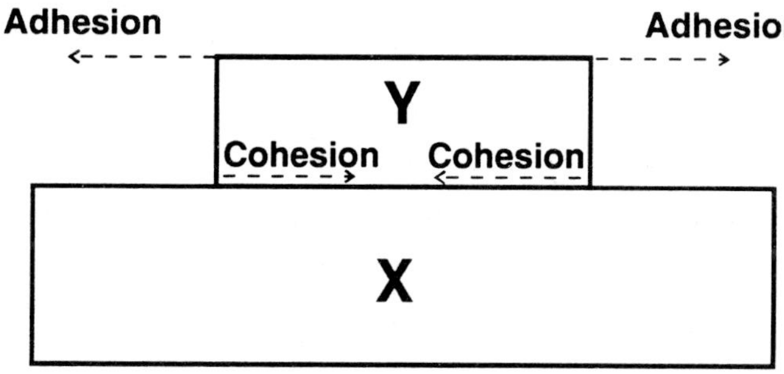

For spreading to occur, the adhesive force between X and Y must be greater than the work of cohesion of Y, mathematically expressed thus,

$$W_{XY} > 2\gamma_Y$$

The function given by the expression $[W_{XY} - 2\gamma_Y]$ is called the initial spreading coefficient, normally represented by the symbol, S. Since from the Dupre equation,

$$W_{XY} = \gamma_X + \gamma_Y - \gamma_{XY}$$

then

$$S = \gamma_X + \gamma_Y - \gamma_{XY} - 2\gamma_Y$$

and

$$S = \gamma_X - [\gamma_{XY} + \gamma_Y]$$

The higher the value of the initial spreading coefficient, S, the greater the tendency for spreading to occur. For spreading to occur $S \geq 0$. If $S < O$, the cohesive force is greater than the adhesive force and the formation of liquid lenses occurs. The presence of impurities in a liquid will affect its surface tension and hence reduce the value of S, making spreading less likely to occur. In some systems, the value of S actually changes after initial spreading has occurred. An example of this is the spreading of hexanol on water. In this case, the value of the initial spreading coefficient, $S_{initial}$, is:

$$S_{initial} = 72.8 - [24.8 + 6.8] = 41.2 \text{ mNm}^{-1}$$

After initial spreading has occurred, some of the hexanol dissolves into the water, thereby reducing its surface tension from 72.8 mNm^{-1} to 28.5 mNm^{-1} and the value of S, such that the final spreading coefficient, S_{final}, is:

$$S_{final} = 28.5 - [24.8 + 6.8] = -3.1 \text{ mNm}^{-1}$$

Therefore the final condition in this system is one of non-spreading. It is therefore important to note that when examining the spreading of one liquid on to another, both the initial and final spreading coefficients should be taken into consideration before a prediction of the behaviour of the system is made.

In cases where a liquid shows a high tendency to spread, the continued addition of the spreading liquid will form a duplex film, in which the spreading liquid behaves independently of the liquid on to which it is being spread, and there is no tendency for the spreading liquid to reduce the surface tension of the bulk liquid. The duplex film so formed tends to be unstable and breaks down into a monolayer with pockets of small liquid lenses formed.

3.3.3 *Emulsions*

An emulsion is a system in which very fine droplets of a liquid are dispersed in a second liquid, the two liquids being partially or totally immiscible. If an emulsion is prepared by high shear dispersion of two pure, dissimilar liquids, phase separation is almost immediate. This is because the adhesive force between the two liquids is very low, and the cohesive forces of each liquid is high. Consequently there is a high interfacial tension. The interfacial tension between two liquids which have low mutual solubility has a value between the surface tensions of the individual liquids. The interfacial tension between two partially miscible liquids usually approximates to the difference between their surface tensions.

To produce stable emulsions, a surface active (surfactant) agent must also be present. Surfactants promote emulsion stability because they are preferentially adsorbed at the liquid–liquid interface, thereby reducing interfacial tension. Emulsions are normally prepared using an excess of surfactant and some form of mechanical agitation.

The two liquid phases in an emulsion are normally classified into the continuous phase, which forms the bulk of the system, and the dispersed phase which describes the phase dispersed in very fine droplet form. Historically, the two emulsion phases are referred to as the "oil" (lipophilic) phase and the "water" (hydrophilic) phase, giving rise to two possible emulsion types:

- Water-in-oil (W/O) emulsions
- Oil-in-water (O/W) emulsions

Generally speaking, the continuous phase of the emulsion is in excess. If equal volumes of the hydrophilic and lipophilic phases are present, then the type of emulsion formed will depend mainly upon the surfactant and the method of formation of the emulsion system. O/W emulsions can be distinguished from W/O emulsions in several different ways. Generally, O/W emulsions will be readily dispersed in water, whereas W/O emulsions will not. O/W emulsions are also readily coloured by water-soluble dyes and will exhibit a higher electrical conductivity than W/O emulsions.

There are many hundreds of surfactants available for stabilising emulsions and methods for choosing the optimum one vary widely. Perhaps the most common method of surfactant selection is the use of the HLB (hydrophilic/lipophilic balance) value. The HLB value scale is generally recognised as running from 1 to 20, with the lower HLB value surfactants promoting W/O emulsions and the higher HLB value surfactants producing O/W emulsions. There are many techniques available for determining the HLB value of a given surfactant but one of the most common methods utilises the following relationship:

$$\text{HLB Value} = 20 \left[1 - \frac{S}{A} \right]$$

where

S = saponification number of the ester
A = acid number of the recovered acid

One limitation of the use of this relationship, is that it is only applicable for certain types of nonionic surfactant. There is no universal method available for determining the HLB values for all types of surfactant. When promoting emulsification, blends of non-ionic surfactants tend to be more effective surfactants, forming more stable emulsions. This is because blends offer better potential for a higher packing density of surfactant molecules at the oil/water interface.

Micelle formation

In solution, surfactant materials exhibit a particular type of behaviour. In very dilute solutions, in the order of a few mMol, surfactants act as ordinary solutes and reduce the surface tension of the liquid containing them. As the concentration of surfactant is increased, however, a variation in the properties of the system occurs, which corresponds to the aggregation of surfactant molecules to form *micelles*. A typical plot of the changes observed in the surface tension, osmotic pressure and turbidity properties of a surfactant solution, as the concentration of surfactant is increased, is illustrated in Figure 24.

FIGURE 24

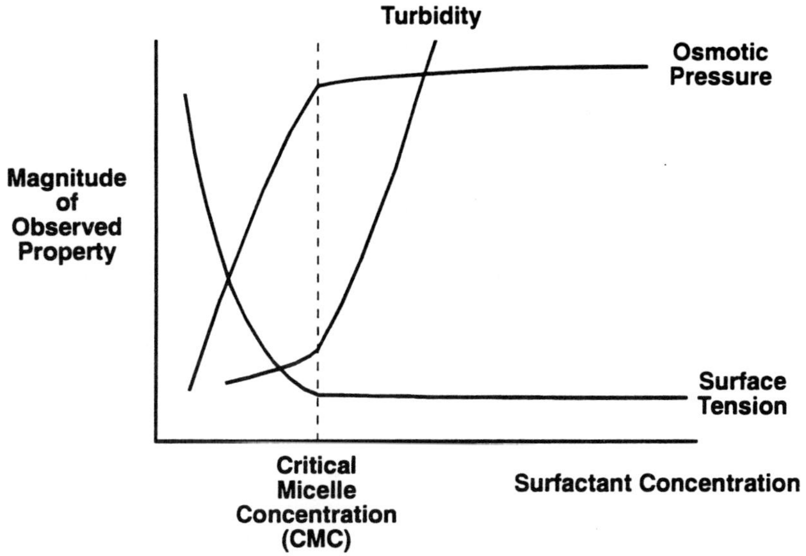

The concentration at which the sudden change in properties occurs (point A on the graph) is called the *critical micelle concentration* (CMC). When micelles are formed, the hydrophobic portions of the surfactant molecules are orientated towards the inside of the micelle, the hydrophilic portions being on the outside. Studies using optical methods confirm the presence of micelles and also of other aggregates known as vesicles (laminar micelles), both of which are diagrammatically illustrated in Figure 25.

FIGURE 25

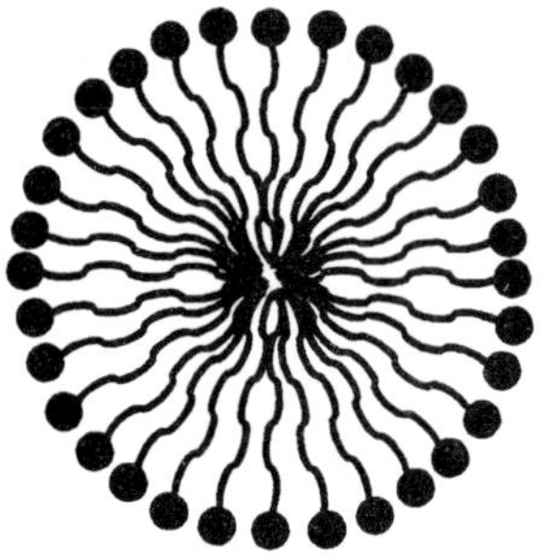

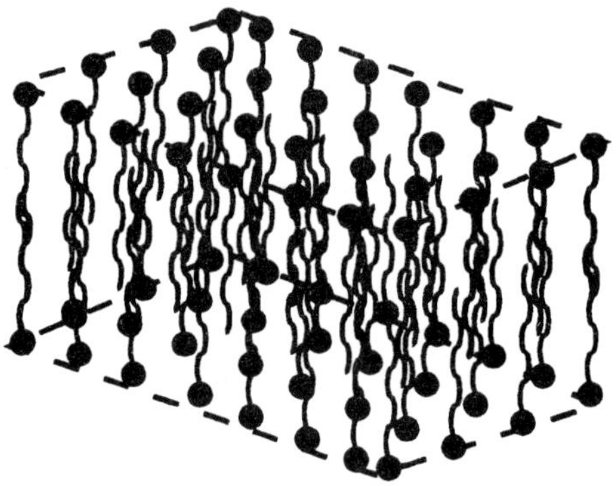

The critical micelle concentration varies for different surfactants and decreases as the hydrocarbon content of the surfactant increases. Therefore, the greater the lipophilic character of the surfactant, the lower its critical micelle concentration. The critical micelle concentration is also reduced as the temperature or ionic content (for aqueous systems) of the dispersion increases.

Micelles themselves behave as colloidal particles and systems containing them are often referred to as *association colloids*. Micelle formation is essential in the process of detergency, where the lipophilic soilage is solubilised by the lipophilic "core" of the micelles, with subsequent dispersion of the soilage taking place. In addition, a number of lipophilic compounds show a greater tendency for dispersion in aqueous media which contains a surfactant at a concentration above its CMC. This process is known as solubilisation. The potential for solubilisation is at a maximum for a surfactant with a high HLB value and industrially this process is very important for dispersing dyes and fragrances into products. Although the critical micelle concentration of surfactant solutions is lowered by temperature increase, there is a temperature, below which, aggregation or micellisation will not occur. This temperature is known as the Krafft point, which is commonly in the range of 10–20°C.

Surface pressure and monolayers

Surfactants are valuable in that, unlike conventional liquids, they do not exhibit adsorption profiles but rather form a monolayer of molecules at an interface. This enables the extent of adsorption at a liquid--gas interface to be investigated. Because surfactant molecules form a true monolayer at the liquid surface they are said to exert a surface pressure, which is given by the difference of the interfacial surface and the surface tension of the pure liquid, thus:

$$\pi = \gamma_0 - \gamma$$

where

π = surface pressure

γ_0 = surface tension of the pure liquid

γ = surface tension at the interface

Effectively, surface pressure (π) is just a representation of lowering of surface tension. The magnitude of π shown by a particular surfactant is useful when studying the associated force/area curves. The force/area curves themselves are determined from surface pressure measurements, using the Langmuir film balance, illustrated in Figure 26.

FIGURE 26

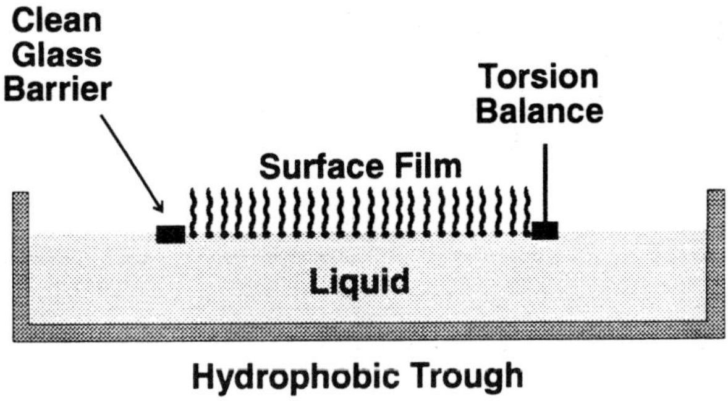

A very small quantity of surfactant is added to the surface of the liquid using a suitable dispersing solvent, which is then allowed to volatilise, leaving a surfactant film. The glass barrier is moved to compress the surface monolayer and a surface pressure reading is taken using the torsion balance. A series of these readings will produce a surface pressure:area curve. Thus, the balance allows the measurement of the surface pressure of the adsorbed film of molecules at the liquid surface. The surface pressure varies with the concentration (expressed as m^2 molecule^{-1}), or number of molecules, of surfactant at the liquid surface.

The surface pressure is inversely proportional to the surface area as shown by the force/area curve in Figure 27.

FIGURE 27

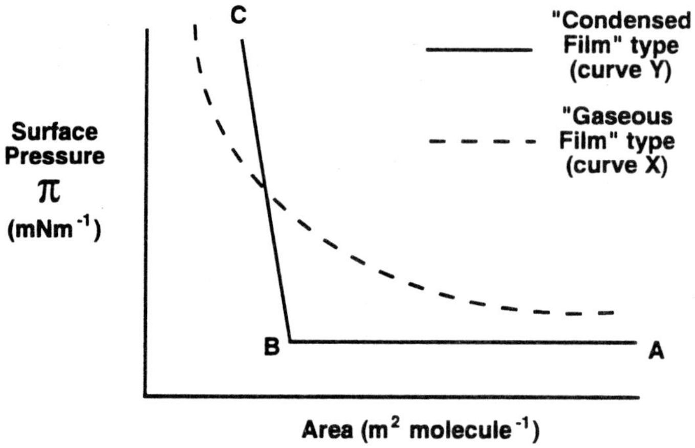

Curve X illustrated in Figure 27 is a gaseous film curve, which is analogous to the behaviour of a compressed gas. Curve Y is a condensed film type, where little change in surface pressure takes place as the surface area is decreased (A to B in Figure 27), until a point (B in Figure 27) is reached where surface pressure increases sharply. This is explained by the fact that over the range A to B surfactant molecules simply undergo rearrangement, whereas within the range of B to C actual compression of the surfactant molecules takes place. In this way, examination of the force/area curves for different surfactants provides information about their structure. It is also possible to investigate the cohesion between surfactants, the effects of pH change and electrolyte additions, the likelihood of micelle formation and the type of reaction kinetics taking place, using force/area curves.

References

1. Sherman P – "Rheology of Emulsions", Pergamon Press, 1963

EMULSIONS

Introduction

Emulsification is the process of dispersing one material throughout another and, in the context of cosmetic applications means effecting a stable state for typically 2 to 3 years.

Emulsification is particularly useful in cosmetics as it promotes cosmetic elegance and allows otherwise impractical combinations of ingredients (i.e. oil soluble and water soluble materials) to be used in the same product. Emulsification also offers great formulation flexibility, enabling modification of such parameters as feel, viscosity and appearance, to be made relatively easily. In addition, emulsions facilitate the "dosing" of active ingredients onto the skin in an aesthetically pleasing and consistent manner. Emulsions are often very cost effective and offer a viable means of producing a commercially successful product.

An emulsion can be defined as "a two-phase system, consisting of two immiscible or partially miscible liquids, one being dispersed in the other as very fine droplets". The emulsion systems considered in this chapter consist of an "oil" phase and a "water" phase only. The phase which is dispersed in the form of very fine droplets is referred to as the dispersed phase and the phase in which it is dispersed is referred to as the continuous phase. The dispersed phase droplets normally possess a minimum diameter of approximately 0.1 microns and a maximum diameter of preferably not greater than 5 microns.

Simplistically, the two basic types of emulsion that can be formed are:

i) water-in-oil (W/O) emulsions
ii) oil-in-water (O/W) emulsions

In reality, emulsion systems are often more complex and cannot be easily classified as either O/W or W/O. In such cases the dispersed phase and continuous phase, instead of being discreet, contain portions of the other and a dual emulsion is formed. Dual emulsions can be classified as either "Water-in-Oil-in-Water" or "Oil-in-Water-in-Oil".

Classical emulsion systems

Historically, three basic groups of emulsion product have been important to the cosmetic scientist. These are:

1. *Cold creams*

These are emulsions with a high oil content and low water content. Originally, cold creams were all W/O emulsions, using lanolin or beeswax plus an alkaline material to emulsify up to 30% of water into a mixture of oils, to give the desired creamy consistency. Modern cold creams are often based on O/W systems for the following reasons:

- they spread and penetrate the skin more easily
- they are aesthetically more pleasing with a less "oily" feel
- they have a pleasing "cooling effect" on the skin after application, due to evaporation of the water phase from the skin's surface

2. *Vanishing creams*

These are O/W emulsions, with a low oil content and high water content. Traditionally, they are based on stearic acid which produces emulsifying soaps capable of holding large quantities of water (60–70% w/w). They rub into the skin very easily, with rapid water loss, leaving a dry, waxy, velvety feel on the skin. The feel is dry and waxy, rather than oily, because the major part of the oil phase consists of the relatively high melting-point wax, stearic acid.

3. *Lotions/milks*

These are O/W emulsions with a low oil content and very high water content (70–90% w/w) and are invariably low viscosity liquids, rather than viscous semi-solids.

Creating an emulsion

Emulsions, in the absence of other influences, are thermodynamically unstable systems and will rapidly separate into the two phases from which they were formed, as illustrated in Figure 1.

FIGURE 1

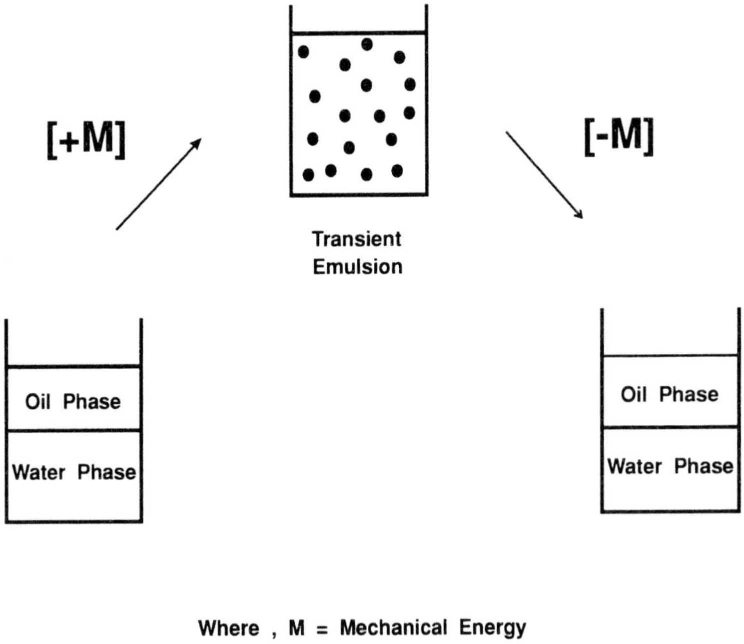

In forming an emulsion, the interfacial area between oil and water phases is increased by several orders of magnitude, the emulsion therefore possessing a much larger interfacial area than the separate phases. Liquid interfaces (surfaces) have a free energy term known as interfacial tension, whose magnitude is directly proportional to the interfacial area, and consequently the massive increase in interfacial area produced when forming an emulsion means that the emulsion has an enormous increase of energy in its system. This energy is lost when the dispersed droplets coalesce. For this reason, an emulsion is a thermodynamically unstable system. The energy required to form the emulsion can be referred to as the creation energy, which is made up of two component parts as indicated in the following relationship:

Creation Energy = Mechanical Energy + Chemical Energy

Surface active materials, or "surfactants" as they are more commonly known, possess the property of reducing interfacial tension/energy and therefore, with reference to the above equation, the addition of a surfactant to the system means that less mechanical energy will be required to form the emulsion. A surfactant used in this way is called an *emulsifier*. The mechanical energy component of the creation energy is the term that is linked to emulsion instability.

Sometimes chemical energy alone is sufficient to form an emulsion without the need for any mechanical aid. This special case of emulsification is referred to as solubilisation and is a perfectly stable state. In cases of solubilisation, the dispersed phase droplets are extremely small. Other types of emulsion system similarly exhibit very small dispersed phase droplet sizes and these are referred to as micro-emulsions, in which the particle size is so small (typically 0.01-0.1 microns in diameter) that the emulsion appears clear. The clarity of these systems is due to the fact that the dispersed phase droplets are smaller than the wavelength of light and therefore do not reflect or refract any light in the visible spectrum.

Emulsion instability

Emulsion instability is brought about by forces that tend to bring emulsion droplets together, enabling them to coalesce. These forces can be classified under the following categories.

1. *Brownian motion*

Brownian motion describes movement of particles due to thermal agitation, and can be observed on particles as small as individual molecules. This motion tends to bring dispersed droplets in the emulsion into contact with each other. The forces of Brownian motion can be described by the following simplified equation:

$$d = \sqrt{\frac{2tkT}{6\pi\eta a}}$$

where

d = distance travelled by the dispersed phase droplet

t	=	time for movement of the dispersed phase droplet
k	=	the Boltzman Constant
T	=	temperature
η	=	viscosity of the continuous phase
a	=	dispersed phase droplet radius

To prevent the dispersed phase droplets approaching each other, with the possible result of coalescence, the temperature must be reduced or the viscosity of the continuous phase of the emulsion increased. When considering the forces of Brownian motion in isolation, an increase in emulsion stability would be predicted by increasing the diameter of the dispersed phase droplets. This prediction, however, is not borne out in reality.

2. Van der Waals forces

These are inter-molecular forces of attraction, which cause particles to gravitate towards each other. The mathematical equations describing Van der Waals forces are complex but, simplistically, the force is described by the following relationship:

$$F \quad \alpha \quad \frac{1}{h^6}$$

where

F	=	Van der Waals force
h	=	distance between two dispersed phase droplets

For an emulsion containing spherical droplets this relationship becomes:

$$F \quad \alpha \quad \frac{a}{h}$$

where

F	=	Van der Waals force	
a	=	dispersed phase droplet radius	$h << a$
h	=	distance between two dispersed phase droplets	

Van der Waals forces predict that an increase in emulsion stability will be achieved by either decreasing the dispersed phase particle size or increasing the inter-droplet distance and, in practice these predictions are found to be valid. In relative terms, Van der Waals are forces more important when considering emulsion instability than those due to Brownian motion.

3. *Sedimentation/creaming forces*

In O/W emulsions, oil droplets tend to migrate to the surface of the emulsion, a phenomenon known as "creaming". Likewise, in W/O systems, water droplets tend to fall with time, a phenomenon referred to as "sedimentation". Both phenomena arise due to the density difference between the two phases in the emulsion system. In each case, the velocity of the dispersed phase droplet can be approximately described by Stokes' law, given by the following equation:

$$V = \frac{2a^2(\rho_D - \rho_C)_g}{9\,\eta}$$

where

V	=	velocity of the dispersed phase droplet
a	=	radius of the dispersed phase droplet
ρ_D	=	density of the dispersed phase droplet
ρ_C	=	density of the continuous phase
η	=	viscosity of the continuous phase
g	=	acceleration due to gravity

When considering the effects of Stokes' Law, an increase in emulsion stability would be predicted by reduction of the dispersed phase particle size, reduction of density difference between the two phases and an increase in viscosity of continuous phase. In practice, all three predictions are observed to be valid.

Emulsion instability is brought about by a combination of all three forces described above. The mechanism of emulsion instability is associated with a progressive occurrence of dispersed phase particle collisions, some of which will result in particle coalescence. When particle coalescence occurs, the larger dispersed phase droplets thus formed have an even greater tendency to collide, again resulting in further coalescence. This process is sometimes referred to as a "circle of instability" and will occur over a period of hours or years, depending upon emulsion stability.

Stabilisation of emulsions

There are two main methods for improving emulsion stability, classified as follows.

1. *Application of electrical charge*

This method relies upon the application of an electrical charge to the dispersed phase emulsion droplet. When the charged droplets approach each other, the forces of mutual electrostatic repulsion significantly reduce the chances of collisions, which could normally give rise to coalescence. This method is only applicable to O/W emulsion systems, as water is required as the continuous phase if any electrostatic interaction is to occur. There are two methods for applying an electrostatic charge to a dispersed phase droplet in an O/W emulsion.

1.1. *Ionisation*

Ionisation will occur in certain systems, such as those containing insoluble proteins as the dispersed phase. A typical equation depicting such ionisation is given below:

$$NH_2 - COOH + H_2O \rightarrow NH_2 - COO^-H^+ + H_2O$$

1.2. *Surface adsorption*

This method involves preferential adsorption of one charge at the interfacial surface. An example of preferential adsorption would be an oil/water system in which is dispersed a soap, $RCOO^-Na^+$. In solution, the more polar end of the soap molecule, the Na^+, will remain in the continuous water phase. The more polar end of the $RCOO^-$ chain (i.e. the $-COO^-$ group) will exhibit an affinity for the water phase, whilst the "R" group (normally a fatty chain) will remain in the dispersed oil phase droplet. The orientation of the soap molecules at the emulsion interface is illustrated in Figure 2.

FIGURE 2

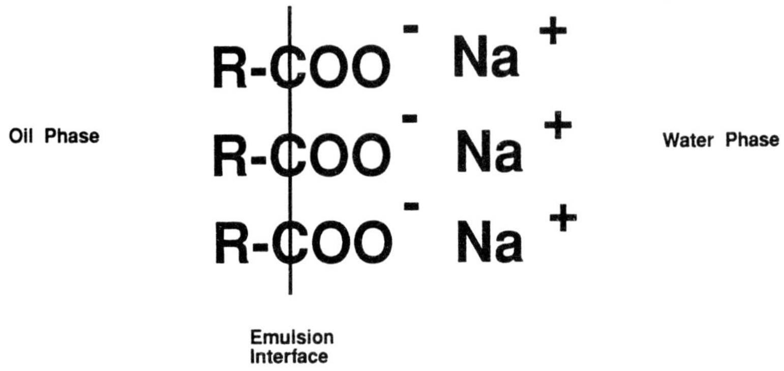

The result of this preferential adsorption is a negative charge at the interface surface, associated with dispersed oil phase droplets. This effect occurs throughout the emulsion system, giving a net negative charge on the surface of the oil droplets, as illustrated in Figure 3.

FIGURE 3

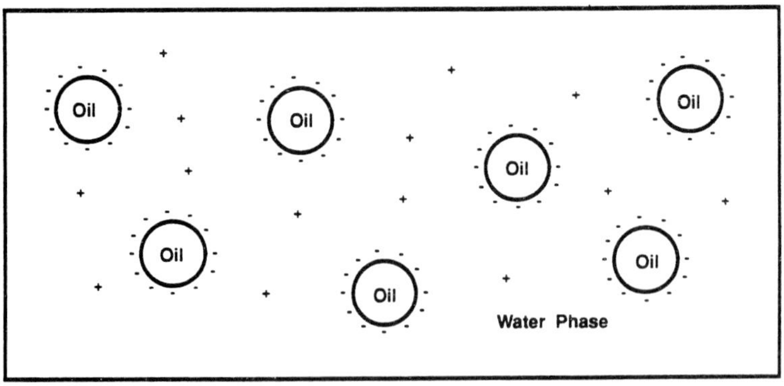

In this emulsion system, the Na ions are known as *counter-ions* and are left distributed somewhat randomly in the water-phase. The $RCOO^-$ ions are referred to as the *co-ions* and they determine the charge on the dispersed droplet itself. The existence of dispersed phase droplets, all carrying a similar net charge within the emulsion system, means that if two droplets approach each other, then electrostatic repulsion will occur. Because of the net negative charge on the dispersed phase droplets, electrostatic attraction will cause some of the Na^+ ions in the continuous phase to migrate towards the negative charges associated with the droplets. The resultant effect is an increase in concentration of Na^+ ions around each dispersed phase droplet such that an electrical double layer is formed. This results in a reduction of the mutual electrostatic repulsion that occurs when two dispersed phase droplets approach each other. This phenomenon is known as the *charge shielding effect.*

This mechanism of repulsion can be more easily understood by examining changes in the electrical potential energy plot that occur as the droplets approach each other.

The electrical potential plot of a negatively charged dispersed phase droplet in isolation (i.e. *in vacuuo*) is represented in Figure 4.

FIGURE 4

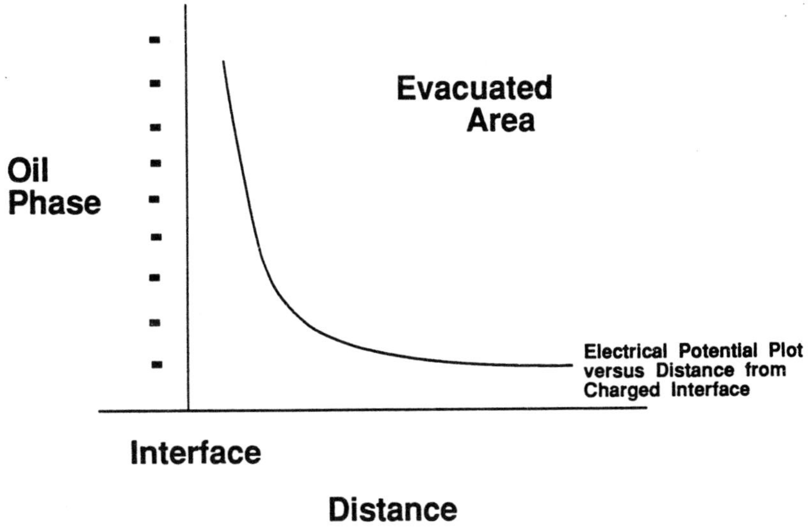

In the emulsion system itself, the charge shielding effect will reduce the forces of mutual repulsion and the electrical potential energy plot will be modified as illustrated in Figure 5.

If further positive ions are added to the emulsion system, as in the case of the addition of small quantities of electrolyte such as sodium chloride, further charge shielding will occur as illustrated in the electrical potential energy plot illustrated in Figure 6.

This increase in charge shielding is due to the increased concentration of Na^+ ions. The Cl^- ions from the added electrolyte are distributed randomly in the system. This "charge neutralisation" or shielding effect is dependent upon the concentration of electrolyte added and the valency of the ion concerned. Thus, for the aluminium ion, Al^{3+}, a much higher charge shielding effect than that observed with the Na^+ ion will occur, for the same concentration of electrolyte.

In summary, if the stability of an emulsion is dependent on charge repulsion, the

FIGURE 5

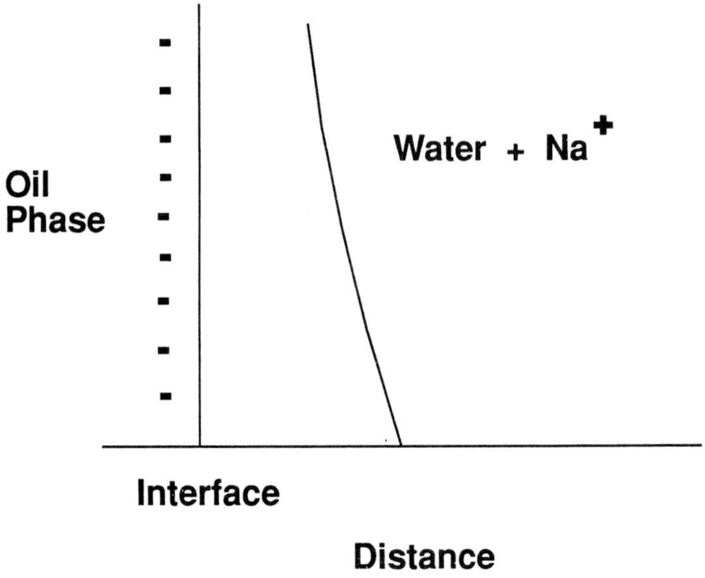

FIGURE 6

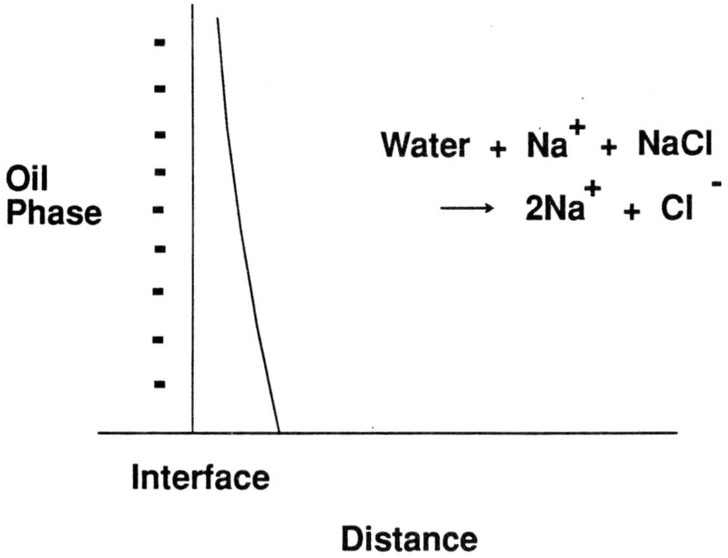

addition of an electrolyte will tend to destabilise it, due to charge shielding. Polyvalent ions give a higher degree of charge shielding than their monovalent counterparts. Clearly, charge shielding only occurs when the species added possesses a charge of opposite sign to that of the one on the dispersed phase droplets in the emulsion.

2. *Interfacial film strengthening*

The stabilisation of emulsions through the use of interfacial film strengthening, is very important for cosmetic and toiletry products. There are three basic methods for improving interfacial film strength.

2.1. *The use of powders*

Any powder considered for use in the stabilisation of emulsion systems must be of much smaller particle size than that of the dispersed phase droplet in the emulsion. Additionally, the powder must not be completely wetted by either phase. If the powder is only partially wetted by either phase, it will migrate to the emulsion interface. The type of emulsion stabilised by any given powder will be dependent upon the characteristics of the powder itself. A hydrophilic powder, that is one which exhibits a higher affinity for the water phase than the oil phase, will stabilise an oil-in-water emulsion system. Conversely, a hydrophobic powder will only stabilise a water-in-oil emulsion.

In either case the mechanism of stabilisation depends on the mechanical strengthening effects of the powder, which resides at the emulsion interface, as illustrated in Figure 7.

FIGURE 7

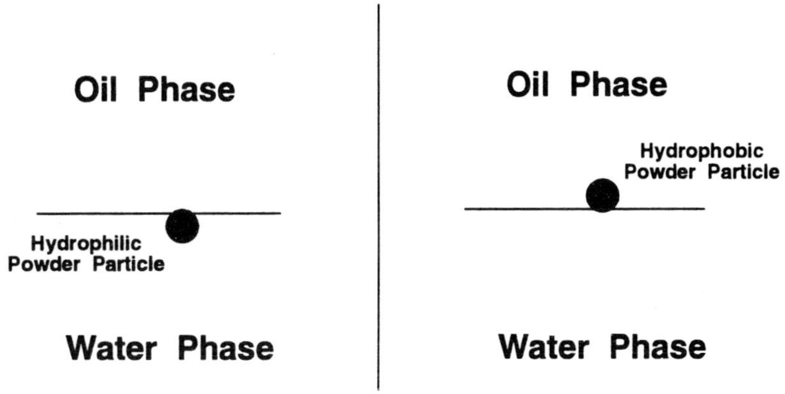

The stabilisation of cosmetic emulsions through the use of powders is not that common and is found only in ccrtain categories of product.

Examples of powders that are used to stabilise cosmetic emulsions are magnesium aluminium silicate and bentonite.

2.2. *The use of polymers*

Certain types of polymer can successfully be used to stabilise cosmetic emulsions. Suitable polymers are normally composed of a high molecular weight, long-chain, hydrophobic backbone, to which is attached a series of hydrophilic polar side groups. Such a structure is illustrated in figure 8.

FIGURE 8

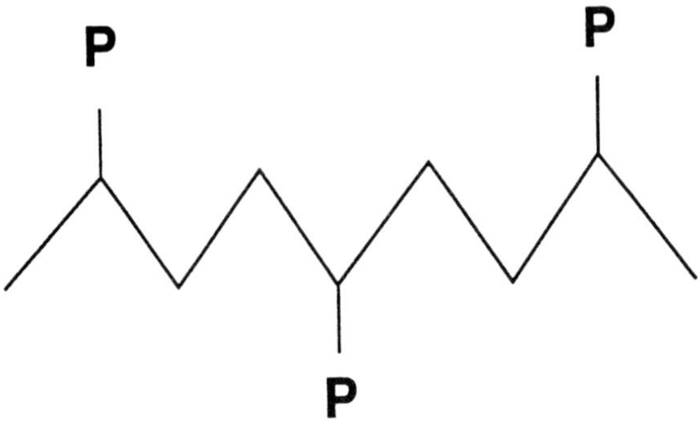

Where;

P = hydrophilic polar side group

When materials of this type are added to an emulsion system, the polar side-groups are orientated into the water phase, whilst the backbone of the polymer chain has an affinity for the oil phase. Thus, the polymeric species is orientated at the emulsion interface, as illustrated in figure 9.

FIGURE 9

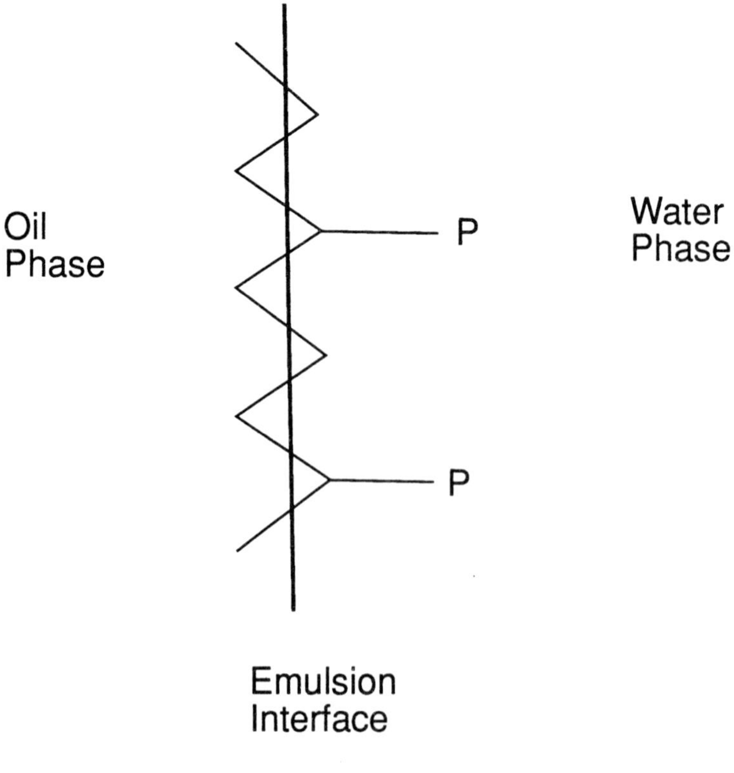

The migration of the polymer to the interface will serve to strengthen it, thus significantly reducing the chances of dispersed phase particle coalescence, in the event of a collision. In practice, polymers are normally added as thickeners (rather than stabilisers) to increase the viscosity of the continuous phase of the emulsion but they also strengthen the emulsion interface. This method of increasing emulsion stability is often referred to as colloid stabilisation and is frequently used in cosmetic and toiletry emulsion products.

Examples of polymeric materials which are used to stabilise cosmetic emulsions are Carbomer 954, Hydroxyethylcellulose and Polyvinyl Alcohol.

2.3. *The Addition of nonionic surfactants*

The use of nonionic surfactants in the stabilisation of cosmetic emulsions is, by far, the most important method utilised today. The use of nonionic surfactants, or emulsifiers as they are more commonly called, allows an enormous degree of flexibility in the formulation and stabilisation of a wide variety of both oil-in-water and water-in-oil emulsion systems.

All nonionic emulsifiers have some degree of affinity for both the oil and the water phases of the emulsion and they therefore migrate to the emulsion interface and reside there, thus stabilising the emulsion through mechanical strengthening. Emulsifiers are often used in combination, rather than singly, such combinations frequently providing a higher degree of emulsion stability due to the increased packing density at the interface that emulsifier blends provide.

Nonionic emulsifiers are sometimes used in combination with anionic emulsifiers, such as the sodium soaps reviewed earlier. In these systems, which are invariably O/W emulsions, the sodium soap acts as the primary emulsifier, whilst the nonionic emulsifier provides secondary stabilisation. The mechanism of stabilisation in this case is interesting. The negative charges associated with the hydrophilic ends of the fatty acid chains will repel each other at the emulsion interface and "spread out" to give the most stable arrangement. If a nonionic surfactant is now added to this system as a secondary emulsifier, the nonionic polar head group will sit at the interface between the positively charged head groups of the primary emulsifier, as illustrated in figure 10.

This produces an interface which is more closely packed and therefore stronger, thus resulting in a more stable emulsion system.

A full review of nonionic emulsifier types is provided later in this chapter.

The selection of an emulsifier system

The most important step in producing an emulsion is selecting the emulsifier system from the many thousands of emulsifiers commercially available. The most commonly used emulsifiers are anionic and nonionic systems, or combinations of the two. The selection of emulsifiers for any particular emulsion system is dependent on many factors.

Primary selection is based on the production of the right emulsion type (i.e. O/W or W/O) and the emulsifier must be compatible with both oil-phase and water-phase components. The primary emulsifier is the main determinant of emulsion type, whilst the secondary emulsifier serves to strengthen the interface and hence improve stability. In

FIGURE 10

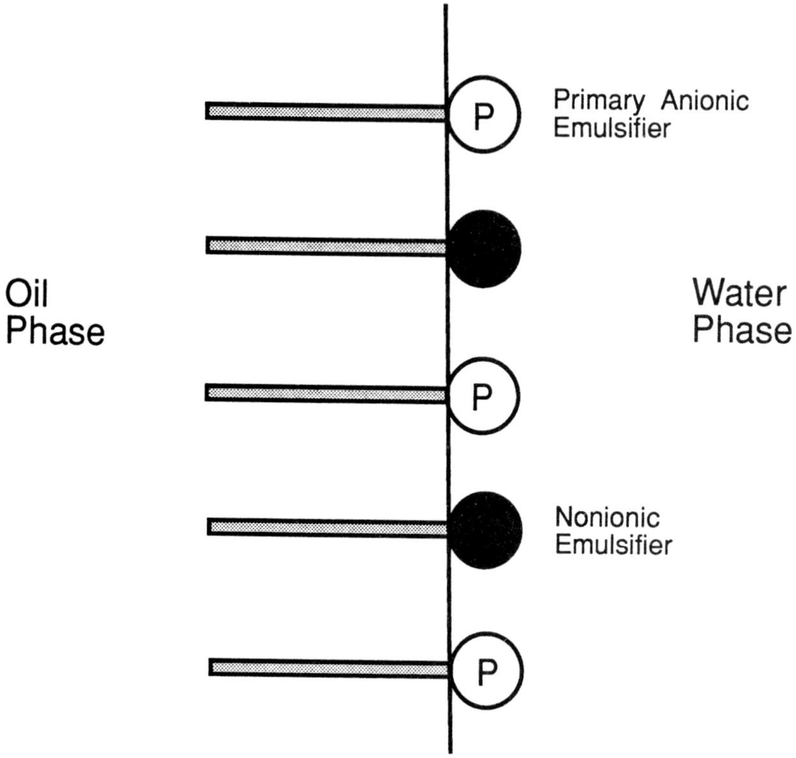

selecting an emulsifier system, the primary emulsifier must first be chosen. This is often a monovalent soap for O/W emulsions, polyvalent soaps being used to produce W/O systems. The use of sodium soaps is often rejected as they are too alkaline when neutralised, producing a hard a soap with limited solubility. Potassium soaps, or more typically alkanolamine soaps (eg. diethanolamine or triethanolamine soaps), are much softer and more soluble. Triethanolamine stearate is one of the most widely used emulsifiers, being cheap and readily available, although it can irritate some people's skin and tends to gel with time. In addition, the concern with the possible formation of nitrosamines in products containing amines of this type has reduced the popularity of this class of emulsifier.

Apart from chemical type, a number of other factors must be considered when selecting an emulsifier system, such as mildness on skin, colour, odour, and lubricity or emollience, as well as the solubility of the emulsifier in each of the two phases. The emulsifier selected must be soluble in at least one of the phases, ideally the continuous phase. If it is not soluble in either phase, no emulsion will be formed. Historically, natural materials such as beeswax or lanolin have been used in anionic emulsifier systems, although both of these often exhibit a tendency to produce W/O emulsions. Such natural raw materials however, tend to vary somewhat in both price and emulsifying ability from year to year and supplier to supplier.

Amongst the range of nonionic emulsifiers available, there is a very wide variety of ethoxylated and non-ethoxylated materials including sorbitan esters (derived from sugars), fatty alcohols, fatty amine ethoxylates, fatty acid ethoxylates and their esters and alkyl phenyl ethoxylates. This choice can be narrowed down for cosmetic applications by considering only materials with good safety data, especially on skin and eye irritation.

One of the most rapid ways of selecting a suitable emulsifier for formulating a new product is to use suppliers literature. Another approach is to use the HLB System.

The HLB (Hydrophilic-Lipophilic Balance) system

The HLB system is one of the most widely used aids in emulsifier selection. The HLB number of an emulsifier is related to the hydrophilic-lipophilic balance of that emulsifier at the oil/water interface.

The system was originally developed for ethoxylated nonionic surfactants, using a scale of 0 to 20, each emulsifier being assigned a number on this scale. A low HLB value means the emulsifier is very oil orientated, whilst a high HLB value means it is very water orientated.

Originally, the HLB value was defined as the percentage of the ethoxylated content of the molecule divided by 5, to bring the theoretical maximum of 100% down to an HLB number of 20. A simple example of this calculation is illustrated below.

$$CH_3(CH_2)_{16}COO(CH_2CH_2O)_{20}H$$

(20 mole ethoxylate of stearic acid)

Molecular weight	=	20 x (CH_2CH_2O)	=	880
		+ 1 x $CH_3(CH_2)_{16}COO$	=	283
		+ 1 x H	=	1
		Total		1164

Therefore HLB = 1/5 (880/1164 x 100) = 15

Surfactants with low HLB values tend to give W/O emulsions, most of the surfactant residing in the oil phase, and require the freedom of the oil phase as the continuous phase. Surfactants with high HLB values tend to give O/W emulsions, most of the surfactant residing in the water phase, and require the freedom of the water phase as the continuous phase.

There is an approximate correlation between the HLB value of a surfactant and the type of emulsion that it will form, if any. This relationship is summarised in Table 1.

TABLE 1

HLB of Surfactant	*Emulsion Type*
0 - 3	None
3 - 8	W/O emulsion
10 - 20	O/W emulsion
>20	Solubilisation

Although the HLB system was originally applied to nonionic ethoxylated surfactants, HLB values can be experimentally determined for other surfactants and can give values much higher than 20. HLB values can also be calculated for mixtures of emulsifiers as illustrated in the example below.

Consider a mixture of:

72% Emulsifier A (HLB = 8.7) + 28% Emulsifier B (HLB = 16.7)
The equivalent HLB of the mixture is given by the calculation:

$$(8.7 \times 72/100) + (16.7 \times 28/100) = \underline{11.00}$$

It is also possible to determine the "required HLB" for a given oil or oil phase. This is equivalent to the HLB value of the emulsifier required to emulsify that particular oil or oil phase mixture. Each oil will have two required HLB values, one for O/W emulsification and one for W/O emulsification. Examples of required HLB values for different oils are illustrated in Table 2.

TABLE 2

	Required HLB	
"Oil" Type	*W/O*	*O/W*
Stearic Acid	6	15 - 17
Mineral Oil	4–5	10 - 12
Cotton Seed Oil	5	7 - 10

Required HLB values for mixtures of oils can also be calculated, using similar methodology to that used for calculating HLB values for emulsifier mixtures.

Experimental determination of HLB values

There are two simple methods for estimating the HLB value of a given emulsifier experimentally. The first is based on visual assessment of the solubility of the emulsifier in water and, by its very nature, can only provide very approximate guidance. The appearance of the emulsifier when added to water is indicative of its likely HLB value as illustrated in Table 3.

TABLE 3

Appearance in water	*HLB value*
Non-dispersed	1 - 4
Poorly dispersed	3 - 7
Milky	7 - 10
Translucent	10 - 13
Clear	13+

The second method available is normally referred to as the "Method of Mixtures". This method requires the use of an oil of known required HLB value and a series of emulsifiers of known HLB value. The steps involved in determining the HLB value of an emulsifier using this method are rather laborious and, once more, can only provide an approximate answer. In practice, the suppliers of modern emulsifier systems will be able to provide the formulator with comprehensive data about their performance, including HLB value. Such information has rendered the use of either of the above methods largely redundant. The HLB system is still widely used however and although it is not the perfect answer for developing stable emulsions, the HLB concept is a very valuable aid. The HLB system does have significant shortcomings however, and these must be recognised by any formulator who attempts to use it. For example, the required HLB value of an oil will change in response to the presence of additives such as perfume, preservatives and secondary emulsifiers. In addition, the HLB system gives no indication of the quantity of emulsifier required to stabilise an emulsion system and HLB values themselves vary with temperature.

Selected emulsifier systems

Although there are many thousands of emulsifiers now available, some find very common use in a wide variety of products. The most well-known of these are described below.

1. *Sugar-based nonionic emulsifiers*

Esterified sugars, based on sorbitan, are a very important group of nonionic surfactants. There are two distinct classes of such materials, the simple sorbitan esters and their ethoxylated counterparts. The sorbitan esters are more hydrophobic and are often oil soluble and therefore tend to be of lower HLB value. The ethoxylated sorbitan esters on the other hand are typically water soluble, hydrophilic materials with higher HLB values.

Sorbitan esters and ethoxylated sorbitan esters are often used in combination to provide optimum stabilisation for emulsion systems. Indeed, many different combinations can be used to give the correct HLB value necessary to stabilise the emulsion but additional factors, other than just HLB value alone, are important in the selection of the best emulsifier system. One of the most important considerations is that of chemical stability of the emulsifier system and its ability to positively affect the stabilisation potential for the emulsion. This is best illustrated by the use of an example. The graph in figure 11 represents a plot of emulsion stability against HLB value, for three different nonionic emulsifier blends, "A", "B" and "C".

FIGURE 11

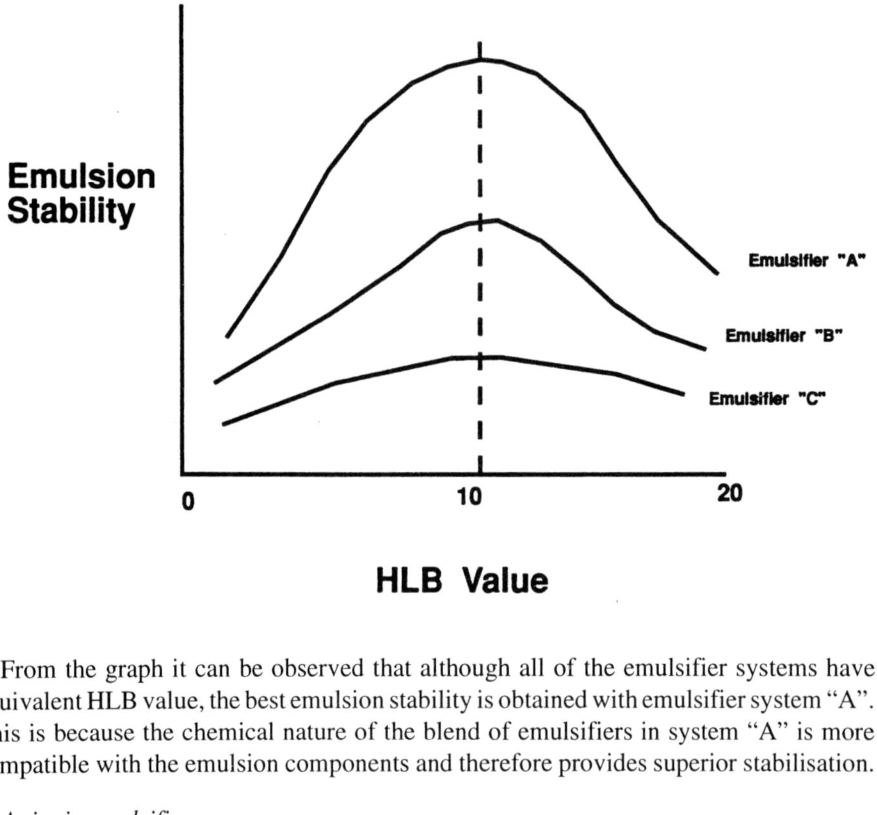

From the graph it can be observed that although all of the emulsifier systems have equivalent HLB value, the best emulsion stability is obtained with emulsifier system "A". This is because the chemical nature of the blend of emulsifiers in system "A" is more compatible with the emulsion components and therefore provides superior stabilisation.

2. *Anionic emulsifiers*

Anionic emulsifiers find widespread use in cosmetic emulsions, the most common of which is triethanolamine stearate, which has an HLB value of approximately 20 and therefore produces O/W emulsions. Triethanolamine stearate is formed insitu in the emulsion system by adding stearic acid into the oil phase and the triethanolamine base into the aqueous phase. Each phase is then heated separately to approximately 70°C, whereupon the hot oil phase is added to the hot aqueous phase and the emulsifier thus created, during emulsification.

This system is always used with a nonionic secondary emulsifier to improve emulsion stability. The secondary emulsifier, a typical example of which is cetyl alcohol, increases the packing density of emulsifier molecules at the emulsion interface, thereby mechanically strengthening it.

3. *Beeswax/borax system*

This system, which is rapidly becoming more of historical interest, is the basis of the traditional "cold cream" type of emulsion product. A beeswax/borax emulsion is produced by adding a solution of borax to a hot mixture of mineral oil and beeswax, thus producing a W/O emulsion. Beeswax consists mainly of fatty acids, straight chain fatty esters, cholesterol esters and hydrocarbon oils. Borax (sodium borate), in the aqueous phase, forms boric acid and sodium hydroxide, whilst the beeswax provides a primary emulsifier precursor (free fatty acids) and a secondary emulsifier (cholesterol esters and some of the straight chain fatty esters). The sodium hydroxide from the borax solution reacts with the free fatty acids, to give the primary emulsifying system.

If the water phase comprises more than about 40% of the formulation, then an O/W emulsion will result. If, however, less water is present, a W/O emulsion will be formed. This is because the long chain fatty esters, with their attendant low HLB values, will stabilise an W/O emulsion, when forced to do so by the phase ratio.

4. *Polyvalent soaps*

Polyvalent soaps, from HLB considerations alone, may be expected to form O/W emulsion systems in a very similar way to their monovalent counterparts. In practice, this is not the case and W/O emulsions are invariably formed. The reason for this is that this type of emulsifier does not function according to HLB concept, but rather their emulsifying action depends upon the emulsifier molecule shape. A typical example of a polyvalent soap is calcium stearate, which is wedged shaped as shown in Figure 12.

FIGURE 12

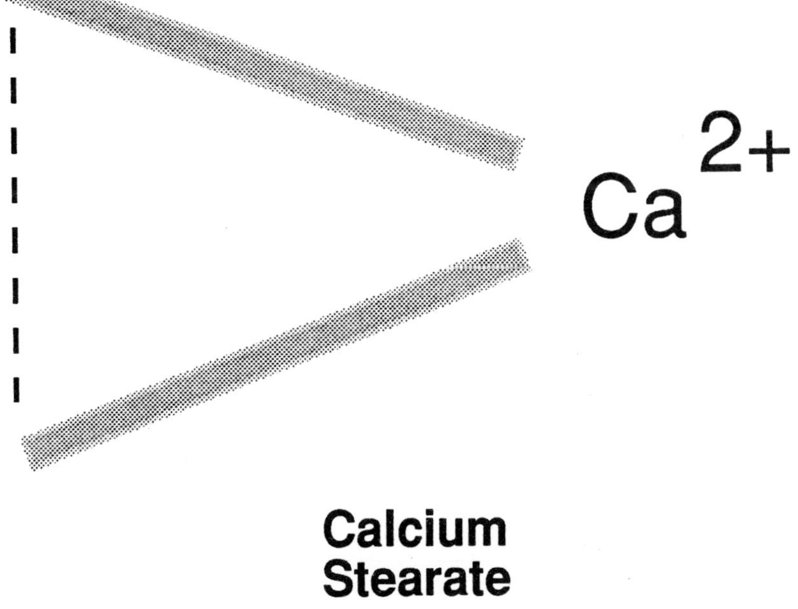

When the emulsifier migrates to the emulsion interface, its wedge-shaped configuration allows greater packing density for a W/O emulsion than that for a O/W system, as illustrated in figure 13.

FIGURE 13

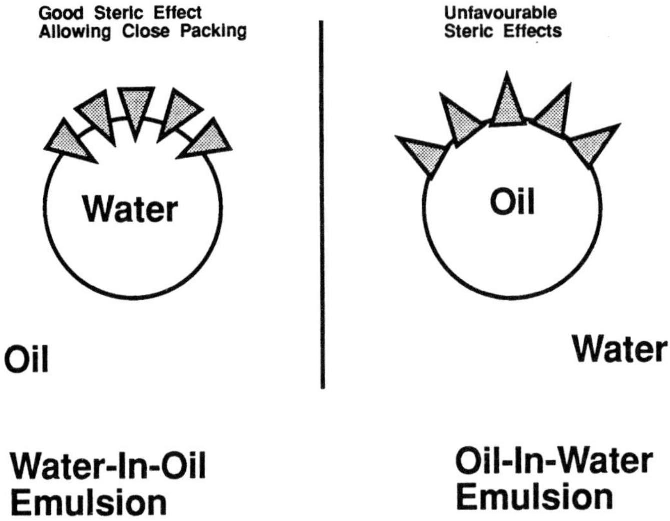

For this reason, polyvalent soaps will form W/O emulsions. Emulsifiers which function by this mechanism are referred to as "stereo-directing emulsifiers" and their emulsifying action cannot be predicted by consideration of HLB values.

The physical properties of emulsions

The physical properties of any emulsion system are determined by a large number of factors, many of which are very complex. Many emulsion characteristics however, are determined primarily by the size, type and charge of the dispersed phase droplets and the nature of the continuous phase in which they are dispersed.

Emulsion appearance, for example, is largely dependent on the size of the dispersed phase droplets and the way in which light is refracted or reflected from them. For this reason, dispersed phase particle size can be approximated from the emulsion appearance, as shown in the table 4.

TABLE 4

Emulsion Appearance	*Particle Size*
milky-white	> 1µ
blue-grey	1 - 0.1µ
translucent	0.1 - 0.05µ
transparent	<0.05µ

The appearance of the emulsion is also affected by the refractive index of the two phases.

Particle type, size and charge are determined by a number of factors and these can be summarised as follows.

1. *Particle type*

The dispersed phase particle type in the emulsion depends upon:

1.1 *The HLB value of the emulsifier*
Using an emulsifier with a low HLB value will normally produce a W/O emulsion with a hydrophilic water dispersed phase, whilst an emulsifier of high HLB value will give an O/W system, with a hydrophobic oil dispersed phase.

1.2 *The order of mixing*
When producing an emulsion product, the phase of the emulsion which is added to the other, often forms the dispersed phase. For example, if the oil phase is added to the water phase, then an O/W emulsion is likely to be formed.

1.3 *The phase-ratio*
The major phase in the emulsion, that is the phase which is present in the largest amount, tends to be the continuous phase. When the emulsifier HLB value dictates the formation of an O/W emulsion, it may dominate the effects of the phase ratio and an emulsion with a minor water continuous phase will result. However, if the emulsifier dictates the formation of a W/O emulsion, but the phase ratio favours the formation of an O/W system, then the phase ratio will be the dominant force and an O/W emulsion will be formed.

2. *Particle size*

Emulsion particle size is dependent upon the following factors:

2.1 *Mechanical energy*
The greater the mechanical energy used to create the emulsion, then the smaller will be the particle size. Thus, an emulsion system which is prepared under conditions of moderate agitation will exhibit larger dispersed phase particle size than an equivalent emulsion, prepared under conditions using higher energy vigorous mixing techniques.

2.2 *Quantity of emulsifier used*
The higher the level of emulsifier used to prepare an emulsion, the smaller the dispersed phase particle size. This observation is only valid if the HLB value of the emulsifier used, is suitable for creating the emulsion in the first place. If the HLB of the emulsifier is unsuitable, then no emulsion will form, irrespective of the quantity of emulsifier used.

2.3 *Order of mixing*
In trying to prepare an O/W emulsion, it is sometimes advantageous to add the water phase to the oil phase, rather than the conventional practice of adding the minor phase (the oil phase in this case) to the major phase (the water phase in this case). When using this

technique to prepare an O/W emulsion, a meta-stable W/O emulsion is formed initially. As further quantities of water are added, the emulsion undergoes a rearrangement to form the desired O/W system. Preparation of emulsions using this technique is known as *phase inversion*. This process is represented diagrammatically in figure 14.

FIGURE 14

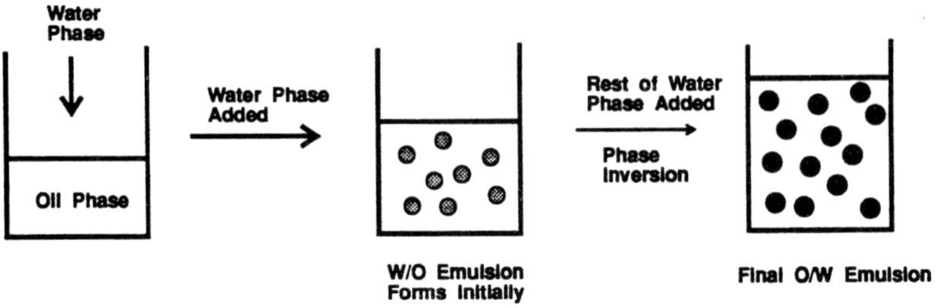

Emulsions prepared using the technique of phase inversion often exhibit a better stability profile than emulsions using classical techniques. This is because the emulsion undergoes two emulsification processes using phase inversion, which ultimately leads to a reduction of the emulsion dispersed phase particle size and, hence, improved emulsion stability.

3. *Particle charge*

Particle charge is relevant for O/W emulsions only, normally those stabilised with anionic emulsifiers such as triethanolamine stearate. In these cases, particle charge is affected by the following factors:

3.1 *Emulsifier type*
An anionic emulsifier leaves a residual negative charge on the dispersed phase droplet, whilst a cationic emulsifier leaves a positive change.

3.2 *Quantity of electrolyte present*
The higher the level of electrolyte that is added to the water-phase and the higher its ion valency, then the greater will be the charge shielding effect, as described earlier.

4. *Phase viscosity*

The viscosity of the continuous aqueous phase in an O/W emulsion can be increased by adding materials which impart a structure in solution. Typical examples of such materials include polymers, clays and thickening gums. The viscosity of the continuous oil phase in W/O emulsions can also be increased by the addition of high melting point waxes.

Reological properties of emulsions

One of the most important properties to consider when producing an emulsion product is its rheology. Emulsion rheology describes the way in which the emulsion flows when external forces are applied to it and, in practical terms, determine characteristics such as appearance, feel and the way in which the emulsion rubs into the skin.

The rheology of any given emulsion system depends on emulsion type and whether the continuous phase or the dispersed phase is present in the greatest amount. If the continuous phase is the major phase of the emulsion, its rheology is likely to be similar to that of the continuous phase.

If, on the other hand, the dispersed phase is the major phase, then emulsion viscosity will rise dramatically and the system will frequently exhibit thixotropic behaviour, with a yield point. The reason for the occurrence of this type of rheology, is that when the dispersed phase is the major phase, a pseudo-structure is set-up in the emulsion. This rheological state in the emulsion is emphasised by the following factors:

- A smaller emulsion particle size will lead to a greater interfacial area and, hence, a higher propensity to form a structure as detailed above.
- The occurrence of electrically charged dispersed phase particles in the emulsion, which produces an additional "drag" due to the *electroviscous effect*. This effect describes the phenomenon that occurs when an electrically charged particle is pulled through a gap between two other adjacent electrically charged particles. The force required to achieve this is increased by the electronegative repulsion effects between the dispersed phase particles, as illustrated in figure 15.

FIGURE 15

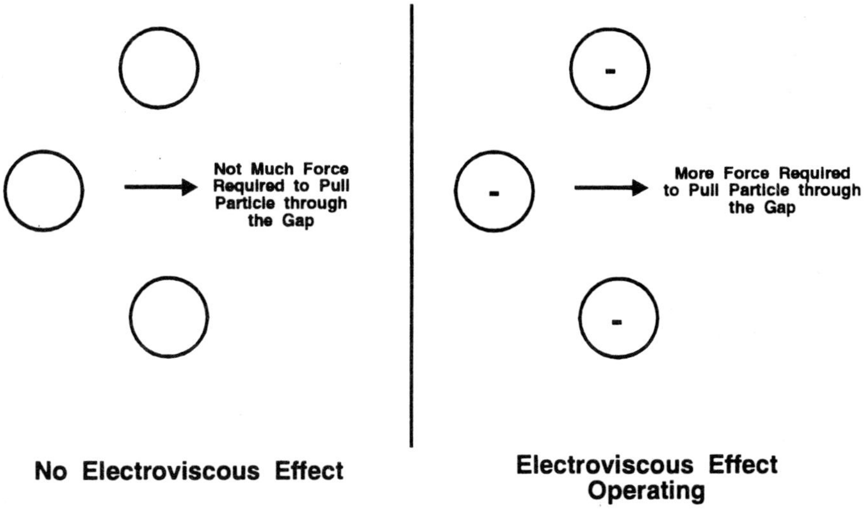

- The presence of a high viscosity dispersed phase, which produces more viscous drag due to a "coupling mechanism" between the dispersed phase particles, as illustrated in figure 16.

FIGURE 16

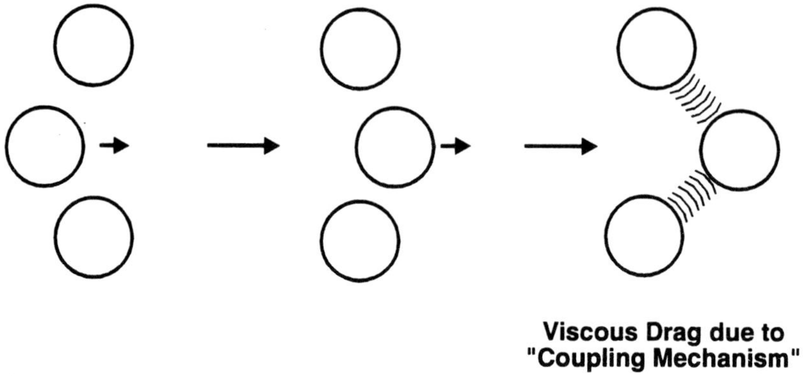

Other properties of emulsions

Other properties of emulsions are often used as a basis for simple identification of emulsion type. The most common of these include the following.

1. *Conductivity measurements*

If the continuous phase is the aqueous phase, the emulsion will conduct electricity, whereas if the continuous phase is the oil phase, it will not.

2. *Dye uptake*

An emulsion with an aqueous continuous phase will assume the colour of a water soluble dye, whereas if the oil phase is the continuous phase, no colour uptake occurs.

3. *Water dispersibility*

An emulsion with an aqueous continuous phase will be readily dispersed in water, whereas an emulsion with an oil continuous phase will not.

The stability of emulsions

The prime requirement for any successful commercial emulsion product, is adequate stability. Apart from the legal implications, products are often 3-12 months old when they are purchased by the consumer and should, of course, still be in good condition at that time. A shelf-life of at least two and a half years is required for cosmetic and toiletry products, with the emulsion being stable with respect to the following properties.

1. *Creaming/Sedimentation*

As described earlier, this is the mechanism whereby oil droplets rise to the surface in O/W emulsions (creaming) or water droplets fall to the bottom of W/O emulsions (sedimentation). The rate of creaming or sedimentation can be retarded by increasing the viscosity of the continuous phase, decreasing the dispersed particle size or increasing the volume of the dispersed phase, thus helping to give the emulsion more structure. In the case of O/W emulsions only, increasing the electrical charge on the dispersed phase particles will also have beneficial effect.

2. *Phase separation*

Phase separation describes the tendency for the dispersed particles to collide and coalesce forming droplets, eventually resulting in complete separation of the two emulsion phases. The rate of phase separation can be reduced by reducing the dispersed phase particle size or physically strengthening the emulsion interface by one of the methods described earlier in this chapter. In the case of O/W emulsions, an increase in the charge on the dispersed phase particles will also have a beneficial effect.

3. *Temperature*

A commercially produced emulsion product will, after manufacture, filling and packaging, be exposed to cold temperatures (eg. warehouse storage in the winter months) and hot temperatures (eg. warehouse storage in the summer months) and must therefore be stable under both conditions.

High temperature emulsion stability can be improved by using a non-melting oil phase system, which normally consists of a blend of high melting-point waxes, or by choosing an emulsifier system with a suitable temperature/ solubility curve. Some emulsifiers can be "leeched out" from the emulsion interface if the are insoluble in the system above a certain temperature.

Low temperature emulsion stability can be improved by using nonionic emulsifiers, rather than anionic emulsifiers, to stabilise the emulsion. The reason for this is that in anionically stabilised systems, as the external water phase turns to ice, the effective electrolyte concentration rises steeply, thus neutralising particle change, causing instability. Low temperature stability can also be improved by mechanically strengthening the interface, thereby reducing the possibility of rupture of the interface by ice crystals during freezing.

Stability testing of emulsions

There are various methods available for predicting emulsion stability, the more important of which are summarised below.

1. *Freeze/thaw testing*

This is a commonly used method in which the emulsion is cycled between a freezer (approximately –10°C) and an oven (usually at 37°C or 40°C) several times and any signs of emulsion break-down noted. When using this technique, 24 hours should be allowed to elapse after preparing the emulsion, before cycling is commenced. A 24 hour period should also be allowed to elapse between cycles, to allow the emulsion to equilibrate at room temperature before assessment is carried out. Successful completion of 5–6 freeze/thaw cycles will not guarantee an adequate shelf-life, but does indicate inherent emulsion stability.

2. *Ageing programmes*

It is also important to carry out constant temperature storage tests, when examining emulsion stability. Storage tests are normally carried out at low temperatures (1–5°C) and higher temperatures (40°C) as well as normal ambient temperatures, since the interfacial film may be weakened or destroyed more easily at temperature extremes. Ambient temperature testing (20–25°C) will normally be continued for up to two to three years. If the principle that first order reactions double in rate for every 10°C rise in temperature is accepted, then an emulsion which is stable for 6 months at 40°C should have a two year shelf-life at 20°C.

3. *Centrifugal force*

The technique of ultra-centrifugation is sometimes employed to assist in the determination of emulsion stability. If an emulsion remains stable after 10 minutes in an ultra-centrifuge, which typical revolves at a speed of 30,000–40,000 revolutions per minute, it has favourable chances of being commercially stable.

4. *Visual assessment*

Visual assessment of emulsion products is carried out both microscopically and by eye. The use of stains, either water or oil soluble, helps to determine dispersed phase particle size. Generally, the smaller and more evenly distributed the particle size, the greater will be the stability of the emulsion under examination.

5. *Low shear rate evaluation*

This technique involves agitation of the emulsion for a period of time, using a laboratory shaker, to accelerate the occurrence of any phase separation which may occur. Any significant signs of separation after a period of approximately three days, indicates inherent instability and the emulsion should be rejected.

A complete emulsion stability testing programme should also examine properties such as the influence of light on the emulsion, chemical incompatibility between emulsion components, potential for atmospheric oxidation of the emulsion and microbial stability.

References

1. Becher P. – Encyclopeda of Emulsion Technology (Volume 1), Marcel Rekken, 1983.

SECTION 4

Decorative Cosmetics

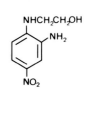

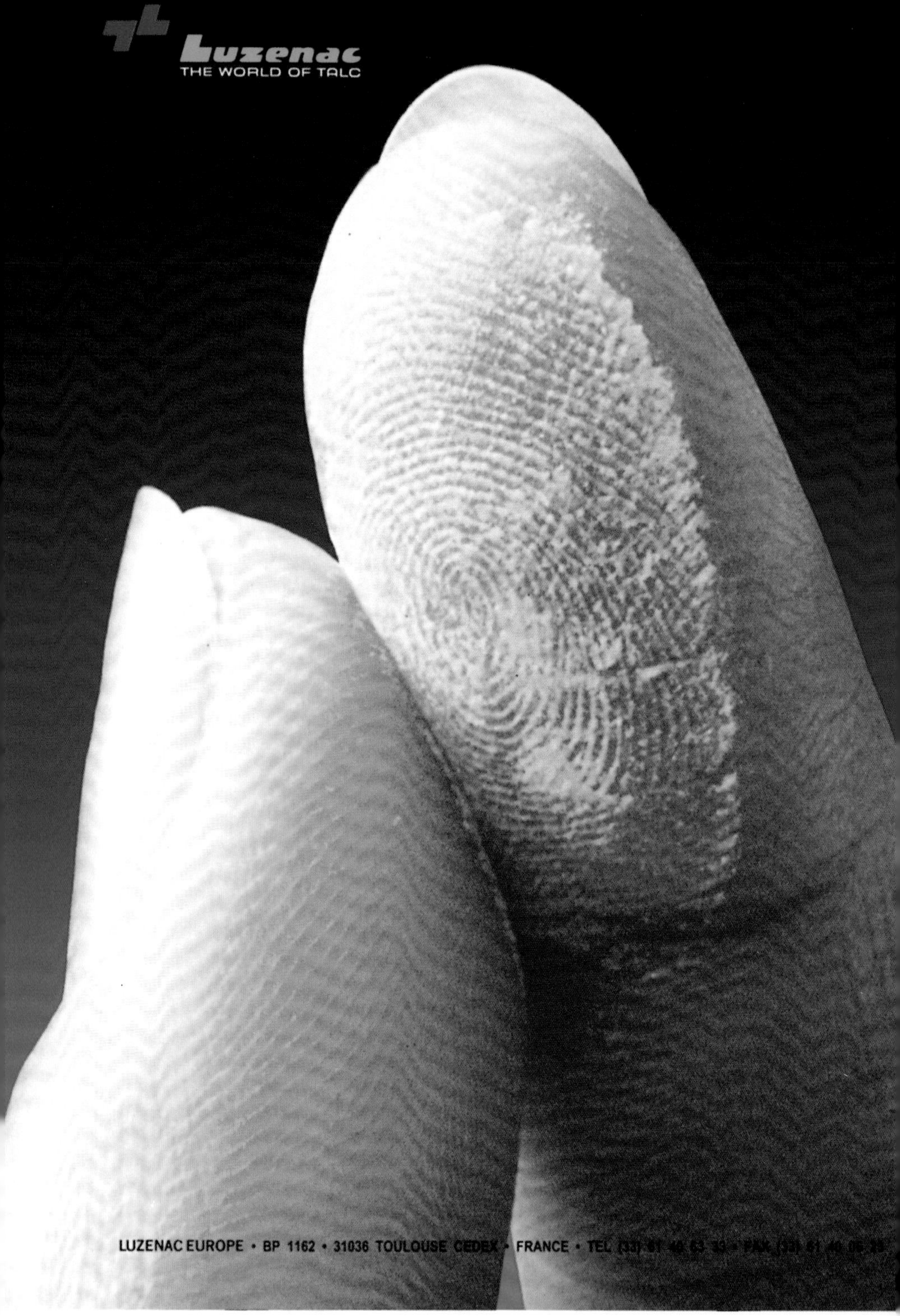
luzenac
THE WORLD OF TALC
LUZENAC EUROPE • BP 1162 • 31036 TOULOUSE CEDEX • FRANCE • TEL (33) 61

DECORATIVE COSMETICS

Introduction

Decorative cosmetics is the most obvious commercially orientated area of cosmetic science, being principally concerned with beautifying and decoration, rather than functionality. In spite of this high consumer profile, the science underpinning the development and production of cosmetics is no less complex and has strong links with both chemical and physical disciplines.

No discussion of cosmetics can be complete without a full understanding of the importance of colour, a prime component of nearly every cosmetic product available. For this reason, this chapter commences with an overview of the science of colour, before discussing the types of colours used in cosmetic products and the development of the products themselves.

Colour theory

Colour is a sensation aroused in the mind by the stimulation of the retina by light. Visible light occupies a small part of the electromagnetic spectrum, with wavelength boundaries from about 380 nm to 780 nm, as illustrated in Figure 1.

The energy of a particular light source is proportional to its frequency, which can also

FIGURE 1

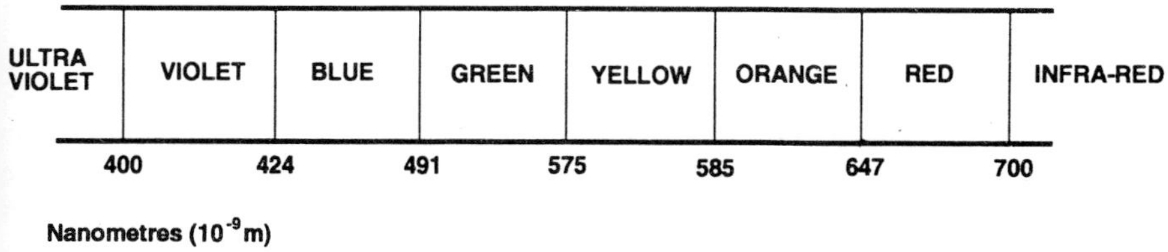

The Visible Spectrum

be expressed as the reciprocal wave-length. This means that violet light, with a wavelength of about 380 nm, has approximately twice the energy of red light at a wavelength of about 760 nm.

Factors which control colour perception are the nature of the source, the nature of the surface onto which the light falls and the response of the eye. Continuous light sources are used to view colour, the standard in the northern hemisphere being diffuse north sky light. This light can be approximated quite accurately by the use of daylight tubes in colour matching cabinets.

Colour vision

When considering the surface of snow, the eye sees predominantly diffuse reflectance from the fine particles of snow and therefore the surface appears white. In the contrasting case of a black carpet, white light is absorbed with no reflectance, and the carpet looks black. Reflectance is also dependent on refractive index, the higher the refractive index the better the reflectance. For example, titanium dioxide with a refractive index of 2.76 appears much whiter than zinc oxide with a refractive index of about 2.01. Talc, which has a larger particle size but low refractive index of 1.589, can be regarded as translucent.

There are two kinds of light receptors in the retina of the human eye. The first type of receptor is composed of rods and is responsible for the phenomenon called *scoptia*. These rods are particularly sensitive to low levels of light, such as moonlight, and may be regarded as monochromatic receptors. This form of visual stimulus cannot differentiate colours. The second sort of receptor is composed of cones, which operate when the level of illumination is of sufficient energy to trigger their response, usually the sun or artificial light. When the cones are activated, the eye is said to be in a condition of *photobia* and colours can be differentiated. There are also intermediate states of illumination in which some cones, as well as rods, are activated and this is called *mesopia*. The different types of colour vision are illustrated in Figure 2.

There are three types of cone and each contains a different pigmentation material. The types are known as an *erythrolabe*, which contains a red-sensitive pigment, a *chlorolabe*, which contains a green-sensitive pigment and a *cyanolabe*, which contains a blue-sensitive pigment. Only a small part of the retina is responsible for detailed vision and this is known as the *fovea*. The fovea contains the highest density of cones in the retina. The angle of the rays of light falling on this area is about 1 or 2 degrees. The eye therefore sees by constant movement so that it scans an area of vision.

1.1. *Defects in colour vision*

About 92% of people have normal colour vision and are referred to as *trichromats*. Such subjects have good colour differentiation and assessments skills and this means that they would be able to colour match in agreement with each other. The remaining 8% of the population, predominately male, have a degree of colour blindness and are referred to as *anomalous trichromats*. This terminology means that their response to red, green and blue is somewhat confused. Genetically, this defect is inherited by male offspring, and less than 1% of females have anomalous colour vision. The number of *monochromats*, that is people suffering from total colour blindness, is extremely low, although it is a more common phenomenon in animals.

FIGURE 2

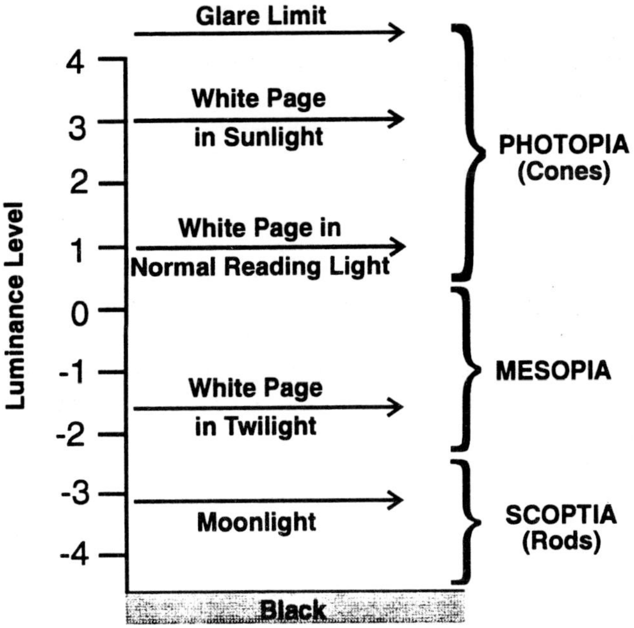

It is critical that anyone involved in colour assessment, or matching, should be tested for colour blindness. A set of test cards called the Ishihara Colour Tests are often used to determine the absence of colour vision defects. In this test, people with defective colour vision do not see the same patterns as those with perfect colour vision. Various other colour vision test can be used but they are often very time consuming and lengthy.

Apart from the physiological defects of colour vision, various other phenomena are sometimes encountered which can lead to incorrect assessment of colour. In some cases, two colour samples can appear identical when viewed under a particular light source but when the light source is changed, a mismatch is observed. This phenomenon is known as *metamerism*. For example, white light can be achieved by a suitable mix of blue light (460 nm), green light (530 nm) and red light (650 nm) only but this would appear identical to a continuous source of white light with components from across the entire visible spectrum.

2. *Colour interference*

Colour interference is an effect which occurs when two light waves travel different paths and arrive at the same point. If the waves are in phase, then they will combine to form a new wave of the same frequency but with double the amplitude. This is known as *constructive interference*. If the waves arrive out of phase, assuming the amplitudes are equal, they will combine and cancel each other out. This is referred to as *destructive interference*. If the waves are only partially out of phase, then some wavelengths will be reinforced and some will be destructively interfered. This is the phenomenon that occurs in the case of pearlescent interference pigments.

Colour interference of this type is best illustrated by the example of a thin film of titanium dioxide covered mica. If the thickness of the titanium dioxide is such that the white light which travels through it is out of phase by one-half wavelength of say red light, then the resulting light which is reflected will be blue-green because of destructive interference of the red component.

The description of colour

Scientifically, the most accurate way of describing colour is to use an objective method, for example quantification using instruments such as a visible spectrophotometer. However, the most common way in which colour is described is by the use of language which is inherently subjective and open to wider interpretation. In order to minimise the ambiguity of this approach, various systems have been introduced, an example of which is the *colour atlas*.

Three variables are normally used to produce these atlases, *hue*, *saturation* (sometimes referred to as chroma) and *luminosity* (sometimes referred to as value). Hue describes a colour in terms of its red, blue or green component, whilst saturation indicates the dominant wavelength. Luminosity is a measure of the brightness of the colour. Given that the eye can see about 3 million different variations of colour, such systems have obvious limitations. In spite of this they are valuable when attempting to describe colour. Examples of systems of this type are the Pantone reference, the Ostwald reference, the Munsell reference and the British Colour Council Dictionary of Colour Standards.

Colour mixing

There are two basic types of colour mixing process, additive and subtractive.

1. *Additive colour mixing*

This is the process of mixing together the three primary colours, red, green and blue. This method can be demonstrated by observation of a white light source through various combinations of primary-coloured optical filters. The different combinations of filters, and the colour observed when looking at a white light source through them, are listed in Table 1.

TABLE 1

Filter combination	*Colour observed*
red/green	yellow
red/blue	magenta
green/blue	cyan
red/green/blue	white

2. *Subtractive colour mixing*

This is the type of colour mixing process that occurs when mixtures of coloured dyes and pigments are used to produce decorative cosmetics. The three subtractive primary colours are yellow, magenta and cyan. Mixing of these colours produces an additive primary colour, as shown in Table 2.

TABLE 2

Colour mixture	*Colour observed*
yellow/magenta	red
yellow/cyan	green
magenta/cyan	blue
yellow/magenta/cyan	black

The legislation and nomenclature of cosmetic colourants

Historically, colourants have been used in cosmetics without any consideration being given to their possible toxicity. Arsenic, lead, antimony and cadmium based pigments were used extensively in the past. At the beginning of the twentieth century, when the understanding of the harmful effects of such compounds was recognised, restrictive use of these cosmetic colourants became necessary and, from that base, legislative regulations grew.

Today, all countries have regulations which control the type and purity of colours that may be used in cosmetics. For example, in the UK there is "The Cosmetic Product (Safety) Regulations 1989", published as a statutory instrument by Her Majesty's Stationery Office. Compliance with this regulation is mandatory for all cosmetic products intended to be distributed and sold in the United Kingdom. Should the cosmetic product be produced in the United Kingdom, for distribution and sale to another EEC country or the USA, then the legislation governing the use of colourants in those countries must also be complied with. The regulations governing the use of cosmetic colourants are continually changing and regular consultation is recommended. The Scandinavian countries and Japan, in particular, have their own specific regulatory requirements concerning the use of coloured materials in cosmetic products.

1. *Colour index and nomenclature*

The need to quote specific colourants in a universally accepted way, for legislative purposes worldwide, gave rise to the development of the Colour Index (CI) system, which catalogues all colourants used in all types of industry. The Colour Index Reference Manual is published jointly by the Society of Dyers and Colourists of Great Britain and the American Association of Textile Chemists and Colorists.

Of specific interest to the cosmetic scientist, is that all colourants are assigned a unique five-digit Colour Index number to describe them. Some cosmetic colourants may also be assigned a sixth digit as a suffix, indicating the salt of the colourant without being specific about the metal ion involved. In addition to describing colours by a specific CI number, the Colour Index also describes colourants generically.

Although cross-referencing using CI number is useful, the cosmetics industry has its own additional classification of cosmetic colourants, which is derived from legislation involving their usage in particular types of cosmetics product. The two major systems used to describe a cosmetic colourant have been designated by the Food and Drug Administration (FDA) of America and the EEC.

1.1. *FDA classification*

The FDA classification of cosmetic colourants splits materials into 3 major categories of use, as indicated below:

use in food, drugs and cosmetics	–	F,D & C Classification
use in drugs and cosmetics	–	D & C Classification
use in externally applied drugs and cosmetics	–	ext. D & C Classification

Coloured materials falling into the last category are not permitted for any application where ingestion may occur, including use in lip products.

This system of categorisation only applies to synthetic organic colourants, which are not permitted for use around the eye area, in the US. Inorganic colourants and natural colourants are not categorised but their purity standards are specified by the FDA.

Apart from the category of use, a sequential number is also used to describe a synthetic organic colourant, for example F,D & C Yellow No. 5 (CI 19140). This dyestuff can be converted into an insoluble lake which has the nomenclature F,D & C Yellow No. 5 Aluminium Lake (CI 19140:1). The suffix :1 in the colour index number denotes an aluminium lake.

In addition to the nomenclature, every batch of colourant intended for use in cosmetic products is tested for purity by the FDA, against a specification for that material, before it is granted FDA batch lot approval. The purity testing, which examines for levels of lead, arsenic and dyestuff intermediates amongst others, is certified and the batch of colourant is then given its own certification documents.

Should a colourant come under suspicion of being harmful, and the fact subsequently proven, the FDA will usually declassify the colour to an external use only classification.

2. *EEC classification*

The European classification system uses the EEC Cosmetics Directive, which is printed in the Official Journal of the European Communities. This classifies colourants in the context of their fields of application, as shown below:

- colourants allowed in all cosmetic products
- colourants allowed in all cosmetic products except for those intended for use in the eye area
- colourants allowed exclusively in cosmetic products not intended to come into contact with mucous membranes
- colourants allowed exclusively in cosmetic products intended to come into contact only briefly with the skin

Colourants are described in these sections by their CI Number and, where applicable, an "E" Number according to the EEC Directive concerning food colourants. In addition to this official method for describing colours, there is also a system of historical nomenclature which has its origins in trade names. For example the colourant CI 19140 (F,D & C Yellow No. 5 in the FDA system) is also commonly known in Europe as tartrazine (E102). Other similar examples are detailed in Table 3.

TABLE 3

CI number	*FDA classification*	*EEC classification*	*E no*
15985	F,D & C Yellow No. 6	Sunset Yellow FCF	E110
42090	F,D & C Blue No. 1	Brilliant Blue FCF	-
45430	F,D & C Red No. 3	Erythrosine	E127
16185	F,D & C Red No. 2	Amaranth	E123
45170	D & C Red No. 19	Rhodamine B	-
47005	D & C Yellow No. 10	Quinoline Yellow	E104

For the sake of clarity, the US FDA system of nomenclature will be used throughout the rest of this chapter, except in the case of natural dyes.

The physical properties of colourant materials

There are literally thousands of different coloured materials used in the field of decorative cosmetics. These can be classified into four categories, synthetic dyes, natural dyes, organic pigments and inorganic pigments, depending on their physical and chemical characteristics.

A dye can be defined as colouring matter which is soluble in a solvent or range of solvents, whilst a pigment can be defined as a coloured or white chemical compound, insoluble in the solvent in which they are used.

1. *Synthetic dyes*

The main use for dyes is in toiletries products such as shampoos and bath foams, in order to impart colour for aesthetic reasons. The use of synthetic dyes is extremely limited in decorative cosmetics because they tend to stain the skin. This feature, however, is used to advantage by formulating water-soluble dyes into artificial tanning creams or gels, where controlled skin staining is desirable. Dyes typically used for this purpose would be a blend of F,D & C Blue No. 1 (CI 42090), F,D & C Yellow No. 5 (CI 19140) and D & C Red No. 33 (CI 17200).

Oil-soluble dyes, commonly referred to as the eosins, are also used in lipsticks. Eosins used for this purpose include D & C Red No.21 (CI 45380), D & C Red No.27 (CI 45410) and D & C Orange No.5 (CI 45370). Caution should be used when formulating these dyes, as they can cause an allergic reaction in some people, when used in lipsticks at high percentages.

2. *Natural dyes*

This group of dyestuffs has come to the forefront of consumer awareness recently, most notably for their use in food. There is, however, no restriction on their use in cosmetics, apart from purity considerations. Generally, the resistance of natural dyes to heat, light and pH instability is much inferior to their synthetic counterparts and these factors should be thoroughly tested when they are used in finished products. A further disadvantage of using natural dyes is that they often tend to exhibit strong odours.

Some natural dyes have been assigned CI numbers, although many have not, due to the fact that they are complex mixtures of more than one colouring principle. For this reason, natural dyes are more often referred to by their E numbers, as assigned by the EEC system.

A list of the more commonly available natural colourants is given in Table 4.

TABLE 4

Colour	*Description*	*Source*	*E number*
yellow	curcumin	turmeric	E100
yellow	crocin	saffron	E160
orange	capsanthin	paprika	E160c
orange	annatto	annatto	E160b
orange	carotenoids	carrots	E160a
red	cochineal	*Coccus cactii*	E120
red	betanine	beetroot	E162
red	anthocyanins	red berries	E163
green	chlorophylls	lucerne grass	E140/141
brown	caramel	sugars	E150

All of the above colourants are of vegetable origin, with the exception of cochineal which is extracted from the crushed insects *Coccus cactii*. This natural dyestuff can also be turned into a pigment by conversion to its aluminium lake, which is known as carmine.

3. *Organic pigments*

These are mostly conjugated cyclic compounds based on a benzene ring structure, although some heterocyclic compounds exist. Apart from cyclic conjugation, which produces the chromophoric system , additional chemical groups known as auxochromes are also required. The auxochromes, typically chemical groups such as –OH or $-NH_2$ are electron donating, whereas halogens (chlorine, bromine and iodine), –C=O, or $-NO_2$, are electron accepting groups. The overall effect is to produce an extensive population of delocalised electrons, which are easily excited by visible light and therefore the compound appears coloured.

There are three main types of organic pigments; lakes, toners and true pigments. Definitions of these classifications are given overleaf but as organic pigments are seldom used without a diluent, all three types are often classified as lakes. The purpose of the diluent, or substrate, is to maintain a consistent colouring strength of the organic pigment, which can vary from batch to batch.

In terms of stability, true aluminium lakes can be affected by extremes of pH, resulting in the soluble dye reforming, a condition known as "bleeding". Also, they are relatively transparent and have poor stability to light. Toners on the other hand, are more resistant to heat, light and pH, although extremes of pH can result in shade changes. True pigments are the most stable, although their existence is relatively uncommon.

3.1 *Lake pigments*

Various definitions exist to describe the word lake, depending on its industrial usage. The definition used in the cosmetic industry is that a lake is essentially an insoluble colourant, produced by precipitating a permitted soluble dye on to a permitted substrate. For example, water soluble F,D & C dyes can be precipitated as aluminium salts onto alumina (aluminium hydroxide), the resultant lakes finding use in both cosmetics and food. Typical examples of aluminium lakes are listed below:

- F,D & C Yellow No. 5 Aluminium lake (CI 19140:1)
- F,D & C Yellow No. 6 Aluminium lake (CI 15985:1)
- F,D & C Red No. 3 Aluminium lake (CI 45430:1)
- F,D & C Blue No. 1 Aluminium lake (CI 42090:2)
- F,D & C Red No. 10 Aluminium lake (CI 16035)

Although of natural origin, carmine (CI 75470) an aluminium lake of the natural dye cochineal, can be made in a grade which is approved for both food and cosmetic use. It is very bright, quite stable and can be used in cosmetics for use around the eye area. Its main disadvantage is that it is very expensive.

Other aluminium lakes which can be used in cosmetic applications, but are not permitted for foodstuffs, include:

- D & C Red No. 21 Aluminium lake (CI 45380:3)
- D & C Red No. 27 Aluminium lake (CI 45410:2)
- D & C Yellow No. 10 Aluminium lake (CI 47005:1)
- D & C Orange No. 5 Aluminium lake (CI 45370:2)
- D & C Red No.19 Aluminium lake (CI 45170:3)

Of these lakes, D & C Red No. 19 is not permitted for use in cosmetic applications in the US.

It is also possible to produce aluminium lakes from dyes which are chemically the same as F,D & C dyes but only meet D&C purity standards. An example of this is D & C Yellow No. 5 Aluminium Lake (CI 19140:1). In cosmetic products, most lakes are based on aluminium, although zirconium lakes are also found.

3.2 *Toners*

Organic pigments can also be made using other cosmetically approved metals, besides aluminium, and these colourants are referred to as toners. Typically, both calcium and barium lakes fall into this category, examples of which are listed below:

- D & C Red No. 6 Barium lake (CI 15850:2)
- D & C Red No. 9 Barium lake (CI 15585:1)
- D & C Red No. 7 Calcium lake (CI 15850:1)
- D & C Red No. 34 Calcium lake (CI 15880:1)

D & C Red No. 9 Barium lake is not permitted in the US, for cosmetic use.

3.3 *True pigments*

A true pigment is an insoluble compound which contains no metal ions, examples of which are D & C Red No. 30 (CI 73360), D & C Red No. 36 (CI 12085) and D & C Orange No. 17 (CI 12075). The latter is not permitted for use in cosmetics in the US.

4. *Inorganic pigments*

The range of inorganic pigments used in cosmetics is made up of several chemical types but, generally speaking, they are all slightly dull when compared to their brighter organic counterparts. In terms of heat and light stability however, inorganic pigments are far superior to organic pigments. Their only disadvantage is that they can react under certain extreme conditions of pH. For example, ultramarines can react with acids to produce hydrogen sulphide and iron blue and manganese violet sometimes react with alkalis to produce shade changes or colourless salts.

Most inorganic pigments can be used in all types of cosmetic products. Subject to purity levels of heavy metals, their greatest use is probably in face and eye make-up. Some of the more common types of inorganic pigment are discussed below.

4.1 *Iron oxides*

These are used in all types of cosmetic products and three basic shades exist, black, red and yellow. By blending these three oxides in the required proportions, browns, tans, umbers and siennas can also be produced. Iron oxides are invaluable in liquid foundations, face powders and blushers, given that an almost infinite range of natural looking flesh tones can be produced by carefully blending them.

Chemically, yellow iron oxide (CI No. 77492) is hydrated iron II (ferrous) oxide, $FeOOH.nH_2O$. It is produced by the controlled oxidation of iron II (ferrous) sulphate. Depending on the conditions of manufacture used, this pigment can range from a light lemon to an orange-yellow shade. Red iron oxide (CI No. 77491), chemically iron III oxide, Fe_2O_3, is the familiar rust-red coloured pigment and is obtained by the controlled heating of yellow iron oxide. Black iron oxide (CI No. 77499) is chemically Fe_3O_4 and is a ferroso-ferric (Iron II/III) mixed oxide, prepared by controlled oxidation of ferrous sulphate in alkaline conditions.

4.2 *Chromium dioxides*

Chromium dioxides are used for most categories of cosmetic preparations but are not permitted for use in lip products in the USA. Two types of chromium oxides exist, Cr_2O_3 (CI No. 77288), a dull olive green anhydrous oxide and $Cr_2O_3.9H_2O$ (CI No. 77289) a blue-green hydrated oxide.

4.3 *Ultramarines*

Ultramarines (CI No. 77007), describes a class of pigment which is a polysulphide sodium/aluminium sulpho-silicate. There are a range of ultramarines varying in shade from blue to violet, pink and even green. The main form encountered in cosmetics, apart from lip products in the USA where it is not permitted, is the traditional ultramarine blue, which has a complex structure and may be written as $Na_8Al_6Si_6O_{24}S_2$. Regardless of shade, all ultramarines are classified under CI No. 77007.

4.4 *Manganese violet*

Chemically, manganese violet (CI No. 77742) is $MnNH_4P_2O_7$, a purple coloured pigment of manganese III and is quite brightly coloured, considering its classification as an inorganic pigment.

4.5 *Iron blue*

Iron blue (CI No. 77510), commonly known as Prussian Blue, is chemically ferric ammonium ferrocyanide, $Fe_4[Fe(CN)_6]_3.nH_2O$. It is a very intense dark blue pigment which is used widely in all cosmetic applications, except for lip products in the USA, where it is not permitted. The use of iron blue is somewhat limited, due to its physical hardness and very strong colouring power.

4.6 *White pigments*

White pigments are commonly used in all types of cosmetic applications. Four main types of white pigment are used in the manufacture of cosmetic products, titanium dioxide, zinc oxide, hydrated alumina and barium sulphate.

4.6.1 *Titanium dioxide (TiO_2)*

Titanium dioxide (CI No. 77891) is perhaps the most commonly used of all white pigments. It has extremely good covering power, or opacity, and is almost totally inert, being extremely stable to both heat and light. Furthermore, it is easily incorporated into all cosmetic products.

4.6.2 *Zinc oxide (ZnO)*

The use of zinc oxide (CI No. 77947) is steadily declining, in many cases being superseded by the use of TiO_2. Zinc oxide is less opaque than TiO_2 and also has a mild antiseptic action on the skin.

4.6.3 *Hydrated alumina*

Hydrated alumina (CI No. 77002), chemically $Al_2O_3.xH_2O$, gives very little opacity and is almost transparent. It is used in products such as lipsticks or mascaras, to give structure without any colour contribution.

4.6.4 *Barium sulphate*

Barium Sulphate (CI No. 77120), $BaSO_4$, is also known as *blanc-fixe*. It is relatively translucent and can be used as a pigment extender. In practice, its use in cosmetics is gradually declining, possibly because of the known toxicity that exists with the soluble barium ions present in some grades of barium sulphate.

Pearlescent pigments (pearls)

The most important requirement for a substance to be pearlescent, is that the crystals from which it is composed should be plate-like and have a high refractive index. The pearlescent effect is then obtained as a result of individual plates lining up and acting like tiny mirrors. Being transparent, each plate reflects only part of the incident light reaching it and transmits the remainder to the plates below. It is the simultaneous reflection of light from these many tiny layers that gives the shimmering lustre called pearlescence. In addition, by controlling the optical thickness of the plates, it is possible to produce colours in certain pearlescent pigments, due to a light interference phenomenon called iridescence.

Various types of pearlescent materials are used in decorative cosmetic products, as detailed below.

1. *Organic pearls*

Organic cosmetic pearls (CI No. 75170) produce a bright silver pearl effect and were only obtainable from fish scales, as either platelets or needle-like highly reflective particles. The natural materials responsible for producing this pearl effect are numerous crystals of a purine called guanine. Although still available, but in short supply, organic pearls are very expensive and have almost exclusively been replaced by synthetic inorganic pearls.

2. *Inorganic pearls*

The first synthetic inorganic pearl to become available was bismuth oxychloride, BiOCl, which has a CI No. 77163 and produces a silvery-grey pearlescent effect. Bismuth oxychloride is synthesised as tetragonal crystals, which produce varying effects depending upon the crystal size. Crystal sizes of approximately 8 μ give a soft, opaque, smooth lustre, larger crystals in the order of 20 μ give a more brilliant sparkling effect. The major disadvantage of bismuth oxychloride is its poor stability to light, causing it to darken and develop a greyish colouration after prolonged exposure. In spite of this problem, which can be overcome by the addition of a UV-absorber in the finished product, it is extensively used in decorative cosmetics. The most frequent applications are as an economical pearling agent in nail enamels, lipsticks, blushers and eye shadows.

Bismuth oxychloride can also be modified in several ways, thus altering the final effect that it has on the finished product. Deposition of bismuth oxychloride on mica produces a silver-grey colouration, whilst bismuth oxychloride bonded to titanium dioxide which is then deposited on mica, produces a whiter, silvery-grey shade. Coloured inorganic pigments bonded to bismuth oxychloride deposited on mica produce coloured pearls and bismuth oxychloride coated talc has a translucent silver appearance. These modifications are not simple mixtures but essentially sandwiches of pearl, or pearl and pigment, made by uniformly coating inner layers of platelets of the mica or talc.

All of these pearls, including bismuth oxychloride, are free-flowing powders which do not agglomerate and are therefore easily mixed into powder, aqueous, or non-aqueous cosmetics. Their particle size determines the lustre and dispersion characteristics. UV-absorbers are recommended to prevent their colour deterioration on exposure to light.

In addition to bismuth oxychloride pearlescents, titanium dioxide coated mica systems, or titanated mica, are also extensively used in cosmetic products. Titanium dioxide coated mica (CI Nos 77891 & 77019) exist in several different forms, as detailed below.

2.1. *Silver pearls*

Titanium dioxide is used to uniformly coat platelets of mica, in much the same way as bismuth oxychloride described earlier. Because titanium dioxide can exist in two crystal forms, either anatase or rutile, these can be used to individually coat the mica. The rutile crystals give a particularly brilliant pearl effect, caused by their higher refractive index.

2.2 *Coloured (interference) pearls*

Titanium dioxide is capable of being coated onto mica with extreme precision, so that different thicknesses can be achieved. At certain thicknesses, interference of light can take place so that some wavelengths of the incident light are reflected and others transmitted. The colours created by this effect are complementary to each other. Each interference colour is determined by the optical thickness of the film, a measurement obtained by multiplying the geometric thickness of the film by its refractive index. If the layers are very thin, a silvery-white reflection colour is produced. As the film thickness is increased, then yellow-gold, red, blue and green reflection colours are observed.

2.3 *Coloured (pigmented interference) pearls*

In addition to ultra-thin layers of titanium dioxide being deposited onto mica, coloured pigments can also be laminated with this interference film. This produces a two-colour effect caused by light rays being reflected, refracted or transmitted at the different pigment layers. The visible interference effects appear like an iridescent lustre, whose intensity is dependent on the angle of view.

2.4 *Coloured (pigmented) pearls*

Coloured pearls consist of sandwiches of titanium dioxide coated mica, with an additional layer of coloured pigment. This produces a combined colour and lustre effect, which is more pure than can be achieved by simply mixing pigment with silvery pearls.

In addition to the bismuth oxychloride and titanium dioxide pearls described above, miscellaneous pigments are also incorporated in cosmetic products. These are basically novelty pigments which nevertheless produce spectacular and eye-catching effects. Polyester jewels are precision-cut regular shapes of brilliant glittering polyester foil, which have been epoxy-coated with light-fast coloured pigments. Their main application is in nail enamels or body make-up. These materials cannot be used in eye make-up because the edges of these jewels are very sharp and may scratch the surface of the eye. Another type of novelty pearlescents are the daylight fluorescent pigments. These are transparent organic resin particles containing fluorescent dyes in solid solution. Their main use is in non-aqueous wax/solvent based products. Compliance with applicable cosmetic regulations should be thoroughly investigated before their use is considered.

Powdered cosmetic products

Powdered cosmetic products is a broad category normally used to described face powders, eye-shadows and blushers. All types of powder cosmetics need to satisfy three requirements if they are to be commercially successful. Firstly the product, when applied to the skin, must be of the desirable shade and must not significantly change as is worn. Secondly the product must feel smooth in use, thereby facilitating easy application to the skin. This property is very much a function of the slip of the product, an important factor when considering formulation. Lastly, the product must remain on the skin for a reasonable period of time, without reapplication. This requirement is a function of the adhesion of the product.

In addition to these requirements, the absorbency of a face powder is also important because, as perspiration take place at the skin's surface oily deposits can be excreted, which may lead to the formation of undesirable shiny patches on the face. Finally, the coverage characteristics of the face powder have to be carefully designed, such that good coverage is achieved without creating an obviously "made-up" appearance. This is normally achieved by carefully balancing the amounts of opaque and translucent raw materials used in the formulation.

1. *Raw materials for powdered cosmetics*

The range of raw materials used in the formulation and manufacture of powdered cosmetic products is extensive and a thorough review of them all is beyond the scope of this text. Some of the more commonly encountered materials are discussed below.

1.1 *Talc*

Talc is the major component of most face powders, eye-shadows and blushers and in some products it can comprise up to 70% of the formulation. Chemically it is a hydrated magnesium silicate, which possesses an outstanding combination of high spreadability or slip, with low translucency and covering power. Talc is principally mined in Italy, France (French Chalk), Norway, India, Spain, USA, Australia, China, Egypt and Japan. Of these, the Italian, French, American, Australian and some Indian and Chinese grades can be used for formulating cosmetic powders. Cosmetic talc should be white in colour and must be absolutely free of asbestos, for safety reasons. Typically, talcs are purified by sterilisation, gamma irradiation frequently being the most successful treatment. Cosmetic talc should also be of the foliated type, consisting of thin plate-like structures for maximum slip, and free from gritty particles. Particle size should be such that it passes through a 200 mesh sieve (75 μ).

Micronised talc is also available and is generally more light and fluffy but very fine grades can feel less smooth on the skin. The absorbency of talc is not particularly high as the material does not have a great affinity for water.

1.2 *Zinc oxide*

Zinc oxide is sometimes used at moderate levels in face powders because it has quite good covering power and is mildly astringent, therefore exhibiting slight soothing properties. Zinc oxide has the unusual property that it can protect against ultraviolet light and, for that reason, is often used as an inorganic sunscreen agent. Only the finest grades of zinc oxide should be used in cosmetic products, as lower quality grades often contain gritty particles. Zinc oxide exhibits moderate adhesion to the surface of the skin.

1.3 *Kaolin*

Kaolin, also known as china clay, is a naturally occurring, almost white, hydrated aluminium silicate. Whilst there are many hydrated aluminium silicates only three, kaolinite, nacrite and dickite can be classified as china clays. Kaolin does not exhibit a high degree of slip, so high levels in formulations can produce a harshness of feel. It is considered to be hygroscopic and therefore exhibits good absorbency, although some opinions suggest that this may be a negative property, which can lead to streaking of product on the face in damp weather. Kaolin is a dense material and is sometimes used to reduce unacceptably high bulk densities in loose powder products. It also provides a matt surface effect and can therefore be used to remove the slight sheen left by the talc in a product, if there is a requirement to do so.

1.4 *Calcium carbonate*

Calcium carbonate, or precipitated chalk, is frequently used because of its excellent absorption characteristics. Like kaolin, it provides a matt finish and can give a bloom effect to the coating on the face. Because calcium carbonate is not particularly smooth, high levels should be avoided otherwise an undesirable dry, powdery, feel can result. It has moderate covering power.

1.5 *Metallic soaps*

Both zinc and magnesium stearates are very important ingredients for imparting adhesion or "cling" to face powder formulations. They are usually incorporated at between 3 and 10 per cent of the formulation. Stearates used in this way must be of high quality because the presence of unsaturated fatty acids can lead to rancidity. Zinc stearate is considered to give the best adherent qualities to a face powder and is also considered to have smoothing qualities. Incorporation of too high a level of the stearates gives a blotchy effect on the skin, although careful formulation will avoid this. The metallic soaps will also add a degree of water repellency to a formulation.

1.6 *Magnesium carbonate*

Magnesium carbonate is available in very light, fluffy grades which absorb well and is often used to absorb the perfume before mixing it into face powders. Magnesium carbonate can also be used to provide bulk in face powders, although over-zealous use will compromise the smoothness of the finished product.

1.7 *Starch and modified starches*

Rice starch was once a very common ingredient in face powders and was considered to give them a "peach-like" bloom. It provides a smooth surface on the skin because the starch particles are smooth in shape and only about 3 μ-8 μ in diameter. One of the main problems when using rice starch is that it tends to cake in the presence of moisture and cling to facial hair. In some circumstances, the moistened product can also become sticky providing an ideal environment for the support of bacterial growth. In view of these disadvantages, the use of rice starch has now been largely superseded by the substitution of talc.

Special hydrophobic grades of modified starches have been developed more recently. These are claimed to possess properties of slip, adhesion, covering power and bloom, without the disadvantages that untreated starches exhibit. The smoothness of starches has a different quality from that of talc and metallic stearates, possibly because it rolls as it moves over the skin.

1.8 *Plastics*

Micronised polythene, polystyrene and nylon microspheres, all impart a smooth feeling to powder products. Plastics are typically used at between 5%-10% of the formulation, higher usage levels being precluded due to commercial considerations.

1.9 *Mica*

Chemically, mica is potassium aluminium silicate dihydrate. It is mined as the ore muscovite and occurs as multi-layered bright translucent sheets. Cosmetic mica is refined and ground to particle sizes of 150 μ or less, before use. Mica imparts a natural translucence to face powders and powder blushers, when used at levels of up to 20% in formulations.

1.10 *Fumed silica*

Fumed silica is basically silicon dioxide which has been processed to give tiny spheres of very low bulk density. It is normally used to decrease the density and bulkiness of very highly pearlised powder products, which would otherwise have an unacceptably high density. Normally only low percentages, less than 1%, are used in the formulation because it can impart a very dry, unpleasant feel to powder products on the skin.

1.11 *Perfumes*

The use of perfumes is extremely important in almost all categories of cosmetics and toiletries and powdered cosmetics are no exception. The choice of perfume for face powders and blushers is extremely important, as powder products often exhibit a rather earthy smell which is better masked. Perfumes used in any powder product should be of high stability and low volatility. It is important to note that powder products designed for use around the eye area should not contain perfume on the grounds of product safety.

1.12 *Miscellaneous additives*

In addition to those raw materials specifically mentioned above, there are many other materials that are often incorporated into powder cosmetics. Materials such as silk powder, silk protein and collagen have all appeared in powders in more recent times. These types of additive are frequently included to make superior product claims on the pack, thus making the product more attractive from the consumers perspective.

1.13 *Preservatives*

Preservation is not usually a problem in powders which are used dry, but the addition of small amounts of biocide are recommended. Typically, the parabens are often incorporated, sometimes in conjunction with imidazolidinyl urea.

1.14 *Binder systems (for pressed powders only)*

Pressed face powders, blushers and eye-shadows are extremely popular because they can easily be carried by the consumer. In order for a free-flowing powder to be pressed into godets, liquid binder has first to be uniformly dispersed throughout. Aqueous binder systems have been used in the past but they suffer from obvious faults such as their eventual drying out, condensation in compacts and, if tin-plate godets are used, the occurrence of corrosion and rusting.

Historically, lanolin and its derivatives were first used as binders, with moderate success. Blends of lanolin with mineral oil and an ester such as isopropyl myristate were the next development and these were far superior to the original lanolin-based binder systems. Indeed, this type of binder system is still used today because of its relatively low cost and effectiveness.

The latest developments in binder technology, made necessary by the trend for high pearl contents in eye-shadows, has been the use of highly branched esters which do not attack the polystyrene packaging in the same way that isopropyl myristate does. Typical examples of such esters are pentaerythritol tetraisostearate (PTIS) and isostearyl neopentanoate (ISNP), which can either be used on their own or as part of a mineral oil/lanolin/ester mixture.

2. *Colourants for powder cosmetics*

Obviously, colourants are an essential ingredient in all powder cosmetics, be they face powders, foundations or blushers.

2.1 *Face powder pigments*

By blending the three basic iron oxide pigments, black iron oxide (CI No. 77099), yellow iron oxide (CI No. 77492) and red iron oxide (CI No. 77491), plus titanium dioxide (CI No. 77891), almost any shade of flesh tone can be achieved. Pearlescent pigments, usually white in colour, can also be added to face powders to give various novelty effects.

2.2 *Blusher pigments*

The three basic iron oxides and titanium dioxide are used with one or more of the lakes listed in Table 5, to achieve various blusher shades for application to the cheeks.

TABLE 5

Colour	*CI number*	*Comments*
D & C Red No. 6 Barium lake	15850:2	
D & C Red No. 7 Calcium lake	15850:1	
D & C Red No. 9 Barium lake	15585:1	Prohibited in the USA and EEC
D & C Red No. 19 Aluminium lake	45170	Prohibited in the USA
D & C Red No. 21 Aluminium lake	45380:3	
D & C Red No. 27 Aluminium lake	45410:2	
D & C Red No. 30 Aluminium lake	73360	
D & C Red No. 33 Calcium lake	17200	Used in blushers
D & C Red No. 34 Calcium lake	15880:1	
D & C Red No. 36 Barium lake	12085	
D & C Orange No. 5 Aluminium lake	45370:2	
D & C Orange No. 17	12075	Prohibited in the USA and EEC
D & C Yellow No. 10 Aluminium lake	47005:1	
F,D & C Yellow No. 5 Aluminium lake	19140:1	
F,D & C Yellow No. 6 Aluminium lake	15985:1	
F,D & C Red No. 3 Aluminium lake	45430:1	
F,D & C Red No. 40 Aluminium lake	16035	

Pearlescent pigments are allowed in all cosmetic products in both the EEC and the USA.

2.3 *Eye-shadow pigments*

The most commonly used approved colours for eye-shadow products are shown in the Table 6. These pigments are also used in mascaras, eyeliners and other eye products.

TABLE 6

Colour	*CI number*
Black Iron Oxide	77099
Yellow Iron Oxide	77492
Red Iron Oxide	77491
Titanium Dioxide	77891
Ultramarine Blue/Green/Pink	77007
Iron Blue	77510
Chromium Oxide	77288
Hydrated Chromium Green	77289
Manganese Violet	77742
Carmine	75470

Pearlescent pigments are allowed in all categories of eye product in both the USA and the EEC. The use of organic colours, normally obtained form coal tar derivatives, for eye products is prohibited in the USA, although they are still permitted for use in the EEC countries. In spite of this however, their use has declined rapidly because of the US ruling.

3. *The formulation, manufacture and filling of powder cosmetics*

There are four main categories of powder cosmetics, loose face powders, pressed face powders, pressed powder blushers and pressed powder eye-shadows. The formulation of each type will be considered in turn, along with some typical formulations for each and the respective manufacturing process.

3.1 *Loose face powders*

This type of face powder is declining in popularity, in favour of pressed face powder products. The smoothness of a loose face powder can be enhanced by using very fine powders, such as talc, metallic soaps, fine pearls and mica in the formulation. Products designed for greasy skin are required to be drier and may contain materials such as magnesium carbonate. Irrespective of formulation type, high quality raw materials must be used if a product is to perform effectively and consistently. Three examples of typical loose powder formulations, aimed at users with normal, dry and oily skin respectively, are illustrated below. All of these products would produce an identical dark brown shade for darker skin types, although the product colour can easily be varied by altering the respective quantities of the iron oxides.

Raw material	*Dry (% w/w)*	*Normal (% w/w)*	*Oily (% w/w)*
Zinc stearate	6.00	5.00	2.00
Zinc oxide	13.00	7.00	5.00
Calcium carbonate	-	4.00	5.00
Magnesium carbonate	-	1.00	5.00
Black iron oxide	0.30	0.30	0.30
Red iron oxide	4.00	4.00	4.00
Yellow iron oxide	3.00	3.00	3.00
Lanolin oil	1.00	–	–
Fragrance	0.80	0.80	0.80
Methyl paraben	0.10	0.10	0.10
Talc	71.80	74.80	74.80

The talc is the base for all of the above formulations and provides the slip on the skin. Zinc oxide is incorporated to improve the covering characteristics of the products, whilst zinc stearate is used as a solid binder. Both calcium and magnesium carbonate provide absorbency characteristics. Lanolin oil is used as a liquid binder in the dry skin formulation. If a higher degree of translucency is required, finely ground mica can be incorporated at between 2% and 5%, at the expense of the zinc oxide. Pearlescent products can be formulated through the use of titanated mica, accompanied by a reduction in the levels of opacifying powders.

In order to prevent streaks, or subsequent colour development during use, all formulations, with the exception of pearls, should be thoroughly ground in the laboratory to fully extend the colours. In practice, this operation is normally carried out using a domestic coffee grinder, which is extremely efficient so long as it is not overloaded. Overloading means that good fluidisation cannot take place in the powder and grinding will be incomplete. Because laboratory grinding in this way is so efficient, problems can arise during the scaling-up process for manufacture, where the grinding efficiency is often much lower. In order to avoid this problem, good correlation must be obtained between the laboratory grinding process and its manufacturing counterpart. Only then can manufacture proceed with sufficient confidence of success. When progressing to full scale manufacture, a better result is often obtained if colours are pre-ground, for example as 20% extensions on talc, in large batches which are then adjusted for colour strength. This approach ensures that less problems are met in colour matching the finished products.

The first stage of the manufacturing process is to combine all the powder ingredients, but not the pearls, in a stainless steel ribbon blender. This is a trough-like mixer with several helical blades which run the length of the trough, very close to the walls. Mixing time can be as long as one or two hours, depending on how quickly the colour becomes even. During the mixing process, the perfume is slowly sprayed into the mix and blended until homogeneous. The entire batch is then pulverised through a hammer mill and subsequently colour checked. If required, colour adjustments are made and the batch returned to the ribbon blender for mixing. After adequate mixing has taken place, the batch is then milled once more. Colour adjustments will be difficult if the original pigment extensions were inadequate because there will be further colour developments in each milling process.

Once the colour of the formula is acceptable, any pearl or mica in the formulation is added and the batch is returned to the ribbon blender once more for final mixing. The next step in the manufacturing process is to run the entire batch through a fine screen and into closed containers, awaiting filling. Storage at this point is important, as the bulk density of the powder mixture is normally reduced during the mixing process. If the product was to be filled in this condition, without the appropriate time allowed for settling, there is a risk that low fill weights may occur.

Loose powders are generally filled through an orifice in the base of the packaging container, which is subsequently sealed with an adhesive disc. If the powder container has a transparent lid, circular patterns can sometimes occur because of static electrical forces and slight classification (separation) of the powder ingredients, particularly if one of the ingredients is very light or fine. In order to identify these problems at an early stage, filling trials during the development process are recommended.

3.2 *Pressed face powders*

Pressed face powders were first introduced in the USA in the 1930's and have overtaken loose powders in popularity because of their ease of application and the reduced danger of spillage when carried in a handbag. Originally, pressed powders were made using a moulding process, based on the inclusion of plaster of Paris. This has now been superseded by methods which rely on the compression of either damp or dry powders.

The basic materials used in pressed powders are the same as those used in loose face powders, the main difference being that in order to produce a cake which is strong, and yet pays-off smoothly, an effective binder must be used. If the powders are to be pressed into a tin-plate godet, then corrosion of the untinned top edge is possible if water-based binders have been used. In such cases the use of aluminium godets should be considered, although this problem can sometimes be overcome by the use of a corrosion inhibitor, mixed at a low level with the binder. If the binder level is too high, difficulty will be experienced by the user in trying to remove the powder with the puff. Excessive binder levels may also lead to glazing of the pressed powder surface making it waxy in appearance, with very little or no pay-off.

Examples of typical pressed face powder formulations are given below. Both produce a light brown shade by virtue of the proportions of iron oxides and titanium dioxide present.

Raw material	*1*	2
Talc	58.60	73.40
Kaolin	20.00	–
Zinc stearate	10.00	10.00
Pentaerythritol tetraisostearate	–	5.00
Calcium carbonate	–	4.00
Magnesium carbonate	–	1.00
Titanium dioxide	4.00	4.00
Yellow iron oxide	0.90	0.90
Red iron oxide	0.50	0.50
Black iron oxide	0.20	0.20
Methyl paraben	0.20	0.20
Fragrance	0.60	0.80
Lanolin/mineral oil/isopropyl myristate binder	5.00	–

The filler in these formulations is either talc or a talc/kaolin mixture. Formulation *1* uses a more conventional zinc stearate/lanolin/mineral oil/isopropyl myristate binder system, whilst formulation *2* incorporates the much newer pentaerythritol tetraisostearate as a co-binder with zinc stearate. The calcium and magnesium carbonates in formulation *2*, provide absorbency and bulk, respectively.

The recent development of silicone/mineral oil treated micas and colour pigments have given rise to compressed face powders which can be applied wet or dry. When dry, they are indistinguishable from regular compressed powders, except perhaps that they are smoother. When a wet sponge is applied to the cake, no water penetrates the cake itself and any water splashed on to the cake surface is repelled. The powder on the sponge blends with the water, giving a more even application to the face and a heavier coat than when it is applied dry. These types of pressed face powder are known as "two way" cakes and they can be used as either a foundation or face powder. The main disadvantage with products using silicone treated powders is cost, the latter being up to three times more expensive than conventional untreated powders.

When formulating pressed face powders, it is important to assess their characteristics carefully during the development process, to ensure optimum performance. Care should be taken to ensure that the cake can be rubbed completely away by the puff, without glazing, even when freshly pressed. Pay-off must be sufficient and the powder should spread in a pleasing way on the skin, without compromising adhesion to the face. The cake consistency should be smooth throughout the usage of the product. Any coarse particles in the ingredients will be left behind by the puff and embedded in the cake surface, giving an exaggerated impression of roughness.

Packaging for pressed face powders should also be carefully considered. As with the loose face powders, the possibility of corrosion of tin-plate godets should be contemplated. Some of the adhesives which are used to stick the godets into the packaging are water-based emulsion systems and these can subsequently cause rusting on storage. Hinges and clasps should be assessed and tested carefully, as they can often break after very little use. If esters such as acetylated lanolin alcohols are used as binders in the formulation, then polystyrene packaging components should be avoided. In such cases, the most appropriate choice of packaging material is styrene-acrylonitrile (SAN) which is not susceptible to attack by short-chain esters. The main disadvantage of SAN is the cost penalty incurred over the use of polystyrene.

During manufacture of pressed face powders, the mixing and colour checking processes are very similar to those used for loose face powders. Sometimes the powder mix is first milled without the binder and then again after its addition. If the product contains a pearl, this is added during the ribbon-blending process. If it becomes necessary to subsequently pass a batch containing pearls through a hammer mill, then this should be carried out with the mill screen removed, a technique known as "jump gapping". More preferable however, is to avoid milling any product containing pearls completely, as the risk of damage to the pearl itself is high. The fine screening process which is carried out with loose face powders is not so essential with pressed powder types.

Powder pressing is sometimes more successful if the powder is kept for a few days to allow the binder system to fully spread over the powders, especially in the case where pearls or mica are present. In the UK, the two most commonly used presses for face powder are the ALITE and the KEMWALL presses. The ALITE is a high-speed hydraulic press, suitable for long manufacturing runs, whilst the KEMWALL is a ram press operated by compression springs. The KEMWALL is a cheaper press to purchase and has a much quicker set-up time, making it more suitable for short production runs. Both machines have a circular table, or carousel, with holes in them into which the godets fit. A piston under the godet lowers it to a particular depth and the resulting cavity passes under a powder reservoir and is filled with powder. The table moves round and at the next position on the carousel the powder mass is compressed against an appropriately shaped punch, across which is stretched a constantly renewed nylon mesh. The mesh gives a neat finish to the surface of the powder and greatly reduces the chances of the punch becoming clogged. It also helps to eliminate any air which may be trapped between the punch and the surface of the pressed powder, during the pressing operation. If air does not escape properly during compression then the compressed mass can flake off. The reason for this is that a pocket of trapped air will prevent the particles from compacting properly, thereby resulting in a laminar layer of weakly adhering particles just under the pressed surface. The pressures used and the speed of pressing depends on the characteristics of the individual formulation.

3.3 *Pressed powder blushers*

Pressed powder blushers are similar to pressed face powder formulations, except that a greater range of colour pigments are used. They are usually applied with a brush and are mostly used for highlighting and shaping the facial features. The formulation of a typical pressed powder blusher, producing a mid-red shade, is given below.

Raw material	*% (w/w)*
Talc	65.70
Zinc stearate	8.00
Titanium dioxide	3.50
Red iron oxide	12.00
Black iron oxide	0.20
D & C Red No. 9 barium lake	0.30
Titanium dioxide coated mica	6.00
Pentaerythritol tetraisostearate	4.00
Methyl paraben	0.10
Imidazolidinyl urea	0.10
Fragrance	0.10

The titanium dioxide and the colours form the pigment system in this formulation, providing the required shade of red on the face. Care should be taken that only non-bleeding pigments are used otherwise, because of the high colour content, skin staining can result. Talc is the filler in the above formulation and the binding system is made up of a combination of pentaerythritol tetraisostearate and zinc stearate. Titanated mica gives the required pearl effect and a methyl paraben/imidazolidinyl urea preservative system is incorporated. The manufacturing process for this type of product is similar to that described earlier for pressed face powders.

3.3.1 *Pressed powder rouges*

Pressed powder rouges are normally sold in small godets and are often quite intense in colour. They are usually applied with the finger. There is a greater tendency for glazing to occur when using finger application and this must be taken into account when the formulation is being developed.

3.4 *Pressed powder eye-shadows*

The technology for producing pressed powder eye-shadows is very similar to that of the pressed powder products already discussed. The permitted colour range is very limited, particularly in USA where no synthetic organic pigments are allowed in the eye area and formulations are generally restricted to the red, yellow and black iron oxides, ultramarine blue, violet and green, chromium oxides, manganese violet, carmine NF, iron blue and permitted white pigments such as titanium dioxide. A surprisingly wide range of effects can be obtained however, due to the many white, coloured, interference and iridescent pearls which can be used.

Typical formulations for a blue/green pressed powder eye-shadow, in both matt and pearly forms, are given below.

Raw material	*Matt (% w/w)*	*Pearly (% w/w)*
Talc	82.30	39.30
Zinc stearate	5.00	8.00
Pentaerythritol tetraisostearate	4.00	4.00
Ultramarine blue	5.00	5.00
Chromium green	3.00	3.00
Yellow iron oxide	0.50	0.50
Titanised mica pearl white	-	40.00
Methyl paraben	0.10	0.10
Imidazolidinyl urea	0.10	0.10

Once again, talc is the filler and zinc stearate/pentaerythritol tetraisostearate forms the binder system. Note the high level of pearl that is necessary in the pearlised formulation, give the desired effect.

A second example of a very highly pearlised pressed powder eye-shadow is given below.

Raw material	*% (w/w)*
Talc	4.20
Bismuth oxychloride pearl	10.00
Titanium dioxide coated mica	65.00
Fumed silica	0.50
Zinc stearate	5.00
Mineral oil	9.75
Isostearyl neopentanoate	1.50
Lanolin alcohol	3.75
Methyl paraben	0.10
Propyl paraben	0.10
Imidazolidinyl urea	0.10

In this formulation, talc is the filler and the binder system is composed of a mineral oil/isostearyl neopentanoate/lanolin alcohol mixture. The very high degree of pearl is provided by a mixture of bismuth oxychloride and titanium dioxide coated mica. The preservative system is a parabens/imidazolidinyl urea combination.

Additional problems occur in eye-shadows which have a high levels of pearl, particularly mica-based pearls. In this type of formulation the binder, as well as holding the cake together, has the additional function of helping the pearl to adhere flat against the skin. If the binder level is too high, the eye-shadow creases easily when worn. It is important to recognise that any changes in binder level affect the colour of a pressed powder eye-shadow. This can subsequently lead to colour matching problems if the binder level is changed in a formulation already on the market.

Another problem associated with high binder levels is uneven flow into the godet-forming dies that are used for manufacture in rotary filling machines. This can result in a pressed godet which is very soft on one side, due to uneven pressing. In manufacture, eye-shadow formulations, particularly those with a high pearl content, should be allowed to settle to remove the entrapped air, before pressing. This process normally takes about 24 hours, over which time the binder will spread fully over the surface of the pearl.

The manufacturing process for pressed powder eye-shadows is similar to that outlined earlier for other pressed powder products, blending first in a ribbon blender and subsequently milling in a hammer mill. An alternative machine for processing powder products, particularly eye-shadow, is known as the Lodige mixer. This mixer, which removes the need for a subsequent milling process, is an adaptation of the ribbon blender and contains a small high speed chopper, as well as larger slow-speed ploughs. Processing time is shortened but pre-ground or extended colours must be used because the chopper cannot break up agglomerates to nearly the same degree as a hammer mill. Pearls can be mixed also but they should be added towards the end of the mixing process, as prolonged processing times can damage them. Lodige mixers are available in sizes from 5 litres to 30,000 litres, covering all the requirements of both small and large-scale mixing.

Micronised extended colours, commonly known as microblends, are available from various suppliers. Although they are more expensive, they are worthy of consideration because a Lodige machine can process them directly. Additionally, the colours are agglomerate-free having already been fully extended on a special grade of calcium carbonate. This means that greater intensity can be obtained from a particular weight of pigment.

4. *Quality assurance tests on powder cosmetics*

In order to ensure the quality of all loose and pressed powder cosmetic products, a series of well-established quality assurance tests should be carried out. A typical quality assurance procedure is given below.

4.1 *Colour*

Correct colour and shade is critical for the vast majority of powder cosmetics. Small heaps of production batch and standard are placed side by side on white paper and pressed flat with a palette-knife. The shades are then carefully compared against each other. The shades of blushers and eye-shadows are also checked on the skin and an applicator, usually a brush or wand, is used to compare the production batch against a known good standard.

4.2 *Bulk density*

This test is carried out on the loose powder only and is designed to ensure that the occurrence of processing problems and incorrect filling weights is minimised. Bulk density tests, which can be carried out using a variety of industry standard methods, detects the presence of any entrapped air in the powder.

4.3 *Fragrance*

Fragrance checks normally only apply to face powders and blushers and are carried out, against a known good standard, using olefactive assessment.

4.4 *Penetration*

Penetration tests are carried out on the surfaces of pressed powder godets, to determine the accuracy of pressure used during the filling operation. The standard method used is to allow a weighted penetration needle to rest on the surface of the pressed powder godet, for a fixed period of time, and measure the distance that the needle has penetrated into the surface.

4.5 *Drop test*

The drop test is carried out on pressed powders only and is designed to test the physical strength of the cake. Normally the godet is dropped on to a wooden floor from a known height (2 to 3 feet) and any signs of damage noted.

4.6 *Glazing*

This test is also for pressed powders only. The pressed cake is rubbed through to the base of the godet with a puff and any signs of glazing that take place during this operation noted.

Lipstick products

Colour for the lips has been used for thousands of years but lipsticks made with a base of oil and wax began to appear just before World War I. Early products were made using the carmine dye, which comes from cochineal. Carmine however, is insoluble and gives a colouring of quite low intensity compared with modern lipstick pigments. Zinc oxide was sometimes used also to brighten the red. Early indelible lipsticks were made by adding a water soluble dye to the carmine. Such sticks were applied to moistened lips. Colour changing lipsticks, which appear orange on the stick but red on the lips, were first formulated by dissolving eosin in stearic acid and melting it with waxes and alcohol. Gelatin/glycerol bases were also used.

Lipsticks became more popular when formulations became more natural looking and in today's society the use of lipstick is almost universally accepted in all civilised countries. Modern lipstick products are available in literally thousands of colours and shades, although fundamentally all lipstick products are very similar. The requirements of a well-formulated modern lipstick are as follows:

- they must be dermatologically safe
- they must be edible
- they should possess a taste and odour that is both neutral and agreeable
- they should be accurately formulated and manufactured to produce a consistently correct colour shade
- they must be easy to apply and pay-off smoothly over a wide range of temperatures
- they should not be subject to breakage, even in hot weather
- they should not change colour on the lips
- they should not smear or run on the lips
- they should not sweat, bloom or produce excrescences
- they should be waterproof
- they should not be greasy
- the finished lipstick should be free of pin-holes and moulding lines.
- their properties should not change over time, for several years after they are manufactured

1. *Raw materials for lipstick products*

A lipstick base must consist of materials which are capable of forming a stick which will pay-off as it is moved over the lips, giving a smooth film. No single material will function in this way and the desired lipstick characteristics can only achieved by combining a selection of different materials, from the enormously large number that is available. Some of the more commonly used lipstick components are discussed below.

1.1. *Lipstick colours*

Up to around 1920, carmine was the predominant colour used in lipsticks. The performance of carmine had been improved by extracting dried cochineal insets with ammonia and then precipitating the soluble salt with alum, to produce a pigment with an intense deep red colour. In the 1920s, eosin, chemically 2,4,5,7-tetrabromo fluorescein, came into use. This material is now known as D & C Red No. 21 (CI 45380). In order to get a stain on the lips, from a water-insoluble waxy base, materials had to be introduced into the formulation which were compatible with the stick materials but which acted as solvents for the eosin.

Other staining dyes, also based on fluorescein, were gradually developed and these gave some variation in the possibilities for lip stain colours. In spite of this, eosin remained one of the best choices, as it gives a fairly natural coloured stain on the lips.

In addition to stains, a wide range of lipstick pigments have been developed by converting dyes into insoluble forms. Colours which may be used in modern lipsticks are regulated by official bodies such as the EEC and FDA. Some of the most commonly used colours in contemporary lipstick products are the oxide pigments, in addition to the various pearling agents. The use of stainers in modern day lipsticks is now much less common.

1.2. *Oils and liquid additives*

The oil in a lipstick product is used for many reasons. In addition to its valuable contribution to the properties of the finished product, it also acts as a dispersing agent for the coloured pigments. The most commonly found oil in lipsticks is castor oil, which was originally used in conjunction with a high proportion of beeswax. It contains a high level of ricinoleic acid, which is responsible for its desirable properties. In its combined form of glyceryl ricinoleate, castor oil is a reasonably good solvent for stainers and is uniquely viscous among vegetable oils, a property which helps prevent pigments settling from the molten lipstick mass during manufacture. Castor oil also has an acceptable taste and is quite resistant to rancidity. Its use in lipsticks helps give a gloss to the applied film, although excessive quantities in the formulation can lead to undesirable drag in use and a somewhat greasy feel. Castor oil is a poor solvent for mineral waxes because it is relatively hydrophilic.

Vegetable oils have also been used in lipsticks but they are inclined to go unpleasantly rancid and act as poor solvents for lipstick stainers. Mineral oils are sometimes used, at fairly low levels in conventional lipsticks, to enhance gloss. They are very resistant to rancidity but are also poor solvents for stainers. Lanolin oils can be used to give a very heavy, rich, feel to the applied lipstick film. They also aid sheen and help pigment dispersion. Acetylated lanolin alcohol is a very smooth, thin, oil which is used at moderate levels to give a greater smoothness to the lipstick film. One disadvantage of this material is that it can undergo partial hydrolysis, producing an unpleasant acetic odour.

Apart from oils, other liquid additives are used in lipstick to improve manufacture and performance characteristics. Amongst these is butyl stearate which is quite widely used, being of bland odour and exhibiting good pigment-wetting properties. Although it is not such a good solvent as castor oil for stainers, butyl stearate will reduce the viscosity of the film on the lips, making the lipstick feel more pleasant in use. Similar effects can be obtained with isopropyl myristate, isopropyl palmitate, diethyl sebacate and cetyl oleate. Fatty alcohols, oleyl alcohol in particular, are often used at quite high levels in lipstick products to modify performance characteristics in use.

Modern lipsticks seldom use stainers, pigment colours now being preferred. In cases where stainers are used, special solvents are sometimes added to enhance the solubility of the stainer in the product. Amongst these additives are carbitol which is an excellent solvent for stainers but tastes rather bitter. The polyethylene glycols are also excellent stainer solvents but are liable to sweat-out of the stick when used at practically useful levels. Phenyl ethyl alcohol and propylene glycol are also good stainer solvents but their respective tastes can be somewhat unpleasant.

1.3. *Fatty materials*

Although lard and other fats were commonplace in lipsticks at one time, true fats are seldom used in modern day products and have been substituted with other additives. Amongst these is cocoa butter, which melts at approximately skin temperature. This can be a disadvantage however, because instability can result in an unacceptable level of blooming. Hydrogenated vegetable oils, frequently used as food shortenings, provide a smooth texture and are relatively stable against autoxidation. Lanolin exhibits some beneficial properties which can be used in lipstick products. Lanolin can aid both sheen and pigment dispersion, although lipsticks with high lanolin levels tend to become rather sticky in use. Allergy problems can also occur, particularly if the lanolin is not of sufficient purity. Modified lanolins, for example acetylated lanolin, also impart good sheen and are slightly less sticky and viscous than lanolin itself. They are also better solvents. The acetoglycerides have also been used for their solvent properties in lipsticks. Other additives include petrolatum and lecithin, the latter providing a rich, heavy, feel to the lipstick film and also helping pigment dispersion.

1.4. *Waxes*

Waxes give structure to lipsticks and keep them solid, even in warm conditions. The rheology of the lipstick structure should be shear-thinning, ensuring good pay-off in use and absence of drag. As such, the particular wax used in a lipstick formulation is very important in determining the performance characteristics of the finished product. Some of the more common waxes used in lipsticks are discussed below.

1.4.1 *Carnauba wax*

Carnauba wax is extracted from the leaves of the Brazilian palm tree, *Copernicia prunifera*. It is a hard vegetable wax, comprising mostly of esters, with a high melting point (83°C). It is used in lipsticks, in modest proportions, to improve the strength and hardness of the finished product. Shrinkage during the stick moulding may be a problem if carnauba wax is used, particularly in excessive amounts.

1.4.2 *Candelilla wax*

Candelilla wax is extracted from *Euphorbiaceae cerifera* plants in Mexico and is a vegetable wax consisting of hydrocarbons and esters. It has a lower melting point (68.5-

72.5°C) than carnauba wax and is used for a similar purpose, that is to provide the stick with increased hardness and strength.

1.4.3 *Beeswax*

Beeswax, extracted from honeycombs, is the traditional stiffening agent for lipsticks and has a melting point of between 62°C and 65°C. It binds other oils, such as castor oil, in the formulation well and provides improved shrinkage qualities on cooling. It is possible to fabricate a lipstick using beeswax as the only wax, although a product of this type is likely to lack gloss and will drag on application.

1.4.4 *Ozokerite wax*

Ozokerite wax is a naturally occurring microcrystalline wax, which was originally obtained from bitumen bearing shales. It is brown in crude form and must be bleached to give the cosmetic grade. Ozokerite wax is often referred to as being amorphous but, in fact, it is more like paraffin wax in character. Ozokerite wax is particularly useful as a high melting-point wax and small quantities added to a lipstick formulation improves performance of the product in hot climates.

1.4.5 *Ceresine wax*

The term cerisine wax is sometimes confused with the term ozokerite wax and, at one time, blends of beeswax and refined ozokerite wax were known as ceresin. In reality, ceresine wax is a microcrystalline hydrocarbon wax that resembles, and is sometimes used in place of, ozokerite. Ceresine waxes are available with a range of melting points between 58°C and 68°C and are characteristically malleable and amorphous.

Other microcrystalline waxes, mostly obtained from petrochemical sources, are also available. They tend to exhibit plastic character and have small crystal sizes with melting points, in some cases, in excess of 90°C. These waxes absorb oil very well and help to prevent the lipstick sweating. A disadvantage of these types of waxes is that they have limited solubility in castor oil.

1.4.6 *Paraffin wax*

Paraffin wax is commonly thought to be too weak and brittle to be of use in lipsticks. Claims have been made, however, that improved gloss can be brought about with small additions of this wax to a lipstick formulation. One particular disadvantage of paraffin wax is its incompatibility with castor oil, thereby limiting its use.

1.4.7 *Synthetic waxes*

Synthetic ester and glyceride waxes are available, many of which are hard and have high melting points. These materials can be added to a lipstick formulation to improve strength and structure, as well as enhancing performance characteristics.

1.5. *Fragrance*

Any fragrance used in lipstick products must have an acceptable taste. Generally, mildly spicy, fruity or floral types are the most preferred.

1.6. *Antioxidants*

Antioxidants must always be added when unsaturated materials are present in the formulation, thus preventing the onset of rancidity. A suitable material that has found extensive use in the past is butylated hydroxytoluene (BHT) at levels of up to 0.1%.

1.7. *Sunscreens*

Oil-soluble sunscreens may be added to lipsticks to help prevent the discomfort that can occur on the lips in very sunny conditions. The growing awareness of the ageing effect of excessive exposure to ultraviolet light has reinforced this trend. Any sunscreen used must be selected from the permitted list of UV filters found in "The Cosmetic Product (Safety) Regulations 1989".

1.8. *Preservatives*

Given that lipsticks are essentially anhydrous systems, there is very little likelihood of bacteria or moulds growing within the body of the lipstick. It must be remembered, however, that during use, the product is brought into regular contact with the mouth and therefore the chance of any naturally occurring micro-organisms on the skin being transferred to the lipstick is very high. Under such conditions, the surface of the lipstick could eventually develop bacterial or mould-like growths. For this reason it is common practice to add a preservative with oil-soluble characteristics into lipsticks. Butyl paraben is sometimes used for this purpose but, because this preservative is very much more soluble in oil than water, its effectiveness against the development of surface micro-organisms must be questioned. 2-bromo-2-nitropropan-1,3-diol, which exhibits good stability in lipstick formulations and exhibits only marginal oil solubility, is a much better choice as more preservative is available to the surface water phase on the lipstick, where bacterial growth is likely to take place.

2. *The formulation and manufacture of lipsticks*

Formulation, particularly of a lipstick with a radically different composition, can be difficult and knowledge and experience of the waxes and oils available, and the effects that they will produce, is invaluable. Slight variations in formulation can exhibit marked differences in product hardness, crystal size and external appearance. Perhaps the most difficult aspect of formulating a good lipstick is to achieve a sufficiently high degree of shine on the lips, without introducing too much fluidity.

When formulating lipsticks at the bench, the ultimate requirements of the manufacturing facility must be borne in mind. The moulded lipstick must release easily from the mould and not be damaged in so doing. In order to fulfil this requirement, the lipstick must harden quickly and shrink to the appropriate degree. If, in the laboratory, the moulds have to be lubricated and chilled in a refrigerator before the salves can be removed, then problems will almost certainly occur in the production environment.

Waxes that produce adequate shrinkage of the lipstick on cooling must be incorporated and it is good practice to include a small amount of wax which melts above the moulding temperature, to give faster nucleation during the cooling process. The hardness and texture of the lipstick can be affected by the quantity and type of colour pigment present. In order to overcome this variability, there is a tendency to use a constant weight of pigment in the formulation. However, differences can still occur in strength because a given weight of a light organic pigment may take-up a much higher level of oil than the same weight of a dense inorganic pigment. In fact, texture variation would be much lower if a constant volume of pigment were to be used but this would be impossibly difficult to manage in production.

It is difficult to produce a lipstick which is stable across a wide range of temperatures. Materials which liquefy or solidify within the stick under different temperature conditions, will very often alter the texture and surface appearance of the stick over a period of time. Cocoa butter, which melts at body temperature, is a good example of a material which can produce this type of effect.

Lipsticks containing powerful solvents, such as acetylated lanolin alcohol in combination with an ester wax, provide an environment in which the latter is partially dissolved by the former, particularly at high temperatures. If the lipstick sweats, then material is sometimes brought to the surface, which crystallises on cooling and cannot then be absorbed back into the structure of the stick. A hard skin then forms over the outside of the stick, which is undetectable until the product is used.

Generally speaking, both oils and waxes used in a lipstick formulation should remain in their original forms within the stick itself and the oils should not easily dissolve the waxes over modest changes in temperature. Naturally, the oils and waxes should be close enough in polarity to readily mix when the lipstick mass is melted, before the stick is formed. Problems can occur if excessively high levels of a microcrystalline wax are used in a high castor oil containing lipstick, the microcrystalline wax only having a very limited solubility in the relatively hydrophilic castor oil.

When formulating lipstick products, it is preferable to use materials which produce a small crystal structure within the product. If larger crystals are produced, then a higher degree of fluidity is necessary to produce the required gloss characteristics in the finished product. Microcrystalline waxes can help overcome this problem because they contain many different branched-chain molecules, which make the formation of large crystals improbable.

The manufacture of lipsticks, particularly the moulding process, can give rise to many problems with associated high reject rates. This is particularly true if the limitations of the production equipment have not been fully borne in mind during the products development stage. In general, the manufacture of lipsticks takes place in three stages.

2.1 *Preparation of the component blends*

The preparation of component blends includes preparation of the colour dispersion, preparation of the wax base and, in some cases, blending of the oils.

2.1.1 *Preparation of the colour dispersion*

For reasons of economy, and to prevent colour change of the pigments during use, the pigments must be blended using a very high shear mixing system. This process is normally carried out by milling viscous pastes of pigment in castor oil, using a triple-roll mill

Adequate mixing is usually monitored by measuring the fineness of grind with a Hegmann gauge.

2.1.2 *Preparation of the wax base*

The wax base is prepared in large batches by heating and mixing all the waxes in the lipstick formulation, plus part of the oils. When the wax base has been thoroughly mixed, it is poured into trays for cooling and is subsequently stored as slabs, awaiting further processing.

2.2 *Preparation of the lipstick mass*

The intermediate components of the lipstick are weighed into a steam-jacketed vacuum vessel, equipped with an efficient mixer capable of producing a moderate degree of shear. Further dispersion of the colour pigments should not take place at this stage because this will affect the colour adjustment process, causing variability of shade in the finished product. Once colour checking and adjustment of the molten mass is complete, a vacuum is applied and the mass stirred for sufficient time to remove any entrapped air. The molten mass is then carefully cooled, under controlled conditions, before being poured into the lipstick salves, or into trays, for final cooling. During this pouring process, it is important to avoid entraining any air bubbles.

2.3 *Moulding of the lipstick*

The simplest lipstick moulds are the split-mould variety, which may be fabricated from either plated brass or aluminium alloy. The lipstick mass is poured into the mould and subsequently allowed to cool fully , before the salves are extracted from the split mould.

3. *Typical lipstick formulations*

As already mentioned the vast array of raw materials available for the manufacture of lipstick products means that the number and type of formulation that can be produced are limitless.

By way of example, the following lipstick formulation is a pearly pink product, based on a synthetic wax base, as shown below.

WAX BASE (a)

Raw material	*% (w/w)*
Glyceryl tribehenate	10.00
C18-36 acid glycol ester	12.00
C18-36 triglyceride	18.00
Isopropyl palmitate	30.00
Castor oil	29.80
Butylated hydroxytoluene	0.10
2-bromo-2-nitropropan-1,3-diol	0.10

SHADE FORMULA

Raw material	*% (w/w)*
Wax base (a)	50.00
Castor oil	30.50
D & C Red No. 3 aluminium lake (1:2)*	1.00
D & C Red No. 21 aluminium lake (1:1.5)*	1.50
Bismuth Oxychloride	15.00
2 hydroxy-4-octyl oxybenzophenone	1.00
Fragrance	1.00

* The ratios referred to in the above formulation are the weights of pigment which are dispersed into a given weight of castor oil.

A second lipstick formulation example, yielding a soft, creamy product based on a combination of carnauba and candilla waxes, is given below.

WAX BASE (b)

Raw material	*% (w/w)*
Carnauba wax	8.00
Candelilla wax	30.00
Multiwax 835	8.00
Lanolin	11.00
Cetyl alcohol	5.00
Isopropyl myristate	37.70
Butylated hydroxytoluene	0.10
Propyl p-hydroxy benzoate	0.20

SHADE FORMULA

Raw material	*% (w/w)*
Wax base (b)	50.00
Titanium dioxide (1:1)*	4.00
D & C Red No. 3 aluminium lake (1:2)*	4.00
D & C Red No. 9 barium lake (1:1)*	0.50
Castor oil	40.50
Fragrance	1.00

* The ratios referred to in the above formulation are the weights of pigment which are dispersed into a given weight of castor oil.

4. *Lipstick formulation acceptability*

The final test of formulation acceptability is to use a panel test. However, an experienced formulator can predict likely product performance by assessment of pay-off characteristics, appearance and the permanence of a film of lipstick, when it is applied to the heel of the thumb or back of the hand.

Other checks can also be carried out on the finished product. Crystal size can be checked by mashing a lipstick sample gently on a tile or glass plate and viewing it under a stereomicroscope. It is also sometimes useful to examine the lipstick structure of a broken stick, which has previously been cooled in a freezer. The viscosity of the liquid film may be assessed by spreading a sample of the mashed lipstick on to the skin. Ideally, the change in viscosity when carrying out this process should be minimal and the viscosity should be as high as possible, consistent with achieving the desired sheen.

It is also useful to measure the drop point and slip point of a lipstick product. Both can be assessed using standard methods that were originally developed for waxes.

Stick strength is an important characteristic of any lipstick. This property is normally measured using the Lauerstein method, which measures the time taken for a lipstick to break under a given load, at a particular temperature. A measure of lipstick strength and pay-off can also be obtained by using the penetration test. Firstly, the formulation is cast and matured under standard conditions. The tapered part of the salve is then carefully sliced away, taking care not to bruise the lipstick surface, and the salve is tested in the vertical position using a small penetrometer cone.

Streaking of the stick and unevenness of colour can be caused by flocculation. Streaking occurs when part of the pigment system tends to flocculate. This can be checked by viewing the lipstick mass under a stereomicroscope and focusing a light on the lipstick surface to melt part of it away. Any tendency to flocculation can readily be observed because areas of clear or partly coloured oil will be noticed. If the lipstick is to be flame-polished, frequent checks should be carried out using the standard flaming system, in order to check that the desired degree of surface gloss will be obtained.

Finally, no testing or evaluation programme is complete without carrying out a storage test. Samples under consideration for commercialisation should be subjected to cycle testing in an oven, with a twice daily change in temperature between 5°C and 25°C, and any changes in product characteristics or performance noted. A testing programme of this type can produce changes in two or three weeks, which would normally take up to a year at room temperature to materialise. Retained samples should be kept and regularly reviewed for some time after the development work is concluded.

Eye mascara products

The main purpose of an eye mascara is to emphasise the eyelashes by making them seem thicker, longer and darker, therefore making the eyes seem warmer, more accentuated and more feminine. There are several types of eye mascara available but, irrespective of type, they should all exhibit the following properties:

- they should be safe, non-irritating and non-sensitising
- they should be well preserved: badly preserved mascaras can cause ulceration of the eye and contamination with the micro-organism *Pseudomonas aerugenosa* can lead to eventual blindness
- they should transfer smoothly from the brush or wand, to the lash, without leaving wet blobs or causing an artificial "sooty" appearance
- they should provide sufficient cover for the lash in a single application
- the formulation should be tear-proof, rain-proof and swim-proof
- the formulation should be resistant to smudging in hot, steamy conditions
- they should resist smudging by greasy skin or eye make-up
- they must be flexible and adhere well to the lash, without flaking. Stiff, difficult to remove mascara can cause lash breakage
- they must dry fairly quickly after application to the lash, to minimise smudging and to facilitate building-up if required. They must not dry so quickly so as to prevent even application.
- they should not be too difficult to remove

There is no universally acceptable mascara product. Most formulations are, to some degree, a compromise between the various competing requirements and the type of mascara bought by the consumer will depend upon personal choice.

1. *The formulation and manufacture of mascaras*

Essentially, there are three basic types of mascara product available, cake mascaras, cream mascaras and waterproof mascaras. Each of these will be discussed in turn.

1.1 *Cake mascara*

Cake mascaras, sometimes referred to as compact or block mascaras were the earliest type of mascara available and were initially sold in the 1920s. Cake mascaras are still in fairly common use although the growing popularity of cream and waterproof mascaras has ensured their steady decline.

Cake mascaras are produced by mixing the oil phase of a soap-based emulsion, together with waxes and other ingredients, plus coloured pigments. During use, a wet brush is rubbed on the cake, so providing the water-phase of the soap-based emulsion and facilitating transfer from the cake to the lash. Application is easy and fairly effective but all cake mascaras suffer from poor waterproofness. They also have a tendency to cause stinging when they get into the eyes.

Early versions of cake mascaras were based on sodium soaps and were particularly irritant. More contemporary formulations are based on triethanolamine soaps, which are less alkaline with a much lower irritation profile. A typical cake mascara formulation based on a triethanolamine soap, is given on page 157.

Raw material	*% (w/w)*
Beeswax	28.20
Carnauba wax	7.60
Stearic acid	23.02
Isopropyl myristate	3.20
Triethanolamine	12.00
Sodium alginate	10.70
Pigments	15.00
Irgasan DP 300*	0.08
Domiphen bromide	0.20

* Tradename of Ciba-Geigy

Manufacture of the above product is carried out by firstly melting the waxes and adding the preservatives. The coloured pigments are then added, followed by the alginate and the triethanolamine. The batch is then placed into a colloid mill for pigment dispersion. Unlike many make-up products, meticulous pigment dispersion is not required because there is very little chance of noticing any colour change when the product is subsequently used. The product is then placed under vacuum for a fixed period of time, before being poured into moulds at between 84°C and 86°C. Depending on the formulation, the product may be cast molten into metal godets or even milled and plodded to give a strip, which is then cut to the required size.

1.2 *Cream mascaras*

Cream mascaras are marketed in special containers, available in a wide range of different materials. Irrespective of packaging composition, the operating principle of all mascara containers is identical. The product is contained in a small barrelled container with a screw-threaded neck. On to the neck fits a cap, containing an integral wand fixed securely to its underside. The wand, which dips into the barrel container, has an applicator at its lower end, normally a comb or brush. The wand is made to fit snugly into a flexible wiper, which fits tightly into the underside of neck of the barrelled container.

In use, the cap is unscrewed and withdrawn from the barrel container. In so doing, the wand and applicator are withdrawn from the container and the product can then be transferred, from the applicator to the eyelash. The wiper on the underside of the container neck removes any excess mascara from the applicator as the wand is withdrawn.

Cream mascaras are emulsion-based formulations incorporating a variety of different waxes in the oil phase, to modify product performance. Beeswax, a common ingredient in cream mascaras, gives a product with a flexible film that adheres well to the lash, without being too difficult to remove. Candilla wax can be used to give a tougher, harder, film on the lash and therefore greater smudge resistance. Microcrystalline waxes give a relatively flexible film with good adhesion properties. The higher melting point microcrystalline waxes can also contribute to the smudge-resistant properties of the product. Finally, wool-wax alcohols can be incorporated into stearate-based emulsion mascaras, to improve emulsion stability.

Polymers can also be added to cream mascaras, to modify their properties. Vinyl acetate/crotonic acid copolymers can be used to modify water-resistance properties. Polyvinyl pyrrolidone is a highly skin-compatible material that will produce tough mascara films, although its water-resistance is modest. Alkylated polyvinyl pyrrolidone is a surface active polymer, which forms highly water-resistant films. Although the films are somewhat brittle, their properties can be modified by plasticisation with non-volatile oils. The main disadvantage of alkylated polyvinyl pyrrolidone, is that it can form gels at low temperatures. The addition of polybutene to mascara formulations improves film flexibility, adhesion and water-resistance.

1.3 *Waterproof mascaras*

Whilst cream mascaras have some degree of waterproofing and smudge resistance, their popularity has been superseded somewhat by truly waterproof products. Waterproof mascaras, which are sold in the same types of container as cream mascaras, are normally based on wax systems dissolved in a cosmetically acceptable solvent. The most commonly used solvents in modern mascara products are those based on cosmetic iso-paraffins, with varying degrees of volatility.

Because the products are based on a wax/solvent system, they dry to a hard film on the lash which is very waterproof. The main disadvantage, however, is that such films are difficult to remove afterwards, even with oil-based cleansing products. This problem can be lessened to some degree by formulating the product to produce a softer film on the lash. Good waterproof performance can still be retained but the product is somewhat easier to remove afterwards. Products of this type can, however, smudge more easily when they come into contact with greasy skin or eye make-up. Smudging can be minimised by reducing the pigment levels in the formulation, although there is then a danger that the depth of colour will be compromised and the covering power reduced.

If the mascara is to be smooth and easy to apply, the wax crystals in the formulation must be small and paraffin wax, which gives large crystals, should be avoided. Microcrystalline waxes (iso-paraffins) are a much better choice, as they produce very small crystals. Some of the best microcrystalline waxes are those that dominate the wax crystal system and reduce the overall crystal size in the finished product.

Because of the importance of crystal size, it is a valuable formulation aid when developing this kind of product, to carry out oven cycling assessments. Samples of the formulation should be cycled between 5°C and 25°C, twice a day, for two to three weeks and the crystal structure of the product examined. This process will very rapidly develop any large crystals, which might have a tendency to form during the shelf-life of the product.

The formulation for a typical waterproof mascara is given on page 159.

Manufacture of this formulation involves mixing the waxes, stearic acid and propyl p-hydroxy benzoate and heating them to approximately 85°C. The iso-paraffin, pigments and triethanolamine are heated to 50°C in a separate container, before being passed into

Raw Material	*% (w/w)*
Carnauba wax	6.00
Microcrystalline wax	10.00
Beeswax	14.00
Stearic acid	2.00
Propyl p-hydroxy benzoate	0.20
Iso-paraffin	56.75
Pigments	10.00
Triethanolamine	1.00
Phenyl ethyl alcohol	0.05

the wax phase, under high shear stirring. When mixing is completed, the batch is cooled to 40°C, before adding the phenyl ethyl alcohol.

The triethanolamine in this formulation combines with the stearic acid to form triethanolamine stearate, which improves ease of removal of the product from the lash, after use. The level of triethanolamine in the formulation should not be too high however, otherwise the water resistance will suffer. A useful ingredient for helping to confer stability in this kind of formulation is Bentone 38, a montmorillonite clay modified with a fatty quaternary salt to be dispersible in aliphatic solvents.

2. The evaluation of mascara formulations

Although a panel test must be the ultimate arbiter of likely commercial success, there are several laboratory-based procedures that will help the formulator discriminate between good and bad mascara formulations.

A useful test for flexibility and adhesion is to coat the mascara on a clean keratin (horn) surface and to allow it to dry. The mascara film is then viewed under a stereomicroscope and, using a pointed object, a circle of approximately 6mm is drawn on the mascara film, penetrating to the surface of the keratin. The disc of mascara is then progressively reduced in size, using the needle, until a small disc of mascara flakes off of the keratin surface. The size of this final disc provides a surprisingly good indicator of how well the product adheres. The removability of the mascara with water, oil or emulsion-based cleansers can also be checked on lashes or a keratin surface.

The viscosity and consistency of any mascara formulation is very important. Low viscosity will produce blobs on the lashes and splashes around the eye area, whilst high viscosity will cause the lashes to stick together. The viscosity should be chosen such that it gives good definition and coverage but will not cause the lashes to stick together. The effects of the product viscosity on ease of application and transfer from brush to lash, should also be checked. If the texture is very short, efficient transfer will be difficult. Conversely if the texture is long and stringy, the lashes will tend to stick together. If the mascara mass leaves small peaks when fingers are placed on the surface and withdrawn, then application characteristics will usually be satisfactory.

3. *Packaging for mascara products*

The importance of the mascara pack has already been mentioned in this chapter. Mascara packs, because they are small, have a high surface area to volume ratio and transmission of water or solvent through the wall of the pack can cause a rapid increase in viscosity, especially if the mascara is designed to dry quickly on the eyelashes.

Alkathene materials like polypropylene should be avoided with solvent-based mascaras and materials like polyvinyl pyrrolidone should be avoided in the case of water-based products. If polystyrene is being considered for the pack design, its incompatibility with some solvents, particularly short chain esters, must be taken into account. Brush wands made of alkathenes will soften in solvent-based mascara products and the result will be a bent or buckled wand.

Wiper design should be such that mascara is not left behind on the brush when it is replaced, otherwise the next time the brush is withdrawn it will be overloaded with an accumulation of product. Brush loading depends on the brush size, product viscosity and wiper size, the latter being probably the most important consideration.

4. *Quality assurance on mascaras*

Quality assurance procedures for mascaras should not only aim to control the quality of the mascara formulation itself but must also address the quality of the packaging and its functionality. There are many tests that can be carried out to check the overall quality of a mascara, some of which are listed below:

- melting point/drop point (cake mascaras only)
- solids content
- preservative(s) assay
- microbiological check
- product viscosity
- colour
- packaging integrity
- functionality
- applicator specifications

Checks on product viscosity should attempt to mimic the high shear conditions that the product experiences in use, as the brush is drawn through the wiper.

Nail lacquers

Nail lacquers are by far the largest and most important group of manicure preparations. Originally, only colourless or pale pink products were acceptable.

A nail lacquer, or nail enamel as it is sometimes called, is essentially a solution of a film former in suitable solvents, capable of supporting colourants and pearls and able to form a smooth, glossy, continuous coating on the nails. The required properties for an ideal nail lacquer are as follows:

- they should be innocuous to skin and nails
- they should be non-staining

- they should be easy to apply
- the drying time should not be excessively long (maximum of about 2 minutes) and the film should dry without blooming
- they should provide even coverage with good wetting and flow
- they should provide a high degree of gloss
- the lacquer film should be tough and flexible, and not prone to chipping
- the lacquer film should have good adhesion to the nail and not easily peel off
- the lacquer surface should be non-tacky and not prone to scuffing
- the colour of the dried film should be uniform
- the dried film should exhibit good stability to light
- the dried film should be permeable to water vapour
- the dried film should be resistant to water and detergent solutions

It is clear that any nail lacquer formulation must be carefully balanced to give the best compromise of performance, for each of the characteristics listed.

1. *Nail lacquer formulations*

Nail lacquer formulations have four major components, film former, solvent, secondary resin and plasticiser. Suitable combinations of these materials are designed to give highly performant lacquer films, with the desired physical characteristics. Coloured nail lacquers also contain pigments and pearls, to give the desired colour effects on the nail, and may contain suspending agents to help stabilise the pigments in the formulation. UV absorbers and other special additives, such as proteins, are sometimes added in order to make more powerful claims and generate a higher level of consumer appeal.

The types and contribution of each of the major ingredients in a nail lacquer product will now be discussed.

1.1. *Film former*

By far the most common film forming material used in nail lacquer is nitrocellulose, obtained by the reaction of cotton or wood pulp with a mixture of sulphuric and nitric acids.

The degree of substitution (nitration) determines the intrinsic characteristics of the nitrocellulose and the degree of polymerisation of the cellulose chain governs the viscosity of the product. Nitrocellulose is normally sold dampened with alcohol.

The properties of a given nitrocellulose can be characterised by three main factors, the degree of substitution, the viscosity and the solubility profile. The solubility characteristics of nitrocellulose in various solvents is determined by the degree of nitration.

1.2. *Solvent systems*

Solvents used in nail lacquer formulations may be classified in three ways, active solvents, co-solvents and diluents.

1.2.1. *Active solvents*

Active solvents are the true solvents for nitrocellulose, normally ketones and esters. Perhaps one of the most commonly used active solvents is ethyl acetate.

1.2.2. *Co-solvents*

Co-solvents are not necessarily solvents for nitrocellulose in their own right but they enhance the solvation power of the active solvents present. Alcohols are often used as co-solvents and are added to reduce the viscosity of nitrocellulose/active solvent solutions, thereby enabling a higher solids content to be used.

1.2.3. *Diluents*

Diluents are solvents that are miscible with the active solvents but are not actually solvents for nitrocellulose themselves. Diluents often act as solvents for the secondary resin, although their main function is to lower and stabilise the viscosity of the nail lacquer. A further advantage in using diluents, is that they reduce the power of the solvent system, which means that application of a second or third coat of nail lacquer is less likely to disturb the previous coat. The addition of diluents also brings a cost advantage, as they are invariably cheaper than the active solvents used. The level of diluents should be carefully controlled, as excessive additions can adversely affect product viscosity or even cause precipitation of the nitrocellulose to occur. Finally, care should be taken that any diluents used should dry at a faster rate than the main active solvents. This ensures that a harsh, rough or cloudy film does not occur through preferential precipitation of the nitrocellulose.

1.3. *Secondary resins*

The main function of secondary resins is to overcome some of the inadequacies of nitrocellulose films. Although several different types of secondary resin have been tried in the past, the most commonly accepted nowadays is the condensation product of p-toluene sulphonamide and formaldehyde, sold under the trade name of Santolite MHP (Monsanto Chemicals). This resin, when included at levels of between 5% and 10%, gives good gloss, good adhesion and improves the hardness of nitrocellulose films. In addition, Santolite MHP confers other benefits to the formulation. It has negligible water solubility and produces a continuous transparent film with good detergent resistance. It also has good light stability and is totally compatible with nail lacquer pigments.

The only major disadvantage with Santolite MHP is that it causes an allergy in some people. In this case, or if a hypoallergenic product claim is required, the Santolite MHP can be substituted with a polyester resin or oil-free alkyd resin, based on polyol dibasic acid ester.

1.4. *Plasticisers*

The function of a plasticiser in a nail lacquer is to control the flexibility and elongation of the film. Nitrocellulose is too brittle when used on its own and even with the incorporation of a secondary resin, will not possess the required flexibility. Plasticisers used in nail lacquers should be of innocuous colour, odour and taste, non-toxic and of low volatility. They must also be compatible with other ingredients in the formulation, including pigments, and must be effective at low concentrations. Very often, plasticisers have negligible water solubility and improve the adhesion of the lacquer on the nail.

The level of plasticiser included in the formulation must be carefully controlled, to achieve the desired effect. Typically, between 25% and 50% of the dry weight of nitrocellulose present is used. If too little is used, the film will retain its brittleness and be prone to chipping during wear. Conversely excessively high levels of plasticiser will leave a film with poor resistance to wear.

Two main groups of plasticisers are used in nail lacquers, solvent plasticisers, which are solvents for nitrocellulose and non-solvent plasticisers, often referred to as softeners. Solvent plasticisers are normally high molecular weight esters with fairly high boiling points and low volatility. The most commonly used example is probably dibutyl phthalate. Non-solvent plasticisers are normally used in conjunction with solvent, to give additional flexibility to the lacquer film. The most commonly used non-solvent plasticiser is castor oil which, when used in combination in a 1:1 ratio with a solvent plasticiser at a level of around 5%, produces a very flexible film. Camphor has also been used as a non-solvent plasticiser in nitrocellulose systems but its plasticising capability is somewhat modest.

When developing a new formulation, it should be borne in mind that the solvent system, in addition to the plasticiser itself, can have a marked effect of the plasticisation of the final film. Solvent substitutions may therefore radically effect the behaviour of the film and each change should be investigated carefully.

2. *The manufacture of nail lacquers*

A typical cream nail lacquer formulation is given below.

Raw material	*% (w/w)*
Nitrocellulose	14.00
Isopropyl alcohol	4.30
Santolite MHP*	9.00
Ethyl acetate	11.00
Butyl acetate	25.00
Toluene	30.72
Dibutyl phthalate	4.00
Bentone 38+	1.00
D & C Red No. 6 barium lake	0.08
Titanium dioxide	0.75
Iron oxide	0.15

* Tradename of Monsanto Chemicals
\+ Tradename of Rheox Inc.

One of the main problems with the manufacture of nail lacquers is that it involves the processing of highly flammable materials, sometimes under conditions of high shear, where localised heating may occur. In view of this fact, all equipment used should be fully flame-proofed and the manufacturing environment should conform to the appropriate safety requirements. The processing of the nail lacquer itself normally takes place in two steps.

Firstly, the pigments for the nail lacquer are prepared by blending them with nitrocellulose, in the presence of the Bentone and plasticiser system. This mixture is then finely ground, usually through a triple-roll mill, to extend the pigments and produce a homogeneous mix. The resultant mix is then dried and broken up into chip form. This stage of the manufacturing process can be quite hazardous, particularly in view of the high shear mixing that is required. For this reason many manufacturers nowadays buy chips in a "ready-made" form.

The second stage of the manufacturing process is to blend the chips into the desired shade, using a high-shear mixing blade such as the Cowles Blender. Once a consistent colour shade has been obtained, the rest of the lacquer ingredients, including any special additives, are added. This stage of the manufacturing process can normally be carried out using a simple propeller stirrer.

3. *Assessment of nail lacquer formulations*

Laboratory assessment and quality control of nail lacquer formulations should be designed to examine several key properties. Of particular importance is the viscosity of the finished product, as this will heavily influence the in-use characteristics of the product. Drying times are equally important and these should be carefully specified and their consistency controlled. Characteristics of the dried film, gloss, hardness, adhesion and wear resistance should all be routinely monitored. Lastly, the process of colour matching is just as important, if not more so, than with other decorative cosmetic products. Incompletely dispersed pigments will not only affect the colour stability of the finished product but may also affect the smoothness of the dried film.

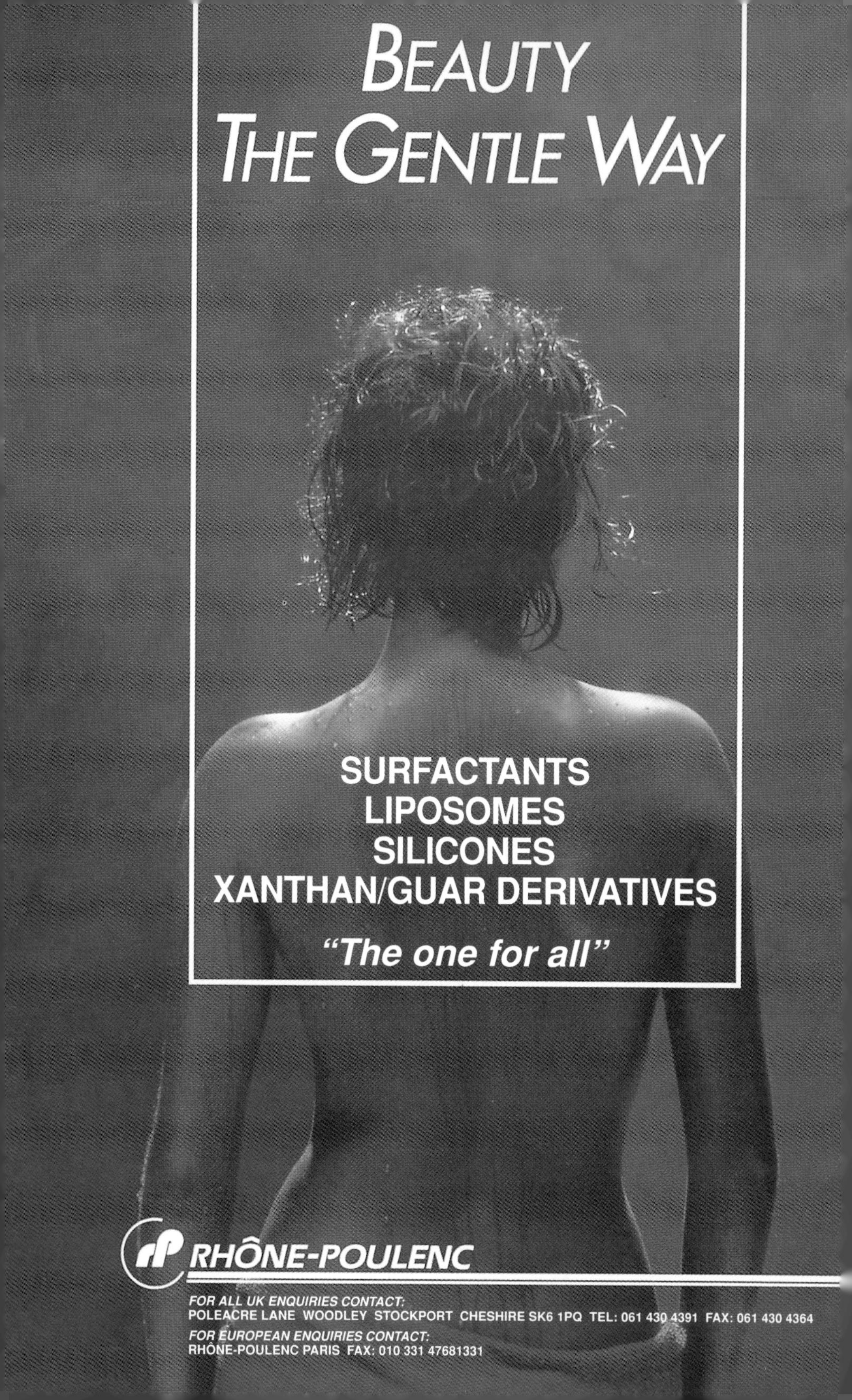

BEAUTY
THE GENTLE WAY
SURFACTANTS
LIPOSOMES
SILICONES
XANTHAN/GUAR DERIVATIVES
"The one for all"
RHÔNE-POULENC
FOR ALL UK ENQUIRIES CONTACT:
POLEACRE LANE WOODLEY STOCKPORT CHESHIRE SK6 1PQ TEL: 061 430 4391 FAX: 061 430 4364
FOR EUROPEAN ENQUIRIES CONTACT:
RHÔNE-POULENC PARIS FAX: 010 331 47681331

SECTION 5

Personal Care

SKIN AND SKIN PRODUCTS
HAIR AND HAIR PRODUCTS
BATH PRODUCTS
THE TEETH & TOOTHPASTE

SKIN AND SKIN PRODUCTS

Introduction

The skin is the largest organ in the human body, comprising approximately 6% of adult body weight. Fundamentally, the skin provides a barrier between the body and its environment and also exhibits a control function over several other factors.

Firstly, the skin helps to maintain a constant body temperature. When the human body becomes hot, heat may be lost in the ambient temperature range of 23-30°C by increased dilation of the blood vessels, giving the visual appearance of what is commonly known as "flushing". Above 30°C, the sweat function begins and evaporative cooling takes place. Similarly, exposure of the body to cold temperatures results in vasoconstriction, thus reducing the flow of blood towards the skin surface, and therefore helping to maintain body heat.

A second and very important function of the skin is to maintain the bodies' water content. Lipids in the outermost layers of the skin provide a barrier to excess loss of water from the body. If these lipids are removed from the skins surface, by detergents for example, the water loss through the skin is increased and the condition known as "dry skin" will result.

A third skin function is to help protect the body from the effects of exposure to light. Exposure of the body to sunlight is controlled by pigments in the skin, which absorb potentially harmful ultraviolet light from the spectrum. The natural level of pigment in the skin varies not only from race to race, but also within individuals of the same race. Pigment levels are also increased with the amount of ultraviolet light falling onto the skin, such changes giving more protection from sunlight, as a direct response to ultraviolet exposure.

Skin also provides some protection from the effects of either chemical or microbial attack. The surface of the skin is covered by a film known as the "acid mantle", which is formed by the sweat and sebaceous glands through secretion. The acid mantle normally has a pH in the range of 4.5-6.5, depending on the site of the body tested, and protection is afforded to the skin by chemical buffering, detoxifying and bacteriostatic functions. Severe changes in the pH of the acid mantle may give rise to unwelcome bacterial invasion, sensitisation and various forms of skin dermatitis. In these circumstances, the benign micro-organisms, normally found resident on the skin at its normal pH, may not survive, leaving the way clear for colonisation of invading transient pathogenic organisms.

Finally, skin provides some resistance to mechanical stress. The outermost horny layer of the skin provides a tough external covering to the body, as a first line of defence to external mechanical pressure and damage. The rheological properties of the skin also give it "shock-absorbing" characteristics.

Physiology of the skin

Simplistically the skin consists of two layers, the *dermis* and the *epidermis*, as illustrated in Figure 1. The epidermal thickness varies over the body from the thickest at 0.8 mm, on the palms of the hands and soles of the feet, to the thinnest at 0.06 mm, on the lips. The epidermis also varies in colour around the body, the concentration of pigmented cells increasing around the nipples and genitals. Although the epidermis is of most importance to the cosmetic chemist, the dermis possesses a number of important functions. The dermis provides support for the epidermis and supplies nutrients to it, via blood capillaries. It also provides support for the sensory nervous system and contains the biological "program" for production of the detailed structure of the epidermis, which varies depending on the part of the body.

FIGURE 1

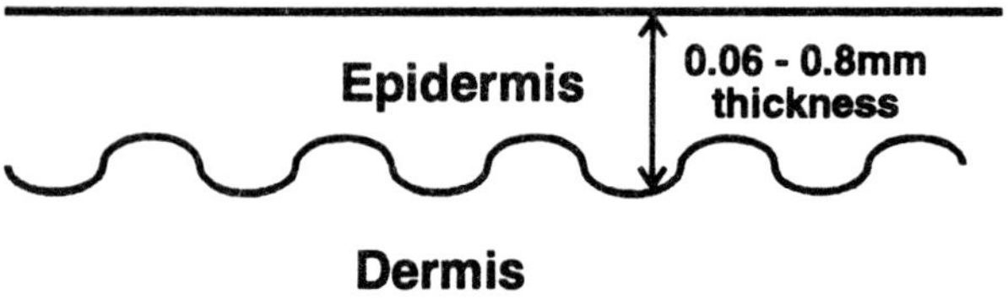

Skin structure changes around the body but some characteristics are common to all sites. All skin appendages such as hair follicles, eccrine sweat glands, apocrine sweat glands and sebaceous glands are extensions of the epidermis.

Hair follicles themselves are continuous with the epidermis. The hairs in the follicles are sloped with respect to the skin's surface and in the obtuse angle formed between the hair and the skin's surface is the arrector pili muscle, which on contraction produces the familiar "goose-pimple" effect as a response to cold, fear or feelings of hostility.

It is important to note that not all follicles contain hair and hair distribution around the body is variable with the occurrence of two main types, terminal hairs and vellus hairs. Terminal hair is found on the scalp, axillae, eyelashes, pubic area, chest, nose, face and eyebrows. Depending on the site of the body, terminal hairs have varying lifespan, growth rate and length. Vellus hair is the fine "down" hair found all over the body.

The sebaceous glands produce a fatty substance known as sebum. These glands are normally attached to hair follicles and found over the entire body, with the exception of the palms of the hands and soles of the feet, and are most concentrated on the face, scalp and back. The sebaceous glands become functional at the age of puberty, leading to the possible occurrence of greasy skin and, in some cases, acne.

The eccrine sweat glands secrete a watery fluid, in response to thermal or emotional stress. These glands are found all over the body, particularly on the axillae, hands and feet.

The apocrine sweat glands feed directly into the hair follicle, normally above the sebaceous gland. They secrete an oily material composed of mainly lipids and proteins, secretion being triggered by nervous and hormonal responses. Unlike eccrine sweat glands, the apocrine sweat glands do not produce any secretions in response to temperature changes. Apocrine sweat glands become functional in the skin after puberty and are found particularly in the underarm area or axillae and around the nipples and genitals. Women have more developed apocrine sweat glands than men and they are also more developed in some races than others.

1. *The structure of the epidermis*

A diagram showing the structure of the epidermis is illustrated in Figure 2. The epidermis itself consists of a number of discrete layers of cells, which are originally produced at the epidermal/dermal junction referred to as the basal layer, using nutrients supplied from the dermis.

FIGURE 2

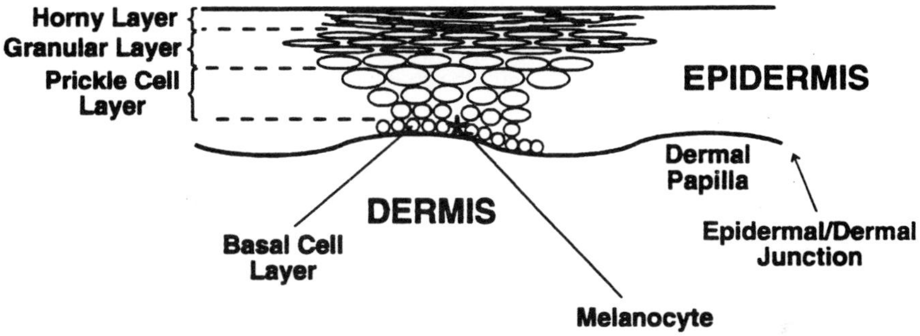

As they move up through the epidermal layer, these living cells lose their spherical shape, becoming rounder and flatter. At the horny layer, they are compressed very flat and they are dead. This process of "cell change" is known as *keratinisation* and Figure 3 illustrates the changes that take place when a single cell undergoes this process.

FIGURE 3

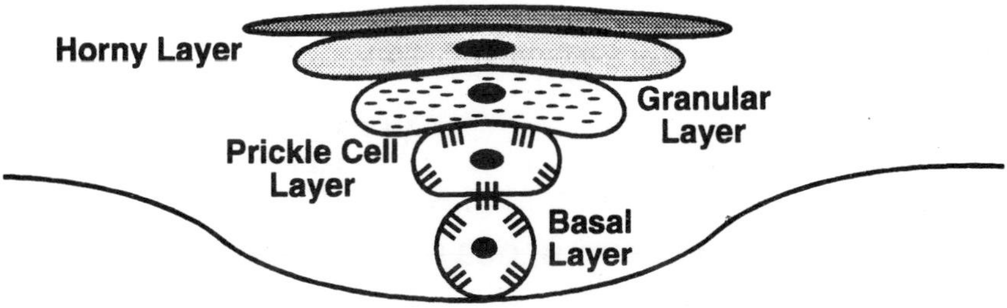

It has been estimated that the normal cell migration time, from the basal layer to the underside of the horny layer, is approximately 13 days, with a further 13 days being spent in the fully keratinised horny layer, before loss at the skin's surface (desquamation) occurs. On average therefore, the human epidermis is renewed approximately every 26 days.

At the start of the keratinisation process the basal layer cells divide, thus reproducing, and move into the layer above. This cell division is known as *mitosis*. The layer above the basal layer is known as the prickle cell layer and this does not reproduce. The prickle cells themselves touch together at the "prickles" which are otherwise known as desmosomes. The desmosomes form intercellular "bridges", growing from which are polypeptide filaments, known as tonofibrils, within each prickle cell. The prickle cells flatten as they move further upward from the basal layer and suddenly become granular. These granules are composed of keratohyalin, which is essentially "three-quarter formed" keratin.

Above the granular layer, breakdown of the cell nucleus takes place, followed by a consolidation process involving the loss of water and the formation of horny layer cells and intercellular cement. Finally, the dead horny layer cells are being continuously lost from the surface of the skin, in the desquamation process.

Cell division at the basal layer is slowed down by a hormone called chalone. The effectiveness of this hormone is reinforced by the presence of adrenalin, which is produced in the body under conditions of stress. In view of the fact that the production of adrenalin is highest during the day, then consequently the chalone is more effective at stopping cell division during this period. The net result of this is that skin tends to "grow" more rapidly during night time.

2. *The structure of the dermis*

A simplified diagram showing the structure of the dermis is shown in Figure 4. The dermis is generally described as an aqueous gelled polysaccharide or mucopolysaccharide. This mucopolysaccharide is strengthened by collagen fibres, collagen being an albuminoid protein and the fibres consisting of long chains of polypeptides. The "bundles" of collagen fibres are thickest in the deeper dermis and finest near the epidermal/dermal junction.

FIGURE 4

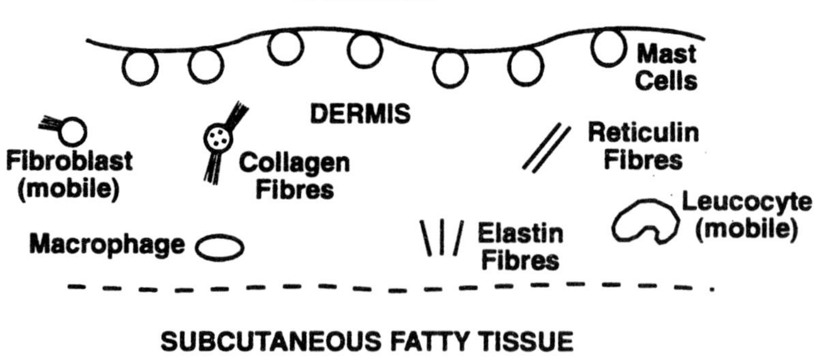

The collagen fibres themselves are wavy, permitting stretching of the skin, but only slightly extensible thus ensuring return of the stretched skin to its original dimensions.

Collagen comprises about 95% of the dry weight of protein in the dermis. Recticulin fibres are present in only small amounts (less than 1%) in normal skin, since they represent "young" collagen fibrils which have not yet aggregated into bundles. Elastin fibres are single amorphous fibrils. The mast cells produce the mucopolysaccharide of the dermis. The mast cells contain lysosomes and the histamine which is released, when the skin is damaged, triggers a "repair program".

Fibroblasts manufacture the collagen fibres and, in carrying out this synthesis, require vitamin C. If insufficient vitamin C is available for this synthesis, defective collagen is produced, resulting in the condition commonly known as scurvy. Macrophages and leucocytes attack invading species such as chemicals and bacteria, when the skin becomes damaged. The ground substance of the dermis occupies the space around the cellular and fibrous elements and consists largely of mucopolysaccharide but also contains glycoproteins, water, inorganic salts and hormones.

The cutaneous nerves are generally found in the deeper dermis, with free nerve endings in the dermal papillae. Sensations of pain, itching and tickling are detected here, the differences between them being one of stimulus intensity and interpretation by the central nervous system. The dermal papillae also contain networks of blood capillaries, with shunts in them which can alter the direction and rate of blood flow. Factors such as heat or emotional embarrassment activate these shunts, causing more blood to flow towards the surface of the skin. This increase in blood flow is commonly observed as reddening of the skin "flushing". Similarly, cold temperatures or emotional shock results in blood being drained away from the capillary network, causing the face and body extremities to whiten. If the body is subjected to very cold temperatures, about 90% of the blood supply may be returned to the circulation, without passing through the skin capillary network. In these circumstances, the requirement for the body to conserve heat takes preference over the skin's need for nutrients and blood supply.

The response of the skin to irritation and damage

Damage to the skin may be caused by any of the following factors:

- mechanical damage
- chemical damage
- bacterial attack
- heat
- light (ultraviolet radiation)

The first effect of mild damage, such as minor abrasion to the skin, is for the epidermal cells to start dividing more rapidly than normal at the basal layer. If the damage to the skin is progressive, the mast cells in the dermis release histamine, causing the blood vessels to dilate and become more permeable. This allows higher levels of nutrients to reach the region where cell division takes place and thus provides more energy for the division process.

As a consequence of this, the "triple response" of skin to damage takes place, with the occurrence of *erythema* or skin reddening, followed by *oedema* or skin swelling and finally *flare*, which is extensive reddening of the skin.

In cases of minor skin abrasion, where only the epidermis is damaged, mitosis continues at the same rate until repair is effected. Tissue repair is stimulated by the temporary removal of chalone from the damaged site. If the damage is more severe and penetrates into the dermis, cells from the blood stream enter the dermis and grow into fibroblasts and mast calls, producing collagen and mucopolysaccharide respectively.

1. *Chemical damage of the skin*

Chemical damage to the skin can result in two conditions, irritation or sensitisation.

Irritation is normally classified in one of two categories, primary irritation or secondary irritation. Primary irritation is caused by exposure of the skin to corrosive chemicals such as acids or alkalis and involves destruction of the epidermis, the extent of damage being dependent on chemical concentration and contact time. Secondary irritation is caused by less corrosive materials, such as anionic surfactants, and has similar characteristics to primary irritation, except that chemical concentration is lower and the rate of destruction much slower. A third form of irritation response, that of phototoxic irritation, is caused by materials which are chemical irritants in the presence of light or ultraviolet radiation.

Skin sensitisation, unlike irritation, is a response of the immune system. The body's normal immunological response describes the way in which it protects itself from attack, based on the history of past attacks made on it, and there are three phases to such a response. Firstly, the attacking species, normally referred to as the *antigen*, attacks the skin. In response to this attack, the body produces antibodies in the blood stream known as immunoglobulins. When the same antigen attacks the body at a later date, it is neutralised by the antibodies previously formed, so that no harm is done.

Sensitisation occurs when this immunological response of the body does not function properly. Only small quantities of chemical are needed to cause sensitisation and only about 0.01% of people are affected by any one sensitising species. The sensitisation problem occurs upon the second application of the sensitising species, provided about 24 hours separates successive applications.

Sensitisation takes place when a normally safe chemical (e.g. a fragrance component) is applied to the skin and antibodies are formed. If the same chemical is applied at a later time, it is now recognised by the immune system as an antigen, which has previously produced antibodies in the blood, and the result is the formation of an antigen/antibody complex. The antigen/antibody complex now acts as an irritant, which sets off a triple response reaction. Antigens are normally high molecular weight chemicals but in some cases low molecular weight chemicals, known as *haptens*, produce the same end result. In these cases, the hapten reacts with skin protein to form high molecular weight antigens.

Photo-sensitisation is a similar process to that described above, except that the hapten/protein reaction requires the presence of light or ultraviolet radiation. A persistent light reaction is said to be present when some of the sensitising chemical remains in the skin after the antibodies have been formed, so that the sensitisation reaction can occur without a second application of the material.

Dry skin and its prevention

Dry skin is a problem associated with the horny layer of the epidermis and its aetiology is therefore an important consideration for the cosmetic scientist, when formulating products such as moisturisers to combat the problem.

Dry skin is most frequently caused when the skin comes into contact with either water or detergent solutions, particularly in cold weather. In some cases, chapping accompanies dry skin, causing superficial cracks to appear in the horny layer of the epidermis. Chapping appears to occur when there is a sudden fall in the moisture content of the air around the body. The mechanism by which water and detergent solutions cause skin roughening or dry skin is not completely understood but initially involves a swelling of the horny layer, as it rapidly absorbs water when the skin is immersed. The inherent stress put upon the skin during this swelling process, causes cracks and splits to form in the horny layer surface. When the skin is removed from the water, the horny layer begins to dry and attempts to return to its original size and form in so doing. However, because of the cracks and splits that have formed in the horny layer during hydration, not all of the skin cells are able to return to their original configuration and some are left protruding above the skin surface, giving it a rough feel. The removal of some of the natural skin lipids, which occurs during the hydration process, contributes to the rough feeling of the skin.

One of the characteristics of dry skin is a feeling of loss of natural oiliness and originally scientists believed that the dead tissue of the skin surface was kept smooth and supple, due to the lubrication supplied by sebum secreted from the sebaceous glands. Current theory suggests that it is water, and not oil or sebum, which makes the skin soft and flexible and the horny layer of normal skin is thought to have as much as 30% of its mass made up of water and water-soluble materials. In the case of a dry skin condition, water is lost from the horny layer faster than it can be replaced and, typically, the water content of the horny layer may be as low as 10% or less. In addition, the water retaining characteristics of the horny layer are reduced. Other factors such as age, ethnicity and physical damage to the skins surface, can increase the likelihood of the occurrence of a dry skin condition.

1. *Keratinisation and the dry skin condition*

To understand further the mechanism for the occurrence of dry skin, the structure of the epidermal horny layer must be considered in more detail. Figure 5 below illustrates the structure of the epidermal horny layer.

FIGURE 5

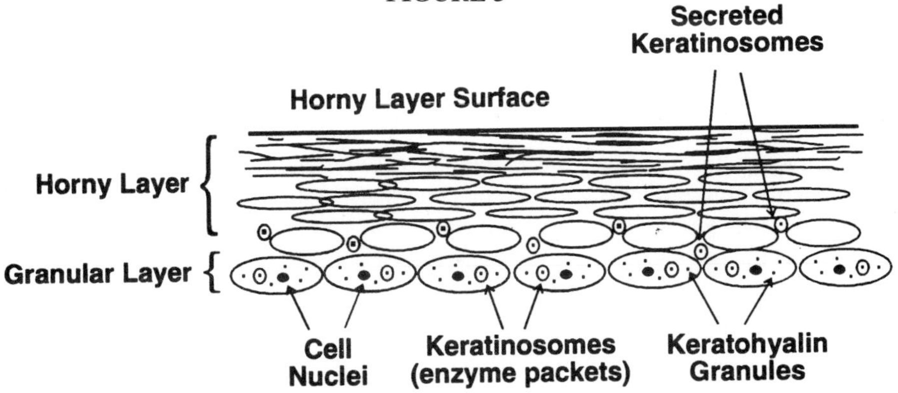

During the keratinisation process, partially formed keratin in the granular layer becomes apparent in the form of keratohyalin granules. This keratohyalin is chemically different from keratin, in that it contains more sulphydryl groups attached to the polypeptide chains than would normally be seen in fully formed keratin. Keratin therefore exhibits a more rigid structure than keratohyalin, due to the presence of disulphide links which join adjacent polypeptide chains together.

Also found in the granular layer are small packets of enzymes known as keratinosomes. These are essentially modified lysosomes that can be broken by the action of UV light. The keratinosomes are unique in that they can be promoted by the granular layer cells into inter-cellular spaces between cells in the horny layer. The hydrolytic enzymes held within the keratinosomes are inactive until the keratinosome is released from the granular cell, whereupon the enzymes become active, performing two major functions in the keratinisation process. The first of these is to induce the final stage of keratin production and promote the formation of the disulphide links. The second function is to promote cell cement degradation when the cells get near the surface of the horny layer, thus allowing the normal process of cell loss from the skins surface to occur. This latter process is normally referred to as desquamation.

Thus it can be seen that the break-down of cells, caused by the action of hydrolytic enzymes, actually takes place in the horny layer, with the resultant loss of the cell nucleus and cytoplasm, and destruction of the keratohyalin granules. A consolidation process accompanies this cell break-down in which water is lost from the cell, a horny matrix is produced from empty cell envelopes and inter-cellular cement is formed.

This process gives rise to the formation of a semi-permeable membrane, constructed from old cell walls which provide the lipid content of the horny layer which, in turn, provides a "fatty film" thus reducing water transfer through the skin. Some water is still lost through the skin despite the protective effects of the lipid layer. This process of water-loss is known as skin transpiration. Amino-acids, derived from the old cell cytoplasm, are also found in the horny layer and, due to their hygroscopic nature, provide a water reservoir which maintains the water content and flexibility of the horny layer. This reservoir of amino-acids and other substances is referred to as the *natural moisturising factor (NMF)* of the skin. The composition of the natural moisturising factor varies but contains lactates, 2-pyrrolidone carboxylic acid, urea, metal ions and sugars, as well as the amino-acids themselves. 2-pyrrolidone carboxylic acid, in particular, is an important natural component of skin and is a very effective humectant, which helps to maintain the skin's moisture balance.

An undamaged epidermal horny layer is a fairly effective barrier to the net gain or loss of water from the body and under normal circumstances water ascends from the underlying dermis, through the viable epidermis and into the horny layer. A proportion of this water is then lost by evaporative diffusion through the horny layer. Loss of water from the body in this way is referred to as *transepidermal water loss*, or TEWL for short.

Transepidermal water loss is normally governed by the laws of diffusion, to give an equilibrium condition in which water is released from the dermis and lower skin, to maintain a balance with the water that is lost from the skin's surface. This results in a moisture concentration gradient through the skin, the moisture content being highest in the dermis and falling away towards the skin's surface.

When the skin is exposed to conditions of low temperature or relative humidity, such as cold, windy weather for example, the concentration gradient is disturbed and moisture is lost faster than it can be replaced. In these circumstances, the horny layer becomes drier than the lower epidermal layers and if the skin is stretched or flexed the surface layers will crack, whilst the deeper layers remain intact. This is due to the loss of the natural moisturising factor in the horny layer causing shrinkage of the keratin, causing the formation of a rough and inflexible skin surface. The NMF itself acts not only as a natural humectant for the horny layer but also as a plasticiser, thus providing a softening effect. Plasticisation is achieved by the relatively highly hygroscopic NMF materials allowing increased hydration and skin flexibility.

When the skin is immersed in water, some of the soluble NMF components are lost and, on drying, the keratin is no longer kept moist. If the skin is immersed in dilute detergent solution, the situation is exacerbated with loss of not only the water-soluble components of the NMF, but also the fat-soluble NMF. Loss of these NMF components deprives the skin of its natural humectancy, with the result that the skin becomes rough and dry. This condition is then known as dry skin.

2. *The treatment of dry skin*

In practice, there are three ways of combating dry skin. The first is by application of a suitable low friction oily or waxy material to the skin. Examples of such materials are stearic acid, cetyl alcohol and silicone oil derivatives, which all give a smooth "velvety" feel to the skin after use. Secondly, humectants such as sorbitol or glycerine can be applied to the skin, in an attempt to compensate for the lost NMF. Commercially available humectants such as sodium lactate, urea and pyrrolidone carboxylic acid can also be used. Lastly, the effects of dry skin can be reduced by the application of occlusive materials, which form an oil continuous film at the skin's surface. This provides a barrier to trans-epidermal water loss from the horny layer, resulting in a higher moisture content. This latter method is probably the most important for combating dry skin conditions.

Skin care products which are designed to reduce skin roughness and increase the water content of the skin, by any of the above methods, are normally referred to as *moisturisers*. Another class of materials used in the formulation of skin care products are *emollients*. These are materials which, when applied to a dry horny layer, will effect a softening of that tissue by inducing rehydration. Water alone cannot be used as an effective moisturiser, as only a thin film of water will adhere to the skin surface and this is very quickly lost by evaporation.

An ideal emollient is a substance which regulates the rate and quantity of water uptake by the skin. Traditionally, emollients tend to be oily substances which, when applied to the skin as occlusive films, provide a protective effect, prevent moisture evaporation from the skin's surface and in so doing, allow depleted moisture to re-accumulate. Such oily materials, when used alone, are aesthetically unsatisfactory and if used over a long period of time can cause swelling and inflammation.

The other traditional treatment for dry skin, that is the application of a humectant, is not wholly satisfactory because in conditions of low relative humidity, a humectant applied to the skin by itself will not only absorb moisture from the atmosphere but will also absorb moisture from the lower layers of the skin, thus exacerbating the dry skin problem that already exists.

The benefits of both oily materials and humectants can be combined, in an aesthetically acceptable form, as an emulsion and the vast majority of skincare products on the market today are in emulsion form. There is, however, an inverse relationship between the value of oily materials as moisturisers and their cosmetic appeal. A good example of this is petroleum jelly which, although an excellent moisturiser by virtue of its occlusive properties on the skin, is unacceptable when used alone due to its feel and physical form. Such materials are therefore formulated into an emulsion to provide an aesthetically satisfactory product, although in this form their moisturising properties are proportionally reduced.

More recently, and in an attempt to overcome this problem, moisturising skincare formulations have utilised volatile oils for both cream and lotion products. During application, these volatile components effectively reduce the viscosity of the product, making it light and pleasant in use. After application, the volatile components evaporate, leaving a thin film of the heavier, more effective, moisturising oils on the skin's surface. The water in the emulsion acts as a plasticiser or softener for the skin, as well as imparting a "cooling effect" on evaporation. The non-volatile emollient oils act both as a lubricants to smooth the skin's surface and as an occlusive film to help the skin retain its moisture.

Illustrated below are three typical moisturiser formulations, each with the functionality of their ingredients explained underneath. The first is a hand cream formulation of the following composition:

Oil phase	
Stearic acid	15.00%
Lanolin	2.00%
Cetyl alcohol	2.00%
Isopropyl myristate	2.00%
Water phase	
Glycerine	3.00%
Sorbitol	3.00%
Triethanolamine	1.40%
Water	71.60%
Dyes, preservatives, perfume, etc.	qs

This is a simple, conventional hand cream formulation, using an O/W emulsion. The stearic acid is a cost effective, readily available emulsifier which reacts with the triethanolamine to produce triethanolamine stearate in the emulsion, acting as the primary stabiliser. It also helps provide skin feel of the product. Cetyl alcohol acts as the secondary co-emulsifier and provides some lubricating properties for the skin. Lanolin and isopropyl myristate are emollients and glycerine and sorbitol assist in providing the moisturising benefits of the product, via their humectant properties. The water content of the product is relatively high at around 70%, making the cream light and easy to apply.

Hand creams often incorporate solid materials that have a higher melting point than the temperature of the skin, producing a relatively dry, non-greasy feel, after application. This property helps to create a "vanishing" effect in the product during use, enhancing easy rub in and aesthetic pleasantness.

The second formulation is that of general purpose skin cream thus:

Oil phase	
POE Glycerol Sorbitan	
Isostearate	10.00%
Beeswax	3.00%
Lanolin (high purity)	3.00%
Caprylic/capric triglyceride	11.00%
Squalane	11.00%
Sunflower oil	3.00%
Water phase	
Glycerine	2.00%
Magnesium Sulphate	0.70%
Water	56.30%
Dyes, preservatives, perfume, etc.	qs

In the above example, the sunflower oil is combined with beeswax, squalane and caprylic/capric triglyceride to give the skin feel and moisturising properties provided by the formulation. The lanolin is included for additional emollience and skin feel. Emulsification is based on a nonionic W/O emulsifier, POE glycerol sorbitan isostearate, with the resultant formation of an W/O product. The magnesium sulphate also helps stabilise the emulsion system.

A final example of a moisturising skincare product is that of a hand and body lotion as detailed below:

Oil phase	
Mineral oil	3.00%
Cyclomethicone	4.00%
Isopropyl myristate	3.00%
Stearic acid	1.80%
Cetyl alcohol	1.00%
Glyceryl stearate	1.50%
Water phase	
Carbomer 984	0.10%
Glycerine	3.00%
Triethanolamine	0.90%
Water	81.70%
Dyes, preservatives, perfume, etc.	qs

This formulation is a very light, aesthetically pleasant lotion, although its properties as a moisturising product would be modest compared to the two previous formulations discussed. The main moisturising benefit is derived from the presence of the mineral oil, whilst both the cyclomethicone and isopropyl myristate provide a "velvety" feel to the skin. The emulsion is stabilised using triethanolamine stearate, formed *in situ* from the triethanolamine and stearic acid, when the emulsion is made. Cetyl alcohol and glyceryl monostearate are present as a secondary co-emulsifiers and also provides some skin lubrication. The emulsion is further stabilised by the inclusion of a small quantity of a acrylic-based polymer, Carbomer 984.

3. *The evaluation of moisturising products*

Methods of evaluating moisturising products vary widely and the choice of raw materials often depends as much on factors such as cost, aesthetic value and compatibility as the ability to provide moisturising benefit. However, as the skin care market becomes more technologically led, there is an increasing requirement for the ability to scientifically evaluate moisturising products for both product claims and competitive performance advantage.

Of the many methods that have been developed to evaluate the moisturising benefits of skincare products, three of the most common are detailed below:

3.1. *Radio-tracer techniques*

Using this method, radioactively labelled water enables the penetration of moisture from a given product, into the horny layer and epidermis, to be monitored with time. The product to be evaluated is prepared using labelled water and applied to the skin. After a period of time, the top layer of the skin is stripped off, using adhesive tape, and evaluated for its radio-activity. This process is then repeated 10-20 times for successive layers of the epidermis.

3.2. *Infra-red spectroscopy*

Infra-red spectroscopy can be used to measure the water-content of the horny later *in vivo*. The human skin exhibits two absorption bands in the infra-red region, one of which is greatly expanded by the presence of water, whilst the other remains unaffected. Thus, the ratio of these two peak heights will provide a measure of the moisture content of skin. Determinations are made before and after application of the moisturising product, using an attenuated total reflectance infra-red cell.

3.3. *Electrical impedance*

The electrical resistance of the skin is a function of the resistance or impedance (reciprocal resistance) of the horny layer. The impedance of the horny layer is proportional to its water content and therefore the water content can be determined by impedance measurement. A pair of dry electrodes are placed on the skin of a non-sweating subject and the impedance of a current passing between these electrodes is measured.

The skin care regimen

Regular practice of a suitable skin care routine is the best way to maintain the skin in good condition and minimise the signs of the skin's ageing process. Characteristics associated with skin ageing become apparent over the years, with the resultant reduction in skin moisture content and elasticity and the ultimate formation of wrinkles. Moisturisation of the skin must be carried out both regularly and conscientiously, if any cumulative benefit is to be observed. There are a wide variety of skin care products on the market today and these can normally be classified by product form and function. Skin type varies from individual to individual and the recognition of any particular skin type is important when selecting a suitable skin care routine. Skin type is normally classified as either dry, normal or greasy, although special cases such as combination skin and sensitive skin must also be acknowledged. The three major steps of any skin care regimen, independent of skin type, are:

1. *Cleansing*

Cleansing products are applied to remove make-up and generally cleanse the skin itself. They are massaged into the skin and wiped-off with a tissue, leaving the skin soft, smooth and clean. The type of cleanser used is dependent upon skin type and subjective attitude of the user. Cleansing products are normally available in two types, creams and milks. Creams usually contain appreciable amounts of oil with higher levels of emollients and are more often used by people with dry skin. Cleansing milk are based on O/W emulsions, with varying levels of oily materials. These products cover the complete skin type spectrum but, in general, more oily types are used by women with dry skin and less oily types by women with greasy skin.

2. *Toning*

Toners are used to complete the cleansing routine, removing the last traces of cleansing product and "freshening" the skin. The products are principally aqueous/alcoholic solutions, the alcohol content itself depending upon the level of astringency required and the type of skin for which the product is designed. Indeed, more recently, alcohol-free toners have become very popular, due to the belief that the presence of alcohol will excessively dry the skin. Toners also contain humectants and esters, at fairly low levels, to improve the feel of the skin after use.

3. *Moisturising*

Moisturisers are normally available in either cream or lotion form and, used on a regular basis, protect the skin from the dryness. Moisturisers can be either O/W or W/O emulsions, with varying oil levels depending on the requirements and skin type of the user.

Flaky skin

Skin cells in the granular layer produce both cell cement and keratin, but each of these is made at the expense of the other. There is a relationship between the rate of cell mitosis in the basal layer and the release of horny cells at the skin's surface, such that the release of horny cells increases with the rate of mitosis. The cells in the horny layer itself are held together with inter-cellular cement, forming a barrier to the passage of substances into and out of the skin. Near the surface of the skin, the horny layer begins to crack, where the cells become loosened in preparation to be shed. This is known as the desquamating zone and is nominally 3-4 cell layers thick. Horny cells are not shed individually, but in aggregates which vary in size. Typically, aggregates are in the order of 200 microns in diameter, compared with a single cell diameter of approximately 40 microns. Larger aggregates, normally visible to the naked eye, are generally referred to as flakes or *squames*.

The aetiology of flaky skin can be better understood by comparing the features associated with the desquamation of normal skin with those of flaky skin. In normal skin the process of desquamation is as follows:

- Epidermal cells are produced in the basal layer
- Keratin polypeptide is synthesised in the prickle layer as tonofibrils
- Keratinosomes appear in the granular layer and tonofibrils aggregate to form granules of keratohyalin
- Some of the keratinosomes are secreted into the cell-cement regions
- Keratinosomes in the granular layer rupture, killing the cell and converting the granules into keratin
- Cell-cement keratinosomes rupture degrading the cell-cement and thus releasing horny cells. The keratinosome rupture can be caused by chemical means, light, or mechanical pressure

Sometimes very large aggregates, up to 2 mm in diameter, containing thousands of horny cells are lost, resulting in a condition of flaky skin or *parakeratosis*. By comparison with the above list, parakeratotic skin is characterised by the following features:

- The normal mitotic rate is approximately doubled, the cells in the basal layer dividing twice as fast as normal
- Only half the normal number of horny layer cells are present
- The horny layer keratin contains more sulphydryl groups than normal skin, indicating improper formation
- There is a higher level of cell-cement in the horny layer
- The horny layer contains less NMF (natural moisturising factor) than normal, causing the skin to be drier
- Some cell nuclei exist in the horny layer
- Very few keratinosomes exist

Clinically, flaky skin is dermatologically characterised by two features, the existence of cell nuclei in the horny layer and the presence of a higher level of sulphydryl groups than normal.

1. *Dandruff*

Dandruff is the most common example of flaky skin and is characterised by a chronic, although non-inflammatory, scaling of the scalp. Dandruff first occurs at the age of puberty, rising to a peak in the late teens and early twenties, gradually declining and falling off rapidly with old age. The occurrence of dandruff also exhibits a seasonal variation, declining in the summer months and rising to a maximum in early winter.

Dandruff is characterised by an increased production of both horny cells and squames at the skin's surface and the viable epidermis is renewed every 7 days, compared with 13 days for normal skin. Although normally a physiological process, desquamation occurs at different rates for different people and the dividing line between a dandruff sufferer and a non-dandruff sufferer is arbitrary.

A possible mechanism for the causation of dandruff arises from irritation of the scalp by fatty acids, particularly oleic and linoleic acids from triglycerides in the sebum, causing flaking. These fatty acids are broken down by a enzyme, secreted from a yeast micro-organism, found on the scalp, known as *Malassezia ovale*. The total amount of fatty acid is dependent on both the amount of sebum produced, and the amount of enzyme present. *Malassezia ovale* is one of the normal yeasts amongst the microorganisms of the scalp and its population is greater where dandruff is present. Originally, *Malassezia ovale* was considered to be a causative factor of dandruff but it is now thought that it's presence is a consequence of it.

1.1. *Anti-dandruff treatments*

Dandruff severity is best assessed by trained observation, using rating scales for degree of scaling on the scalp. Any anti-dandruff treatment is judged to be significantly effective if the degree of scaling drops by 50% or more. The most important actives used in the treatment of dandruff are:

- Selenium disulphide
- Zinc pyrithione (ZPT)
- Piroctone olamine ("Octopirox")

All of these materials are effective anti-microbial agents and reduce the presence of dandruff after the second or third application. The mode of action of these agents is cytostatic, a suppressive effect being exerted on the rate of cell mitosis and, consequently, on the rate of scaling.

Two other actives, which have been used in anti-dandruff preparations, are sulphur and salicylic acid. Both of these materials are keratolytic, dissolving the cell-cement and causing the large dandruff flakes to disintegrate. Coal tar is also an effective anti-dandruff agent with an unknown mode of action.

Anti-dandruff agents are most conveniently applied to the scalp in shampoo form. Selenium sulphide is normally used at levels of approximately 2% and is rapid in its action. Zinc pyrithione is incorporated at levels of between 0.5% and 2%, although some difficulties in formulating stable shampoos with this material exist. Piroctone olamine is normally used at levels of 0.5% to 1% and has the distinct advantage of being suitable for clear shampoo systems, due to its solubility characteristics.

A typical anti-dandruff shampoo formulation, based on zinc pyrithione, is detailed on page 182.

Sodium Lauryl Ether Sulphate	15.00%
Lauric Isopropanolamide	2.00%
Zinc Pyrithione	1.00%
Bentonite	0.50%
Sodium Chloride	1.00%
Water, preservatives, dyes, etc.	to 100.00%

In the above example, the cleansing properties are provided by the sodium lauryl ether sulphate, the resultant foam being stabilised by the lauric isopropanolamide. The active agent, zinc pyrithione is not soluble in water and must therefore be held in suspension using a suspending agent, typically a synthetic clay, such as bentonite. The sodium chloride is added to the formulation to give viscosity control. Zinc pyrithione is also unstable in the presence of light and product packaging must therefore be opaque.

Piroctone olamine has advantages over zinc pyrithione, in that it is soluble in water and stable to light, therefore enabling it to be used in clear products with transparent packaging.

The ageing of skin

The ageing associated changes that take place in the skin, are caused by changes in the skin's rheology, mainly the dermis rather than the epidermis.

The dermis is a network of essentially non-elastic fibres in a polysaccharide gel, the rheology of which can be described as having viscoelastic properties. These properties can be analysed with the aid of a model, such as the one shown in Figure 6.

FIGURE 6

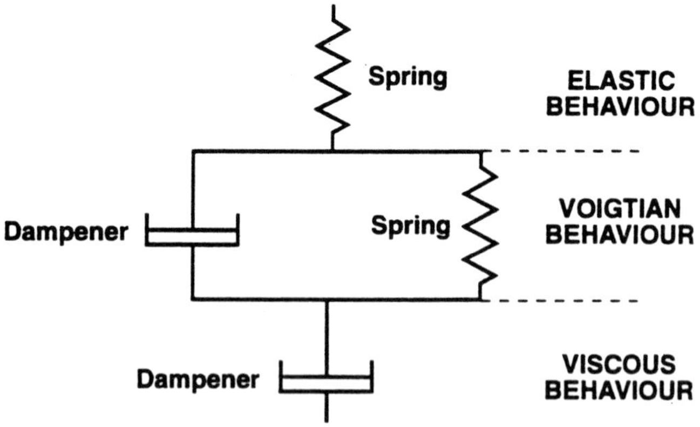

This model provides a representation of viscoelastic behaviour and contains three components, elastic behaviour, Voigtian behaviour and viscous behaviour. Elastic behaviour describes instantly recoverable deformation, whilst the Voigtian component describes deformation that is still reversible but with dampening. The viscous behaviour describes non-reversible deformation. There is also a fourth component present in the rheological make-up of the skin, known as the slack component which describes skin found around the knuckles and joints.

The elastic component maintains the tension in the skin, whilst the Voigtian component is responsible for better overall accommodation of objects in contact with the skin. Lastly, the viscous component accommodates any rapid growth in the skin, such as the growth of pimples or boils.

The rheological behaviour of the skin is provided by the non-elastic collagen network in the dermis, which itself is loosely joined and flexible. The environmental polysaccharide gel, moving through this lattice, provides Voigtian behaviour and the adjacent dermis maintains the elastic property of restorative pressure.

Collagen in the dermis is a fibrous protein, produced by fibroblasts, that slowly dissolves after manufacture. Free radical cross-linking can occasionally occur between collagen fibres, such that the collagen can no longer be dissolved. With age, the amount of cross-linking increases, giving rise to decreased flexibility within the skin. Other changes, occurring with age, in the epidermal appendages include:

- Atrophy of the apocrine and eccrine glands, but not of the sebaceous gland
- A retardation in hair growth rate and a reduction in the number of hairs and hair follicles
- A loss of lustre and change of colour on nail plates
- A decrease in the volume of blood circulating in the dermis
- A decrease in the amount of sebaceous fat below the dermis

Occurrence of the above changes leads to a marked change in the skin, as age increases. Quantitatively, there is less collagen than normal found in aged skin, with a higher proportion than of this being cross-linked. Although the amount of collagen is decreased, the quantity of elastin, the precursor to collagen, remains the same. There is also less mucopolysaccharide in the dermis, due to the reduced rate of synthesis. Older skin is characterised by the occurrence of a thinning of the epidermis, associated with a reduction in mitotic rate, and a reduction of fat in the dermis. Water transpiration in older skin is lower, leading to a higher water content in the dermis and a drier horny layer. Finally, the pigmentation processes in the epidermis frequently malfunction, leading to blotching or hyperpigmentation of the skin.

1. *Anti-ageing products*

Some drugs, applied topically, can affect the age induced changes in the skin. Traditionally, hormone creams containing oestrogen have been used for this purpose although current interest revolves mainly around the use of vitamin A and its analogues.

As a therapeutic drug vitamin A, or retinoic acid, has been used to treat a number of skin disorders, particularly psoriasis where it effects the liberation of hydrolytic enzymes in the granular layer. Much lower doses of Vitamin A are used as actives in cosmetic products, but these low concentrations do provide noticeable improvements in the skin with maximum benefit occurring some time after the treatment is commenced. Beneficial effects of topically applied vitamin A include an increase in the rate of cell mitosis, leading to thickening of the epidermis. Collagen and blood vessel formation in the dermis is also stimulated and improvements are observed in dry skin. Lastly, there is a noticeable improvement in the occurrence of skin pigmentation disorders.

Hair growth and its control

Hairs are dead structures, composed of keratinised cells compactly cemented together, growing out of epidermal tubes, or hair follicles as they are more properly called, sunken deep into the dermis. A diagrammatic representation of a hair, within its follicle, is shown in Figure 7.

FIGURE 7

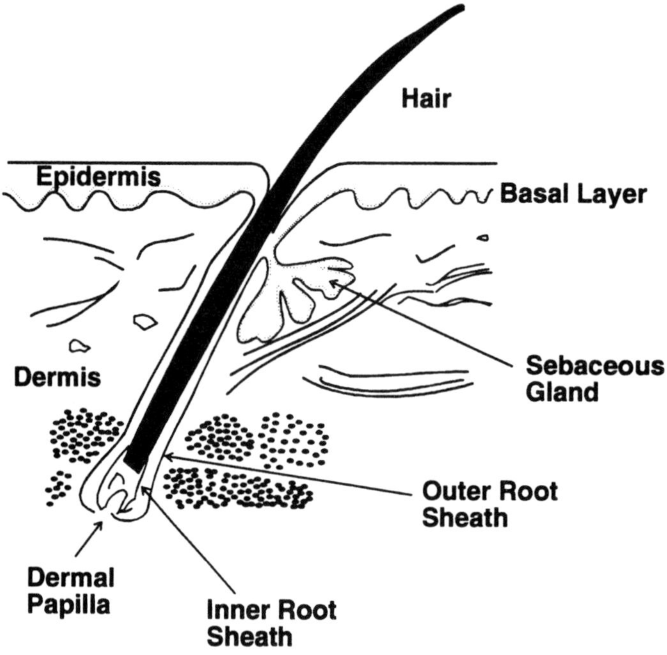

The keratinising zone extends along the lower part of the hair and two supporting sheaths, the inner and outer root sheaths, surround the hair shaft from the root, through the follicle, to the skin's surface. The inner root sheath is interlocked with the cuticle cells of the hair and grows upwards to the surface with the hair, sliding against the outer root sheath which is stationary. Above the level of the sebaceous gland, the outer root sheath becomes identical in structure to that of the surface epidermis. The dermal papilla provides a source of nutrients, from the blood stream, for hair growth.

1. *Depilatories*

Depilatory products are used to soften unsightly body hair, so that it can be easily removed by rinsing, or wiping with a tissue a short time after application. Depilatory products should be as cosmetically elegant as possible and should be non-toxic and non-irritating to the skin. They should be easy to apply and economical in use, converting human hair to a soft mass within 2-5 minutes, so that it can be easily removed.

The major problem with depilatory products is that because both skin and hair are composed of keratin, both are therefore subject to the same degree of attack from the same chemicals. Keratin is rich in sulphur-sulphur bonds which link the polypeptide chains, giving the hair fibre its stability and flexibility. The aim of a depilatory product is to break just enough sulphur-sulphur bonds, by chemical reduction, for the hair structure to disintegrate.

The most commonly used active ingredients used in depilatory products are strontium sulphide and calcium thioglycollate. The former is faster acting than the latter although it is also more irritant. Most modern products are based on thioglycollates.

A typical depilatory formulation, based on strontium sulphide, is given below:

Strontium sulphide	20.00%
Talc	20.00%
Hydroxyethyl cellulose	3.00%
Glycerine	15.00%
Water, preservatives, dyes, etc.	to 100.00%

Depilatory products are usually formulated as creams or pastes because they allow sufficient contact time to be effective and are cosmetically elegant. This consistency is normally achieved using thickeners, such as hydroxyethyl cellulose. Non-reactive powders, for example talc, are used as product fillers, reducing the irritation profile of the product and allowing ease of removal of the cream after application. Glycerine is included in the formulation for its emollient properties.

A contemporary formulation, based on calcium thioglycollate, is detailed below:

Calcium thioglycollate	7.00%
Calcium hydroxide	7.00%
Calcium carbonate	20.00%
Cetyl alcohol	5.00%
Sodium lauryl sulphate	1.00%
Water, preservatives, fragrance, dyes, etc.	to 100.00%

This formulation has a lower irritation profile than the first and can be safely used in the facial area. Depilation time is reduced by raising the pH of the product to between 10 and 12 pH units, by the addition of calcium hydroxide. This causes the body hair to swell and react with the active ingredient more quickly. The pH range of 10 to 12 is very important. Above pH 12 there is a high risk of skin damage, whilst below pH 10 depilatory action falls off rapidly. Contact with the skin is enhanced by the use of a wetting agent, in this case sodium lauryl sulphate. Good wetting action in the product allows better contact with the hair shaft, thus increasing the rate of swelling and softening of the hair. Calcium carbonate, in the form of chalk, is used as a filler, with emolliency being provided through the use of cetyl alcohol.

2. *Shaving products*

Wet shaving products are normally based on soaps and can take the form of sticks, aerosols or lathering creams. Aerosols are, by far, the most popular form of wet shaving aid, this being largely due to their convenience and ease of use.

Shaving products function by surrounding each hair with an array of very small air bubbles which separate and present the hair in erect form, ready for blade cutting. Shaving products must therefore produce copious quantities of foam in a short space of time. Apart from being stable and of low irritancy, wet shaving products must promote hydration of the beard hair, thus causing it to swell and soften and consequently reduce cutting resistance. The foam produced should also be durable and lubricious, so that the cutting blade slides easily over the face throughout the shaving process.

Originally, all wet shaving products were in the form of soap. These were composed of short-chain fatty-acids (typically lauric acid) which provided the rapidly forming, copious lather, essential in an effective product. In order to increase the foam viscosity and stability, long-chain fatty-acids (typically C_{18}) were also included in the formulation. In practice therefore, a blend of long and short chain fatty-acids is used to optimise product performance. The ease of lathering also depends upon the solubility of the soap formed in the product. Sodium soaps, for example, are hard with relatively low solubility, whilst their potassium counterparts are softer and more soluble. The softest and most soluble soaps are those based on triethanolamine salts. The effect of various components, on both the soap and the foam properties of a shaving product, are summarised in Table 1.

TABLE 1

Ingredient	*Effect on soap*	*Effect on foam*
Short chain fatty acids	Softening. More irritant to the skin	Produces large quantity of loose bubbles quickly
Long chain fatty acids	Hardens the soap	Slowly produces small quantity of thick foam
Triethanolamine	Produces a very soft soap	Produces a quick copious foam
Potassium hydroxide	Produces a soft soap	Produces a copious foam
Sodium hydroxide	Produces a hard soap	Produces less foam
Glycerine	Produces a soft soap	Stabilises foam and stops drying out

For any given formulation, blends of the above ingredients are chosen to give the correct properties, depending on physical form. Some typical formulations are illustrated below:

	Stick	*Cream*	*Aerosol*
Lauric acid (C_{12})	18.00%	10.00%	0.95%
Stearic acid(C_{18})	65.00%	30.00%	5.70%
Glycerine	5.00%	10.00%	1.90%
Triethanolamine	—	—	2.85%
Potassium hydroxide	4.30%	2.40%	—
Sodium hydroxide	1.70%	0.60%	—
Water	6.00%	47.00%	83.60%
Butane/propane	—	—	5.00%

A further category of shaving products are the brushless shaving creams, which are normally oil-in-water emulsion products that function by substitution of oil bubbles for the air bubbles in normal shaving soaps. They tend to give a more comfortable shave, due to their skin lubricating qualities, and leave the face feeling softer. The disadvantage with this type of product is that the beard must first be pre-softened with soap and water. An example of a brushless shaving cream formulation is illustrated below:

Oil phase	
Stearic acid	20.00%
Mineral oil	2.00%
Cetyl alcohol	0.50%
Water phase	
Water	76.00%
Triethanolamine	1.20%
Dyes, preservatives, perfume, etc.	to 100%

Sunlight and the skin

1. *Skin pigmentation*

The colour of the skin is determined by a number of physiological characteristics. The most important factor is the quantity of melanin in the skin, the melanin itself being of two different types, *phaeomelanin* (reddish-brown in colour with a high cysteine content) and *eumelanin* (blackish-brown). All of the melanin pigment in the skin is found in the epidermis. The next most important factor in determining skin colour is the presence of beta-carotene, which is found distributed throughout the epidermis, dermis and subcutaneous fatty layers. Finally, the amount of blood in the dermal and subcutaneous blood vessels, and whether or not the blood haemoglobin is oxidised, will also have a marked effect on skin colour. The visible appearance of the skin is also affected by the way in which incident light is reflected or scattered at the skin's surface. A diagrammatic representation of the skin, showing the way in which light is scattered or reflected at the surface, is illustrated in Figure 8.

FIGURE 8

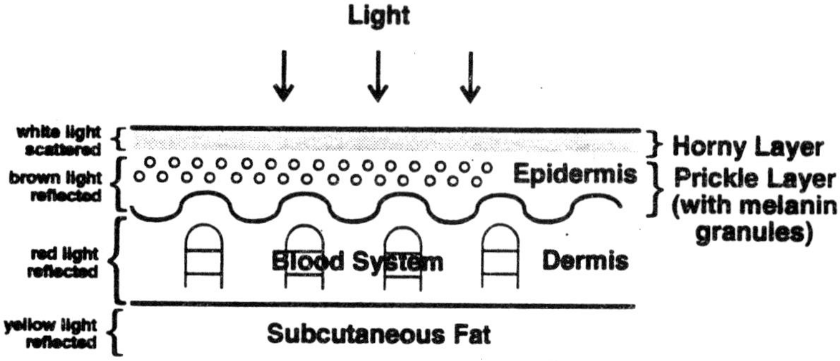

The melanin content of the skin varies widely with race and even within different parts of the body. More melanin is produced following exposure to sunlight, consequently turning the skin brown. The horny layer of the skin can become dry or flaky thus scattering light and appearing whiter, whilst the superficial blood system of the skin may make the skin appear red in colour.

The melanin pigment itself is formed in the skin by dendritic cells called melanocytes, which are located in the basal layer of the epidermis. The melanin is synthesised in several stages, starting with the synthesis of pre-melanosomes in the golgi body of the melanocyte. The endoplasmic reticulum of the melanocyte contains the amino acid tyrosine, which is oxidised by copper containing enzymes known as tyrosinase found in the pre-melanosomes, through a sequence of several steps to produce melanin as the final product.

The colourless pre-melanosomes become part of the cell dendrites and as the dendrites grow longer so the pre-melanosomes are promoted further away from the cell nucleus. During this promotion they become more identifiable as melanosomes, which darken as melanin is formed within them. The dendrites, containing the synthesised melanin, become absorbed by surrounding prickle-layer cells and melanin passes into these cells, from the tips of the dendrites, thus pigmenting them. The melanocytes and their associated epidermal cells are called melanocyte units and the amount of pigment in the cells depends on both the activity of the melanocyte and the associated cells.

There is a similar number of melanocytes in both white and negroid skin, the darker colour negroid skin being a result of increased melanocyte activity associated with the production of melanosomes which are larger than those found in white skin. Melanosomes in white skin are also usually complexed with surrounding cell protein, whereas in negroid skin they are individually dispersed. Oriental skin contains relatively little melanin and substantially more beta-carotene, accounting for the yellowish colour observed.

2. *Skin lightening products*

Skin lightening products contain depigmenting agents which may function by either destroying the melanocytes, by interfering with melanin biosynthesis, by inactivating the enzyme tyrosinase, by interfering with melanin transfer to adjacent epidermal cells or by discolouring melanin itself, producing a lighter reduced form.

Historically, many very toxic and dangerous materials have been used in skin lightening products but the most recent products all contain hydroquinone as the depigmenting agent. Although hydroquinone is a primary skin irritant, it is relatively innocuous at levels of less than 5%, although depigmenting ability is modest at this level. The upper limit of hydroquinone allowed in the EEC is 2%. Cosmetic skin lighteners are normally used where the skin is undesirably dark or mottled due to age or drug therapy.

3. *The effects of sunlight on the skin*

Sunlight contains visible, UV and IR radiation which have wavelength distributions as detailed in Table 2.

TABLE 2

Ultraviolet Radiation	< 400 nm
Visible Light	400-800 nm
Infra-red Radiation	> 800 nm

Ultraviolet radiation is subdivided into UV-A light in the wavelength region of 400-320 nm, UV-B light in the wavelength region of 320-290 nm and UV-C light with a wavelength of less than 290 nm, the latter being absorbed by atmospheric ozone. UV-B is present only in bright sunlight, whereas UV-A is present even in ordinary daylight on dull, cloudy days.

Originally it was thought that only UV-B light was responsible for erythema and tanning of the skin but it is now recognised that UV-A light also has very important implications in its reaction with skin. UV-A light is also capable of producing erythema of the skin but, more importantly, it penetrates deep into the dermis causing damage to the skin's structure. Indeed, it is now recognised that UV-A radiation is a major contributory factor of skin ageing.

When the skin is exposed to sunlight, it undergoes a series of complex biological and physiological changes. Firstly, melanin granules are promoted above the prickle cell layer of the skin to help protect the cell nuclei from radiation damage. The melanin granules then darken within a few hours of exposure to sunlight resulting in a tanning of the skin to offer further protection from UV light. As more melanin granules are produced, the skin continues to darken producing a more protective tan. Some protection from the harmful effects of sunlight is also provided by urocanic acid, which is a component of the eccrine sweat secreted from the skin in response to heat. Continued exposure to sunlight activates the synthesis of vitamin D at the top of the prickle cell layer, by the action of UV-B radiation on 7-dehydrocholesterol in the epidermis. Vitamin D synthesis, in turn, stimulates the activity of vitamin A, causing an increase in the rate of cell production in the basal layer. This increase in cell production in the basal layer produces an increase in the rate of production of pigmented cells and promotes a thickening of the horny layer of the skin, to provide protection from further damage.

Despite these natural protective mechanisms, the skin undergoes considerable damage when exposed to sunlight. Lysosomes in the mast cells of the dermis are broken by the effects of the UV radiation and the subsequent release of histamine produces the classic triple response in the skin. Fibroblast lysosomes are also broken by the effects of sunlight, releasing enzymes into the dermis where collagen is synthesised. The collagen in the dermis is then permanently damaged, eventually leading to premature skin ageing. The horny layer of the skin can also become flaky due to the release of enzymes from keratinosomes in the granular layer, which cause a breakdown of cell cement.

UV-B radiation is a major cause of both tanning and burning of the skin. Short term exposure to UV-A radiation has an augmentative action on these effects but, over longer periods, is primarily responsible for skin ageing. Symptoms of sunburn are a painful reddening and swelling of the skin, often followed by blistering and peeling. In the absence of further exposure, these effects usually subside within 7-10 days. The rate of tanning depends to a large extent on race and skin type and, in some cases, never takes place at all. Prolonged exposure of the skin to sunlight causes peeling and subsequent loss of the protective melanin.

Long term effects of exposure of the skin to UV radiation include premature skin ageing, characterised by thinning and wrinkling of the epidermis, loss of elasticity and pigmentation defects. The dermis of damaged skin contains a lower level of collagen and higher level of mucopolysaccharide than normally found in undamaged skin. Damage of the collagen in the dermis results in a reduction of dermal elasticity, causing loss of restorative capability in the supportive tissue.

A very serious effect of long term exposure to UV radiation, is that of melanoma, or skin cancer. There is no doubt about the correlation of exposure to UV radiation and incidence of the more common types of skin cancer. Skin cancer is more prevalent in people with fair skin and particularly those of Celtic origin.

4. *Sunscreens and sun protection factors*

As the long-term effects of exposure to sunlight have become widely recognised, the use of effective sun protection products has increased sharply. Sunscreens are added to skin care products in order to control the amount of UV light that reaches the skin. These sunscreens function by chemically absorbing UV radiation, thus helping to protect the skin from the effects of sunlight exposure. The level of protection can be altered by varying the type and level of sunscreen used in the product although, until recent years, only protection against UV-B radiation was provided by commercially available products. The protective efficacy of the product against UV-B radiation can measured by its *sun protection factor (SPF)*, defined as the number of times longer that can be spent in the sun in achieving the same degree of tanning or burning as that achieved without the product.

Clinically, the degree of efficacy of a sunscreen product, as defined by its sun protection factor, is defined by the following relationship.

$$\text{Sun protection factor} = \frac{\text{MED}^* \text{ with sunscreen product}}{\text{MED}^* \text{ without sunscreen product}}$$

* The MED, or Minimum Erythemal Dose, is defined as the amount UV-B radiation necessary to cause first visible reddening of the skin.

The SPF value for any particular sunscreen product is given by the average value of a logarithmically normal distribution of individual SPF values, as measured on a number of test persons, normally approximately 20. Evaluation is carried out by exposing a horizontally positioned skin area on the subject's back to uniform irradiation from an artificial source of UV light, normally referred to as a solar simulator. For each subject, the protective effects of the product under test are compared with a control product and with unprotected skin. A series of UV light exposures is administered to 1 cm^2 test sites on each of the subject's backs. One series of exposures is administered to unprotected skin to determine the subjects inherent MED. A second series of exposures is given to sites treated with the control product and a third series to sites treated with test product. The exposure time given to each successive site is normally 25% greater than that of the previous site. Approximately 24 hours after the exposure, the test sites are uncovered and the MED for each site determined.

Historically, two different adaptations of this method have been used in different parts of the world. The most popular is the ASA method which utilises a xenon arc lamp as the UV radiation source. The second method, which conforms to the German DIN standard, uses mercury vapour lamp and the resulting SPF values are expressed on a different scale.

More recently, and in light of the knowledge of the damaging effects of UV-A radiation, commercially available sunscreen products have sought to provide some UV-A protection, as well as the more normal protection from UV-B light. Unfortunately, the determination of protective efficacy against UV-A radiation is more difficult to measure. The principal reason for this is that doses of UV-A radiation required to produce erythema are impracticably large and the fact that such levels would potentially cause permanent skin damage raises ethical issues.

In recent years, several methods have been proposed for the measurement of UV-A sunscreen efficacy. Whilst many of these are still under development at the time of writing, it is appropriate to briefly review the methods being investigated. Several methods have been developed around the use of topically applied photosensitizers, the most notable being the psoralens and 8-methoxy psoralen (8-MOP) in particular. These methods measure the ability of the UV-A filter under test to protect the skin against phototoxic reactions. The activity of the filter can then be calculated as the ratio of the minimal phototoxic dose of the protected skin, to the minimal phototoxic dose of the unprotected skin. The efficacy, expressed by the UV-A light protection (LPF) value, expresses the extent by which the risk of skin injuries might be reduced by use of a UV-A filter under test. The main drawback of this method is that measurements are carried out on presensitised skins and the relevance of such protection factors to normal unsensitised skin is open to question.

Another method that has been used for assessing the effectiveness of UV-A sunscreens is Immediate Pigment Darkening (IPD). IPD involves measuring the pigmentation of the colourless melanin precursor existing in epidermal keratinocytes, which is photo-oxidized to darken by UV-A and visible light. IPD occurs within 5-10 min. of exposure to noonday sun, as a gradual darkening which is confined to exposed skin. Darkening increases unti it reaches its maximum after about one hour. IPD is optimally produced by UV-A but i is also produced by visible light.

The third *in vivo* method of assessing UV-A sunscreen efficacy is that using ornithine decarboxilase enzyme. This method involves the measurement of the effectiveness of the sunscreen in its ability to inhibit an increase in the level of the enzyme.

Apart from the *in vivo* methods reviewed thus far, there are also invitro methods tha have been developed to measure UV-A sunscreen efficacy.

Whilst conventional spectrophotometric methods are being developed, perhaps the most widely accepted *in vitro* method, at the time of writing, is the Diffey method. This technique uses a commercially available transpore tape as a substrate through which the spectral transmission of UV rays is measured, at 5 nm intervals, with and withou sunscreen. The Diffey method provides information which can be computed in a variety of ways, to measure the ability of a given sunscreen composition, applied on an artificia surface, to filter UV-A and UV-B radiation.

5. *Sun-tanning and sunscreen products*

Whether or not a product is classified as a sun-tanning or sun-protection product, depends largely on its SPF value. A "total sunblock" is considered to be provided by a product with an SPF rating of 15 or over, although many products on the market have an SPF value fa in excess of this. The correct SPF for any particular skin depends entirely upon skin type although the emphasis is now biased much more towards sun protection than sun-tanning The type of protection required to prevent burning, as a function of skin type, is detailed in Table 3.

TABLE 3

SPF	*Category description*	*Protection Level*
15+	Maximum protection, providing a total block	For highly sensitive skin which burns easily and never tans
8-15	High protection	For skin which burns easily and only tans slightly
4-8	Medium protection	For skin which tans fairly easily and only burns slightly
2-4	Minimum protection	For skin which tans easily and never burns

The most important classes of chemical sunscreens that either have been, or are being used in commercial sunscreen products are listed on page 193.

- Para-amino benzoic acid (PABA) derivatives. At one time these materials enjoyed widespread use but more recently they have fallen out of favour because of their potential to cause sensitisation reactions on the skin
- Homomenthyl salicylates
- Benzophenone derivatives, particularly 2-hydroxy, 4-methoxy benzophenone
- Cinnamic acid derivatives. These materials are the most widely used in Europe, a typical example being ethylhexyl-p-methoxy cinnimate

Commercially available sunscreen products vary widely in formulation and are often quite complex. A para-amino benzoic acid free, water-in-oil sunscreen lotion, with an SPF value of approximately 11, is given below:

Oil phase	
Glyceryl sorbitan oleostearate	3.70%
PEG-7 hydrogenated castor oil	3.30%
Decyl oleate	3.00%
Isopropyl lanolate	4.00%
Petroleum jelly	7.00%
Dimethicone	3.00%
Lanolin alcohol	0.80%
Octyl methoxycinnamate	6.00%
Benzophenone-3	1.50%
Water phase	
Glycerine	3.00%
Magnesium Sulphate	0.50%
Water	63.20%
Dyes, preservatives, perfume, etc.	to 100%

Most sunscreen products nowadays contain UV-A sunscreens, in addition to the more conventional UV-B sunscreens described above. Unfortunately, very few commercially acceptable UV-A sunscreens are available at the time of writing. Those materials that are available can be subclassified into organic and inorganic UV-A sunscreens.

The organic sunscreens include menthyl anthranilate, benzophenone-3, benzophenone-4, benzophenone-5, benzophenone-10 and butyl methoxydibenzoylmethane. Outside of the US, where its general use is not permitted, the latter is the most widely used substance.

Inorganic UV screens are all pigments which are used as physical blockers. These materials have been used in decorative cosmetics for many years because of their ability to cover or hide skin blemishes. The most common materials used as UV-A sunscreens are titanium dioxide and zinc oxide, both normally incorporated in micronised form. The light scattering effectiveness of the pigment particles depends primarily on the refractive index and particle size. Micronised pigments have particle sizes so small that they cannot be seen, since they no longer reflect visible light.

6. *Artificial tanning products*

With the increasing concern about the harmful effects of sunlight on the skin, artificial tan products or "skin bronzers" are becoming more popular. These products, normally sold in emulsion form, usually contain dihydroxy acetone at levels of up to 5%. Dihydroxy acetone reacts with the free amino-acids in the horny layer of the skin to form a reddish-brown colour, the intensity of which is proportional to the thickness of the horny layer. This artificial tan does not screen UV light and offers no protection from sunlight.

The eccrine and apocrine sweat glands

1. *Eccrine sweat glands*

Eccrine sweat glands are by far the most numerous of the sweat glands (in excess of 2,000,000 around the body) and are found all over the body, with relatively more on the hands, feet and axillae. Racially, caucasian people have more than negro types who in turn have more than oriental races. Additionally, females have more eccrine sweat glands than males. The primary function of eccrine sweat glands is to control body temperature by producing sweat, largely composed of water, at temperatures of about 30-32°C and above. The sweat evaporates from the skin, the heat of vaporisation removing heat from the body. Mental stress also causes an eccrine response, resulting in sweating on the palms of hands and soles of the feet, both areas in which thermal sweating does not occur. A condition known as gustatory sweating can also occur on the lips, forehead and nose. This condition is activated by stimuli such as hot and spicy foods.

Under normal conditions, the body secretes 1 to 2 litres of eccrine sweat per day produced by the eccrine sweat glands in pulses of approximately 6 per minute. The exact composition of eccrine sweat depends on the rapidity of secretion through the sweat duct, but is mostly water with a solids content of between 0.3-0.8%. The solid matter is largely composed of sodium chloride, with trace amounts of lactic acid and urea. The pH of eccrine sweat is normally between 4.5 and 6.5. A diagrammatic representation of the eccrine sweat gland is given in Figure 9.

The mechanism of sweat production starts with the transportation of sodium ions, by carrier molecules, from the outside to the inside of the secretary coil. Water then follows by osmotic pressure to dilute the high concentration of sodium ions now inside the secretary coil. The increase in water volume inside the secretary coil causes water to move into the proximal duct via the ampulla. The wall of the proximal duct is permeable to sodium ions but not to water and consequently sodium ions leak back into the dermis where they are available for more sweat production. The remaining water rises up the distal duct and out onto the epidermis via the spiral duct. The concentration of sodium ions in the secretion increases with the rate of sweating, as less time is available for their re-absorption.

The control of eccrine sweating is normally achieved through the use of astringent metal salts. Those most commonly used today are aluminium and zirconium salts. The mode of action of sweat inhibition is a subject of much debate and many theories have been proposed over the years. One proposal is that the presence of aluminium ions on the skin's surface causes sweat to leak back into the dermis where it is reabsorbed into the blood

FIGURE 9

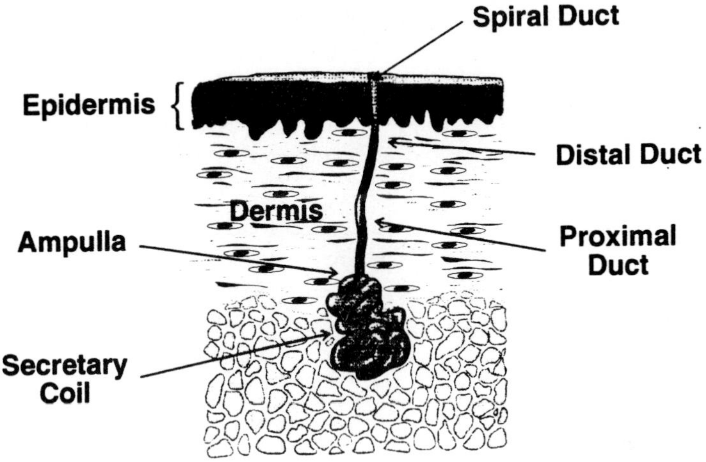

stream. Alternatively, other theories suggest that the that metal ions bind to the walls of the sweat duct, gradually blocking the duct and causing pressure to rise to a point where it stops secretion. The closure of the duct is then maintained until the effected keratin is lost by normal desquamation. A modified version of this theory suggests that the acid metal ions are buffered to the skin's pH after entering the spiral duct, causing formation of a gel which blocks the duct.

1.1. *Antiperspirant products*

Early antiperspirant products used aluminium chloride in solution to prohibit sweat production. These products were very acidic with pH values of around 2-3 and were thus often irritant to the skin. Since the late 1940s aluminium has been used in a complexed form, which is stable at higher pH values. The most commonly incorporated complex is aluminium chlorhydrate, which is used in aqueous solution at pH 4, or just above. It can also be used in alcoholic solution when complexed in a 3:1 ratio with propylene glycol, to make it alcohol soluble. Aluminium chlorhydrate is used in fine powder form in aerosols at levels of approximately 3-10% and in roll-ons and sticks at levels of approximately 15-20%.

A more recently developed group of actives are those based on zirconium or aluminium/zirconium chlorhydrate mixtures, with varying ratios of the two metals being available. These products are more effective than corresponding aluminium compounds and are enjoying use in the development of roll-on and stick antiperspirant products. They cannot, however, be used in aerosol based antiperspirant products, due to their inhalation toxicity. In use, aluminium/zirconium salts are normally buffered with a glycine because of their higher acidity.

A typical antiperspirant aerosol formulation, based on aluminium chlorhydrate, is shown on page 196.

Aluminium chlorhydrate	6.00%
Bentone gel	8.00%
Alcohol	2.00%
Isopropyl myristate	2.50%
Cyclomethicone	6.00%
Perfume	0.50%
Propellant (Butane)	75.00%

The bentone gel in this formulation acts as a suspending agent for the aluminium chlorhydrate powder. Both the isopropyl myristate and the cyclomethicone improve the skin feel after use of the product. The cyclomethicone also helps skin adhesion of the spray and, in part, reduces skin irritancy.

A simple antiperspirant roll-on formulation is illustrated below:

Aluminium chlorhydrate	15.00%
Cellulose gum	1.50%
Alcohol	20.00%
Isopropyl myristate	1.00%
Water, fragrance, etc.	to 100.00%

This is a simple solution product. The cellulose gum gives the product viscosity and feel, the alcohol improves drying time and the isopropyl myristate acts as an emollient.

A higher efficacy stick formulation, based on aluminium/zirconium chlorhydrate, is shown below:

Aluminium/zirconium chlorhydrate	20.00%
Cyclomethicone	51.00%
Stearyl alcohol	20.00%
PEG 200 trihydroxysterin	5.00%
Pyrogenic silica	1.50%
Glyceryl monostearate	1.00%
Talc	1.00%
Perfume	0.50%

In this formulation the stearyl alcohol acts as the main stick-forming agent. Emolliency and pay-off is provided by the volatile cyclomethicone and some slip is provide by the talc. The finely divided silica acts as a suspending agent.

1.2. *Antiperspirant efficacy*

Antiperspirants are generally used to reduce under-arm sweating, as total inhibition of body sweating is neither necessary or desirable, given the vital nature of body temperature control. The test method by which antiperspirants are evaluated in most laboratories is a simple gravimetric one, in which subjects are treated (axillae) with test products and placed in a hot room environment, typically 40°C, with a relative humidity of 50-60%. After a "warming up" period, the sweat is collected over a fixed (normally 40 minutes) period by taping pre-weighed absorbent pads, under the axillae. The pads are removed at the end of the period and re-weighed to determine the amount of sweat produced. Normally, one axilla is treated with the test product and the other with a control. Axillary sweating is asymmetric and this must be taken into account in the experimental design. This method is rapid, adaptable for a large number of subjects and requires a minimum of special equipment.

2. *Apocrine sweat glands*

In contrast to the eccrine sweat glands, the apocrine sweat glands become functional only at the age of puberty and they are found only in a few well defined areas of the body, namely the axillae, the ano-genital area, and around the nipples. Their development and function parallel with the growth of body hair. A diagrammatic representation of the apocrine sweat gland is shown in Figure 10 (sebaceous glands omitted for clarity).

FIGURE 10

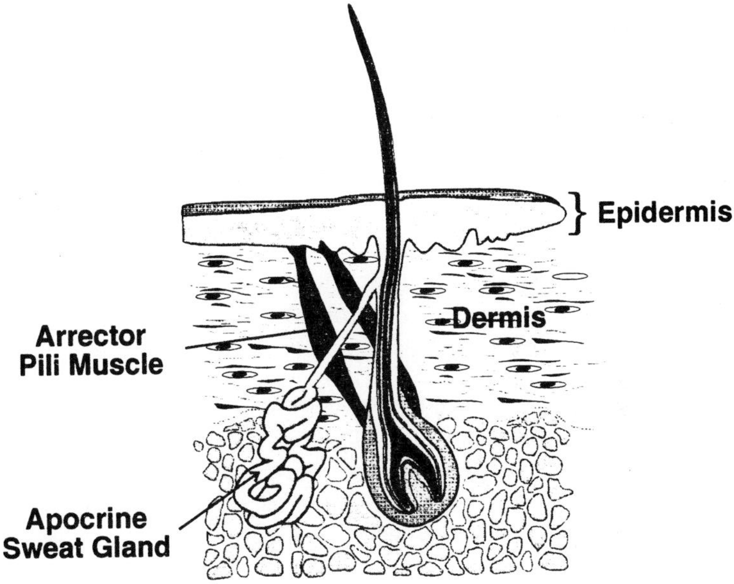

Like the eccrine sweat gland, the apocrine sweat gland consists of a coiled secretory portion and a tubular duct which empties into the hair follicle. In contrast to the eccrine sweat gland, however, the apocrine sweat gland secretes by a process called *picnosis*, which is under hormone control. Whereas eccrine sweat is produced rapidly and in large quantities, the apocrine sweat gland secretes sweat gradually and continuously at a slower rate, forming only small quantities of sweat within the coiled gland. The apocrine duct is muscular and operates by nervous control, responding to emotional rather than thermal stimulus. Apocrine sweating may take place at a much faster rate than the secretion and once emptied the apocrine sweat gland takes some time to refill. The sweat itself is a viscous, turbid liquid containing proteins, lipids, carbohydrates and salts and appears in the hair follicle orifice as a globular droplet.

2.1. *Body malodour*

Body odour is most apparent in areas of the body where apocrine sweat glands are found. Both eccrine and apocrine sweat, when delivered to the skin surface, are sterile and odourless. The odour of stale axillary sweat is caused by subsequent bacterial decomposition of apocrine sweat on the skin's surface. The presence of skin bacteria, apocrine sweat and water from eccrine sweat in the warm enclosed environment of the axilla, rapidly leads to the formation of break-down products of apocrine sweat, largely composed of volatile fatty acids and protein derivatives with an offensive odour. The feet can be particularly prone to malodour production. The epidermis of the soles of the feet has a larger number of eccrine sweat glands than normal skin and the horny layer is much thicker than normal with less cell degradation and a higher availability of lipids. Foot odour arises by bacterial degradation of the fats and amino-acids in the moist, enclosed region of the foot.

2.2. *Deodorant products*

Antiperspirants help reduce body odour by reducing the amount of available water on the skin, whereas deodorants stop odour production by killing skin bacteria and masking malodour through the use of fragrance. The aluminium and zirconium salts used in antiperspirants have some anti-microbial and deodorant activity. The purpose of a deodorant is to mask or remove the source of body malodour, or merely mask the odour with a fragrance.

Most deodorant formulations function through antimicrobial action in the presence of a perfume. The most widely used antimicrobial agent is triclosan, marketed under the trade name Irgasan DP 300 by Ciba-Geigy. This material is insoluble in water and substantive to the skin and is therefore not removed by sweat. It is soluble in most organic solvents including alcohol, which is often used to incorporate it into product formulations.

Other agents are available which absorb or chemically neutralise body malodour. These include zinc recinoleate which is typically used at levels of up to 5%, although the disadvantage with this material is that it is expensive. Sodium carbonate is an effective odour neutraliser, which functions by converting the volatile, malodourous fatty acids to their non-volatile sodium salts. This material is cheap and effective but is only soluble in water and is therefore difficult to formulate with.

The principle solvent used in liquid, aerosol and stick deodorants is ethanol, which itself has a measurable bactericidal and deodorant effect. Some perfume components also have deodorant efficacy.

Products for the feet also contain anti-fungal materials, such as calcium or zinc undecylinates. These are used against fungal infections such as athletes foot.

2.3. *Deodorant efficacy testing*

The best method of assessing malodour and product efficacy is the human nose, used directly, sniffing the axillae at close quarters. Typically, a panel of at least 3 assessors, chosen for sensitivity and consistency, assess test products under one arm with a control product under the other. Products are normally tested unperfumed.

A typical aerosol deodorant formulation is illustrated on page 199.

Alcohol	50.00%
Triclosan	0.30%
Perfume	0.50%
Propellant (propane/butane)	49.20%

A typical formulation for an alcohol-based deodorant stick is shown below:

Sodium stearate	7.50%
PEG-8	10.00%
Triclosan	0.30%
Zinc recinoleate	3.50%
Perfume	0.50%
Alcohol	70.00%
Water, dyes, etc	to 100.00%

The sodium stearate in this formulation is the main stick former and the PEG-8 modifies the feel of the product.

Sebaceous glands

Sebaceous glands are found all over the body except the palms of the hands and the soles and backs of the feet, the greatest density occurring on the scalp, face and back. The sebaceous glands are normally associated with a hair follicle and become functional at the age of puberty, with maximum activity occurring at about 14 years of age in girls and 16 years of age in boys. Sebaceous glands are under hormone control. Estrogen, the female sex hormone, decreases sebaceous activity whilst androgen, the male sex hormone, increases it.

Secretion from the sebaceous gland is composed of whole cells shed from the walls of the gland. These cells degenerate before being secreted, by capillary action, onto the skin's surface and the hair shaft. This secretion, known as holocrine secretion, when resident on the skin is known as *sebum*. Sebum on the skin is a mixture of triglycerides, unsaturated free fatty acids, waxes, paraffins, cholesterol, squalene and other minor components. In newly produced sebum there are no free fatty acids, as they are all bound up in the triglycerides. Skin sebum has several functions. It helps to provide an emollient film on the skin, assisting in reduction of the rate of water evaporation from the skin's surface. It also provides protection against de-fatting of the skin by external substances and helps prevent bacterial infection.

Excessive sebum production is responsible for the problems of greasy skin and acne. Astringent lotions or tonics are available to reduce the greasiness of the skin, although they do not control the flow of sebum. These products are normally based on aqueous solutions of alcohol, witch-hazel or metal salts and tone the skin, causing pore constriction. A typical formulation for such a product is illustrated below:

Alcohol	40.00%
Glycerol	3.00%
Zinc phenol sulphate	1.00%
Water, fragrance, etc.	to 100.00%

1. *Acne*

Acne is associated with high sebum secretion and is more prevalent in boys than girls. Acne is not just related to the quantity of sebum secreted but also to the quantity of free fatty acids that are liberated from triglycerides it contains. Sometimes free fatty acids are liberated inside the sebaceous gland and irritate and puncture the sebaceous gland walls, causing foreign body and triple response reactions to occur. Consequently, the region around the gland swells and reddens as the blood vessels expand and often a little pus results from the arriving white blood corpuscles. If the bodies defence mechanisms operate successfully nothing is really noticeable on the skin's surface, except that a small comedo may have formed. Sometimes a group of sebum cells agglomerate and block the duct from the sebaceous gland. If this blockage occurs below the skin's surface it is called a closed comedone or whitehead, whereas if the plug enlarges or rises out of the duct, it is called an open comedone or blackhead. The tip of the open comedone is black due to localised build-up of melanin pigment. The formation of either open or closed comedones are known as non-inflammatory acne.

If, however, the body's defence mechanisms do not operate efficiently, the comedones become inflamed and material from the duct leaks into surrounding tissues. Subsequently, pimples and pustules develop, resulting in the condition known as inflammatory acne.

1.1. *Anti-acne products*

The most effective ways of controlling acne involve changing the free fatty acid concentration available. One method of achieving this is to prescribe low levels of oxytetracyline for a period of up to 12 months. This reduces the conversion of triglycerides to fatty acids, thus reducing the occurrence of irritation. A second method of treatment is to use estrogen to deactivate sebum gland activity, thus reducing the amount of free fatty acid available. This treatment normally takes several months to become effective.

Cosmetic methods of reducing acne involve increasing the epidermal cell turnover rate using materials such as vitamin A (retinoic acid). This increases cell turnover in the follicle and reduces the cohesiveness of the cells shed from the sebaceous gland walls. This treatment normally reduces plug and comedone formation by more than 50%. The most widely used methods of increasing cell turnover incorporate keratolytic agents such as sulphur, resorcinol, benzoyl peroxide and salicylic acid. These materials also function by keeping the sebaceous ducts open. Antiseptics and bacteriostats, such as triclosan or cetrimide, are often used in combination with these materials to improve efficacy.

The formulation for a typical anti-acne product, based on a combination of keratolytic and bacteriostatic agents, is illustrated below:

Colloidal sulphur	8.00%
Resorcinol	2.00%
Bentonite	2.00%
Propylene glycol	6.00%
Triclosan	0.30%
Water, fragrance, etc.	to 100.00%

HAIR AND HAIR PRODUCTS

Introduction

Hair is a filament-like extension of the epidermis found in mammals and is composed largely of a proteinacious substance called *keratin.* The outer layer of the hair shaft is covered in a hard layer, known as the cuticle. The inside of the hair is known as the cortex which has, running up the middle of it, the medulla. Hair itself is not a living organ and therefore contains no nerves, blood vessels or viable cells.

Hair possesses a multi-functional role in mammals, giving some protection and enhancing the sensitivity to touch. In animals, hair also helps keep the skin dry and provides, in many cases, camouflage against predators.

The structure of hair

There are three types of hair (vellus, terminal and bristle) found on the human body and, irrespective of type, the hair is always associated with a hair follicle, which is an extension of the epidermis. Vellus hairs are very fine, normally 1-2 mm in length and cover the entire body. Terminal hairs are the long, course hairs that cover the scalp, whilst bristle hairs are those that cover the eyebrows etc. and are up to 5 mm in length. From the cosmetic scientist's viewpoint, by far the most important type of hair is the terminal type.

1. *Terminal hair and the hair growth cycle*

Terminal hairs have a definite cyclic activity, known as the growth cycle. Hair production originates at the bottom of the hair follicle, which exhibits periodic activity giving rise to the hair cycle itself. Follicle activity can be classified as either "active", during which hair growth takes place, or "passive" during which growth is static. The hair growth cycle itself occurs in three stages referred to as *anagen*, *catagen* and *telogen* as illustrated in Figure 1.

The initial growth of a hair in its follicle is activated by natural hormones or physical stimulus. Epidermal cells located at the base of the hair follicle divide and the cells surrounding the dermal papilla then proliferate, pushing a column of keratinising cells outwards towards the skin's surface. This column of keratinising cells forms the hair shaft, invested in an inner root sheath and, as growth continues, a hair and hair canal are formed.

FIGURE 1

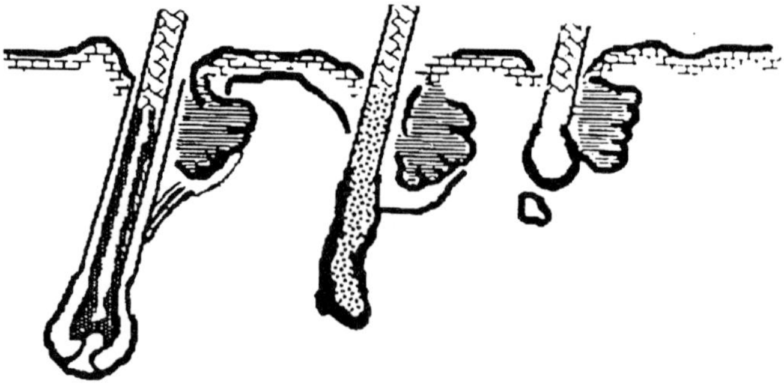

This growth period, referred to as the anagen phase, continues for a period of between 2 and 5 years and is often subdivided into two distinct stages. The first stage is referred to as the proanagen stage and describes hair growth that continues under the skin's surface. The second stage, metanagen, is used to describe the stage of growth after the hair has emerged from the skin's surface. At any one time, 80-90% of human terminal hair is in a stage of anagen. When anagen has ceased, the cycle goes into a transition stage, known as catagen, which normally only lasts a few weeks. During this period the hair follicle shortens and starts to move towards the skin's surface, as illustrated in Figure 1. When the catagen phase subsides, the growth cycle moves into the telogen or "passive" stage, when the cuticle scales on the outside of the hair bind on to the inside of the follicle, thus keeping the hair firmly in place. The hair remains in telogen for some time, on average approximately 100 days. Eventually, the anagen period starts once more and the hair follicle becomes longer again, moving down deeper into the dermis. The hair growth cycle then begins once more, with new hair growth commencing, sometimes whilst the original hair is still in place.

Hair dimensions vary widely depending on hair type, stage of growth, race and location on the body. Typically, terminal hair exhibits diameters in the range of 40-90 μm and, on average, the healthy human scalp contains approximately 1.2×10^5 hairs. Hair cross-sectional shape is dependent on race. Negroid hair, for example, is elliptical in shape whilst caucasian hair is much more regular and cylindrical in shape. In order to describe hair cross-sectional shape, a term referred to as the hair index is often used. This is defined by the following relationship.

$$\text{Hair Index} = \frac{\text{least diameter of the hair}}{\text{greatest diameter of the hair}} \times 100$$

2. *Hair and follicle physiology*

The structure of the hair and its associated follicle is illustrated in Figure 2.

The hair follicle itself is made up of the outer root sheath, which is merely an extension

FIGURE 2

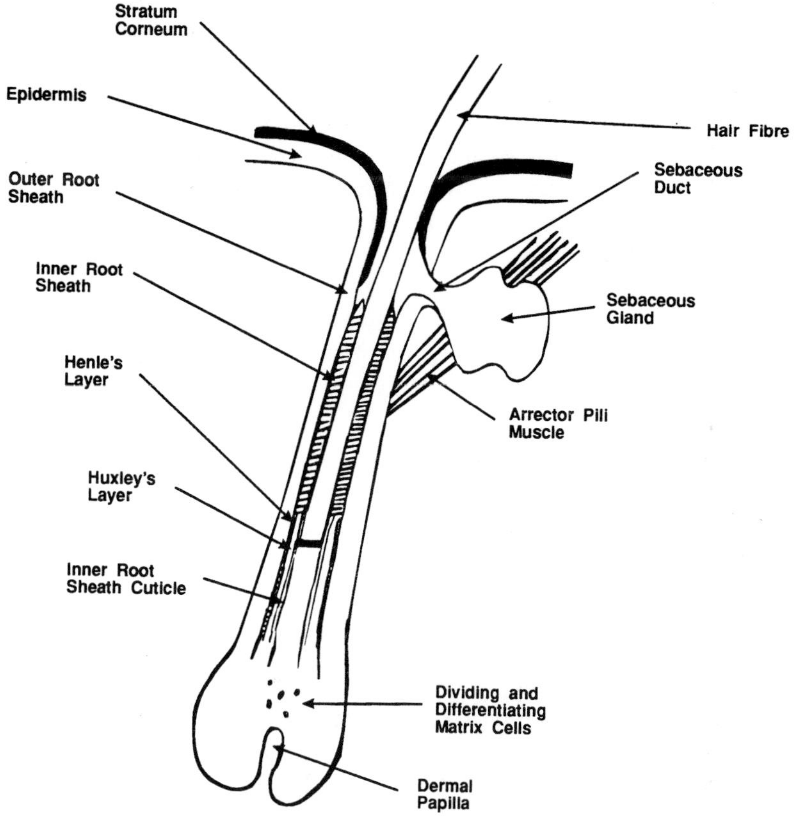

of the epidermis, the outermost layer of which is a heavily keratinised layer known as the stratum corneum. The inner root sheath, in the metagen phase, is the layer immediately adjacent to the hair. It is a hard, rigid, heavily keratinised tube, which determines the hair cross-section as it grows. The lower portion of the inner root sheath is divided into three parts, known as *Henle's layer*, *Huxley's layer* and the inner root sheath cuticle. Keratinisation of the inner root sheath occurs before keratinisation of the hair itself, thus enabling it to guide the hair during growth. The inner root sheath and the hair shaft grow together, both originating from the proliferating cells at the base of the hair follicle. As the inner root sheath nears the skin's surface it desquamates and emerges from the follicle in pieces.

The dermal papilla contains blood-vessels and nerve-endings and is the area in which the epidermal cells at the base of the follicle divide. These epidermal cells do not all reproduce in the same way but differentiate to form the different components in the hair shaft (i.e. cuticle, cortex, medulla), with the outermost cells forming the inner root sheath. At this stage the cells are soft and not keratinised.

The sebaceous glands' function is to provide the secretion of sebum onto the hair. Sebum produced in the gland is forced out of the sebaceous duct and as the hair grows past the sebaceous duct, it carries sebum out with it. Sebum provides natural protection for the hair, giving it a lubricious waterproof coating, a factor which is of more importance to animals than man.

The arrector pili muscle is a very small muscular organ which is attached to the epidermis and is able to expand or contract depending on the physiological environment. Contraction, which can be activated by low temperature or by the hormone adrenalin, pulls the follicle into a more upright position causing the hair to stand upright on the skin's surface.

3. *Hair morphology*

Figure 3 illustrates the cross-sectional profile of a hair fibre.

FIGURE 3

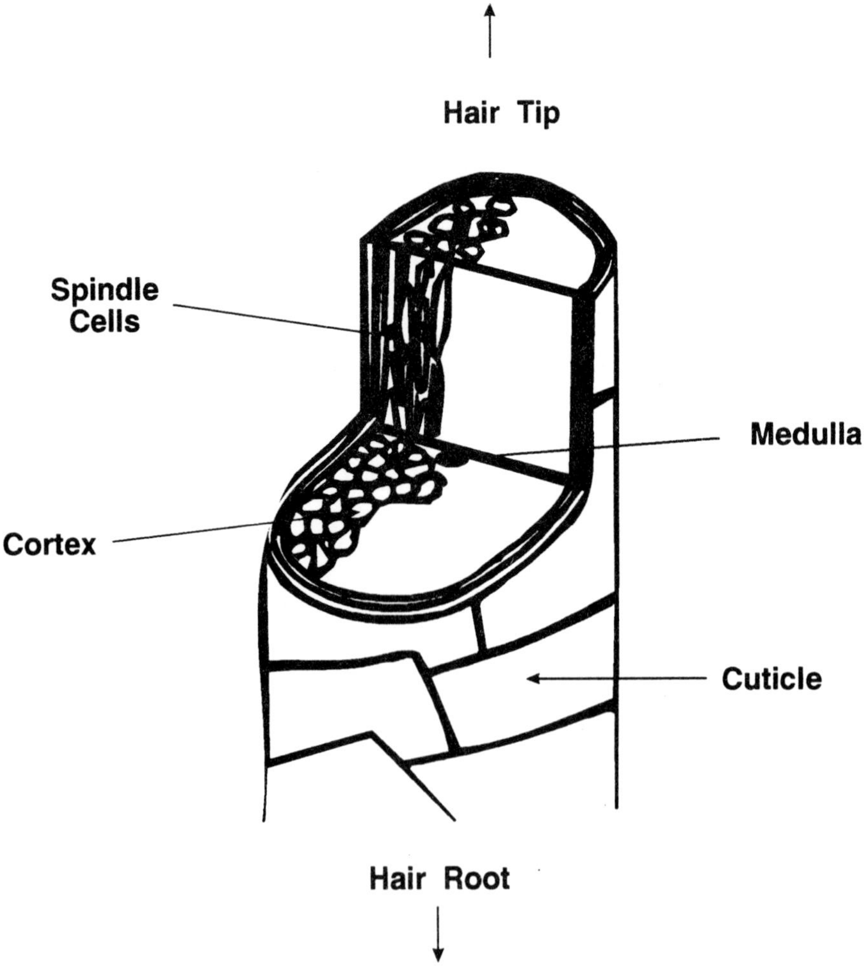

The *cuticle* is the hard outer layer of the hair shaft which is composed of flattened plates of heavily keratinised material. Each cuticle plate is approximately 50 μm square and of 0.5 μm thickness and they are wrapped around the hair shaft in several layers, overlapping from the root end of the hair to the tip. Cuticle keratin is amorphous and very hard, providing a chemically inert protective layer around the hair shaft. When newly formed, cuticle keratin scales have a very smooth, lustrous, surface which gives the hair its characteristic shiny property. The composition of each cuticle scale can be further subdivided in to three layers, the *endocuticle*, *exocuticle* and *epicuticle*, as illustrated in Figure 4.

FIGURE 4

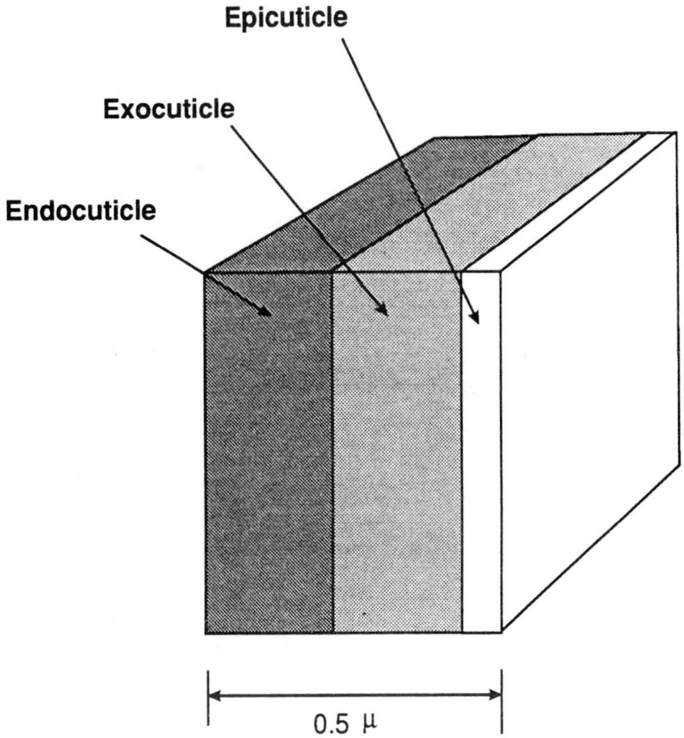

The exocuticle provides the outer surface of the cuticle scale and is relatively rich in the amino acid cystine, which contains the disulphide linkage, -S-S-, so important to the physico-chemical properties of hair. The exocuticle is therefore much harder than the endocuticle, which possesses a much lower cystine content. The epicuticle is the outermost layer of the cuticle scale and is approximately 25 angstroms in thickness. The epicuticle is a heavily keratinised proteinaceous structure, associated with lipids, which confers the property of water-repellence to the hair shaft.

If the hair surface suffers mechanical attack, such as that caused by washing, combing and brushing, the cuticle scales become damaged and dislodged, exposing the fibrous keratin of the cortex, which quickly breaks up resulting in split ends. Hair in which the cuticle is damaged also appears dull and in poor condition when visually examined.

The *cortex* consists of fibrous *spindle cells* bound together with a mucopolysaccharide. This mucopolysaccharide binding material is very strong and the spindle cell itself will break before rupture of the mucopolysaccharide. A cross-sectional analysis of each spindle cell reveals that it is composed of *macrofibrils*, the structure of which is illustrated in Figure 5.

FIGURE 5

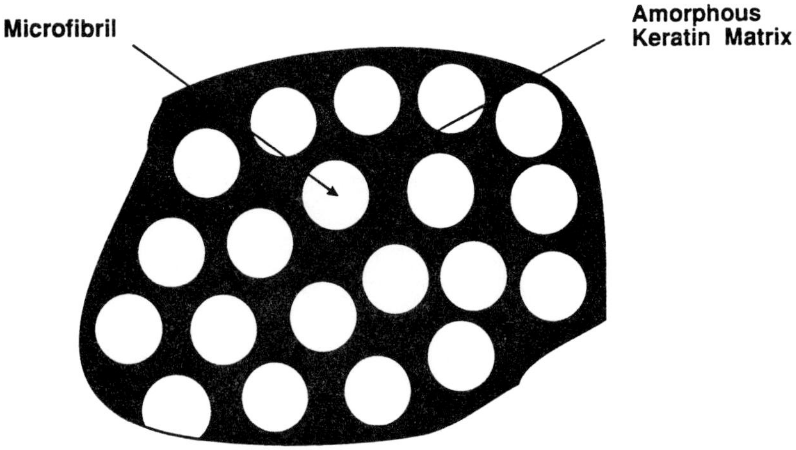

Structure of a Macrofibril

Each macrofibril is between 0.2 μm and 0.9 μm in diameter and is composed of highly organised fibrous *microfibrils*, which are embedded in an amorphous keratin matrix. The microfibrils themselves contain little or no sulphur-containing amino acids, unlike the matrix in which they are embedded. Each microfibril is made up of 11 *protofibrils*, each containing three alpha-helix structures.

The cortex therefore consists of a composite structure of fibrous material, embedded longitudinally in an amorphous matrix. It is the cortex which is responsible for the behaviour of hair when it is stretched under load, giving a high longitudinal tensile strength.

The medulla is only present, if at all, in small amounts in human hair and may be either continuously, or more commonly discontinuously, distributed longitudinally along the hair fibre. The properties of the medulla contribute little to the physical or chemical properties of the hair fibre and, as a consequence, it is of relatively little importance.

Apart from the different types of proteinaceous keratin, there are other components within hair that are important in determining its behaviour. Hair is very hygroscopic and contains significant amounts of water, which is drawn from the surrounding environment. The moisture content of hair will vary with relative humidity, rising from about 6% at a relative humidity of 30%, to about 25-30% when the hair is immersed in water. Hair also contains small quantities of lipids, found both in the body of the hair fibre and at the surface. These lipids are derived from the presence of sebum, distributed onto the hair shaft by the sebaceous gland.

Finally, hair contains various types of melanin pigment, which are responsible for hair colour, in addition to low levels of metals such as iron, zinc and calcium. The latter are thought to be absorbed into the hair when it comes into contact with water and some toiletries preparations.

The chemistry of keratin and hair

Keratin is an insoluble proteinaceous complex, found in the epidermal tissue of vertebrates, which comprises greater than 85% of the hair. Keratin is often referred to as a protein but, more correctly, is composed of protein or polypeptide chains in certain configurations. These polypeptide chains are composed of amino acids, the properties of the chain being dependent on the number, type and order of amino acids present. The amino acids are linked through a series of peptide links, as illustrated in Figure 6.

FIGURE 6

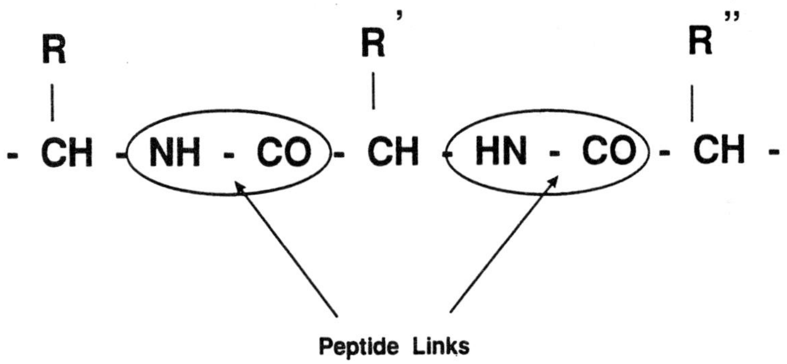

The side groups in the amino acid chain, depicted by the letter "R" in Figure 6, can be very reactive and this is very important when considering the chemistry of hair. The polypeptide chain structures found in hair are composed of 18 different types of amino acid.

Two fundamental types of keratin exist, crystalline keratin and amorphous keratin, each of which is of vital importance to the structure and behaviour of hair.

1. *Crystalline keratin*

Crystalline keratin, unlike amorphous keratin, has a crystalline structure which will give characteristic patterns when using X-ray diffraction techniques. Crystalline keratin is commonly found in two different forms, alpha-keratin and beta-keratin.

Alpha-keratin is normally found in the form of an alpha-helix, so called because of the pattern it produces when examined using X-ray diffraction techniques. The configuration of the alpha-helix coil is held in place by hydrogen bonding between adjacent coils in the helix itself. The protein chains in the alpha-helix are shorter than those found in beta-keratin but if subjected to a combination of moisture and physical load, they will extend to similar dimensions of the latter. Beta-keratin is normally found in the form of pleated sheets, with a ribbon or sheet-like structure, held together with hydrogen bonds. Alpha- and beta-keratin forms can be differentiated by the use of X-ray diffraction techniques.

Crystalline or fibrous keratin is found in the spindle cells of the hair cortex. The alpha-helix is the main "building block" of the cortex, each protofibril consisting of three alpha-helices which form its structure. Hydrogen bonding plays a key role in the configuration of each alpha-helix and normally the maximum allowable number of hydrogen bonds will be present, so as to allow the minimum energy within the structure.

2. *Amorphous keratin*

Amorphous keratin is, as its name suggests, keratin which exhibits no crystalline structure. In hair, amorphous keratin is found in the cuticle and in the interstices between the spindle cells of the cortex. Amorphous keratin is composed of macromolecules consisting of approximately 100 proteins, each containing 18 amino acids.

3. *Keratin configuration*

The configuration of the protein chains in keratin is determined by several factors, as discussed below.

3.1. *Hydrogen bonding*

In proteins, hydrogen bonding takes place between adjacent –NH and >C=O groups, as illustrated in Figure 7.

FIGURE 7

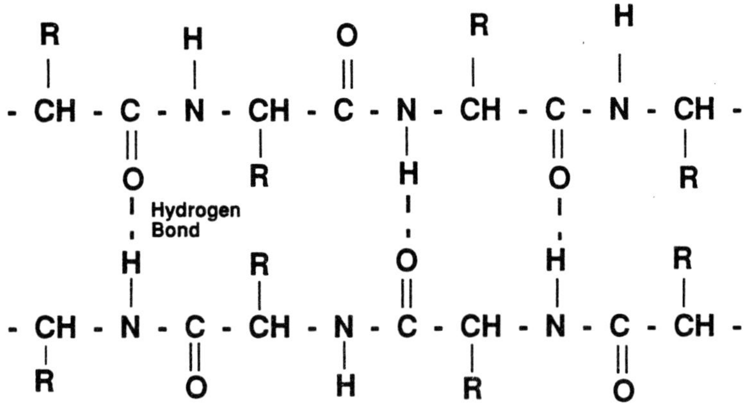

Hydrogen bonds are generally fairly weak but are nevertheless important to the macromolecular structure of keratin proteins, by virtue of their number.

3.2. *Disulphide links*

These bonds, which are very strong, are crucially important to the macromolecular structure of keratin protein. Disulphide links are derived from two units of the highly reactive, sulphur-containing, amino acid cysteine, which combine to form the sulphur-sulphur linkage, as shown in Figure 8.

FIGURE 8

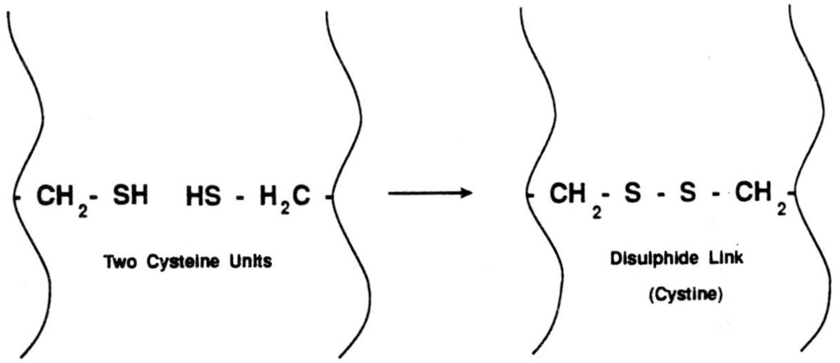

This reaction takes place in the zone of keratinisation during the hairs development and forms the basis of the keratinisation process. The resulting dimeric amino acid formed between the two chains is known as cystine. The existence of disulphide links in keratin is the main reason for their resistance to enzyme and chemical attack. The disulphide link is, however, susceptible to rupture by one of the following methods:

- ultraviolet light
- oxidising agents
- reducing agents
- strong acids and alkalis
- prolonged exposure to boiling water

The disulphide link is also responsible for the very hard nature of amorphous keratin, which is proportional to the quantity of disulphide linkages present.

3.3. *Salt linkages*

Salt linkages derive from natural attraction between carboxyl and amino groups on favourably placed side-chains, within the keratin structure. This reaction causes electrostatic attraction giving a relatively strong bond. Many salt linkages occur in the chain matrix, their nature being affected by the presence of acid or alkali.

3.4. *Peptide bonds*

Although peptide bonds are commonly found within protein chains, they can also sometimes occur between two adjacent chains, giving a relatively strong, covalently bonded, cross-link. These bonds are, however, readily broken under the influence of acids or alkalis.

3.5. *Van Der Waals forces*

Van der Waals forces are the forces of attraction which exist between adjacent molecules and are associated with the oscillating dipoles found in each molecule.

Whilst many of the above factors (3.1–3.4) are important in determining the structure of amorphous keratin, its fundamental character is dictated by the high proportion of disulphide linkages present.

Physical properties of the hair

The physical properties of the hair in bulk are of paramount importance to the cosmetic scientist and attributes such as lustre, feel, extent of fly-away, style-retentiveness, manageability and ease of combing, are all important factors when designing products for consumer or professional use.

Before considering these attributes, it is important to understand the single fibre properties of hair and what resultant effect they will have on the hair in bulk.

1. *Properties of single hair fibres*

There are many single fibre properties which will have a direct impact on the appearance and behaviour of hair in bulk. The most obvious properties, length and diameter, require little explanation although their variability with the surrounding environment is interesting. Hair length varies immensely but is almost completely independent of the temperature and humidity of environment. Hair diameter and cross-sectional shape is of critical importance in the determination of the bulk properties of hair. Fibre diameter will influence the stiffness, rigidity (resistance to twist) and surface area:volume ratio of hair, the latter decreasing as hair diameter increases. The surface area:volume ratio is important as it affects the way in which cleansing, conditioning and styling products interact with the hair. Both hair length and diameter are conventionally measured using microscopic techniques, although other methods such as gravimetric determinations are also used. In the case of the latter, the density of the hair must be known to enable the fibre dimensions to be calculated. Care is needed if these techniques are used, as the density of hair will vary considerably with relative humidity of the environment. Hair density decreases with an increase in relative humidity and typical density values of just over 1.3 at 0% RH, down to approximately 1.27 at 100% RH, are observed.

Other single fibre properties of hair require more examination, in order to determine the effect on the bulk hair properties. The most important of these single fibre properties are discussed below.

1.1. *Colour*

The colour of hair relies on pigment distribution throughout the cortex. Pigments are distributed in the form of melanin granules, approximately 1 micron in diameter, which can be classified into two main types. The first type, true melanins or *eumelanins*, are brown-black in colour and have a very complex macromolecular structure, based on indole derivatives, such as those shown in Figure 9.

The most probable explanation for the presence of this type of melanin structure is that

FIGURE 9

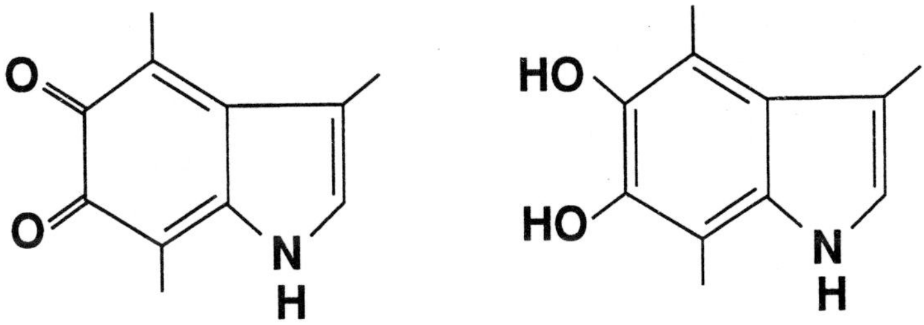

Indole Derivatives

they are synthesised from tyrosine, which is present in keratin in the hair. The second type of melanin pigment found in hair is the *pheomelanin* type, formerly known as the trichosiderins. These are also complex aromatic structures but are much lighter in colour than the eumelanins and generally impart a yellow-red colouration to the hair. The final colour of the hair will depend upon the ratio of eumelanins and pheomelanins. The chromophores produced by both are readily destroyed by reducing agents, such as hydrogen peroxide, resulting in bleaching and loss of colour.

1.2. *Lustre*

Hair fibre in good condition is very shiny due to the very smooth, amorphous cuticle platelets on the outside of the hair shaft. This smooth surface reflects light in a regular fashion, with little scatter, producing a characteristically high degree of shine. As the hair grows, however, it becomes physically damaged and the cuticle scales become chipped or even dislodged completely. This causes a much higher degree of light scatter at the hair surface, resulting in a dull appearance.

1.3. *Electrostatic properties*

The electrostatic behaviour of hair is directly connected with the bulk property of fly-away. Dry hair is not a good electrical conductor and when the fibre is subjected to forces of mechanical friction, such as combing, an electrostatic charge is generated on the hair surface. On combing, hair is nearly always left with a positive residual charge, although this can be affected by factors such as the direction and rate of combing. As the relative humidity of the surrounding environment increases, the potential for the accumulation of an electrostatic charge diminishes significantly, due to the resultant increase in the electrical conductivity of the hair fibre.

1.4. *Frictional properties*

The frictional behaviour of the hair surface is very important and determines the characteristic behaviour when hair fibres rub against each other. The frictional behaviour of two hair surfaces in contact can be measured using a rotating mandrel composed of many hair fibres, over which is laid a single hair fibre under tension, as illustrated in Figure 10.

FIGURE 10

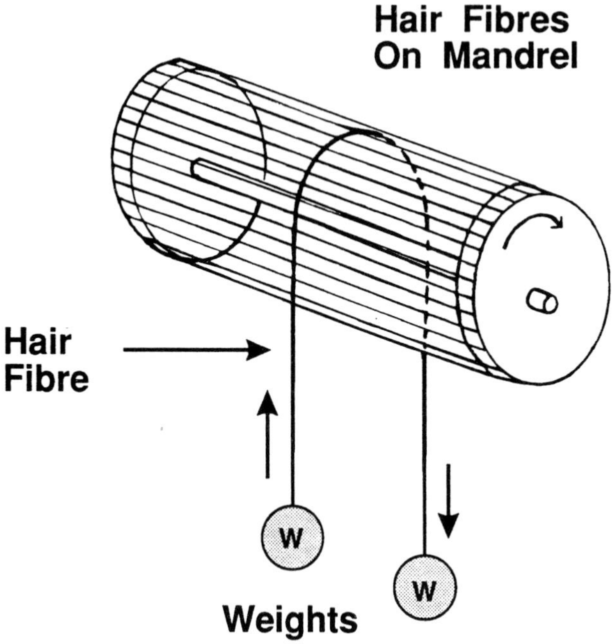

The mandrel rotates at a constant speed in the direction shown and the hair-on-hair frictional forces cause the weights to rise or fall in the direction of the movement. A known external force, which is equivalent to the force of friction, is then applied to counterbalance the movement of the weights. This method of measurement does have the limitations of only measuring the hair-hair frictional coefficient at right angles and the frictional constant obtained will depend upon the direction in which the hair is tested (i.e. hair to root or vice versa).

1.5. *Tensile and elasticity properties*

The tensile and elastic properties of hair are generally described by examining the relationship between stress, the force applied and strain, the resultant deformation caused. In practice, the tensile and elastic properties of a single hair fibre can be measured by stretching the hair, at a constant rate, and recording the stress/strain curve. The actual shape of the stress/strain curve will depend upon the relative humidity in which the hair is tested but a typical stress/strain plot, at a relative humidity of 65% RH, is illustrated in Figure 11.

In the region between points O and A, the hair fibre exhibits true elastic properties which obey Hookes law. Physically, this represents fully reversible bond distortion in the hair fibre as it is stretched. Point A on the curve is known as the yield point, which represents physical disruption of hydrogen bonds in the alpha-helices and possible

FIGURE 11

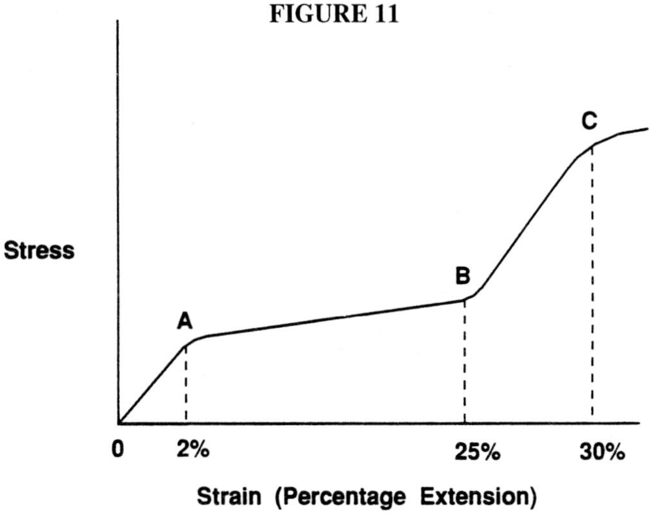

disruption of some sulphur-sulphur bonds within the fibre. Between points A and B, there is very little, if any, increase in stress over a relatively large increase in strain. This represents the rearrangement of alpha-helices into beta-keratin pleated sheets. Between points B and C, Hookes Law is obeyed once more, accompanied by an apparent stiffening of the hair fibre. Finally at point D the hair fibre breaks.

If the same curve is generated at a higher relative humidity, an increase in elasticity and decrease in tensile strength is noted, although the shape of the curve remains largely unchanged. Other factors that affect curve characteristics are hair treatments, such as dyeing and permanent waving, which can considerably weaken the hair fibre. In hot water, hair will extend by up to 100% of its original length, before it breaks. This is due to ingression of a higher level of water into the hair fibre, which lubricates it, increasing the extension. In practice, only the region of the curve between 0% and 5% strain is important, as this represents all practical stress that the hair experiences in day to day grooming. A value often used in relation to stress/strain studies is the 30% work index. This is the work required to produce 30% strain and is calculated from the area under the curve.

Two other types of elasticity are important in characterising the behaviour of a single hair fibre. The first is the elasticity associated with the force required to bend the hair fibre, given by the following equation:

$$\text{Bending force} = N \times \text{Deformation}$$

where

$$N = \text{modulus of stiffness of the hair fibre}$$

The second is the torsional elasticity of the hair fibre, which describes the behaviour due twisting forces. This type of elasticity is related to the stiffness of the hair fibre and can be measured in several different ways.

2. *Bulk fibre properties of the hair*

Compared to the single fibre properties, the bulk properties of hair are much more difficult to measure, as they often rely upon subjective assessment. Therefore the handling and measurement of bulk fibre properties are often achieved by the use of sensory methods, which may be either psychophysical or psychometric in type. Psychophysical methods give a mental judgement of an attribute which can also be measured physically, whilst psychometric methods refer to mental judgements for which no physical methods apply. Nowadays, the term psychophysics is often used to describe both psychophysical and psychometric methods. Psychophysical assessments on hair can involve the use of rating scales which give a numerical value to a particular attribute, rank order judgements which put a ranking order on different samples and difference judgements which use subjective or sensory methods to assess differences between samples.

The bulk fibre properties of hair reflect the types of attribute which will concern the end user of hair care products. Each of these attributes, and the way in which they are assessed, will be discussed in turn.

2.1. *Ease of combing*

The ease of combing is related to the resistance encountered as the comb is moved through the hair. The single fibre properties which affect the ease of combing are the coefficients of friction between the hair fibres and the hair fibres and the comb, hair diameter, hair stiffness and the presence of any electrostatic charge. The evaluation of ease of combing often employs psychophysical methods, using manual combing of hair tresses and a rating scale. Statistical methods can be used to give a score for each tress tested. In cases where quantitative measurement of ease of combing is required, tensile testing machines, such as the one illustrated in Figure 12, are normally used to measure the force required to pull a comb of known dimensions through a hair tress. Care must be taken when using this

FIGURE 12

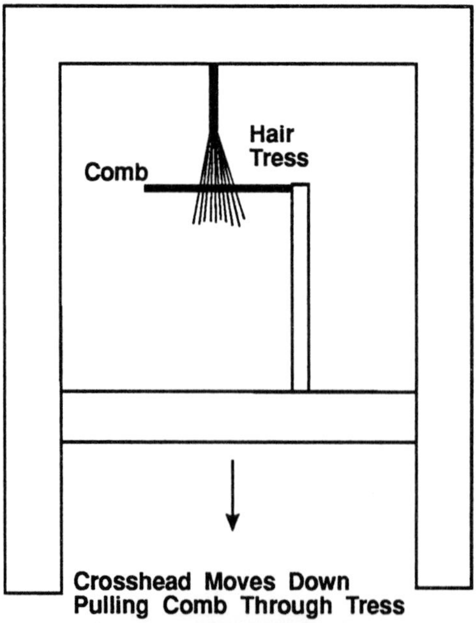

technique to avoid errors, such as those arising from tangles in the tress, and temperature and relative humidity must be held constant.

2.2. *Style retention*

Style retention is the ability of hair to stay in place after styling and is therefore a function of time. The single hair fibre properties of hair curvature, hair-hair friction, hair density and hair stiffness all have an impact on this bulk property, in addition to any pretreatment, such as hair-sprays or setting lotions, that the hair may undergo. Techniques for evaluating style retention normally involve putting a hair tress into a given configuration (i.e. setting the "style"), for example wrapping it around a glass rod of known dimension. The hair is treated appropriately whilst on the glass rod and then removed when the "style" has set. Style retention can then be measured by deforming the hair with a known force and measuring the loss of the "style" as a function of time. The longer the hair tress takes to lose its style, then the better the style retention. Whenever style retention measurements are being made, the temperature and relative humidity must be carefully controlled.

2.3. *Fly-away*

The extent of fly-away is largely dependent on the presence of static charge on the hair which, in turn, is dependent on temperature and relative humidity, in particular. During testing, static charge is produced by pulling foreign material, for example a comb, through the hair. The extent of charge depends upon area of contact and the pressure between the foreign material and the hair. The simplest way to measure fly-away is to comb the hair tress in a controlled manner and measure the ballooning of the tress after combing, as illustrated in Figure 13.

FIGURE 13

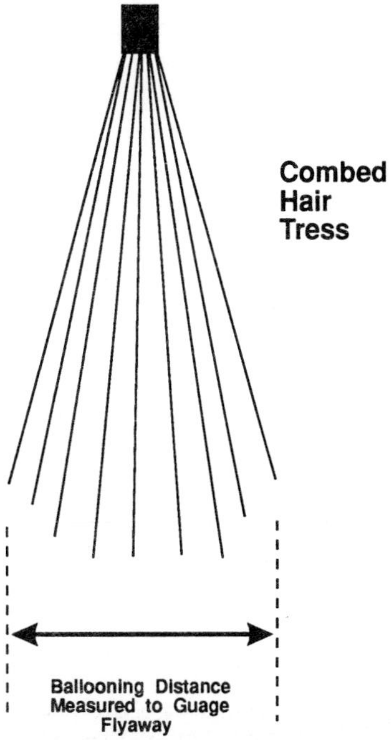

In practice, this procedure is carried out in a controlled environment to minimise errors and the ballooning of the tress is normally measured using a travelling microscope.

A more sophisticated method for measuring fly-away employs the use of a gold-leaf electroscope. The hair tress is combed and subsequently the comb is brought towards the electroscope until a deflection occurs. The distance of the comb from the electroscope is then measured to give a quantitative indication of the extent of static charge present.

2.4. *Lustre*

Lustre depends, to some extent, upon the smoothness of the single hair fibres, but other more complex bulk properties have to be considered also. The reflective properties at the surface of a single hair fibre are represented, in Figure 14.

FIGURE 14

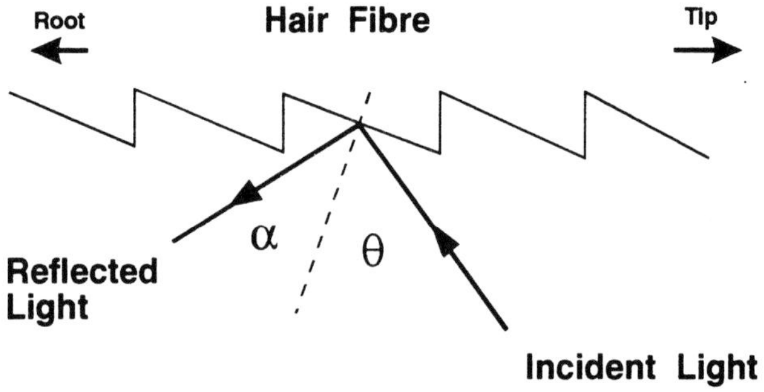

The angle of incidence (Θ) and the angle of reflection (α) are not equal, even on hair in very good condition, due to the angle of the cuticle scales on the outside of the hair shaft. When light is reflected from hair in bulk, the degree of lustre is much higher longitudinally along the hair shafts, than it is across the hair. This is because in the latter case a significant amount of light is lost due to the scatter across an irregular surface.

Psychophysical methods for the assessment of lustre are fairly satisfactory but errors do occur due to the difficulty in assessing lustre by eye, the dependence of gloss on the alignment of the tress and the fact that handling the tress will change the reflective properties of the hair surface. Lustre can be measured instrumentally using a goniophotometer, represented in Figure 15.

The light source sends a beam of light towards the hair surface at an angle of incidence Θ, whereupon some of the light is scattered and some reflected at an angle of (Θ+δ). The detector is then moved round in an arc so it that picks up firstly the scattered light, then the main reflected beam and finally scattered light once more. A plot of light intensity versus distance travelled (as a function of angle) can then be made to produce a graph of the type shown in Figure 16.

Lustre is then measured as the difference between the peak light output and the background scattered light level. The degree of lustre is quantified by the relationship:

$$\text{Lustre} = \frac{S}{D}$$

FIGURE 15

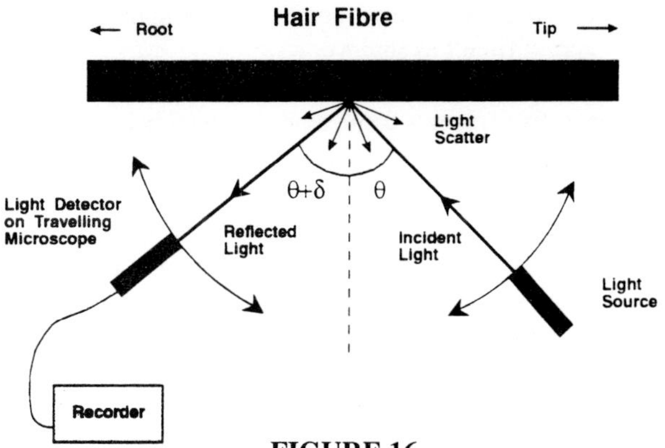

FIGURE 16

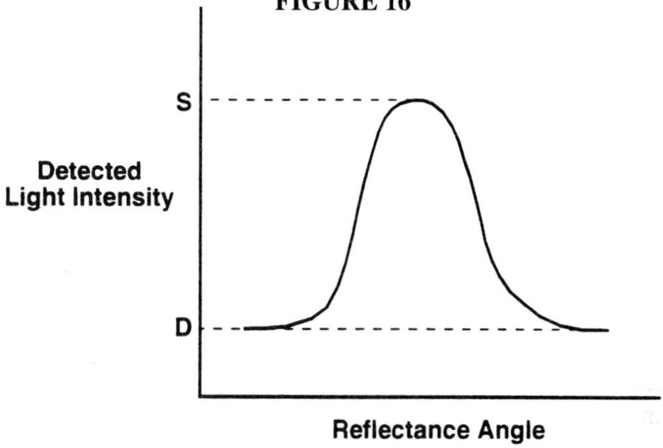

S = Specular Light Level
D = Diffuse/Scattered Light Level

A better way of characterising lustre is to use the equation:

$$L_c = \frac{S - D}{S} = 1 - \frac{D}{S}$$

where

$$L_c = \text{contrast lustre}$$

For a highly reflective surface, L_c will approach unity. Measurement of the angle Θ, and comparison with the angle $(\Theta+\delta)$, will enable determination of the angle of the cuticle scales on the hair shaft.

2.5. Manageability

Manageability can be defined as the ease with which hair can be arranged and temporarily kept in place. Manageability of the hair is influenced by the single fibre properties of static charge, kinetic friction and static friction, hair curvature and the weight of the hair. Fo optimum manageability low kinetic friction, combined with high static friction, is require which, in practice, is very difficult to achieve.

Methods for evaluating manageability are normally subjective techniques employin tresses or, in some cases, live subjects using a professional hair salon.

2.6. *Body*

Body is a highly subjective concept but can normally be related to hair volume an thickness. It is difficult to precisely define body, although reference to hair thickness o apparent volume, as judged by visual and sensory methods, is relevant. Other descriptor associated with hair body are springiness, structural strength, resilience, stiffness an volume.

Measurement of body is largely undertaken using sensory evaluations, although th resilience of a hair mass may be measured instrumentally using a compression testin device. In the latter technique, the force of compression required to distort a hair tress i determined. The greater the amount of the work (compression force x distance travelled necessary, then the more body the hair tress exhibits.

Environmental effects on the hair

An understanding of the influence that the external environment and various chemica treatments have on hair, is very important in the design of hair care products. Factors whic the cosmetic scientist should consider are described below.

1. *Environmental effects*

Hair fibres suffer from general wear and tear, due to mechanical damage and air pollutio but both air and sunlight can adversely affect the condition of the hair. Sunlight has bleaching effect on hair and, in cases of severe exposure, rupture of the cystine disulphid links can occur, as shown in Figure 17.

Breakage of the disuphide links causes significant weakening of the hair structure

FIGURE 17

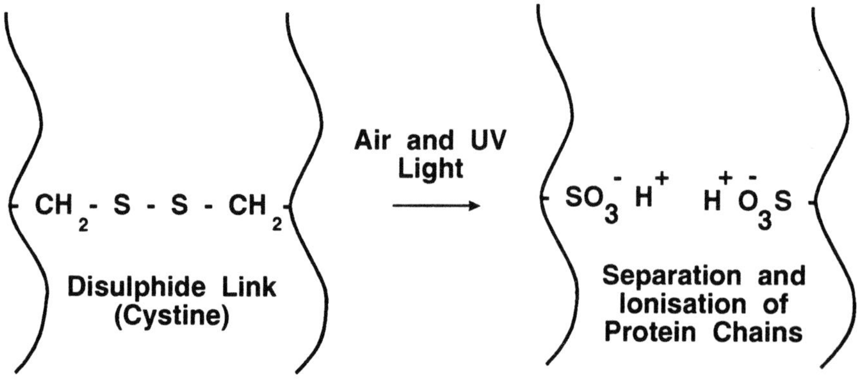

eventually resulting in hair damage. Hair damaged in this way contains a higher amount of ionised -SO_3H groups and is therefore more acidic in character.

2. *Effects of heat on hair*

The resistance of hair to dry heat is very good and little effect is observed below temperatures of around 100°C. Temperatures of up to 200°C can be tolerated before the keratin structure suffers significant damage.

3. *Effects of water on hair*

The effects of water on hair are very important. Hair is very hygroscopic and can absorb water at quite high levels, as discussed earlier in this chapter. When water is absorbed by the hair, it goes into the keratin matrix destroying hydrogen bonds and causing the hair to swell. Swelling occurs principally in the diameter of the hair which increases by approximately 16% for a rise in relative humidity from 20% to 80%. Very little increase in hair length is observed as the hair absorbs moisture. Increase in hair diameter with moisture content is best followed using a travelling microscope.

Exposure of the hair to water can also provide a temporary "set". If a length of hair is immersed in cold water, its length stretched by a few per cent, and then subsequently dried under stress, it will retain its increased length. The type of set obtained with water depends largely on the temperature. If cold water is used a cohesive set, which is temporary and easily removed by rewetting the hair, is obtained. If hot water is used, a temporary set is obtained which is more permanent, relative to the set obtained with cold water. Removal of this type of set cannot be achieved with cold water but only with water at a temperature the same as that with which the set was made. Treatment of hair with boiling water produces a permanent set.

It is obvious therefore, that for each type of water set, a different reaction in the hair is taking place. In the case of cold and hot water sets, hydrogen bonds in the hair matrix are destroyed, allowing the hair to stretch. In the case of a cold water set, rewetting of the hair allows these hydrogen bonds to reform. In the case of the hot water set, the increase in temperature increases the penetration of water into the hair shaft. Treatment of hair with boiling water causes rupture of the disulphide links, resulting in irreversible damage.

If hair is boiled in water, under stress, for only a few minutes and then removed and dried, reimmersion in water will cause a contraction of its original length. This phenomenon is known as super contraction and is due to shrinkage of protofibrils, microfibrils and alpha-helices, by rearrangement.

4. *Effects of acid and alkali on hair*

The changes in hair properties in the presence of either acid or alkali are largely determined by the accompanying changes in pH. Hair contains both acidic carboxyl groups and basic amino groups, both of which are attached, as side groups, to the protein chains. Hair will therefore react with both acids and alkalis and, as such, is amphoteric in nature. The amphoteric nature of hair can be quantified by reaction with dilute acids and alkalis and measuring the quantity of either acid or alkali taken up by a known quantity of hair. The extent of uptake obviously depends upon the pH of the solution. Between pH values of 2 and 9, hair takes up very little acid or alkali, suggesting that a wide pH range exists in which very little damage to hair occurs. In practice, hair at all pH values will still react with anionic and/or cationic materials, taking them up accordingly. This observation is very important in the study of the behaviour of shampoos and conditioners on the hair. There are two terms commonly used when examining the effects of pH on the hair. The *isoionic point* is used to describe the pH at which there is an equal number of positively and negatively charged groups on the hair and normally occurs at pH values of between 5.5 and 6.0. The *isoelectric point* is defined as the point at which the hair will not migrate in an electrical field and normally occurs at pH values between 3.5 and 4.5.

5. *Effects of anionic and cationic materials on the hair*

Anionic materials, such as the sodium alkyl sulphates (R-O-$SO_3^-Na^+$), will be substantive to ionised basic groups, $–NH_3^+$, on the hairs surface, whilst cationic materials, such as the quaternary ammonium salts ($R_4N^+X^-$) will be substantive to the corresponding ionised acidic groups, $–COO^-$. Therefore, whilst relatively small pH changes have little effect on the condition of the hair per se, a large variation in the substantivity of both anionic and cationic substances, with changes in pH, is observed. This substantivity is directly related to the number of $–COO^-$ and $–NH_3^+$ groups that exist on the hair at a given pH value. Other factors, such as Van der Waals forces and partition coefficients, also affect the substantivity of anionic and cationic materials to the hair surface, to a lesser degree.

6. *Effects of reducing agents on the hair*

Some reducing agents will react with the disulphide links in hair keratin, causing temporary or permanent rupture. It is this process which forms the basis of the design of permanent waving products, dealt with in greater detail later in this chapter.

Shampoo products

Although shampoo function has become more sophisticated in recent years, its original purpose was to clean the hair by removing sebum and associated foreign debris. Apart from flexibility in functionality, the fundamental requirements of a shampoo product remain unchanged. Shampoo products must be safe and non-irritant to the skin and eyes and should be pleasant and convenient to use. They must also exhibit good foaming characteristics and leave the hair in good condition, as well as being adequately preserved and stable throughout their shelf-life. Finally, the marketability of any shampoo product will be enhanced if it is attractive in appearance and has a pleasant fragrance.

1. *Shampoo formulations*

In order to achieve the properties described above, most shampoo formulations contain the following types of ingredient:

- primary surfactant (cleanser)
- secondary surfactant (foam booster/stabiliser)
- thickener
- colour
- opacifier (provides pearlescent appearance)

- perfume
- preservative
- conditioners
- performance additives (where applicable)
- sequestering agent

Each of these components are discussed in further detail below.

.1. *Primary surfactants*

The function of the primary surfactant is to confer cleansing properties to the shampoo formulation, through the process of detergency. Although there are many different types of primary surfactant used in shampoo formulations, they are almost always anionic in nature to give optimal detergent effects. The use of true soaps is avoided in shampoo formulation, because of their propensity to form a precipitous scum in hard water areas. By far the most common types of primary anionic surfactants used in shampoo systems are the alkyl sulphates and the alkyl ether sulphates or, sometimes, combinations of the two.

.1.1. *Alkyl sulphates*

The most commonly used alkyl sulphate is the sodium salt derived from lauric acid, sodium lauryl sulphate (SLS). Industrially, this is manufactured by the reaction of either sulphur trioxide or chlorosulphonic acid with lauryl alcohol, the resultant lauryl sulphate being neutralised with sodium hydroxide to give the sodium salt. SLS has the advantages of being a soluble, high-foaming surfactant which provides excellent cleansing, irrespective of water hardness, and is commercially available in pure forms. Disadvantages of SLS are that it is fairly irritant and cannot be used at pH values below about 4.5, due to its tendency to hydrolyse and become unstable in acidic media. Another alkyl sulphate commonly found in more recent shampoo systems is ammonium lauryl sulphate. This material possesses all the advantages of sodium lauryl sulphate but is milder to the scalp and leaves the hair in better condition. Formulation utilising this material must be kept at about pH 6.5 or below, because of the possible liberation of ammonia at higher pH values. More recently, magnesium alkyl sulphates have become available. These are claimed to posses all the advantages of their sodium-based counterparts and to exhibit significant improvements in mildness. Disadvantages, however, include commercial cost and the need to use magnesium chloride for electrolyte thickening.

.1.2. *Alkyl ether sulphates*

Alkyl ether sulphates are more commonly found in shampoo systems than straight alkyl sulphates. Alkyl ether sulphates are prepared by reaction of the corresponding alcohol with ethylene oxide, before sulphonation takes place. The most common alkyl ether sulphates used in shampoo formulations correspond to the formula $R{-}[OCH_2CH_2]_nOSO_3^-\,M^+$, where n is an integer between 1 and 3 and M^+ is a sodium, magnesium, ammonium or triethanolamine ion. Of these, probably the most popular alkyl ether sulphate is the 2 mole ethoxylated sodium salt of lauric acid, sodium lauryl ether[2] sulphate (SLE[2]S).

Like SLS, SLE[2]S is an effective foaming agent, with good cleansing properties. SLE[2]S is, however, more soluble than SLS, due to the presence of the hydrophilic ethylene oxide moieties and exhibits better properties in hard water. SLE[2]S also reacts with electrolytes more effectively than SLS, enabling the formulator to more readily thicken shampoo formulations in which it is incorporated. Finally, in view of its increased molecular size and its interaction with other secondary surfactants, SLE[2]S is milder than SLS to the skin and eyes, a property that is of significant importance when formulating frequent wash shampoos or products for babies and young children. The mildness of alkyl ether sulphates increases with an increase in the ethylene oxide content of the molecule, although cleansing properties are inversely proportional to ethylene oxide content. Like SLS, SLE[2]S must not be incorporated in formulation below a pH of about 4.5, as acid hydrolysis will occur, leading to instability. Magnesium-based alkyl ether sulphates are also now available and these have the relative advantages and disadvantages of the magnesium alkyl sulphates, discussed earlier.

1.1.3. *Other anionic surfactants*

There are many other classes of anionic surfactants that can be used in shampoo formulations. Amongst these are the sulphosuccinate half-esters, n-acyl taurines, alkoyl sarcosinates and alpha-olefin sulphonates.

The sulphosuccinate half-esters conform to the structure $ROCOCH(SO_3^-M^+)CH_2COO^-M^+$ and exhibit similar characteristics to the alkyl sulphates and the alkyl ether sulphates. They are, however, much more resistant to hydrolysis in acidic conditions and exhibit far greater mildness, an important factor in the formulation of shampoos for babies and young children. N-acyl taurines conform to the structure $RCONR'CH_2CH_2SO_3M^+$ and have the advantages of not only being very mild but also exhibiting good stability over a wide pH range. The alkoyl sarcosinates, which conform to the structure $RCON(CH_3)CH_2COO^-M^+$ produce a desirable combination of rich foam and excellent mildness. They are also extraordinarily good hydrotropes and are remarkably compatible with a wide range of other surfactants, including some cationic conditioning agents. The main disadvantage of incorporating sarcosinates into shampoo systems is that they do not respond to attempted thickening with electrolytes and, very often, it is necessary to use an external thickener to achieve the desired shampoo viscosity. The alpha-olefin sulphonates, which have only been commercially available over recent years, are also relatively mild and exhibit good stability over a wide pH range. Once again, cost prevents wider use in favour of the alkyl sulphates or alkyl ether sulphates.

1.2. *Secondary surfactants*

The functions of the secondary surfactant in a shampoo system are to stabilise the foam and mitigate the irritancy of the primary cleansing surfactant, although they are often selected to enhance the viscosity of the product, both with and without the presence of electrolyte thickeners. The secondary surfactants used in shampoo formulations are normally drawn from two classes, non-ionic surfactants and amphoteric surfactants.

1.2.1. *Non-ionic surfactants*

The most commonly used non-ionic secondary surfactants incorporated in shampoo systems are the fatty acid alkanolamides, which conform to the structure $RCONH(CH_2CH_2OH)_n$, where n is normally 1 (monoethanolamides) or 2 (diethanolamides). Industrially, these materials are manufactured by reacting the appropriate fatty acid with either monoethanolamine or diethanolamine, depending on the desired end product. The fatty acid alkanolamides provide excellent foam stabilising properties, producing a rich, creamy stable foam, particularly when used in combination with alkyl sulphates or alkyl ether sulphates, optimal performance normally being obtained when they are used in the ratio of between 1:3 and 1:4 with the anionic material. The alkanolamides also provide excellent enhancement of shampoo viscosity, particularly in the presence of electrolytes, and are commercially cost effective.

A second class of non-ionic surfactant used in shampoo formulation is the amine oxides, which conform to the structure $RN(CH_3)_2 \rightarrow O$. These materials exhibit different characteristics depending upon the pH of the system in which they are used. In alkaline media, they remain as uncharged species but in acidic conditions they assume cationic character. Amine oxides are compatible with a wide range of other surfactants and they offer foam stabilising properties and the ability to mitigate irritancy in anionic surfactants, as well as offering a mild conditioning benefit in systems formulated at lower pH values.

The last class of nonionic surfactant that deserves mention is the polyethylene glycol diesters, which conform to the formula $RCO[OCH_2CH_2]_nOOCR$, where n is an integer. Whilst these materials have little or no foaming power, they provide a means of thickening shampoo systems when used in combinations of anionic and amphoteric surfactants with low total activity. The most important property of the polyethylene glycol diesters, however, is their remarkable ability to mitigate the irritancy of other surfactants in the shampoo system, particularly important in the formulation of products for babies and children.

1.2.2. *Amphoteric surfactants*

Amphoteric surfactants are those containing both positive and negative character in the same molecule and, as such, their behaviour will be dependent to some extent on the pH of the system in which they are used. Amphoteric surfactants provide foam stabilisation in combination with the ability to mitigate irritancy of other materials and, in some cases, will modify product viscosity. There are three main types of amphoteric surfactant that are important in shampoo systems, the n-alkyl betaines, the n-alkylamido betaines and the alkyl carboxyamphoglycinates (carboxylated imidazolines).

Both the n-alkyl betaines, which conform to the structure $RN^+(CH_3)_2CH_2COO^-$, and the n-alkylamido betaines, which conform to the structure $RCONH(CH_2)_3N^+(CH_3)_2CH_2COO^-$, exhibit similar properties and are zwitterionic at all pH values. The betaines exhibit good foam stabilising character, even in hard water, and are compatible with anionic surfactants, except at low pH values. Betaines also possess the ability to mitigate the irritancy of anionic surfactants with which they are combined, resultant combinations showing excellent viscosity building characteristics which can be enhanced through the use of electrolytes.

The alkyl amphocarboxyglycinates, which conform to the formula $RCONH(CH_2)_2$-$N(CH_2CH_2COO^-M^+)CH_2CH_2OCH_2CH_2COO^-M^+$, exhibit similar properties to the betaines, with the exception that they are not zwitterionic at all pH values. Although the alkyl amphocarboxyglycinates are compatible with both anionic and nonionic surfactants over a wide pH range, at lower pH values they assume cationic character, which may result in an increase in irritancy of the system.

1.3. *Thickeners*

The mechanism used to thicken a shampoo formulation depends largely upon the type of surfactants used. In shampoo systems where the primary surfactant is an alkyl sulphate or alkyl ether sulphate, viscosity enhancement can be achieved through the use of electrolytes, which function by modifying the micelle structure. This type of viscosity modification is particularly effective in the presence of a fatty acid alkanolamide foam stabiliser. In shampoo systems where alternative primary surfactants, such as the sulphosuccinates or sarcosinates, are used, thickening with electrolytes is not successful. In these cases, thickening is normally achieved through the use of external thickeners, such as cellulose derivatives or polyethylene glycol diesters.

1.4. *Colour*

The consumer appeal of a shampoo may be significantly increased by the use of a suitable dye. Dyes used in shampoo formulations must be readily water soluble and are generally anionic in nature. Compatibility problems are rare when using dyes of this type, except in the case where the shampoo formulation contains a significant quantity of cationic material. From a legislative point of view, dyes are controlled substances and care must be taken to ensure that they conform to the EEC Cosmetic Directive. Dyes used in shampoo formulations should also be carefully evaluated for light and chemical stability.

1.5. *Opacifiers*

Opacity or pearlescence in a shampoo can enhance consumer appeal and can be provided by various materials, most of which are waxy and crystalline in nature. The most common type of opacifier used in shampoo systems is ethylene glycol distearate (EGDS). This material is incorporated into the shampoo system, typically at levels of between 1% and 2%, by heating to temperatures in excess of 55°C causing dissolution. When the shampoo is subsequently cooled, the EGMS recrystallises to give a pearlescent effect. The pearlescence obtained will depend upon the size, shape, distribution and reflectance of the opacifier crystals formed which, in turn, is largely influenced by the rate of cooling. Pearlescence can also be obtained through the use of "inorganic pearls" such as micronised titanium dioxide, although the shampoo system must be reasonably viscous, in order to avoid settling of the pearl, at high temperatures, under gravity.

1.6. *Perfume*

The perfume or fragrance used in a shampoo formulation is often critical in determining its success in the market-place. Perfumes in a shampoo are typically incorporated at levels of between 0.3% and 1.0% and they must be both compatible and stable with respect to the other shampoo ingredients.

1.7. *Preservative*

Historically, the most commonly used preservative in shampoos was formaldehyde although this is now less popular due to concern over the possibility of sensitisation reactions. Nowadays the choice of preservative is largely dependent upon toxicological data available and the materials used in the shampoo formulation concerned. Common choices of preservatives include parabens, 1-(3-chloroallyl)-3,5,7-triaza-1-azoniaadamantane, isothiazolinones and 2-bromo-2-nitropropane-1,3-diol.

1.8. *Conditioners*

The most commonly used conditioners in hair care are the quaternary ammonium salts, covered in greater detail later in this chapter. Generally these materials are incompatible with the primary anionic surfactants used in shampoo systems, due to the possible formation of a salt, resulting in precipitation of the conditioning active. The degree of incompatibility is dependent upon the size and configuration of the hydrophobic chains in the quaternary ammonium salt and, in some cases, this type of material can be incorporated into shampoos without compatibility problems. Other types of conditioning agent incorporated into shampoo systems include cationic polymers and water-soluble silicones. More recently, shellac has been suggested as a hair conditioner suitable for shampoo systems. This material, which was originally used as a setting resin in early hair spray products, can impart good condition and wet comb characteristics to the hair and provide it with volume and hold.

1.9. *Performance additives*

Performance additives are normally incorporated into shampoos that are designed to perform special functions, such as anti-dandruff shampoos. Historically, coal tar and selenium sulphide were used as anti-dandruff actives but modern formulations use materials such as zinc pyrithione, incorporated at levels between 0.5% and 1.0%. or "Octopirox" (Hoechst) which is effective against dandruff at similar levels. Zinc pyrithione is insoluble in water and is therefore normally suspended in the shampoo formulation using a stabiliser such as bentonite clay. Octopirox, on the other hand, can be used to formulate totally clear shampoo systems due to its solubility in water, although care must be taken to prevent it from reacting with ferric ions, which may cause a yellow discolouration in the shampoo.

1.10. *Sequestering agents*

Sequestering agents are incorporated into shampoo systems to combat the effects of hard water, which normally results in poorer foam quality and dull appearance of the hair after shampooing, due to precipitation of insoluble calcium salts on the surface of the hair. The most commonly used sequestering agents are based on EDTA (ethylenediaminetetraacetic acid) which preferentially complexes with any metal ions in solution, thereby removing their availability for precipitation on the hair in the form of insoluble salts.

Other hair care products

Although shampoos are primarily functional, there are many other categories of hair care products that are designed to improve the condition, style and appearance of the hair. A brief overview of these products will now be discussed.

1. *Hair conditioners*

Hair conditioners are normally applied after shampooing to improve the condition and appearance of the hair. In order to maximise their performance, the conditioner is normally left on the hair for between 1-2 minutes, before being rinsed off. Some of the newer types of hair conditioner are designed to remain on the hair after being applied, although this method of use is more common with products intended to provide a high level of conditioning benefit.

Hair conditioners are normally formulated to provide a combination of the following benefits:

- to increase shine and lustre of the hair
- to reduce the level of static charge on the hair, thus minimising "fly-away"
- to provide good lubrication to the hair shafts, thereby easing combing of the hair in both the wet and dry states
- to give the hair body and manageability
- to make the hair soft and pleasant to the touch

Conditioners normally contain, as actives, cationics surfactants which are substantive to the hair, via attraction to the anionic charges on the hair shafts that remain after washing with shampoo. The most popular class of cationic surfactants incorporated in modern conditioners are the quaternary ammonium compounds, which comply to the general formula $(R_4N)^+X^-$, where R is a fatty chain and X is a halide, commonly chloride. The quaternary ammonium salt is normally formulated into a high water content, oil-in-water emulsion, which is stabilised using external thickeners, secondary emulsifiers and emulsifying waxes. Typical conditioning actives include distearyl dimethyl ammonium chloride, $[CH_3(CH_2)_{16}CH_2{-}N(CH_3)_2{-}CH_2(CH_2)_{16}CH_3]^+Cl^-$ and stearyl dimethyl benzyl ammonium chloride, $[CH_3(CH_2)_{16}CH_2{-}N(CH_3)_2{-}CH_2{-}C_6H_6]^+Cl^-$, use of the former being illustrated in a typical hair conditioner formulation, shown below.

Distearyl dimethyl ammonium chloride	2.40%
Hydroxyethyl cellulose	1.00%
Cetyl alcohol	2.00%
Emulsifying wax	0.50%
Perfume	0.50%
Preservative	qs
Water	to 100.00%

The pH of hair conditioning formulations is normally adjusted to fall within a pH range of 3.0-5.0. This ensures that the active remains in its cationic form, thus maximising its substantivity on hair. Other conditioning actives which are also incorporated into hair conditioners include ethoxylated alkyl amines, hydrolysed proteins and cationic polymers.

2. *Hair sprays*

Simplistically, hair sprays consist of a polymer resin, is solution, which "holds" the hair by bonding hair fibres together at points of mutual contact. The general requirements for hair spray products are:

- they must exhibit good holding properties, without appearing too rigid
- they must not detract from the gloss of the hair
- they must be easily removed by shampooing
- they must not be susceptible to absorption of moisture from the atmosphere
- they should brush out of the hair in the form of a fine powder, without flaking

Historically, hair spray products were based on shellac, a natural resinous secretion from insects, but this produced hard, insoluble films that were not easily washed out of the hair. Modern hair sprays are formulated with synthetic polymeric resins, specifically designed to give the required properties described above. Initially, the most popular polymers were those based on PVP (polyvinylpyrrolidone) which is non-toxic, substantive to hair and a good film-forming agent, producing clear films. The main disadvantage of PVP lies in its hygroscopic nature, which causes it to absorb moisture from the atmosphere making the hair sticky. Modern synthetic polymer resins are much more sophisticated, incorporating more than one type of monomeric unit to achieve the desired properties. Typical examples of modern hair spray resins are PVP\vinyl acetate copolymers, vinyl acetate\crotonic acid copolymers (Resyn 28-1310 from National Starch), vinyl acetate\crotonic acid\vinyl neodecanoate terpolymers (Resyn 28-2930 from National Starch) and octylacrylamide\acrylates\butylaminoethyl methacrylate terpolymers (Amphomer from National Starch). In some cases, where the polymers contain acidic carboxyl groups, it is necessary to neutralise the resins for use in the hair spray product. Neutralisation enables control to be exercised over the degree of hardness and solubility of the hair spray films, thus improving their holding and washing out properties. This neutralisation process is normally carried out with organic amines such as triethanolamine, aminomethyl propanol (AMP) or aminomethyl propandiol (AMPD). A typical hair-spray formulation, based on Resyn 28-2930, is given in below.

Resyn 28-2930	1.60%
Aminomethyl propanol	0.15%
PEG-75 lanolin	0.20%
Ethanol	68.05%
Butane 30	30.00%

Evaluation of hair sprays can be carried out in several ways. Initial evaluations can be made by spraying the product onto a clear glass plate and examining the resultant film for clarity, flexibility and tackiness. The drying time of the film, and the ease with which it is washed off of the glass plate can also be estimated using this method. After prototype screening, tests on hair tresses are normally carried out, to determine the effectiveness of the holding properties of the product, before moving to full-head tests in the final stages of product development.

3. *Setting lotions*

Setting lotions are very similar to hair-sprays, except that they are bottled liquids which are applied to wet hair. The setting lotion deposits a polymeric film onto the wet hair, which forms a flexible film when the hair dries, thus providing the hold. Setting lotion formulations are similar to those of hair sprays, employing the same resin types, except that the base is approximately 60% water and 40% alcohol. A typical setting lotion formulation is shown in below.

PVP\vinyl acetate copolymer	2.00%
Silicone oil	0.20%
Perfume	0.30%
Ethanol	37.50%
Water	60.00%

Nowadays, setting lotions are much less popular due to the developments of styling mousses, discussed in more detail below.

4. *Styling mousses*

Styling mousses provide an effective and convenient method for applying a setting lotion type formulation to the hair, using an aerosolised container, in the form of a semi-stable foam. After application to the wet hair, the foam collapses, providing an even dispersion of small quantities of setting fluid onto the hair. Hair mousse formulations normally contain the following components:

- a setting resin
- a cationic conditioner
- a plasticiser
- a surfactant for emulsification of the concentrate which, in turn, provides the foam
- ethanol
- water

A typical hair mousse formulation, based on a mixed resin system (polyquaternium-11 and polyquaternium-4) is illustrated below.

Polyquaternium-11	3.00%
Polyquaternium-4	1.50%
Hydrogenated tallow trimonium chloride	0.20%
Nonoxynol-10	0.35%
Ethanol	10.00%
Water	74.95%
Butane-48	10.00%

5. *Bleaching products*

Hair is pigmented by melanin, which has a very complex structure based on the indole derivatives. Reaction with hydrogen peroxide bleaches melanin, thereby destroying hair colour. The basis for a hair bleaching product is therefore a dilute solution of hydrogen peroxide. One of the undesirable side reactions of the bleaching process is the cleavage of disulphide links, which weakens the hair. In a normal bleaching operation approximately 15-20% of disulphide links in the hair are broken. Bleaching products typically contain hydrogen peroxide solutions at a strength of 3-4%, adjusted to approximately pH 9.0-9.5 with ammonia. The bleach is applied to the hair and rinsed off when the required degree of bleaching has taken place.

6. *Hair waving products*

In addition to being cleaved by oxidation processes, disulphide links can also be broken by reduction, a reaction that forms the basis of the permanent waving process. The process of permanent waving involves the cleavage of the disulphide links, by a reduction process, whereupon the hair is reconfigured into the desired style (using rollers, etc.) and the disulphide links reformed by oxidation. This process is represented diagrammatically in Figure 18.

FIGURE 18

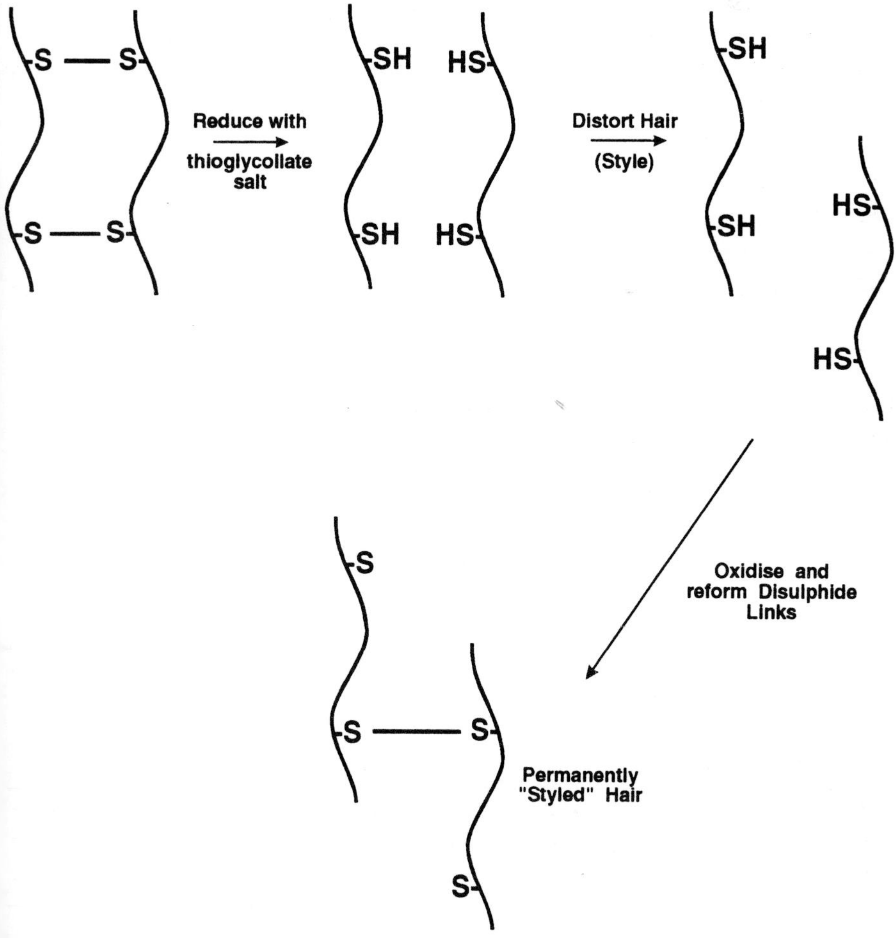

When the hair is in the distorted state, the disulphide links will reform in different places, reaction occurring between the two closest –SH bonds on oxidation. This happens throughout the matrix of the hair cortex, thus maintaining the distorted state or style.

Hair reduction can be carried out with many different materials but the most widely used reducing agents are those with a mercaptan structure, the most common example of which is thioglycollic acid which is readily obtainable in pure liquid form. This universally used material possesses the correct level of activity for the permanent waving process and is small in molecular size, allowing penetration into the hair. Another material which is also used in permanent waving systems is thiolactic acid, although this does not enjoy the wide usage of thioglycollic acid. A typical reducing solution for permanent waving contains the following components:

- thioglycollic acid, normally at a level of approximately 6.0%
- wetting agent
- thickening agent
- oils or protein derivatives for conditioning the hair
- ammonia for pH adjustment to approximately pH 9.0.

The pH of the reducing solution must not be too high, otherwise permanent destruction of the hair will result. A typical permanent waving formulation is illustrated below.

Ammonium thioglycollate (50% active)	10.00%
Ammonia (25%)	5.00%
Wetting agent	0.80%
Conditioning agent	0.70%
Opacifier	0.20%
Perfume, dyes, etc.	qs
Water	to 100.00%

The pH of this formulation would be adjusted to pH 9.0-9.5.

In the permanent waving procedure, the hair is firstly shampooed to ensure maximum penetration, before being styled by winding on to curlers close the scalp. The waving solution is then applied and left to react with the hair for up to 30 minutes, before being rinsed off with water and left for a further 30 minutes. The hair is then oxidised with a dilute solution of hydrogen peroxide and subsequently dried to give permanent set.

7. *Hair colourants*

Hair colourants are a very important type of hair product, commonly applied in the form of a shampoo with the following additional requirements:

- the hair fibre should be coloured in a relatively short time, between 10-30 minutes
- the colouring reaction should occur at ambient temperatures
- the product should be safe and not injure eyes or scalp
- the product should produce a minimum of scalp staining
- the coloured hair should be natural in appearance

- each dye component should fade from the hair to the same degree when washed
- the dye components must be adequately light stable and fade at the same rate
- the dye shade should not change when applied to bleached or permed hair, or with subsequent use of conditioners
- the dye should not sublime when heated
- the dye should not be removed by mechanical abrasion when the hair is wet or dry
- the product should satisfy all cosmetic legal requirements

There are essentially three classifications of hair colouring products, temporary, semi-permanent and permanent.

7.1. *Temporary hair dyes*

These are colouring products which give a colour to the hair, which washes out on the first shampooing. These products use colourants of high molecular weight that are deposited on to the surface of the hair, without penetration. The most common types of temporary hair colourant are anionic in nature, and are normally applied in slightly acidified form. Examples of temporary dye materials include acid violet, tartrazine yellow and eosin red. Dye application can be made directly from aqueous solution, but it is more common to apply these dyes in the presence of a non-ionic surfactant.

7.2. *Semi-permanent hair dyes*

This is a very common type of hair dye, with a permanence between that of temporary and permanent products. Typically, hair coloured using semi-permanent colourants will tolerate between 3 and 6 shampoo washings, before significant colour fading takes place. The greater degree of permanence occurs because the dye penetrates into the hair, by the process of diffusion, due to its small molecular size. The great majority of semi-permanent dyes belong to the chemical classes of nitrophenylenediamines, nitroaminophenols or aminoanthraquinones. Perhaps the most commonly used of these is the nitrophenylenediamine class of material, which conforms to the general formula shown in Figure 19.

FIGURE 19

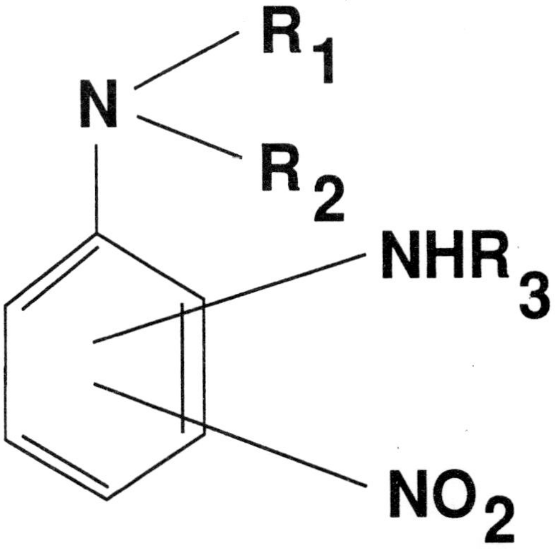

The dyeing process depends on the degree of diffusion which, in turn, depends upon dye concentration, temperature and contact time. The higher the temperature and the longer the contact time, the more permanent the colouring process will be. All of these dyes are nonionic in nature, enabling them to be added to shampoo bases without any compatibility problems.

7.3. *Permanent hair dyes*

Permanent hair dyes provide colour to the hair which resists shampooing, mechanical wear and tear (brushing, combing, etc) and light. Permanent hair colourants are based almost exclusively on oxidation dyes. These are small molecules that diffuse into the hair, whereupon they are coupled to form large highly-coloured dye molecules, which cannot diffuse back out of the hair shaft. These small molecules, or bases as they are sometimes known, are commonly based on aromatic p-diamines or aromatic p-aminophenols, in the para-configuration, a typical example of which is p-phenylene diamine. The internal coupling reaction is brought about by oxidation with hydrogen peroxide, inside the hair, and the reactions that occur are very complex. Secondary agents in the formulation couple with the p-phenylene diamine molecules to form more complex high molecular weight structures. These secondary agents are known as couplers or modifiers and are generally aromatic m-diamines, aromatic m-diphenols (resorcinols), or aromatic m-aminophenols. Typical examples of couplers are m-aminophenol and resorcinol.

References & further reading

1. The Physiology and Pathophysiology of the Skin (Edited by A. Jarrett). Academic Press 1977. Vol. 4.
2. Keratins, their Composition, Structure and Biosynthesis. R. D. B. Frazer, T. P. MacRae, G. E. Rogers. Charles C. Thomas 1972.
3. Keratin and Keratinisation. E.H. Mercer, Pergamon Press 1961.
4. Chemical and Physical Behaviour of Human Hair. Clarence R. Robbins. Van Nostrand Reinhold Company 1979.
5. Wool: Its Chemistry and Physics (2nd Edition). P. Alexander, R. F. Hudson, C. Earland. Chapman and Hall Ltd. 1964.
6. Cosmetics Science Vol. 2. (edited by Breuer). Academic Press 1980. (Contribution by F. J. Ebling "The Physiology of Hair Growth").
7. "Hair" by Michael L. Ryder 1973. Institute of Biology's Studies in Biology No. 41. Edward Arnold Ltd.

BATH PRODUCTS

The bath products sector has increased steadily over the last decade and now constitutes a fairly large segment of the total personal care products market. Although bath foams still occupy, by far, the biggest part of the market, the range of bath products available is quite diverse. Not all bath products are designed to provide a cleansing function and many just deliver pleasant skin feel and the aura of pleasantness, through fragrance.

In this chapter, a brief discussion of the many types of bath product available is given, along with an overview of their function and formulation. For completeness, shower gels and after bath preparations are also included.

Bath salts, bath cubes and bath tablets

Bath salts may be used in powder, crystal and tablet form, with the primary function of softening the bath water, particularly in hard water areas. They are often attractively coloured and carry a refreshing or relaxing fragrance. Bath salts are also used in cases where water requires no softening, simply to enhance the aesthetic effect of the bath by supplying fragrance and colour.

Bath salts are commonly based on sodium bicarbonate, or borax in combination with trisodium phosphate. Sodium sesquicarbonate is a particularly stable salt, which is readily soluble in water and can be easily perfumed and tinted. It does not effloresce or require any special precautions in packing and is therefore one of the most popular materials used in the preparation of bath salts. Sodium sesquicarbonate is attractive in appearance and is available in the form of fine, white, needle-shaped crystals. It also gives less alkaline solutions than sodium carbonate, which, because of its high alkalinity, is difficult to colour and perfume.

Borax does not affect the constituents of perfumes normally employed and is easy to tint. It is, however, less soluble in water. Trisodium phosphate is used as a water softener but has the disadvantage of being difficult to tint, and is, therefore, used in combination with borax.

Sodium tripolyphosphate and tetrasodium pyrophosphate, both sequestering agents, are often used in combination with the above mentioned bath salts constituents, to increase their efficacy. Sodium tripolyphosphate is stable and has water-softening properties.

Another useful sequestering agent, used in bath salt preparations, is sodium hexametaphosphate, commercially available as "Calgon".

Once the components of a bath salt have been selected, they are mixed together, then tinted and perfumed. The tinting and perfuming process is normally carried out by spraying, or immersion of the bath salt constituents in an alcoholic solution of a permitted dyestuff.

The ingredients found in bath cubes and tablets are the same as those used in bath salts, except that the salts are powered and mixed with starch or, sometimes, low levels of sodium lauryl sulphate. This prevents hardening of the product and facilitates their disintegration. In addition, various mucilages may be included as binders. The salts, after being mixed, tinted and perfumed, are powdered and granulated with starch and a binder, before finally compression into the cube or tablet shape.

For rapid disintegration in bath water, effervescent tablets are frequently employed; the gas bubbles produced by these tablets also producing a tingling sensation on the skin. The most commonly used ingredients for this purpose are sodium bicarbonate and citric or tartaric acids. To prevent possible deleterious effects of atmospheric moisture, such as deliquescence or efflorescence, manufacture of such tablets should be carried out in a dry atmosphere. To prevent premature gas release, a small amount of soluble starch may be included in the formulation to coat the component organic acid, preventing direct contact with the alkaline constituents of the formulation.

An example of an effervescent bath tablet formulation is given below.

Sodium bicarbonate	55.00%
Tartaric acid	15.00%
Citric acid	5.00%
Sodium sesquicarbonate	20.00%
Perfume, colour, etc	5.00%

In terms of the total bathcare market-place, this group of products represents the smallest proportion.

Bath oils

Bath oils are, after foam baths, the second largest group of bath preparations. Their function is to produce a pleasant fragrance in use and to impart pleasant skin feel, making the skin soft and supple. Bath oils can be divided into two main groups, floating or spreading bath oils and emulsifying bath oils. The latter can be further sub-divided into soluble and dispersible bath oils.

Apart from their use as additives to bath water, bath oils can also be used directly on the skin, as an emollient, after showering. After application, they are gently rinsed off and the skin patted dry.

1. *Floating bath oils*

Floating bath oils are hydrophobic in nature and, by virtue of their specific gravity, will float on top of the bath water, depositing a film on the bather's skin on emergence from

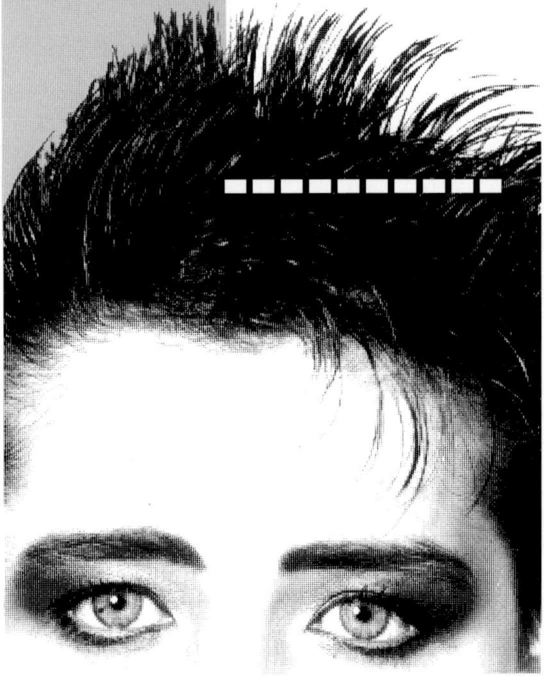

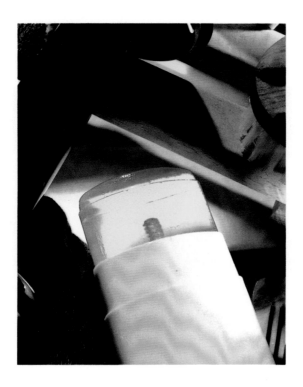

the bath. Many of the oils used in these products do not spread rapidly and, if no detergent is present, may form large globules instead of a continuous film on the surface of the water. This would be cosmetically unpleasant and deposit oil, in patches, on the bather's body. For this reason, an oil-soluble surfactant is often added, at levels of up to 5% of the composition. This will lower the surface tension of the bath water and enable the oil to form a continuous film on its surface. The spreading capacity of a bath oil formulated in this way can be directly related to the HLB of the surfactant used, the higher the HLB, the greater the spreading capability of the product. However, as the HLB of the added surfactant is increased, so its solubility in the oily ingredients decreases. When formulating a product of this type therefore, a compromise between oil solubility and spreading efficacy must be made. Typically, chemically suitable nonionic surfactants with an HLB value in the range of 8-10 provide optimum results, giving workable solubility in the oil, whilst providing adequate spreading capability for the product. A typical example of such a surfactant is PEG-40 sorbitan peroleate, which has an HLB value of approximately 9.

Floating bath oils often contain a fairly large amount of perfume, typically 5-10%, enabling the fragrance to diffuse throughout the bathroom, enhancing the pleasant sensation of the bathing experience. However, because of their tendency to leave an oily scum along the sides and on the bottom of the bath, which cannot readily be removed, floating bath oils are not as popular as dispersible bath oils.

Light mineral oil and isopropyl myristate are amongst the most frequently used ingredients of commercially available floating bath oils. However, if mineral oil is used on its own, it gives an excessively greasy product with poor perfume solubilisation properties. It is therefore common practice to use mineral oil, or other oils, in conjunction with another emollient or blend of emollients. A variety of other materials may be used to produce floating bath oils, such as fatty acid esters, lanolin derivatives, and vegetable oils. A frequently used ingredient is castor oil, which can be easily diluted with alcohol to give the desired viscosity and to assist in the solubilisation of the perfume. A typical floating bath oil formulation, based on mineral oil and isopropyl myristate, is shown below.

Mineral oil	71.00%
Isopropyl myristate	24.00%
Laureth-2 benzoate	1.00%
Perfume	4.00%

A second formulation, based on mineral oil, lanolin oil and isopropyl myristate is illustrated below.

Lanolin oil	8.00%
Mineral oil	30.00%
Isopropyl myristate	53.00%
PEG-40 sorbitan peroleate	2.00%
C12-15 alcohols lactate	2.00%
Perfume	5.00%

So-called "functional bath oils", which are usually mixtures of mineral and/or vegetable oils, are used to relieve dry skin, as well as being the source of fragrance. The presence of an oily film on the skins surface impedes skin dehydration in conditions of low relative humidity.

2. *Emulsifying bath oils*

These types of bath oil are intended to form a homogeneous dispersion within the bath water, rather than float on its surface. They deposit less oil on the body than floating bath oils. Two types of emulsifying bath oils are commercially available, soluble and dispersible. The former will produce a clear solution in the bath after dispersion, whereas the latter produces a milky emulsion. The effect produced will depend on the type and level of surfactant employed and the HLB value of the components used.

Soluble bath oils are transparent emulsions, containing a sufficient level of surfactant to solubilise the oily constituents, through micellar solubilisation. They are either concentrates, consisting of perfume, surfactant and, optionally, alcohol, or solubilised products, comprising perfume, surfactant and water. Soluble bath oils generally contain a perfume at levels between 3-5% and nonionic, oil-soluble, water-dispersible surfactants with an HLB value of between 12 and 18. These surfactants may be included at levels of up to 15%, in order to solubilise the perfume in the bath oil and eventually assist in dispersing the oil in the bath water, producing clear dispersions. The ratio of surfactant to perfume required, for effective solubilisation, is normally between 2:1 and 5:1.

This class of bath oils also includes bath essences, designed principally to perfume the bath. Alcohol is sometimes added to these formulations to give the perfume "lift" when the product is added to the bath water. A typical soluble bath oil formulation is shown below.

Light mineral oil	65.00%
Isopropyl myristate	20.00%
PPG-15 stearyl ether	10.00%
Perfume	5.00%

Dispersible, or "blooming", bath oils normally contain the oil, combined with an emollient, a perfume and a surfactant. They produce a milky effect when poured into bath water and, with mild agitation, produce a uniform dispersion of the oil in the bath, in the form of very small droplets. This effect is due to the emulsification of the oil in the bath water. To create such an effect, surfactants with a low HLB value are employed, a typical example of which is polyoxyethylene-2 oleyl ether, which is an oil-soluble surfactant with an HLB value of approximately 5.

A typical example of a dispersing bath oil formulation is given below.

Light mineral oil	70.00%
Isopropyl myristate	15.00%
POE-5 oleyl ether	10.00%
Perfume	5.00%

3. *Bath oils based on silicones*

Silicones have a number of advantages over traditional fatty oils when used in bath oil formulation. They form non-tacky, non-greasy films on the surface of the skin, as well as reducing the tacky feel of greasy emollient ingredients. Silicone glycol copolymers are used in dispersible bath oil preparations because of their emollient properties and their tendency to depress surface tension assists in the dispersion of the product in the bath.

Silicone glycols are also used in the formulation of soluble bath oils, at low levels, where their properties have beneficial effect on product performance.

Low viscosity, polydimethyl cyclosiloxanes exhibit good compatibility with many cosmetic ingredients, in products where clarity is important. They also possess good solubility in many cosmetic solvents, including mineral oil, and reduce the tackiness of other constituents, imparting a silky feel to the skin. Polydimethyl siloxanes are therefore excellent materials for inclusion in bath oils and other bath preparations.

Amongst the disadvantages of the use of polydimethyl siloxanes is their tendency to inhibit instantaneous blooming in the bath and, in the case of floating bath oil formulations, it is preferable to use straight-chain dimethyl polysiloxanes at levels of up to 5%. Polydimethyl siloxanes can be successfully incorporated into dispersible bath oils and are often used in combination with mineral oil and emollient esters in these formulations.

4. *Foaming bath oils*

Foaming bath oils appear to combine the conventional properties of a foam bath with the benefits of an oil and are often marketed as premium priced products. They combine fragrance with foaming action, as well as reducing or eliminating scum in the bath.

The term "foaming bath oil" is somewhat inaccurate, as these products are often foam baths, which give a high degree of emolliency on the skin as a result of the formulation design. Foaming bath oils contain foam producing surfactants such as the sodium, ammonium or alkanolamine salts of alkyl sulphates and alkyl ether sulphates, as well as foam stabilisers such as fatty acid alkanolamides which, in addition to stabilising the foam, tend to increase the viscosity of the product.

In view of the conflicting requirements of good foaming capability and good emolliency, foaming bath oils are often difficult to formulate. The use of an effective nonionic solubiliser, possessing a high degree of emollient character, at an adequate level, minimises these problems and the required product characteristics may be achieved. It should be borne in mind, however, that the inclusion of large amounts of emollient oils will have a depressant effect on the foam. As such, foaming bath oils do not combine the best properties of bath oils and foam baths and are essentially a compromise of the two.

In addition to using nonionic surfactants as solubilisers for the perfume, use is made of sequestrants, to prevent the formation of insoluble soaps which give rise to scum. If a product of relatively high viscosity is required, cellulosic thickeners can be incorporated into the formulation to work synergistically with any fatty acid alkanolamides present.

More recently, low levels of water-soluble or water-dispersible silicone glycols have been used in foaming bath oils. These materials enhance both the foaming and emolliency of such products.

The skin benefits provided by foaming bath oils arise in two ways. Firstly, the physical deposition of oily emollients on the skin subsequently inhibits loss of water from the skin at low relative humidities. Secondly, the deposition and absorption of certain ingredients, such as triglycerides, also confer skin moisturisation benefits in their own right.

A typical foaming bath oil formulation is illustrated below.

Triethanolamine lauryl sulphate	40.00%
Lauric/myristic diethanolamide	15.00%
Polyoxyethylene lanolin alcohols	5.00%
Modified coconut triglyceride	5.00%
Citric acid	1.00%
Perfume	5.00%
Water	29.00%

5. *"Special effect" bath oils*

These products do not provide any additional functional benefits and are primarily formulated to maximise commercial appeal. Product types include two-layer and three-layer dispersible and foaming bath oils, fluorescent bath oil formulations and those which produce a colour change when diluted by the bath water.

An example of a two-layer foaming bath oil formulation is given below.

Sodium laureth sulphate	30.00%
Propylene glycol dicaprylate-dicaprate	15.00%
Mineral oil	19.00%
Hexylene glycol	4.00%
Perfume	2.00%
Preservatives, dyes and water	to 100.00%

Foam baths

Of all bathroom products, foam baths constitute by far the largest and most lucrative sector. Their requirements have changed little over the years and a typical foam bath formulation should conform to the following fundamental criteria:

- it should produce a copious and instantaneous foam at practical product concentrations
- the foam should be stable within fairly wide limits of temperature, in both hard and soft water, in the presence of soil and fatty impurities
- the foam should not collapse immediately on the addition of soap, nor should it be too stable to make its removal after bathing difficult
- it should have adequate detergency and emolliency to cleanse the body and condition the skin, providing some compensation for the loss of sebum during the bathing process

- the product should be non-irritant to both skin and eyes and non-sensitising to the skin

In order to maximise potential for commercial success, the foam characteristics and stability must be carefully balanced to satisfy the requirements listed above.

1. *Detergents used in bath foams*

When choosing the best detergent for a bath foam, account should be taken of its water solubility, dispersibility, flash foam characteristics at different dilutions, rapidity of foam collapse in the presence of soap and its viscosity dependence on the presence of electrolytes.

Preferred detergents for liquid foam baths are the alkyl ether sulphates which are cost-effective, readily available and satisfy most of the above mentioned requirements. They are also considered to be milder on the skin than the corresponding alkyl sulphates. Alkyl aryl sulphonates have also been used, but do not foam as readily in hard water and their foam tends to collapse in the presence of soap.

Sulphosuccinates are considered to be very mild detergents with good foaming properties. In addition to being non-irritant to skin and mucous membranes, these materials are also claimed to increase the skin's tolerance to other detergents. Amphoteric surfactants, such as disubstituted imidazolines and alkyl betaines, are sometimes included to improve the mildness of finished products.

Other surfactants employed are fatty acid condensates and protein/fatty acid condensates. The former includes the methyl taurides and sarcosides. Methyl taurides are water soluble materials with good calcium soap dispersing properties. The second group of protein/fatty acid condensates are, like the sulphosuccinates, considered to be very mild anionic detergents capable of improving the skin's tolerance to other surfactants. Their foaming properties, however, are not as good as alkyl sulphates and alkyl ether sulphates and they are therefore not used on their own, but in combination with other primary detergents.

2. *Foam bath formulation*

Foam baths come in many forms, ranging from powders to liquids, and their formulations are many and varied. In the case of clear liquid formulations, in transparent bottles or sachets, it is necessary to ensure that the product retains its clarity at low temperatures and over periods of prolonged storage. This can be achieved using surfactants with good water solubility at high active matter concentrations (15-35%), over a wide temperature range. The detergents selected should also have good foaming capacity, given the high dilution ratio when poured into the bath. Transparent, coloured preparations are often exposed to sunlight, either at the point of sale or during their period of use, and should therefore be stable to ultraviolet and visible light. Products should therefore be formulated using light-stable dye systems or, optionally, should contain low levels of a UV absorber to prevent discolouration and fading of the product.

The viscosity of a bath foam has little effect on its functionality, provided it disperses easily in the bath water. Aesthetically, however, it is advisable to thicken bath products of this type, to enhance their consumer appeal. Bath foams based on alkyl sulphates or alkyl-ether sulphate can normally be thickened by the addition of an electrolyte, typically sodium chloride, at levels ranging between 0.5% and 2%. Other thickening agents include PEG 6000 distearate and PEG-55 propylene glycol oleate. The inclusion of fatty acid alkanolamides will also increase the viscosity of many detergent systems.

Foam bath preparations often contain as the main component a sodium or alkanolamine salt of an alkyl ether sulphate, although numerous other anionic surfactants have been employed, either singly or as mixtures, to generate the foam. A further component of these compositions is a foam stabiliser, commonly a fatty acid alkanolamide, such as coconut or lauric acid dialkanolamide.

Other ingredients include dyestuffs and a thickener, often of cellulosic derivation. Cellulosic thickeners have the additional benefits of providing some foam stabilisation and helping to prevent the redeposition of dispersed soil back on to the skin. Modern foam baths are mostly "conditioning" products and therefore contain water-soluble skin emollients to enhance post-use skin feel. A classic ingredient of this type is PEG-7 glyceryl cocoate.

Fragrance is an important part of many cosmetic and toiletry products and nowhere is this more true than in the case of bath foams. Fragrance not only enhances initial consumer appeal but, together with the product's performance, will be an important factor in its ultimate commercial success. Fragrance level usually varies between 2% and 5%, depending on product class. In the case of clear products, fragrance will often be added in the presence of a nonionic solubiliser such as POE-20 sorbitan monolaurate or PEG-40 hydrogenated castor oil.

A typical, clear, foam bath formulation is illustrated below.

Sodium laureth sulphate	50.00%
Coconut diethanolamide	4.00%
PEG-7 glyceryl cocoate	3.00%
Dimethicone copolyol	1.00%
Sodium chloride	1.50%
Citric acid	0.50%
Perfume	3.50%
Preservatives, dyes and water	to 100.00%

More recently, there has been an increase in the popularity of products sold on a "natural" platform and it is common nowadays to find bath foam products which contain herbal extracts, claimed to enhance skin benefits or the well-being of the user. Extract of marigold, sage, burdock, wild chamomile and lime tree are all claimed to provide therapeutic benefit for a variety of minor skin disorders. Others, such as horse-chestnut, clematis and arnica, are claimed to stimulate blood circulation. In general, herbal extracts are added to create a feeling of well-being, which can be further enhanced by the inclusion of essential oils and refreshing fragrances of the woody, mossy type.

The visual appeal of foam baths, in common with other product types, is critical to commercial success and, in addition to clear liquids, pearlescent and opaque liquid preparations are also marketed on a "luxurious" platform. These are similar in composition to their clear equivalents, except that they contain pearlescents or opacifiers to produce the desired visual effect. The most common types of pearlescent are those based on ethylene glycol distearate. Such products must have adequate viscosity to uniformly suspend the particles of the pearlescent ingredient. Traditionally the pearlescent agent is added to the formulation, before fragrance addition, at temperatures of 65-70°C. On cooling, with gentle agitation, the pearlescent crystallises to produce a high degree of pearlescence, the final effect being dependent on many factors such as type of pearl and the rate of cooling. More recently, there has been an increase in the availability of "cold" pearlescent products, which are pre-formed and added to the bath foam at room temperature.

An example of a "luxurious" pearlescent bath foam formulation is illustrated below.

Ammonium lauryl sulphate	35.00%
Disodium laureth sulphosuccinate	8.00%
Cocoamidopropyl hydroxysultaine	5.00%
Lauric diethanolamide	3.00%
Hexylene glycol	2.00%
PEG-7 glyceryl cocoate	4.00%
Ethylene glycol distearate	0.70%
Tetrasodium EDTA	0.30%
Perfume	3.00%
Preservatives, dyes and water	to 100.00%

Finally, it is important that bath foams are well preserved, to prevent any subsequent problems in the market-place. Commonly used preservatives include the esters of p-hydroxybenzoic acid, 2-bromo-2-nitro-1,3-propanediol, isothiazolinone derivatives and 1-(3-chloroallyl)-3,5,7-triaza-1-azoniaadamantane chloride. These materials are often used in combination to provide synergistic preservative effects.

Bath gels, powders and tablets

Gel foam bath products are similar in composition to liquid foam baths, except that they have a higher viscosity. These viscosities are normally produced using higher active detergent levels, in combination with sodium chloride, or by inclusion of thickening agents like PEG 6000 distearate or PEG-55 propylene glycol oleate. A typical foam bath gel formulation is shown below.

Triethanolamine lauryl sulphate	40.00%
Lauroyl diethanolamide	10.00%
Linoleic diethanolamide	7.00%
PEG-75 lanolin oil	5.00%
Tetrasodium EDTA	0.50%
Perfume	3.00%
Preservatives, dyes and water	to 100.00%

Powders and tablets consist essentially of a mixture of foaming detergents, a sequestering agent, a perfume and a filler. Use of non-hygroscopic, free-flowing filler ensures that no caking of individual constituents or the final product will result, upon exposure to a humid atmosphere. Common fillers are sodium chloride, sodium tripolyphosphate and sodium sesquicarbonate. Starch can also be used as a filler and exhibits a high absorption capacity. The sequestering agent, for example sodium hexametaphosphate, is incorporated to reduce or prevent formation of sticky lime soaps and will also increase foam stability in the presence of soap.

The most frequently employed surfactant in foam bath powders is sodium alkyl benzene sulphonate, because of its excellent foaming properties and cost-effective performance. This material is used alone, or in conjunction with other surfactants, such as sodium alkyl sulphates. The perfume in a powder formulation, may be absorbed on to an inert carrier such as starch, which is then blended with the remaining, already mixed ingredients. A typical foam bath powder formulation is illustrated below.

Sodium alkyl benzene sulphonate	55.00%
Sodium sesquicarbonate	31.00%
Corn starch	10.00%
Tricalcium phosphate	2.00%
Perfume	2.00%

Foam bath powders can also be successfully compressed into tablets at low humidities.

After-bath preparations

After-bath preparations include body or dusting powders, after-bath emollients and after bath splashes. Talcs and dusting powders absorb moisture, cooling the skin and imparting a smooth feeling. Talc, hydrated magnesium silicate, is the major ingredient that provides the smooth, velvety, feel on the skin, although products sometimes contain metallic stearates to improve adhesion properties, as well as magnesium carbonate, starch, kaolin or precipitated chalk, to improve absorbency. They also act as carriers for applying perfume to the skin.

After-bath emollients are designed to combat the effects produced as a result of sebum loss, during bathing, and to prevent dry skin. The most popular types are oil-in-water emulsions and hydro-alcoholic emulsions. Although available in a variety of forms, the most popular is the lotion type, sometimes referred to as a moisturising body lotion. These products are usually formulated to rub-in quickly and to provide a soft, smooth, feel. Hydro-alcoholic lotions serve a similar purpose and are sometimes referred to as "bath satins". They also deposit an oily film on the skin but, in addition, produce a more pronounced cooling effect, as the alcohol evaporates from the skin. These products sometimes contain a pearlescent pigment to give the skin a smooth lustre, while also enhancing the appearance of the product. An alcohol-compatible gum may also be included in the formulation, to improve the stability of non-hydro-alcoholic emulsions.

Shower gels

Shower gels are a relatively new addition to the bath products market, although their growth, over recent years, has been dramatic. This type of product was first developed in France and Germany in the mid-seventies but have now spread to most parts of Europe.

Fundamentally, shower gels are highly viscous liquids which, in composition and function, largely resemble conditioning shampoos. Their cleansing function extends to the whole body and the ingredients they contain impart a soft, silky-smooth feel to the skin, after use.

Products typically have viscosities of between 2000-8000 cP, in order to stay on the palms of the hands or sponge in use and not be washed away too quickly under the shower. Like shampoos, shower gels consist essentially of a primary surfactant, which provides detergency and foam, and secondary surfactants which enhance the detergency and modify foam characteristics. Also included are foam stabilisers, conditioners, viscosity enhancers, preservatives, perfume and colourants.

The primary surfactants used are usually of the anionic type, because of their superior foaming properties and cost-effectiveness. Amphoteric surfactants, such as betaines or imidazoline betaines, are often used in shower gels because of their good detergency, mildness and performance as foam boosters/stabilisers. They also have a tendency to produce close-textured foams, a highly desirable attribute for a shower gel product, as well as being capable of reducing the irritation potential of anionics.

A typical conditioning shower gel formulation is shown below.

Sodium laureth sulphate	35.00%
Sodium lauroyl sarcosinate	5.00%
Cocoamidopropyl betaine	10.00%
Cocoamidopropyl hydroxy sultaine	5.00%
Glycerine	2.00%
Polyquaternium-7	3.00%
Citric acid	0.60%
Tetrasodium EDTA	0.25%
Perfume	0.50%
Preservatives, dyes and water	to 100.00%

References

1. R Harry, 'Harry's Cosmeticology', 6th Edition, revised by J B Wilkinson, Leonard Hill, London, 1973, p 239.
2. P Becher & D L Courtney, American Perfumes & Cosmetics, 1965, 80, (11).

THE TEETH AND TOOTHPASTE

Introduction

Teeth are an important factor in personal appearance and the need for well formulated oral hygiene products cannot be disputed. A good oral hygiene regimen will not only enhance the appearance of the teeth, but also reduce the likelihood of premature tooth loss and the accompanying unpleasant facial changes that may take place as a result of it. Dental products therefore play an important part in the maintainance of tooth health and the prevention of tooth disease.

Tooth damage can arise in three fundamental ways. The physical forces exerted on the teeth during the chewing process are surprisingly high and lateral forces, in particular, can ultimately result in tooth damage. Abrasive effects can also play a part in tooth damage, as food is chewed. The microbiological environment within the mouth is also important, in consideration of tooth health. Bacteria found in the mouth are being continually inoculated by the favourable environment and metabolic by-products formed may attack teeth and gums. Thirdly, chemical attack from particularly acid foodstuffs can damage teeth by dissolution of tooth material. Any oral hygiene regimen should aim to minimise or prevent these factors wherever possible.

The tooth and its structure

A diagram illustrating the structure of the tooth is shown in Figure 1.

The enamel is the outermost layer on the crown of the tooth and disappears just below the gum margin. Enamel is the hardest material found in the body and is composed of 95% *hydroxyapatite* ($Ca_5(PO_4)_3OH$), 4% bound water and 1% organic material, principally keratin. The strength of the main component, hydroxyapatite, is derived from its crystalline structure. It can, however, undergo an ion-exchange reaction in the presence of the fluoride ion, producing a material known as fluorapatite, which is much stronger and more acid resistant. This reaction takes place topically at the enamel surface when substances (foodstuffs, oral care products, etc.) are introduced into the oral cavity. Enamel, to some extent, protects teeth against physical damage and chemical or microbial attack. It can, however, be abraded and is eroded at pH values of 5 or less. It is important to recognise that once enamel has been lost from the tooth surface, it cannot be reformed in its original structure, although some remineralisation may occur.

Henkel
Plantaren
A new generatio
of surfactant
from
Cospha
Plantaren stands for
polyglucosides, a new
tion of surfactants from
Nonionic surfactants for
spectrum of applications in
cosmetic industry. Based on
(e.g. maize starch) and fatty
alcohols (e.g. from coconut oil).
Doubly close to nature through
renewable starting materials.
Plantaren is distinguished by:
● very good dermatological compatibili
● good cleaning capacity
● good foaming capacity
● stability across a wide pH range
● viscosity-increasing effects
● outstanding ecological characteristics
As co-surfactants and special surfactants, Plantaren types reflect the trend towards greater compatibility in the cosmetic industry.
Plantaren products, new surfactants with a sophisticated performance profile. They are manufactured in commercial quantities and qualities at the oleochemical centre of the Henkel Group in Cincinnati/USA.
Henkel KGaA · Cospha
Postfach 101100 · W-4000 Düsseldorf 1 · Germany
Represented in GB: Henkel Ltd.
Henkel House · 292-308 Southbury Road
Enfield · Middlesex EN1 1TS
Telephone 01-804-3343
Telex 922708 · Fax 81-443-4392
Cospha
COUPON
I am interested in Plantaren. Please send me, without obligation:
product information ☐
a free sample ☐
Please state full address or simply attach your business card.
COS 392e-2

FIGURE 1

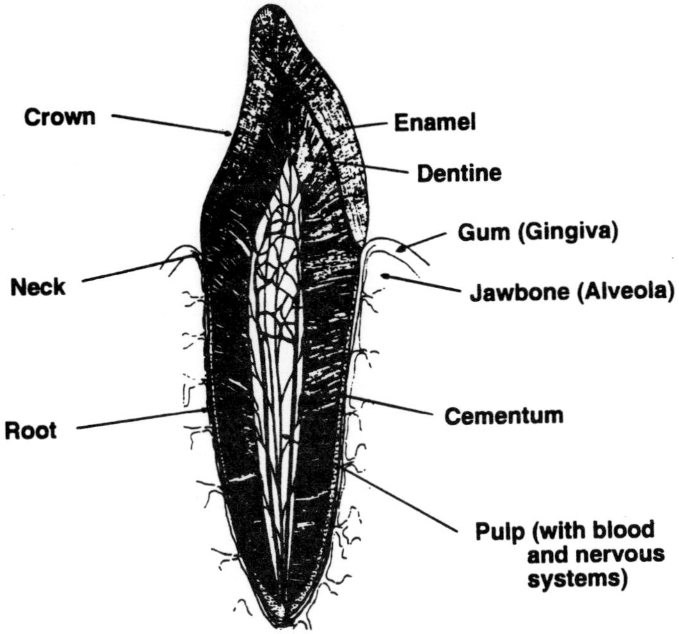

The *dentine* is the layer of the tooth found beneath the enamel surface. This is similar enamel, in that it contains hydroxyapatite but at a lower level. The crystalline structure much less defined and the material more porous. Typically, the composition of dentine 70% hydroxyapatite, 15% organic material, comprising of keratin and collagen, and 5% water. The dentine matrix contains *dentine tubules*, which gives the material its orosity. These tubules can be exposed, particularly in older people, due to enamel orasion and gum recession. Dentine is relatively soft and not particularly resistant to orasion or acid attack and any oral hygiene regime should aim to protect exposed dentine, far as is practicable. The reaction of dentine with topical fluoride is similar to that escribed with hydroxapatite.

The *pulp* is the only part of the tooth that is biologically viable and contains systemic ood vessels and nerve cells. Attack of the pulp results in the occurrence of pain, often ferred to as "toothache".

The root of the tooth is covered by cementum, which has a rough surface. The cementum attached to the jawbone through the periodontal membrane, which acts as a shock osorber for the tooth. The gums provide a lining for the mouth and protection for the tooth here there is no enamel. They are synonymously referred to as the *gingiva*.

Any discussion on the structure of the tooth should not omit mention of saliva which, hilst not being part of the mouth, is of vital importance in oral hygiene. Saliva contains ater, mucoprotein and enzymes. The mucoprotein gives saliva the ability to lubricate the outh, whilst the enzymes not only help the digestion process but also exert a natural anti-bacterial effect on mouth bacteria. The presence of calcium and phosphate ions in the saliva also helps reduce the solubility of enamel in the mouth's environment. Salivary function is critical in controlling levels of bacteria and nutrients in the mouth, thereby providing a measure of protection. Loss of salivary function results in severe tooth decay and the onset of poor oral health.

Oral accumulations

There are several different types of oral accumulation which can occur in the mouth, each of these having an important bearing on overall oral health. The most important types of accumulation are listed below.

1. *Pellicle*

This is an acellular, bacteria-free coating, formed on the surface of teeth, within seconds of contacting saliva and is extremely difficult to remove. It is implicated in the staining of teeth but, but more importantly, provides a substrate for bacterial colonisation leading to the formation of plaque.

2. *Dental plaque*

The importance of dental plaque can not be overstated, being the most significant underlying factor of dental disease. Plaque is a mucous coating composed of bacteria, their by-products and food debris, which reside in the mouth.

3. *Calculus (tartar)*

Calculus is the mineralised deposit on teeth and is formed by the calcification of plaque. There are two types of calculus that occur, supra-gingival (above the gum line) and sub-gingival (below the gum line). Calculus is difficult to remove and normally requires physical removal by a dental practitioner.

4. *Materia alba*

This is the name given to the white, diffuse, loosely attached material on the teeth.

Dental disease

1. *Dental caries*

Dental caries is the condition often referred to as tooth decay. It occurs as a result of the formation of plaque, which consists of proliferating anaerobic bacteria, their by-products and food debris. The bacteria produce several metabolic by-products in the plaque, amongst which are extra cellular polysaccharides. These are sticky substances which enable the bacteria to adhere to the teeth to an even greater extent, any fissures in the tooth surface providing a protected site, making plaque removal very difficult. These bacteria also produce acids (mainly lactic acid) amongst their metabolites, which are in intimate contact with the enamel at the tooth's surface. These acids attack the enamel and so the onset of dental caries occurs. The levels of acid in the mouth, as a function of time, can be investigated using *Stephan curves*, which plot the pH of the mouth environment with time. An example of a Stephan curve is illustrated in Figure 2.

FIGURE 2

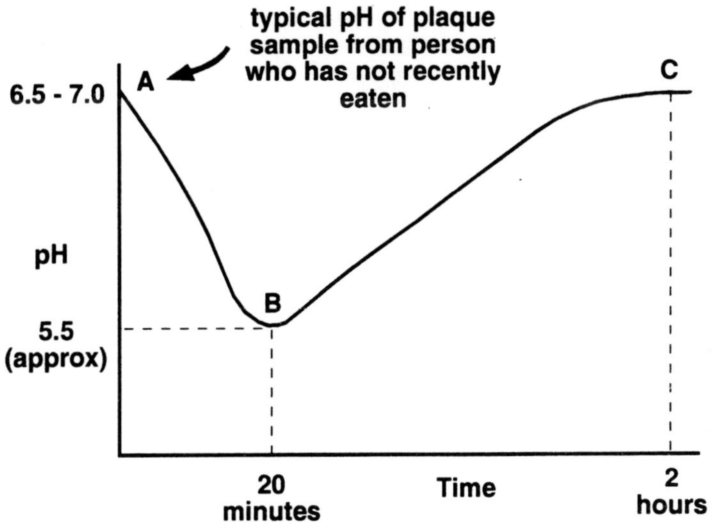

Stephan curves are often used to predict the cariogenicity of different foodstuffs, which is a measure of their potential ability to cause dental caries. In the example above, at time zero, the pH of a plaque sample from somebody who has not recently eaten is measured. The person is then given the food under test and the metabolisis is monitored, as a function of pH versus time. Between the points A and B, the resultant drop in pH is associated with metabolic acid formation. The rate at which the pH drops depends upon the decline in nutrient level, the rate of salivation, the types and level of saliva enzymes present and the presence of calcium and phosphate ions in the saliva, which provide a buffering effect on the mouth. At point B, the minimum pH is recorded and this is the point at which most enamel damage takes place. The curve between points B and C, represents re-equilibration of the mouth, over time, back to its normal state.

The solubility of hydroxyapatite increases significantly at below values of approximately pH 5.5 and the acids produced at this pH will have the most marked effect in areas where the plaque remains undisturbed. The process of tooth decay is gradual, and is accompanied by loss of calcium and phosphate, as illustrated in Figure 3.

Plaque formation in the mouth, and the associated decrease in pH, causes demineralisation of the enamel layer through loss of calcium and phosphate just below the enamel surface. This leads to a change in enamel opacity, often known as a *white-spot lesion*, which is indicative of an incipient carie. The demineralisation below the tooth surface penetrates deeper into the tissue and spreads, finally penetrating the dentine which then undergoes demineralization process. Finally, the enamel surface breaks and a cavity is produced,

FIGURE 3

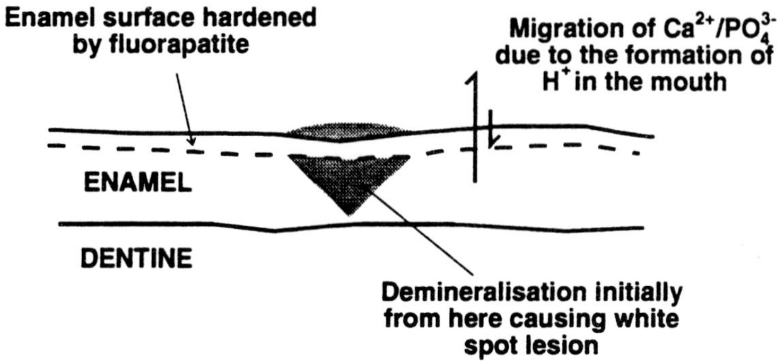

exposing the dentine. Once the dentine has been exposed, decay is rapid and progressive.

The factors affecting the incidence of dental caries are wide and varied but the following are certainly implicated:

- enamel thickness and extent of gum recession
- pattern of eating
- hereditary factors
- the pH of the mouth environment
- the level of plaque formation
- age
- the level of available fluoride
- oral hygiene regimen (plaque removal)

2. *Periodontal disease*

Periodontal disease is a disease of the tissues around the teeth, which gives rise to damage of the gum and jaw-bone. The underlying cause of periodontal disease is also plaque. The mechanism resulting in the occurrence of periodontal disease is illustrated diagrammatically in Figure 4.

With reference to figure 4(a), plaque becomes trapped at the tooth/gingiva junction. Plaque occurring at this point is not easily removed by brushing and will therefore accumulate over time. Not only do the microbial organisms in the plaque produce acids, which are implicated in tooth decay, but also other metabolic by-products known as *toxins*. With reference to figure 4(b), these toxins attack the gums causing reddening and inflammation, ultimately resulting in gum enlargement at the tooth/gingiva junction. This enlargement traps further amounts of plaque in the pocket thus formed in the gum. Over time, this pocket can progress down the gum margin resulting in a condition known as *gingivitis*. In figure 3(c), the enlarged pocket of plaque has produced further amounts of toxins and, in so doing, becomes much bigger. A condition in which the gums become more inflamed occurs, eventually resulting in gum detachment and erosion. This condition

FIGURE 4

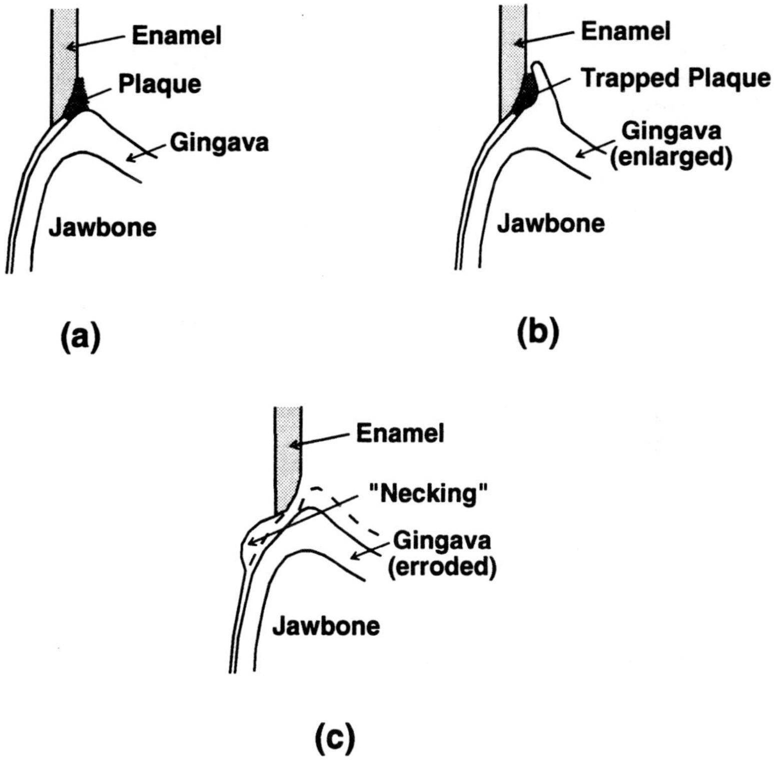

commonly referred to as *gum regression.* Progressive gum erosion results in a more rious oral health condition, in which the dentine exposed starts to erode very rapidly and e tooth starts to neck. Eventually, the lack of attachment causes loss of the tooth.

A further problem associated with plaque formation is the occurrence of calculus or rtar. Calculus is plaque which has been mineralised by the calcium and phosphate ions resent in the mouth and is very hard in nature. The calculus thus formed is very rough, ausing gum abrasion and inflammation. It also acts as a reservoir for plaque, bacteria and eir associated toxins. As such, the presence of calculus also encourages gingival etachment.

Halitosis (mouth odour)

he possible occurrence of halitosis, or mouth odour, is the principal motivation for onsumer use of dentrifices. Most mouth-derived malodours are high in sulphur-contain-g compounds, which are more likely to occur where significant amounts of plaque are und. The origins of the malodour are varied but principally result from some, or all, of e factors listed on page 250.

- production of malodour by the stomach
- production of malodour by the lungs
- lack of salivation during sleep
- food residues
- the presence of plaque

Of these factors, only the last three can be effectively controlled through the use of a goo oral hygiene regimen.

4. *Dental hypersensitivity*

Dental hypersensitivity can be experienced when the root surfaces become exposed as result of gum recession. The exposed dentine provides very limited insulation for th nerves in the pulp against external stimuli, such as temperature extremes. Exposed denta tubules allow the efficient transmission of the stimulus to the nerves, resulting in th condition commonly referred to as "sensitive teeth".

Toothpaste

Toothpastes are cosmetic products, as well as having therapeutic action. As a consequenc of this, toothpastes have been marketed on almost a commodity platform which, in turr has led to relatively low awareness of the benefits that they offer.

Advertising spend on the toothpaste sector is often substantial, as the market is not onl competitive but also very profitable. Because the market is lucrative, manufacturers inves a great deal of money in promotional advertising and special offers. This investment ha been reflected in the fact that high capital spend is often invested in toothpaste researcl

There are many brands and types of toothpaste available, including childrens tooth paste, anti-plaque/gum protection toothpaste, tartar control toothpaste, toothpaste fc sensitive teeth and stain-removing toothpaste, as well as convention fluoride-based famil products.

1. *The Functions of a toothpaste*

The basic functions of a toothpaste are listed below:

- When used properly, with efficient brushing for a suitable period (approximate 1-2 minutes), the toothpaste should clean the mouth and teeth adequately, removin food debris, plaque and stains
- The toothpaste should leave the mouth with a clean, fresh sensation
- The toothpaste should strengthen the teeth and exhibit prophylactic action
- The toothpaste should encourage regular use, leading to good oral hygiene

2. *Toothpaste formulations*

Although many different types of toothpaste are available in the market-place, most o them contain similar types of ingredient which have a certain function to fulfil as part overall product performance. A typical toothpaste formula consists of the following typ of ingredient, at the indicated percentage levels shown on page 251.

abrasive	20-50%
humectant	20-30%
gum/thickener	0.5-2%
foaming agent (surfactant)	1-3%
flavour	1-3%
noncariogenic sweetener	0.1-0.5%
therapeutic agent (optional)	1-10%
preservative	variable
water	to 100%

Each of the components listed above are discussed in further detail below:

2.1. *Abrasives*

Clean teeth very quickly develop a proteinaceous layer known as the *acquired pellicle.* The absence of an effective abrasive in the toothpaste would result in the teeth very quickly becoming non-uniformly discoloured through stain absorption by the acquired pellicle. The toothpaste abrasive must therefore clean the teeth and remove the pellicle but, at the same time, should not abrade the enamel or any exposed dentine in the mouth.

The abrasivity of any material used in toothpaste formulation is therefore critical. When considering the type of abrasive to be used, the following factors should be borne in mind.

2.1.1 *Hardness*

Hardness can be expressed using the Moh scale of hardness, which is a logarithmic scale graded from 0-10. Using this scale, enamel has a hardness value of 4-5, dentine a value of 2-2.5 and the acquired pellicle a value of approximately 1. Therefore for an abrasive to be effective on stained pellicle, yet exert minimum damage to the dentine and enamel, it must have an accurately pitched Moh value of approximately 2.0.

2.1.2 *Particle size*

To work effectively, the abrasive particle size must be within a given range. If the particles are too small, they will not abrade the stained pellicle. If they are too large, insufficient particles will be available around the toothbrush bristles to abrade the stained pellicle efficiently. Most abrasive materials exhibit a normal distribution curve of particle sizes. A particle size of $<2\ \mu$ will not produce effective abrasive action, whilst a particle size of 20 μ or more will result in the toothpaste feeling gritty in the mouth. Ideal toothpaste abrasives should therefore have a particle size distribution within the range of 2-15 μ and this should be reflected in the normal distribution curve of any abrasive used in a toothpaste product.

2.1.3 *Particle shape*

In addition to particle size, particle shape is also an important factor in the determination of toothpaste abrasivity. A fairly regular, smooth, spherical particle will move over the tooth surface without abrading the acquired pellicle effectively. An abrasive particle with an irregular surface will be much more effective at removing acquired pellicle through the resultant "ripping" action which takes place during brushing.

2.1.4 *Compatibility with fluoride*

The interchange of the fluoride ion, F^-, with enamel to form fluorapatite is not a kinetically favourable reaction. Some abrasive minerals, particularly those based on calcium phosphate, can compete for the available fluoride in the mouth, thus rendering the toothpaste less effective as a fluoride tooth protectant.

2.1.5 *Cost and availability*

Any material used as an abrasive in a commercial toothpaste product must be readily available in the right quality and at an economically viable cost.

Several different types of material are typically used as toothpaste abrasives. One of the most commonly used materials is dicalcium phosphate dihydrate (DCPD) which has the chemical formula, $CaHPO_4 . 2H_2O$. It is white in colour and exhibits good flavour stability. However, two main disadvantages occur when using calcium phosphate dihydrate. Firstly, it has an abrasive hardness similar to that of dentine and can therefore result in dentine damage, particularly in the case of over-zealous brushing. Secondly, it does not exhibit good fluoride compatibility and sodium monofluorophosphate is the only source of fluoride that can be practically used.

Calcium carbonate ($CaCO_3$) is a very cheap abrasive and is readily available in many grades and qualities. Although its alkaline pH is beneficial for resisting acid formation and build-up in the mouth, its chief drawback is the availability of calcium ions which reduce the effectiveness of any available fluoride in the formulation. Once more the most suitable source of fluoride, if calcium carbonate is to be used as the abrasive, is sodium monofluorophosphate.

One of the most widely used abrasives in the past was hydrated alumina ($Al_2O_3 . 3H_2O$). This material is readily available as a white amorphous solid and is very cost-effective. Some compatibility problems exist in the presence of the fluoride ion but these are not significant and are fairly easily overcome.

The newest abrasive material to find general use in toothpaste products is silica (SiO_2) either as is or in modified form. Silicas can be manufactured by many different methods offering the formulator a wide variety of particle sizes in very pure form. Silica can also be made porous, causing particles to break-down upon the application of shear, which effectively results in a self-adjusting particle size distribution during use. It also exhibits good compatibility with fluoride ions and all types of therapeutic agent. One unique property of silica is that it has a refractive index which is very low, enabling the formulator to match the refractive index with that of the liquid phase in the product. This results in a clear-gel type of toothpaste, which has become more popular over recent years. The main drawback with silica is its price, putting it at a significant commercial disadvantage when compared with materials such as dicalcium phosphate dihydrate and calcium carbonate.

2.2. *Humectants*

The liquid phase of a toothpaste consists of a humectant and water. The presence of th humectant is to prevent excessive drying out of the product, which would occur if only water were used as the vehicle. Excessive drying out of the product causes a viscosity increase and eventually solidification of the product. The humectant must have an affinity for water, thus helping the toothpaste formulation to retain water. Typically used humectants are glycerin, sorbitol and propylene glycol. Although materials of this type are highly effective at preventing excessive dry out in a toothpaste formulation, they are also expensive. In practice these ingredients are mixed in specific ratios with water, to give the optimum performance/cost ratio. Humectants do not affect the efficacy of a toothpaste whatsoever, although they can sweeten the flavour slightly.

2.3. *Foaming agents*

All toothpastes contain detersive surfactants, or foaming agents, at low levels. The function of these materials is not to cleanse the teeth but rather to give the product a pleasant feel in the mouth during use. The relatively low amount of foam produced also helps create a stable suspension of abrasive in the mouth, thus making it available for cleaning the teeth effectively.

Examples of commonly used foaming agents are sodium lauryl sulphate ($C_{12}H_{25}OSO_3Na$) and sodium N-lauroyl sarcosinate ($C_{11}H_{23}CON[CH_3]CH_2COONa$) which is said to possess anti-enzyme properties.

2.4. *Thickeners*

Some form of thickening agent is essential in a toothpaste formulation for two reasons. Firstly, the product would be of too low a viscosity, resulting in separation of the liquid and solid phases. Secondly, in the absence of a thickening agent, the toothpaste viscosity would be totally dependent on inter-particulate spacing, resulting in a moisture-loss sensitive product which would exhibit variable viscosity on storage. The function of the thickener is therefore to increase product viscosity and bind the abrasive particles together. Thickener levels must be very carefully controlled if product separation or excessive viscosity build-up is to be avoided.

Originally, natural products were often used as toothpaste thickeners and materials such as alginates and carrageenates were often incorporated. Modern toothpaste formulations use more sophisticated thickeners such as refined bentonite, silica and sodium carboxymethylcellulose. The latter can be used to give a variety of rheological characteristics to the toothpaste formulation.

2.5. *Flavour*

Flavour is a very important ingredient in toothpaste, providing brand signature and representing an important factor in motivating consumer purchase. Originally, only peppermint and spearmint oils were used to flavour toothpaste but modern flavours are much more complex with individual subtleties being created.

The importance of a toothpaste flavour lies in its ability to give fresh breath or a pleasant mouth sensation after brushing. Most flavours are based on peppermint or spearmint. Both contain menthol at significant levels, which provides the toothpaste with a cooling aftertaste. Peppermint and spearmint oils still form the basis of most toothpaste flavours.

Peppermint oil is commonly produced from two different sources. *Mentha arvensis* is grown in large quantities in South America and China and provides a crude peppermint flavour. *Mentha piperita* is grown in fewer parts of the world, mainly in North America, and forms the basis of many oils used in toothpastes. *Mentha piperita* contains menthone, menthyl esters and menthofuran. Spearmint oil is mostly produced in North America and contains carvone, limonene and other minor ingredients. Both peppermint and spearmint oils are extracted by a crude distillation process and refined using more sophisticated distillation techniques.

Toothpaste flavours are inherently slightly bitter in taste and commonly a non-cariogenic sweetener is added to compensate this effect. The most common sweetener used is sodium saccharin, normally incorporated at levels of 0.2%-0.4% (w/w). Other sweeteners such as sodium cyclamates are also used.

2.6. *Preservatives*

Toothpaste formulations must be microbially stable products, although the likelihood of microbiological insult during use is not as high as with some other types of product. To some extent, toothpaste is self-preserving, due to the presence of the humectant in the product. The humectant binds the available water in the product which, by a mechanism of adverse osmosis, will inhibit the growth of microbes.

Some of the ingredients in toothpaste do, however, support microbial growth. A good example is the thickener, sodium carboxymethylcellulose. During manufacture of the toothpaste, this material is normally added in the form of an aqueous pre-mix, which is highly susceptible to microbial growth. Bacterial attack on this pre-mix, prior to its inclusion in the product, can result in the formation of an enzyme by-product known as *cellulase*. If a pre-mix containing this enzyme is added to the product during manufacture, subsequent enzymatic attack on the thickening system in the toothpaste can occur, ultimately resulting in complete product spoilage. In order to minimise occurrence of this problem, cellulose pre-mixes of this type are often preserved, particularly if they are to be kept for any length of time.

For these reasons, it is important that not only the toothpaste but also any microbiologically susceptible ingredients are effectively preserved against microbial challenge.

2.7. *Therapeutic agents*

Therapeutic agents are added to toothpaste formulation to address one or more of the following problems:

- the formation of dental caries
- periodontal disease
- sensitive teeth

Each of these problems, and the ways in which they can be addressed, are discussed below.

2.7.1. *Prevention of dental caries*

Therapeutic agents used for the prevention of dental caries are mostly fluorides. Early research in the 1950s enabled dental workers to establish a definite correlation between the level of fluoride in drinking water and the incidence of dental caries. These discoveries lead to the inclusion of fluoride in toothpaste formulations to provide similar benefits.

Historically, three main types of fluoride have been used in toothpaste formulation, annous fluoride, sodium fluoride and sodium monofluorophosphate (Na_2MFP). Early othpaste products often incorporated stannous fluoride, which provided a convenient ay of avoiding the compatibility problems that exist between sodium fluoride and some orasive materials. Nowadays, stannous fluoride is hardly ever used. The most commonly sed therapeutic agent in modern toothpaste formulations is sodium monoflourophosphate, s it avoids many of the incompatibilities that occur when using sodium fluoride.

The concentration of fluoride ion in toothpastes is normally around 1000 ppm, which ields a fluoride contact concentration on the teeth of just over 300 ppm, assuming a othpaste:saliva ratio of 1:2 during brushing. A representation of the equilibrium that xists in the mouth, during brushing and immediately after brushing, is illustrated in igure 5.

FIGURE 5

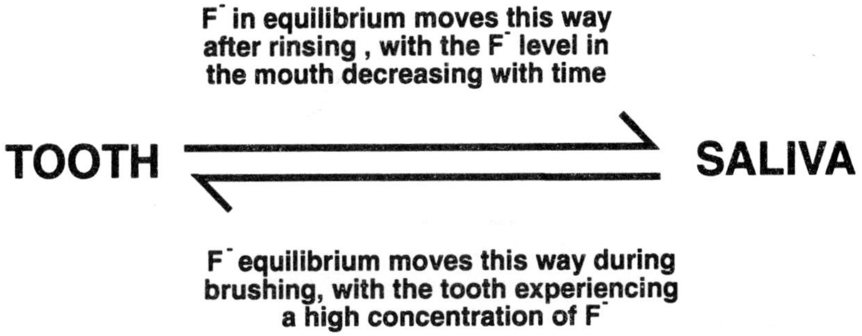
F⁻ in equilibrium moves this way after rinsing , with the F⁻ level in the mouth decreasing with time
TOOTH
SALIVA
F⁻ equilibrium moves this way during brushing, with the tooth experiencing a high concentration of F⁻

Many toothpaste products in the UK make therapeutic claims and therefore require a oduct licence. The maximum allowable fluoride concentration in toothpaste products is 500 ppm and the requirement of a product licence necessitates clinical trials during oduct development. An outline of clinical trial methodology is given later in this chapter.

7.2 *Prevention of periodontal disease*

Most modern toothpaste products claim benefits in the reduction of periodontal disease d its precursor plaque. Plaque is harmful to the teeth for two reasons. Firstly, the bacteria the plaque produce harmful toxins as by-products and, secondly, plaque is readily ineralised to produce tartar or calculus on the teeth.

The most obvious method for retarding the damaging effects of plaque, is to include an ti-microbial agent in the toothpaste, thereby preventing plaque formation. In practice, owever, this is extremely difficult as very few toxicologically safe anti-microbial agents ith the required properties exist. Any anti-microbial agent used must be compatible with her ingredients in the toothpaste formulation and should be substantive to teeth to ensure ficacy.

Chlorhexidene, normally incorporated as the gluconate salt, is a particularly suitable anti-plaque agent for toothpaste formulations. It is substantive to the tooth surface, thereby enhancing its efficacy. Despite this fact, chlorhexidene exhibits many disadvantages in use. It is incompatible with many other toothpaste ingredients, possesses an extremely bitter taste and can result in staining of the teeth.

Another substance which is used as an anti-plaque agent is zinc citrate trihydrate, which has been shown to inhibit plaque growth, thus preventing bacterial colonisation. Zinc citrate trihydrate itself, however, is not an anti-microbial agent. The efficacy of zinc citrate can be greatly enhanced by combining it with the antimicrobial agent 2,4,4'-trichloro-2'-hydroxydiphenylether, more often referred to as triclosan. Triclosan is effective against a broad spectrum of yeasts and bacteria, including those implicated in the occurrence of gingivitis.

The other commonly used anti-plaque ingredient is cetyl pyridinium chloride (CPC) which exhibits both anti-bacterial action and substantivity to teeth.

Enzymes have, in the past, also been used as anti-plaque agents. The enzyme enhance the production of saliva in the mouth, thereby interfering with the metabolism of plaque formation. Despite this, very few commercially available anti-plaque toothpastes, based on enzyme action, have been produced.

Although there are a variety of methods for evaluating materials as anti-plaque agents they all essentially fall into two types.

2.7.2.1. *In vivo studies*

Plaque forms very quickly in the mouth and is therefore relatively easy to monitor *in vivo*. Initially, the subjects teeth are treated by a professional clinician to bring the plaque down to a known base level. The subjects are then required to use the test product for a period of time, after being given cariogenic foods, and the rate of growth of plaque is regularly monitored over time. Plaque monitoring is carried out with the aid of plaque disclosant tablets, by measuring the area of stain on the teeth. Results obtained for the test product are compared with a placebo control.

2.7.2.2. *Invitro studies*

These are all essentially laboratory techniques, based on various methods of microbiological analysis.

2.7.3. *Relief of sensitive teeth*

People with sensitive teeth are usually those who have exposed dentine, due to gingival recession. The pain arises because of the transmission of stimuli, such as variation in temperature, at the exposed dentine surface to the pulp nerves, via the dentinal tubules. This mechanism is illustrated in Figure 6.

This pain sensation can be prevented by the dentist putting a lacquer over the exposed dentine, to block the dentinal tubules. There are, however, various ingredients which can be added to toothpastes which provide benefits to sensitive teeth sufferers. Early formulations used formaldehyde at levels of approximately 0.5% but these have been superseded due to the toxicological profile of formaldehyde and its doubtful side-effects.

Modern sensitive toothpastes typically use either strontium chloride at levels of approximately 10.00% (w/w) or potassium nitrate at levels of around 5.00% (w/w). Th

FIGURE 6

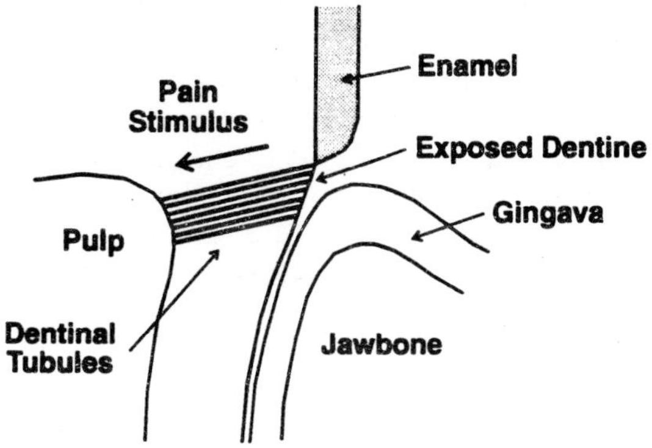

mechanism of desensitisation is not fully understood but a "blockage" of the dentinal tubules is thought to occur when these materials are incorporated into the formulation.

. Measurement of toothpaste abrasivity

Toothpaste abrasives must remove the stained pellicle without removing enamel or dentine.

Historically, many different *in vitro* methods for measuring toothpaste abrasivity have been developed, although currently the two most widely accepted are the surface profilometry and radio-tracer methods.

.1. Surface profile method

This involves controlled brushing of teeth, *in vitro*, and subsequent monitoring of the loss of tooth material. Using this method, a healthy tooth is firstly sectioned to give a dentine or enamel surface, depending on the study. The tooth is held by mounting it in epoxy resin, the exposed surface being ground to an optical flat. The surface of the tooth is then measured using a *profilometer*, to measure the smoothness of the ground surface. The surface of the tooth is then brushed in a controlled manner, as illustrated in Figure 7 below.

FIGURE 7

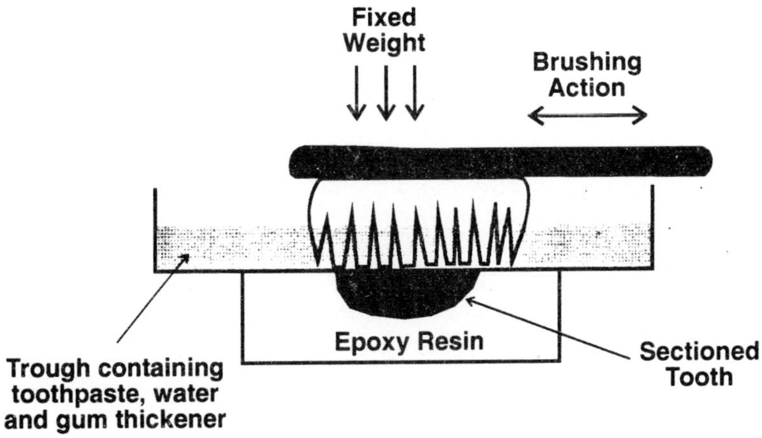

After brushing for a fixed period of time, often designed to represent exaggerated brushing conditions, the tooth surface is re-examined with the profilometer for smoothness. Examination of the profilometer chart enables calculation of the amount of tooth material lost, to be made. Comparison of the test product with a standard compound of known abrasivity, enables the abrasivity of the test formulation to be quantified.

The advantages of this method is that the actual change that takes place at the tooth surface can be measured, by comparison of the surface profile before and after brushing. The main disadvantage is that it is extremely difficult to locate the same area of the tooth for measurement, both before and after the brushing operation.

3.2. Radio-tracer method

This method also involves determination on a section of tooth, as described earlier. The sectioned tooth is irradiated with a neutron source, causing a proportion of the phosphorus in the hydroxyapatite to become radioactive, forming ^{32}P isotope. The tooth is then brushed, in a controlled fashion, as described above. After brushing, the slurry from the brushing is collected and its radioactivity measured. The amount of tooth material removed by brushing is directly proportional to the level of radioactivity detected. Quantification is normally made by comparison with a toothpaste formulation of known abrasivity.

The main advantage of this method is that it is not necessary to locate the same point on the tooth, before and after brushing, to make the measurement. Disadvantages are that the method gives no indication as to the nature of the changes in the tooth surface and the treatment of the tooth with radiation can change the hardness of the tooth material itself

3.3. *In vivo measurement of toothpaste abrasivity*

Sometimes abrasivity is measured using *in vivo* methods. Firstly, a replicate of the subjects' teeth is taken and, using these moulds, a positive impression of the teeth is made using special materials. This process can be repeated over a period of time and tooth wear can be monitored using a profilometer on a series of positive impressions. In this way, gum recession, dentine exposure and other oral characteristics can also be monitored.

Measurement of toothpaste efficacy

Toothpaste efficacy can either be measured *in vivo*, through the use of clinical controlled studies, or *in vitro* using several different methodologies.

1. *Clinical trials*

The only method for satisfactory substantiation of claims for toothpaste efficacy in the prevention of dental caries is a *clinical trial*. Such trials normally last up to three years. The test subjects must show maximum prevalence for dental caries and, in practice, this normally means using school children between the ages of 7 and 11 years old. The number of subjects used in the trial must also be large, to guarantee statistical significance, and typically between 1000 and 1500 children are used.

The methodology for performing a clinical trial firstly involves the recruiting of a panel of subjects, normally taken from local schools, and carrying out careful examination of their teeth. Examination is carried out firstly by a dental practitioner, who assigns a DMFS (diseased/missing/filled/surfaces) score for the teeth, and secondly by examination using clinical radiographs to reveal incipient caries. The test subjects' are then randomised into two groups, one of which is asked to use the test toothpaste product containing fluoride. The other group is given a non-fluoride containing control product. Both groups are then monitored, with examinations of the teeth taking place at regular intervals, over the period of test.

Clinical trials are very expensive to conduct and are becoming more difficult to carry out from an ethical standpoint. Historically, the result obtained from clinical trials have, however, shown up to 30% fewer incidence of caries when using a fluoride containing product.

2. *Invitro methods for studying toothpaste efficacy*

Whilst a clinical trial provides an *in vivo* method for studying the benefits of fluoride toothpastes, there are also *in vitro* methods available which are often used to pre-screen products before embarking on expensive *in vivo* studies. Two methods commonly used are:

2.1 *Chemical analysis*

In this method, pure hydroxyapatite mineral is subjected to controlled contact with the toothpaste product, the contact time being carefully selected to give good correlation with *in vivo* studies. The uptake of fluoride ion is then monitored analytically, whereupon it can be related to the formation of fluorapatite.

2.2 *Enamel solubility*

A compressed disc of hydroxyapatite is ground to an optical flat, before being exposed to the test toothpaste product, in slurry form, using a carefully controlled contact time. The optical flat is then rinsed and treated with a fixed amount of dilute acid. The acid solution is then decanted and its phosphorus content determined. This analysis provides a measure of the amount of hydroxyapatite dissolved from the flat by the acid. Results from the test product and a non-fluoride containing control are then compared.

Toothpaste packaging

Although aluminium tubes were historically used for packaging almost all toothpastes, these have now been totally superseded by laminated tubes. The exact types of laminate used vary but are typically aluminium sheets that have been paper-printed and sandwiched between sheets of polyethylene. Polyethylene alone cannot be used for tube manufacture, due to the so-called "memory" of the plastic, which describes its tendency to revert to its moulded form when distorted. The aluminium in the laminate overcomes this difficulty and also provides a barrier to prevent moisture and/or flavour loss.

Laminated tubes are sealed thermally, the polythene on the inside surface of the laminate providing two adjacent surfaces which can be sealed, under pressure, in this way. The polythene on the outside of the tube increases aesthetic appeal and provides protection for the pack graphics, which are printed on to the paper layer in the laminate.

In more recent years, toothpaste dispensers have grown very popular, as an alternative method to tubes for packaging of toothpaste. Dispensers can be obtained in several different types but most of these operate using a ratchet and piston dispensing system.

NOVOSPRAY / SCHA LER

THE TECHNOLOGICAL LEADER IN COMPRESSED GAS AEROSOLS

To dispense your products as:

GAS · SPRAY · LIQUID · PASTE

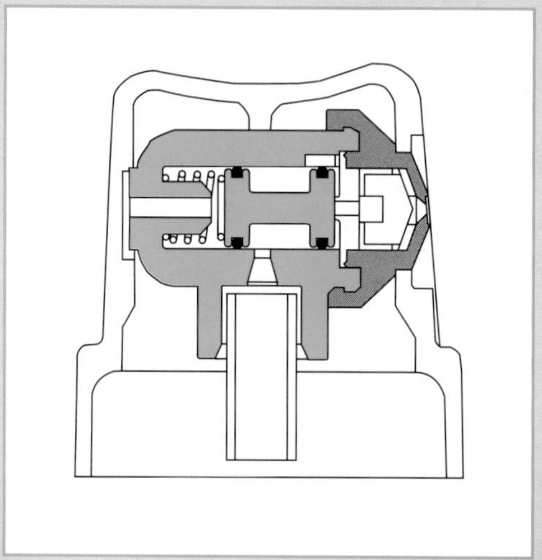

OFFERS TECHNICALLY SUPERIOR PERFORMANCES :

- Constant flow rates
- Constant particle size
- 100% of your product dispensed
- No contamination of your product
- Entire system fits inside the actuator button

PROVIDES CUSTOMER BENEFI

- New range of active ingredients
- Soft spray characteristics
- Cannisters can be used in any pos
- No cold effect from evaporation of volatile gases
- Non toxic, environmentally pure, not inflammable gas

A Compressed Gas Aerosol System using a NOVOSPRAY Regulator offers flexib solutions to help you improve the delivery performance of traditional products opens up new application opportunities for new products.

SECTION 6

Packaging

AEROSOLS
PACKAGING

AEROSOLS

Aerosol - a definition

An aerosol or pressurised product may be defined as “a liquid, solid, gas or mixture thereof, discharged by a propellant force of liquefied and/or non-liquefied compressed gas, usually from a disposable dispenser, through a valve”. The term *concentrate* is used to describe all of the components of an aerosol product, with exception of the propellant.

A liquefied propellant is defined as a liquefied gas, with a vapour pressure greater than atmospheric pressure, at a temperature of 105°F. The term compressed gas propellant is used to describe compressed gas systems and is used in addition to the term liquefied gas propellants.

The history of aerosols

The concept of aerosols was developed in the mid-1920s, although the aerosol did not become a realistic proposition until after World War II. From the end of World War II through to the early 1960s, much work was carried out on the development of new aerosol containers and valve systems and gradually aerosol packaging components became available in reasonably large numbers. The growth and development of the aerosol, as a consumer product, continued through the 1950s and 1960s and by the early 1970’s aerosols had become an established consumer product, particularly in the USA where the largest market, by far, existed.

In the late 1960s and early 1970s the overwhelmingly popular choice of aerosol propellant was the chlorfluorocarbons, commonly referred to as CFC’s. At that time, some researchers had already raised concerns about the effects of CFC’s on the environment, particularly their impact on the ozone layer in the Earth’s atmosphere. This concern continued throughout the early 1970s, until in 1974 a paper published by two scientists at the forefront of this investigation, Professor Rowland and Dr Molina, had a drastic impact on the US aerosol market. The Rowland and Molina paper claimed that CFC’s could rise into the Earth’s atmosphere and, under the influence of short wavelength UV radiation, react with the atmospheric ozone, thus causing its depletion. At the time, public reaction in the US to this theory was significant and, in an attempt to retain consumer confidence in the aerosol concept, many manufacturers moved away from CFC propellant systems into other alternatives, principally hydrocarbons. Unfortunately, in the absence of properly controlled development work with these alternative propellant systems, many products appeared on the US market which were, at best, of poor quality and, at worst, potentially dangerous to the user.

This event was reflected in consumer purchase of the aerosol and for the first time since its inception the US aerosol market in 1975 declined significantly from the previous year's volumes. Nevertheless, as the functionality of products with alternative propellant systems improved, consumer confidence was regained and the aerosol market started to grow once more. By 1979, the use of CFC propellant systems in cosmetic aerosols had been banned in the US and hydrocarbon propellant systems became the prime choice for the aerosol formulator.

In Europe, the use of CFC propellant systems continued until the mid to late 1980's. In 1987, the so-called Montreal Protocol was signed by over 20 participating countries, including those from Europe, agreeing to restrict the use of CFC's as aerosol propellants in cosmetic products. The decline in the use of CFC's in the UK came rapidly and by 1990

TABLE 1

Aerosol category	*Volume (10^6 Units)*
Insecticides	27.5
Paints & Lacquers	20.0
Air Fresheners	59.0
Waxes & Polishes	47.5
Oven Cleaners	7.0
Starches	12.0
Other Household	10.0
Hairsprays	144.0
Hair Styling Mousse	51.0
Colognes & Perfumes	35.0
Deodorants, Deo Colognes, AntiPerpirants	183.0
Shaving Foam	42.0
Other Personal Care	6.0
Medicinal & Pharmaceutical	62.0
Automotive Products	19.0
Industrial	18.0
Miscellaneous	28.0
Total 1990 Volume	771.0

UK Aerosol Production Volume Figures for 1990

almost all consumer aerosol products were propelled by alternative systems, principally combinations of butane and propane hydrocarbons.

In the UK, the aerosol market is still very buoyant, as indicated by the aerosol production figures for 1990, shown in Table 1.

Perhaps one of the reasons that the aerosol remains popular, is that no truly acceptable alternative to this form of product delivery has been developed. At the time of writing, concern is growing over the use of volatile organic compounds (VOC's) and the environmental impact that such materials have. The vast majority of aerosol propellants in use today, principally butane and propane hydrocarbons, fall clearly into this category. What effect any future legislation on the use of VOC's has on the aerosol industry, remains to be seen. Perhaps the future of the aerosol is under question but currently the only well-established alternative to the aerosol, the pump spray, has not yet enjoyed full consumer acceptance.

Classification of aerosols systems

Fundamentally, spray aerosols fall into two distinct classes, two-phase aerosols and three-phase aerosols. A third type of aerosol product, the mousse, is now commonly available, particularly in the hair care market. This can be classified as a special type of three-phase aerosol system.

1. *Two-phase aerosols*

Two-phase aerosols refer to systems where the propellant and the concentrate are totally miscible with each other, the two phases being defined as the propellant/concentrate mixture and the propellant vapour in the head space. A diagrammatic representation of a two-phase aerosol is illustrated in Figure 1.

FIGURE 1

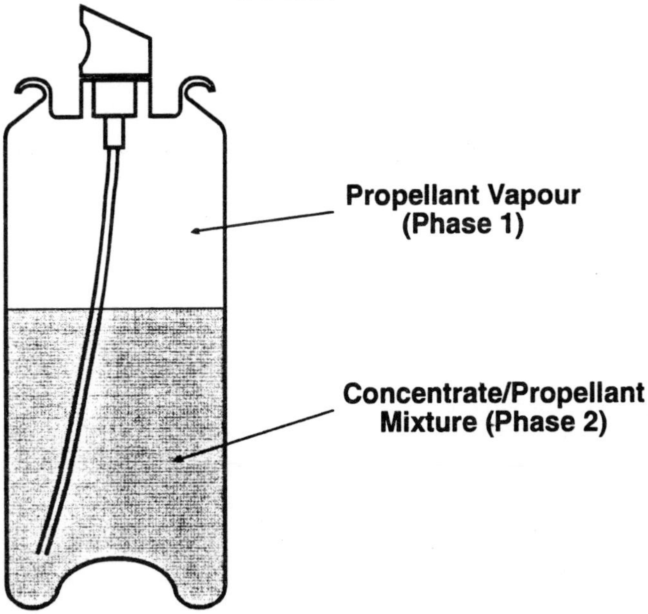

FIGURE 2

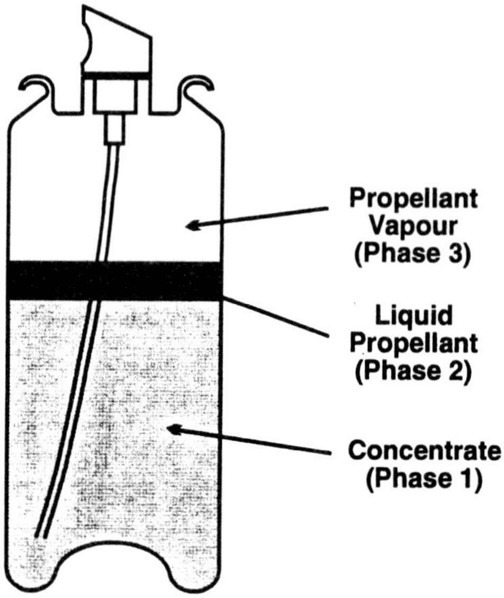

2. *Three-phase aerosols*

Three-phase aerosols refer to systems where the propellant and the concentrate are either totally immiscible or only partly miscible with each other. In this case, the three phases are defined as the concentrate, the liquid propellant phase and the propellant vapour in the head space. A diagrammatic representation of a three-phase aerosol is illustrated in Figure 2.

3. *Aerosol mousses*

Aerosol mousses are essentially three-phase aerosol systems, in which the concentrate is an oil-in-water emulsion. The propellant in the aerosol is immiscible with the external water phase of the concentrate but is soluble in, or miscible with, the internal oil phase. This system is represented diagrammatically in Figure 3.

When the aerosol is activated, the emulsion concentrate is dispensed through the aerosol valve and the propellant dissolved in the internal oil phase of the emulsion, expands rapidly upon exposure to atmospheric pressure. The expansion of propellant, in the presence of the surfactants present in the emulsion concentrate itself, causes the immediate and rapid formation of a foam, or mousse. The mousse is a convenient and clean way of dispensing certain types of product and the foam characteristics can be readily modified to suit the individual purpose.

For example, if concentration or type of propellant is changed, foam of varying stability is formed. Generally, the foam will become firmer and drier with increasing quantities of propellant and higher propellant pressures. The appearance and stability of the foam can be modified by adding various types of thickening agent, solvent or surfactant to the mousse concentrate. The addition of solvents, such as ethyl alcohol, will produce a looser

FIGURE 3

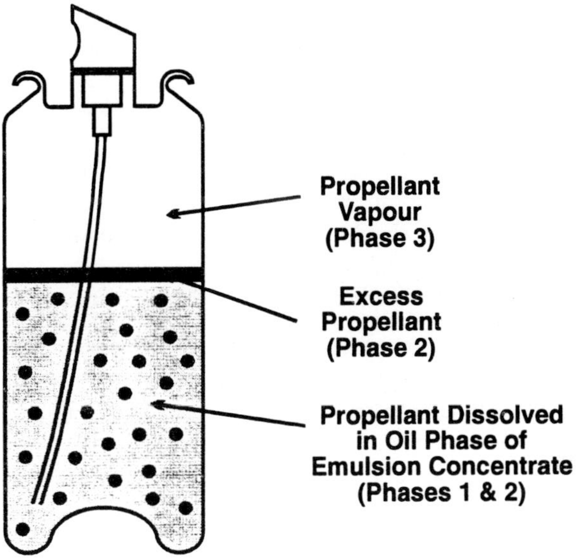

oam structure which collapses fairly quickly, a feature used to advantage in the ormulation of some aerosol mousses. Conversely, increasing the level of surfactant in the nousse concentrate will generally produce a thicker, creamier foam structure, with a much igher degree of stability.

Typically, mousse compositions contain between 85% to 95% of the emulsion concentrate and between 5% and 15% propellant.

Water-in-oil concentrates can also be delivered in aerosol form but such products do not orm mousses. With this type of product, a fine spray pattern is normally obtained, as a arge proportion of the propellant is dissolved in the external oil phase of the emulsion. Vater-in-oil emulsion aerosol sprays have very little application in the field of cosmetics nd toiletries and perhaps the most commonly known use of this technology is in furniture olish.

erosol propellants

efore reviewing the specific types of propellant in use today, it is instructive to look at e various requirements of an ideal propellant system, as follows:

- the boiling point and vapour pressure characteristics of any propellant must be such that the pressure inside the aerosol can is sufficiently high to expel the contents, when the system is opened to atmospheric pressure
- the propellant should be safe, with low contact and inhalation toxicity
- the propellant should ideally, but not essentially, be non-flammable. Highly flammable propellants require special handling during manufacture and the use of flame-proof equipment throughout. High flammability also poses a potential hazard to the end-user, particularly under conditions of mis-use

- the propellant should have an innocuous odour - this is particularly important in the field of toiletry and personal care products
- the propellant should be obtainable in grades of high purity and should be of consistent quality
- the propellant should have good solvent power and compatibility, in a wide range of raw materials
- the propellant should exhibit a high degree of chemical stability and yet be fully biodegradable, without causing harm to the environment
- ideally, the propellant should be available in a wide range of pressures, giving maximum flexibility of formulation
- the propellant should be cost-effective and available in commercial quantities

Since the demise of chlorofluorocarbon propellants in cosmetic products, by far the most common type of propellant systems used today are based on hydrocarbons. Other propellants, including dimethyl ether and some selected fluorocarbons, enjoy some usage, but volumes are still very small. In more recent years, the interest in compressed gas propellants has grown, in light of the increasing concerns over the use of hydrocarbon VOC's. Each of these propellant types will now be discussed in more detail.

1. *Hydrocarbon Propellants*

By far the most commonly used hydrocarbon propellants are propane (C_3H_8) which has a vapour pressure of 728 kNm^{-1} (105.6 psig) at 20°C and n-butane (C_4H_{10}), with a vapour pressure of 106 $kNm^{-1}C$ (15.4 psig) at 20°C. Somewhat confusingly, "Butane" is the industry term used to describe mixtures of butane and propane, mixed in different proportions, to give the required vapour pressure. These mixtures are commonly available with vapour pressures of 30 psig, 40 psig and 48 psig, these grades normally being referred to as Butane 30, Butane 40 and Butane 48 respectively. The compositions of these three propellant blends are given in Table 2.

From this data, it will be noted that as the propane/butane ratio is decreased, the vapour pressure of the propellant mixture also decreases. The very high vapour pressure of propane normally precludes its use as a propellant in its own right and it is frequently mixed with other propellants to obtain acceptable delivery pressures.

TABLE 2

	Butane 30	*Butane 40*	*Butane 48*	*B.Pt.*
Propane	10.6%	22.1%	30.8%	–42.1°C
iso-Butane	28.6%	24.0%	22.9%	–11.7°C
n-Butane	60.2%	53.5%	45.8%	–0.5°C
iso-Pentane	0.6%	0.4%	0.5%	27.9°C

Compositions and Boiling Points of Commercially Available Butane Propellants

Perhaps the reason that the hydrocarbon propellants are so popular, is because of their ·w toxicity and high chemical stability, combined with the fact that they can be readily ixed with other organic solvents, without any compatibility problems. Hydrocarbons are so relatively cheap to produce, making them a cost-effective choice for the formulator.

One of the major disadvantages of hydrocarbon propellants is their flammability and otentially explosive nature. Whilst this is not a major problem in the consumer market-lace, except in the case of mis-use, the handling and safety requirements for a manufac-ring environment can pose severe problems.

More recently, environmental concerns have been expressed about the use of hydrocar-ons and other compounds which may be classified as VOC's. VOC's evaporating into ie Earth's atmosphere are said to be a major contributory factor in the so-called greenhouse effect", a term used to describe the phenomenon of global warming. lthough no firm legislation regarding the use of hydrocarbon propellants exists at the me of writing, it is likely in the future that their use in aerosol products will steadily ecline.

Major suppliers of hydrocarbon propellants include Calor, Shell and BP.

. Dimethyl ether (DME)

lthough dimethyl ether (DME), CH_3–O–CH_3, has been available for many years as an erosol propellant, it is still only used in very limited volumes, except in Holland where s use is more widespread. DME is a gas, with a freezing point of –24.8°C and a vapour ressure of 460 kNm^{-1} (66.7 psig) at 20°C. DME has a good toxicity profile and, like the ydrocarbon propellants, it is highly flammable. Perhaps the most unique property of ME is that it is totally miscible with water, a property that can be used to modify ropellant pressure and reduce flammability simultaneously. As such, fine particle sprays ith low flammability levels can be achieved with DME/water mixes. DME shows no ropensity for damage of the ozone layer.

It is difficult to understand why DME has not become more popular as an aerosol ropellant, particularly since the restriction on the use of CFC's. One factor may be cost, ME being much more expensive than hydrocarbon propellant types. DME is also lassified as a VOC and any future legislation relating to the restriction of these materials ould preclude its future use in cosmetic applications.

Major suppliers of DME include Air Products, Du Pont and Aerofako.

. Fluorocarbon propellants

Inlike CFC's, fluorocarbon propellants, containing carbon, hydrogen and fluorine only, o not show a propensity for damaging the ozone layer. The remaining properties of the uorocarbons are very similar to those of the chlorofluorocarbons. They have excellent olvent power, show good chemical stability and compatibility and are available in high urity grades. Many of the fluorocarbons have safe toxicity profiles, as indicated by their pproval for use in food products in the USA.

Fluorocarbon propellants are named numerically, by applying the rules derived for aming chlorofluorocarbons propellants, thus:

- the first digit on the right of the propellant number corresponds to the number of fluorine atoms in the compound
- the second digit from the right is one integer higher than the number of hydrogen atoms in the compound
- the third digit from the right is one integer lower than the number of carbon atoms in the compound. When this digit is zero, it is omitted from the number
- the number of chlorine atoms in the compound (not relevant for the naming of fluorocarbons) is found by subtracting the sum of the fluorine and hydrogen atoms from the total number of atoms that can be connected to carbon

Examples of the application of this naming system are given in Table 3.

TABLE 3

Chemical name	*Chemical formula*	*Propellant nomenclature*
tetrafluoromethane	CF4	Propellant 14
trifluoromethane	CHF3	Propellant 23
1,1-difluoroethane	CH_3CHF_2	Propellant 152

Despite their promise for use as aerosol propellants, the fluorocarbons have never been widely adopted. Propellant 152, chemically 1,1-difluoroethane (CH_3CHF_2) has been used in some toiletries applications, principally mousses, in the USA. It is doubtful, particularly with the rising concerns over VOC's, that fluorocarbons will ever be fully exploited as aerosol propellants, particularly in view of their high cost.

Suppliers of fluorocarbon propellants are ICI and Du Pont.

4. *Compressed gas propellants*

The attraction of using compressed gases, principally nitrogen and carbon dioxide, as aerosol propellants has always been extremely high. In many ways, compressed gases can be considered to be closer to the "ideal" propellant than any other system. They are odourless, have very low toxicity and are non-flammable. Compressed gases are also very economical to produce and are totally acceptable from an environmental perspective being classified as neither ozone-depleting or VOC's. Furthermore, the pressure of a compressed gas is not significantly affected by temperature, in the same way that a liquefied gas propellant is.

There are, however, several major disadvantages in using compressed gases as aerosol propellants and these account for their virtually non-existent use. Firstly, compressed gases have very limited solubility in most products and generally do not provide good spray characteristics. The most significant disadvantage is that compressed gases suffer a pressure drop, throughout the usage life of the aerosol product. In turn, this leads to continual variation of spray pattern and particle size as the aerosol is used and very often all of the compressed gas propellant is spent before the contents of the can have been fully expelled. This latter point, in particular, has made the compressed gas propellant commercially unacceptable.

With the rising concern over environmental issues, much research has recently taken place on new aerosol valve designs, which would make the use of compressed gases a viable alternative. One of these, the "Novaspray" valve system, operates by regulating the compressed gas pressure throughout the life of the aerosol, thus minimising or eliminating the pressure drop normally observed. Although in principal this idea seems to work fairly well, the cost is prohibitively high and commercialisation of the concept, at the time of writing, has not taken place.

Aerosol packaging components

An aerosol pack consists of three basic components, the *container*, the *valve* and the *actuator*. Careful selection of each should be made, depending on product to be filled, the propellant used and the type of spray characteristics required. Each of these components is discussed in further detail below.

1. *Aerosol containers*

The purpose of the container is to provide a safe environment for accommodation of the concentrate and propellant. Effectively, the container is a pressurised vessel and a high degree of confidence for the safety of the end-user is required. There are basically three types of container, tin-plate, aluminium monobloc and plastic. Glass containers are also available but the potential hazards associated with the explosion of a glass container limit their use to low-pressure systems, typically alcoholic perfumes and colognes.

1.1 *Tin-plate containers*

Tin-plate containers are the cheapest and most commonly used type of aerosol container. They are fabricated from three separate pieces, a body plate, the base and the top. Firstly, the body plate is printed with the external pack design and it is then rolled, the adjoining edges being welded to form a cylinder. The base and top is then applied to the cylinder by a sealing process known as *swageing*.

One major drawback of tin-plate containers is their potential for corrosion, particularly in the presence of concentrates containing water. In order to prevent this, all tin-plate cans are lined, on the inside, with a protective lacquer coating to stop the concentrate contacting the tin-plate surface. A variety of internal lacquers are used, including phenolic and epoxy-phenolic resins, vinyl resins and vinyl-organosols. Epoxy-phenolic lacquers are commonly used but where extra protection is required vinyl-organosols, which can be applied in heavier film weights, can be employed. A further option for extra protection is to use double-coated internal lacquers, normally combinations of epoxy and vinyl resins or epoxy and phenolic double organosols. Whatever the internal lacquer, it is essential that the film is of uniformly high quality, with no pin-holing, otherwise can corrosion will occur.

Construction of tin-plate containers must be of sufficient quality to withstand a given pressure, without risk or distortion of explosion. Typically, tin-plate cans are available with recommended maximum working pressures of 12 bar or 18 bar, the latter utilising a thicker gauge of material.

The selection of a tin-plate container for any product is normally based on considerations of cost and product compatibility, although their cosmetic appearance is not as pleasant as that of the aluminium monobloc.

1.2 *Aluminium monobloc containers*

Aluminium monobloc containers are made from a single slug of aluminium, using an impact-extrusion process. Aluminium monoblocs have many advantages over tin-plate containers. Being of one-piece construction, they are cosmetically more acceptable than tin-plate cans, having no seam or welds present. This also means that the cans can be decorated around the whole circumference of their body, normally by a screen-printing process. Aluminium monoblocs are much less susceptible to can corrosion than tinplate cans although, in spite of this, they are invariably coated with an internal lacquer.

Perhaps the major disadvantage is cost, aluminium monoblocs being comparatively expensive to produce. This is particularly true of the larger sizes and cost considerations, combined with difficulty in manufacture, prohibit the availability of can sizes greater than 200 ml. Aluminium monobloc containers are selected in circumstances where the concentrate has a high corrosivity potential or where cosmetic appearance is of paramount importance.

1.3 *Plastic containers*

At one time the concept of a plastic aerosol container was unheard of, in view of the lack of suitable materials for their fabrication. With the development of polyethylene terephthalate (PET) containers and associated moulding processes, in the early to mid 1980s, the plastic aerosol container became a reality. Plastic aerosol containers are made by an injection stretch blow-moulding process to form a one piece unit, upon which a specially designed aerosol valve can be crimped.

The PET container has many attractions. It is inert, with good chemical compatibility, and is completely free of any corrosion problems. It is light to handle in manufacturing and, unlike tin-plate containers, is not subject to minor denting or consequential damage. Because it is made by a moulding process, the container can be formed into decorative shapes, thus enhancing cosmetic appeal.

The major disadvantages of the PET container are cost, the restricted availability of sizes and the relative difficulty in container decoration. The latter is normally achieved by the use of a thermally-sensitive shrink-wrap sleeve. PET containers also require special filling equipment which has restricted their popularity to some degree.

1.4 *Specialist containers*

The driving force of environmental concerns, combined with the need for customised containers for some types of product delivery, has brought about the development of containers with specialist applications. Perhaps the best known commercial example of this is the post-foaming shaving gel, which is sold in the so-called "bag-in-can" container. Essentially, this is a standard tin-plate container, inside which is a aluminium/plastic laminated bag attached to the underside of the valve assembly. The post-foaming gel is filled, through the valve, into the bag. The propellant, normally a standard hydrocarbon, is injected into the space between the bag and the inside of the tin-plate container, through a rubber septum in the base of the can itself. When the valve is actuated, the propellant pressure around the bag inside the can, expels the product.

Another recent development in container technology is the "Airspray" system, manufactured by Airspray BV in Holland. The "Airspray" is a three-part plastic container, consisting of a container body, a lid, and a pump/valve system insert. The concentrate is placed into the container, which is then fitted with a pump insert. The lid contains a cylindrical protrusion on its underside, which slides into the pump/valve insert in the container body. The product contains no propellant as such, the pressure inside the container being generated by the end-user, through a pumping action of the lid. This pumping action compresses the air inside the container and, when the valve is released, the product is expelled.

Although this system is environmentally very sound, it is also expensive and is only suitable for certain types of product. There is also a question mark regarding consumer convenience, because of the need to pump the container before use. At the time of writing the "Airspray" system has been successfully used by one or two major manufacturers but its acceptance is not widespread.

2. *Aerosol valves*

There are many types of aerosol valve available, both in tin-plate and aluminium, and the main criteria for selection are product category and the type of spray characteristics required. Aerosol valves are made up of three components, the valve cup/gasket, the valve assembly itself and the dip tube. The valve assembly is held in place by the valve cup which is crimped on to the top of the aerosol container. The valve cup gasket ensures that a good seal is obtained between the cup and the aerosol body, thereby avoiding significant leakage. Valve cup gaskets are available in a wide range of materials including neoprene rubber, viton rubber and nitrile rubber. Selection of the correct gasket material is essential, as any incompatibility between this and the concentrate will cause the gasket to swell and the risk of unacceptably high weight loss will be increased. The underside of the valve cup is lacquered, using similar materials to those described earlier for containers.

The valve assembly consists of a valve housing, inside which is a valve stem and gasket. The valve return action is provided by a spring located under the valve stem, inside the valve housing. The valve stem and housing are normally fabricated from plastic and valve gaskets are optionally available in the types of material described earlier for cup gaskets. The valve return spring is normally fabricated from grade 316 stainless steel, thus avoiding the risk of corrosion.

The dip tube is attached to the lower end of the valve housing and effects transfer of the concentrate from the container into the valve assembly, when the valve is actuated. Dip tubes are commonly made from plastics such as high density polyethylene. A cross-

FIGURE 4

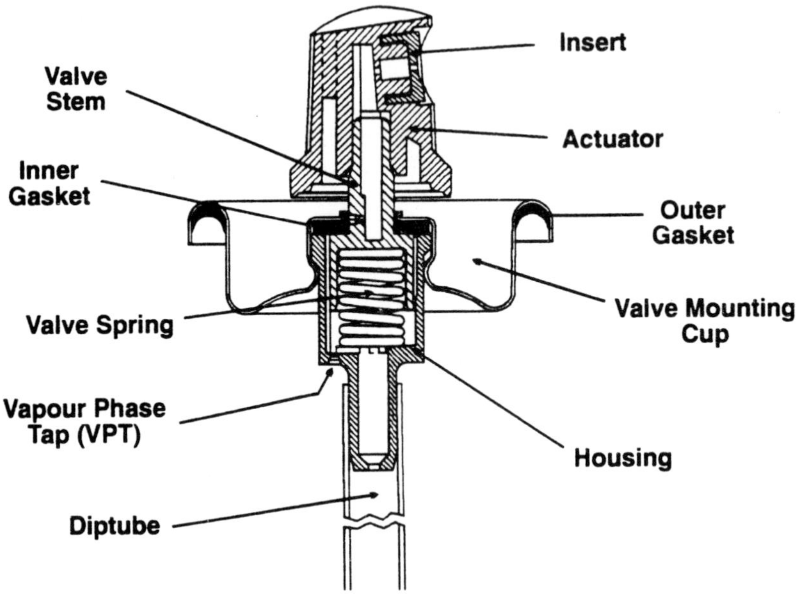

PCA 39F Powder Valve

Illustration courtesy of Aerosol Research & Development Ltd, Hampshire, England

section of a valve assembly, clearly showing the composite parts, is shown in Figure 4. The particular valve illustrated is a powder type, typically used in aerosol antiperspirants.

In some cases, valve performance can be improved by the addition of a vapour-phase tap (VPT). This is a small orifice located in the valve housing, connecting the headspace of the aerosol container with the inside of the housing itself. This allows propellant vapour to pass directly into the valve housing, when the valve is actuated. The use of a VPT has two advantages. Firstly, the aerosol can be operated in both upright and inverted positions and, secondly, the particle size of the spray is reduced, thus giving finer spray characteristics.

3. *The actuator*

The actuator, which sits on the top of the valve stem as shown in Figure 4, is critical in determining the characteristics of the spray pattern and will directly influence the discharge rate, the spray pattern and the particle size. Actuators are made of plastic and are available either as one piece mouldings or as two-part assemblies consisting of the actuator button and an insert.

Essentially, there are two types of actuator, those that produce mechanical break-up of the aerosol spray and those that produce non-mechanical break-up.

3.1 *Mechanical break-up actuators*

These are usually of the "swirl-arm" type and are typically used with propellant loadings of about 60% or less. Mechanical break-up actuators function by mechanically disintegrating the concentrate inside a swirl-chamber, located in the actuator housing, before the product is dispensed. This enlarges the cone angle of the resultant spray and reduces the particle size. Deodorants, hairsprays and colognes are all products that typically use this type of actuator.

3.2 *Non-mechanical break-up actuators*

These are used for products where the propellant loadings are high and further assistance in disintegrating the aerosol spray is not required. The generation of the spray in these cases is provided by the propellant pressure. Antiperspirants and air-fresheners are typical of products using this type of system.

Aerosol spray characteristics

The final spray characteristics of any aerosol are affected by many factors, the choice of propellant, valve specification and actuator design all playing a critical role. The mechanism of spray formation is governed by the change in a physical state of the propellant, from a liquid to a gas. When the propellant flashes, it breaks up the product concentrate into droplet particles, the higher the proportion of propellant that flashes, the smaller the particle size of the spray. Smaller spray particle sizes will occur with propellants that have high specific heat capacities, low heats of vaporisation and low boiling points (high vapour pressure). Other factors such as the surface tension and viscosity of the concentrate, droplet coalescence and the frictional forces between the spray droplets and air, will also affect the particle size.

The process of evaporation of particles continues as they continue to move away from the button orifice. The boiling point of any solvents used in the propellant concentrate then becomes a critical factor in determining the particle size. High boiling point solvents, with a relatively low rate of evaporation, will produce sprays with larger particle sizes, whilst low boiling solvents, with a higher rate of evaporation, will produce finer spray particles.

In designing the spray characteristics for an aerosol product, there are three important factors that must be specified.

1. *Discharge rate*

This is the volume or weight of product expelled from the container in a given time, when the valve is fully opened. Discharge rates are normally expressed in units of grams per second ($g\ s^{-1}$) and vary according to product type.

2. *Particle size*

This refers to the size of the spray particles, just after they have been expelled from the aerosol container. There is, of course, for any aerosol spray, a range of particle sizes and the value quoted is normally an average, expressed in microns (10^{-6}m). Particle size will directly influence the wetness of the spray, the rate of evaporation and, in some cases, the potential inhalation toxicity of the aerosol.

A list of typical discharge rates and particle sizes, for various product categories, is given in Table 4.

TABLE 4

Product Type	*Product/ Propellant Ratio*	*Discharge Rate (gs^{-1})*	*Particle Size (μ)*
Hairspray	70/30	0.70	50
Deodorant	40/60	0.50	30
Anti-Perspirant	25/75	0.70	30
Perfumes/Colognes	50/50	0.50	30
Shave Foams	93/7	1.00	—

Typical Discharge Rates and Particle Sizes for Various Hydrocarbon Propelled Products by Category

3. *Spray cone angle*

This is the angle made by the spray cone as it leaves the aerosol dispenser and is expressed in degrees. The size of the cone angle required is largely dependent upon the product type. Where a high degree of directionality is required, for example in a deodorant spray, then a narrow cone angle is preferred. In the case where a wide spray coverage is required, a more obtuse cone angle is selected.

Specific aerosol products

Having reviewed the various types of propellant and packaging components available for the production of aerosol products, some specific types of aerosol products will now be discussed.

1. *Hair sprays*

Hair sprays are one of the largest categories of aerosol products in the market-place today. Fundamentally, hairsprays consist of a setting resin, which is dissolved in an alcoholic solution, in the presence of other additives to improve the condition and hold of the hair. A perfume is invariably included to give the product a pleasant fragrance, both during and after use. Since the restriction on the use of CFC's, nearly all modern hair sprays are propelled using hydrocarbon propellant systems. The properties of an ideal hair spray are as follows:

- the spray must be very fine and should not cause obvious matting of the hair
- the sprayed product should exhibit good adherent and coherent strength
- residual product on the hair should be easily removed with shampoo

- the film should dry clearly, without becoming powdery or flaky on brushing and combing
- the film should be flexible enough to allow movement of the hair, without breaking
- the film should not be tacky or sticky to the touch, when dry
- the film should enhance the lustre of the hair, leaving it shiny in appearance
- the film should not make the hair feel heavy, oily or greasy

When a hair spray is sprayed onto the hair, the propellant evaporates rapidly and a viscous polymer/resin solution falls on to the styled hair. The solvent system evaporates and the resinous residue locks the individual hair shafts together through a series of "spot welds".

One of the earliest hairspray resins used was shellac but this was not particularly satisfactory as it produced hard, sticky, brittle films, that were not easily shampooed out. Later resins were based on polyvinyl pyrrolidone (PVP) and, whilst they were reasonably effective film-formers, they suffered from the major disadvantage of being hygroscopic and became sticky very quickly.

The next development in hairspray resins was to combine the positive attributes of PVP with other materials, in the form of copolymers, to improve overall performance. In so doing, the polyvinyl pyrrolidone/vinyl acetate copolymer (PVP/VA) was developed. This material, which is available in various grades, produces films that are clear, non-hygroscopic and exhibit good hair-holding characteristics, with the required degree of flexibility. PVP/VA copolymers still enjoy fairly widespread use today, particularly in lower-priced hairsprays on the market.

In more recent years other synthetic copolymers have been commercially developed with the aim of further improving the "naturalness" of the hold obtained. Amongst these are vinyl acetate/crotonic acid copolymer (Resyn 28-1310 from National Starch), vinyl acetate/crotonic acid/vinyl neodecanoate copolymer (Resyn 28-2930 from National Starch), butyl ester of polyvinyl methyl ether/maleic acid copolymer (Gantrez 425 from ISP) and octylacrylamide/acrylates/butylaminoethyl methacrylate copolymer (Amphomer 28-4910 from National Starch). Each of these materials are designed to give a more sophisticated hold to the hair and provide certain attributes, depending upon the target market.

Some of these modern synthetic resins, those containing free carboxylic acid groupings, require neutralisation before they can be successfully used. Materials commonly employed for this purpose include 2-amino-2-methyl-1-propanol (AMP), 2-amino-2-methyl-1,3-propandiol (AMPD), triisopropanolamine (TIPA), and 2-amino-2-ethyl-1,3-propandiol (AEPD). The degree to which the hairspray resin is neutralised, allows the formulator much more control over the final performance. Water solubility, film hygroscopicity and ease of shampooing all increase, with an increase in the degree of neutralisation. At the same time, both hydrocarbon tolerance and film hardness are decreased. Care must be taken by the formulator to ensure that the degree of neutralisation is carefully balanced, to give the optimum compromise of overall performance.

The most commonly used solvent in hairsprays is denatured alcohol, commonly referred to as DEB-100 or Ethanol B. This is principally ethyl alcohol, which has been denatured with low levels of "Bitrex" (trademark of McFarland-Smith), to comply with

HM Customs and Excise requirements. An example of a modern hairspray formulation, using Amphomer 28-4910 resin, is given below.

Raw Material	*% (w/w)*
Amphomer 28-4910*	2.50
AMP	0.38
Cyclomethicone	0.15
Dimethicone Copolyol	0.05
Ethanol B	68.72
Fragrance	0.20
Butane 40	28.00

* Tradename of National Starch

In this formulation, the Amphomer resin is neutralised with AMP to give the desired hold and wash-out characteristics. Cylcomethicone has been included at a low level to reduce the tackiness of the formulation and dimethicone copolyol is present to provide slight plasticisation of the resin film. The product is propelled using Butane 40.

One of the key factors in hairspray stability, is to keep the resin in solution at all times and ensure that it does not precipitate, particularly at low temperatures. Before their use was discontinued, CFC's acted as good co-solvents for resin systems and resin precipitation was not a significant problem. In the move to hydrocarbon propellants, the solvation power of the CFC's was lost and high ethanol levels were needed to ensure complete solution of the resin at all temperatures. With the increasing concern over the use of VOC's, of which alcohol is clearly one, the pressure on resin manufacturers to produce resins with higher solvent solubilities and better hydrocarbon tolerance is increasing.

Finally, care should be exercised when adding fragrances to hairspray formulations. If the levels are too high, some of the fragrance components can act as plasticisers on the resin film, thereby inhibiting its performance.

2. *Styling mousses*

Styling mousses first became commercially popular in the early 1980's and provided the consumer with the ideal vehicle for applying a setting-lotion type product evenly on to the hair. Styling mousses are oil-in-water emulsions, typically propelled by hydrocarbon propellants, normally included at levels of between 5% and 15%. The "oil-phase" of the emulsion is normally made up of cationic styling resins, not dissimilar to those described above for use in hairsprays, plus some additional conditioning components.

As the product is dispensed, the concentrate expands into a foam which can then be applied to the hair with either the hands or a comb. As the application is taking place, the foam breaks on the hair, leaving a thin, aqueous film of styling product coating the hair shafts. When the hair is dried, the water vaporises and leaves a residual film of styling resin on the hair, holding the style. A typical styling mousse formulation, propelled by Butane 48, is given on page 279.

In this formulation, the combination of polyquaternium-11 and polyquaternium-4 give the required set and style to the hair. The emulsifier system is based around nonoxynol-

Raw material	*% (w/w)*
Polyquaternium-11	2.00
Polyquaternium-4	1.20
Amodimethicone (and) Tallowtrimonium Chloride (and) Nonoxynol 10	0.20
Nonoxynol-10	0.50
Tallowtrimonium Chloride	0.50
Fragrance	0.20
Ethanol B	10.00
Water	75.40
Butane 48	10.00

10, with tallowtrimonium chloride acting as a co-stabiliser. The amodimethicone is a substantive hair conditioner, which further enhances product performance. The alcohol is required to produce a foam of the correct stability, which will remain intact on the hand after it is dispensed, but break quickly as it is applied to the hair. Generally speaking, the higher the level of alcohol, the quicker the foam will break.

Being water-based products, containing cationic surfactants, styling mousses can be very aggressive to aerosol cans, with the associated high risk of can corrosion. For this reason, styling mousses are invariably packed into aluminium monobloc containers that have been coated internally with a particulrly resistant lacquer. The most common lacquer used for this purpose is micoflex which, whilst relatively expensive, affords the required degree of protection.

3. *Antiperspirants*

Aerosol antiperspirants are one of the largest aerosol personal care categories, particularly in the US, UK and Northern Europe. Antiperspirants are less common in the Mediterranean countries, where personal deodorants are preferred.

The most common type of antiperspirant sold in the UK is the dry powder suspension aerosol, so called because it contains a suspension of aluminium chlorhydrate powder in an anhydrous stabilising matrix, normally quaternium-18 hectorite (Bentone 38 from Rheox Inc.) or fumed silica. A volatile silicone is normally added to improve the feel of the dry powder on the skin, particularly in the underarm area where the usage is most extensive.

The formulation for a typical dry powder suspension aerosol antiperspirant is given below.

Raw material	*% (w/w)*
Aluminium Chlorhydrate	10.00
Cyclomethicone	12.00
Quaternium-18 Hectorite	1.00
Ethanol B	0.80
Fragrance	1.20
Butane 48	75.00

Although zirconium salts are known to provide a higher degree of antiperspirant efficacy, it is important to note that their use in aerosol antiperspirants is not permitted, because of problems associated with inhalation toxicity.

Being totally anhydrous systems, aerosol antiperspirants are very often packaged in tin-plate containers. One of the major problems in developing this type of product is to ensure careful selection of the valve and actuator system, so that blockage does not occur when the suspended aluminium chlorhydrate powder is dispensed. The propellant content is kept deliberately high, normally between 70% and 80%. This ensures that the spray is dry when it reaches the underarm area and minimises the feeling of wetness or stickiness.

4. *Deodorants*

Next to antiperspirants, deodorants, or deo-colognes as they are sometimes known, are the second largest category of personal care aerosol products.

Simplistically, aerosol deodorants are alcoholic solutions of a deodorant active, most commonly 2,4,4'-trichloro-2'-hydroxy diphenyl ether, sold under the trade name of Irgasan DP-300 from Ciba Geigy. Other additives such as isopropyl myristate, glycerine or volatile silicones are sometimes included to promote skin feel. The mode of action of these products is based around the fact that Irgasan DP-300, a proven bactericide, kills the skin organisms responsible for the breakdown of apocrine sweat into malodorous by-products.

A typical aerosol deodorant formula, based on the use of Irgasan DP-300, is given below.

Raw material	*% (w/w)*
Irgasan DP-300	0.10
Cyclomethicone	0.50
Ethanol B	38.60
Fragrance	0.80
Butane 40	60.00

Some deodorant aerosols do not contain a deodorant active and just rely on a higher level of fragrance to mask any body odour that is produced. Such products are often referred to as deo-colognes and fragrance levels of approximately 1.50% are not atypical.

5. *Perfumed aerosols*

Perfumed aerosols which includes colognes, perfume sprays and toilet waters, are alcoholic solutions of perfume oil, propelled by a hydrocarbon propellant system. Ethyl alcohol is used, being mixed with 2% to 10% of perfume oil, depending on the nature of the aerosol product.

This type of product is often packed into a glass aerosol container, so particular attention must be given to the safety of the pressurised product in use. For this reason, it is not unusual to find that glass-packed perfumed aerosols have fairly low pack pressures, well below the burst-point for the container itself. In cases where higher propellant pressures are needed, the glass is externally coated with an invisible film of plastic. This reduces the chances of any breakage of the aerosol and contains any glass fragments, should breakage take place. The appearance of a diptube through the transparent containers is somewhat unattractive and this is often concealed by fluting or decorating the glass bottle.

Products of this type are normally prepared by dissolving the perfume in the alcohol, cooling the product to approximately 5°C and then filtering cold. This allows the removal of trace amounts of precipitated solids that may form, when the product is cooled. Although perfumed aerosols are fairly easy to formulate and manufacture, care should be taken that the propellant used does not adversely react with the perfume oil, which can sometimes produce a colour change. Colour changes can also occur when the product is exposed to ultraviolet light, the glass container providing little protection in this case. Perfume ingredients particularly prone to discolouration include indole, vanillin, nitro-musk, and phenols such as eugenol or iso-eugenol. All perfumes to be filled in glass aerosols must therefore be tested for discolouration, or precipitation, at low temperatures. Some typical formulations for perfumed aerosols are given below.

Raw material	*Perfume % (w/w)*	*Cologne % (w/w)*	*Toilet water % (w/w)*
Ethyl Alcohol	52.00	63.00	52.00
Perfume Oil	8.00	2.00	3.00
Butane 30	40.00	35.00	45.00

In order to avoid the possibility of flocculation or precipitation of some of the perfume oil ingredients, particularly at low temperatures, the propellant content should not usually exceed 50% of the total aerosol fill.

Aerosol filling

Aerosol filling is a specialised operation, carried out on dedicated filling lines fabricated for this purpose. It must be remembered that the filling of most modern aerosol products involves the handling of significant amounts of flammable, and potentially explosive, materials. This places stringent requirements on all aspects of safety in the manufacturing environment. A thorough review of the safety requirements is beyond the scope of this text but the following key points should be noted:

- propellants should be properly stored away from the manufacturing and filling area
- all storage vessels and transfer systems, for propellants and flammable liquids, should be properly earthed to prevent build-up of static electricity
- all electrically-powered manufacturing equipment should be spark-proofed and flame-proofed, in accordance with the relevant safety standards
- efficient extraction systems should be located in the filling area, particularly where build-up of propellant vapour is likely, such as around the propellant filling apparatus

Full details on the safety requirements for the manufacture and filling of aerosols can be obtained from the British Aerosol Manufacturers Association (BAMA).

The first stage in filling the aerosol, is to add the concentrate to the aerosol container. This is often carried out using conventional liquid filling techniques, recognising that additional precautions need to be taken in the case of flammable products. Next, the propellant is added by one of three different methods, cold filling, pressure filling and under-cup filling.

1. *Cold filling*

The product concentrate is cooled to about 0°F, depending on the product and propellant type, and added to the open aerosol container. The propellant is cooled to below its boiling point, usually between –20°F and –40°F, and metered into the open container on top of the concentrate. If product is not cooled sufficiently, then propellant losses which occur as the cold propellant mixes with the concentrate in the can, could be excessive. Immediately following the addition of propellant, the valve assembly is crimped on to the container.

Cold filling of aerosols has many disadvantages. The equipment required to cool the concentrate and propellant is expensive and cooling the concentrate to such an extent can cause precipitation of active ingredients, breakdown of emulsions and increase in viscosity. Ice crystals that sometimes form on the filling head nozzles can contaminate the product with excessive moisture, causing can corrosion at a later date. For these reasons, the cold filling method is only used on a very limited scale, normally for perfumes and pharmaceuticals that employ metering valves.

2. *Pressure filling*

Pressure filling of aerosols is, by far, the most popular method used. The product concentrate is added to the container at room temperature, air is removed from the head space by purging with propellant or vacuum-purging and the valve assembly is crimped on to container. The propellant is then pressure-filled through the aerosol valve using high injection pressure, normally between 600 and 1200 psig. Single stage (up to 50 cans min^{-1}) and rotary multi-stage (up to 350 cans min^{-1}) filling machines can be employed.

One of the advantages of pressure filling is that no refrigeration equipment is required and the product does not have to be cooled, therefore removing any low temperature concentrate sensitivity problems. Pressure filling is particularly suitable for filling hydrocarbons or other flammable propellants and propellant losses are very low.

3. *Under cup filling*

Nowadays, under-cup filling of aerosols is very rare. Firstly, the product concentrate is filled into the open aerosol container at room temperature. The valve assembly is then loosely placed on to the container, the head space purged and propellant injected, under pressure, around the valve cup and into the container. The valve assembly is then immediately crimped on to the container.

Under-cup filling has many attractions, combining the advantages of cold filling and pressure filling. Disadvantages of this method include the high capital cost of equipment and high propellant losses on filling.

When the aerosol product has been filled, it is passed through a heated water bath, set at between 50°C and 55°C. This process pressure tests the aerosol containers individually and identifies any leaking units. Any leaking or faulty cans are rejected at this stage.

Following water-bath testing, the aerosol cans are dried, normally with the assistance of a warm-air blower, packed into cardboard trays and shrink-wrapped for distribution.

Aerosol corrosion

Corrosion may be defined as the destruction of metal, by chemical or electrochemical action. Chemical corrosion normally involves the gradual accumulation of corrosion products on a metal surface. Electrochemical corrosion involves a reaction where the formation of corrosive products is not necessarily observed, or where the corrosion products are precipitated at a distance from specially separated electrodes. Overall, the mechanism of the corrosion process is electrochemical in nature, involving the transfer or displacement of electrons. When a metal changes to its metallic ion, and the ions flow from one electrode to another, then corrosion will take place. The anode is the electrode that loses electrons and is most often the element that is subject to corrosion. In this case, the anodc is called a sacrificial anode.

In tin-plate aerosol containers, the tin-plating can behave as a sacrificial anode and the base steel of the tin-plate container forms the cathode. This condition prevails when the oxygen content of the environment is low. In cases where the oxygen content is high, the polarity can be reversed, the tin-plating becoming the cathode and the base steel the anode.

The setting up of electrochemical reactions of this type, means that tin-plate aerosol cans are particularly susceptible to corrosion, particularly if the inner tin-plate surfaces are not protected and isolated from the concentrate. The nature of the concentrate will also affect the likelihood of can corrosion. In totally anhydrous systems, ion formation is impossible and corrosion cannot take place. Care must be exercised in these cases, however, as apparently anhydrous raw materials containing low levels of water as an impurity, will be enough to initiate a corrosive reaction. Generally, concentrates which are acidic solutions are more corrosive than neutral or alkaline ones, although it is impossible to predict a likely corrosive reaction on the basis of pH alone.

Another form of corrosion can also take place in aerosol containers. Corrosive attack takes place in parts of the container where high tensile stress can be observed, for example where the tin-plated metal has been sharply distorted during can manufacture. This type of corrosion is normally referred to as stress corrosion.

The best way of minimising or eliminating aerosol can corrosion, is to interfere with the electrode function, by applying a protective coating to the inside of the aerosol can. The types of protective lacquers used for this purpose have already been reviewed earlier in this chapter.

It is also possible to add corrosion inhibitors to aerosol products, although this is not commonplace in toiletries and personal care products. Typical general-purpose corrosion inhibitors include sodium nitrite and sodium benzoate, or oxygen scavengers such as sodium sulphate.

The assessment of testing of aerosol products

During the development and manufacture of aerosol products, there are a number of tests that should be carried out to ensure that the aerosol will be stable, safe and effective, before going into the market-place. Some of these tests can be applied to finished products as quality assurance procedures, although many are aimed at the formulator, to assist in the development process.

1. *Storage testing*

Like any other personal care product, aerosols should be fully storage tested before they are deemed to be suitable for commercial sale. Testing procedures are not dissimilar to those used for other toiletry and personal care products and high temperature accelerated storage tests are normally employed. Cans are normally stored both in the upright and inverted position, in order to determine whether any incompatibilities exist between the concentrate and the materials in the valve assembly. Storage temperatures used are normally 4°C, room temperature (20°C), 35°C and 45°C. The latter temperature is particularly important if the product is to be marketed in tropical countries.

As with any storage study, any significant changes in product characteristics or performance should be noted. It is difficult to provide a comprehensive list of the properties that should be measured throughout the storage test period but the following list should represent the minimum number of parameters tested.

- aerosol function
- can pressure
- appearance of concentrate (where relevant)
- odour
- pH of concentrate (where relevant)
- spray pattern
- discharge rate
- weight loss
- occurrence of can corrosion

Acceptance or rejection of a product at this stage, depends upon the degree of change in any of the above parameters throughout the period of testing, which should be a minimum of 3 months. With respect to can corrosion, some of the aerosol containers should be examined internally after at 2 months, 4 months and 8 months. The 8 month test at a temperature of 35°C is a good guide to the likely condition of the aerosol after 2 years at ambient conditions. It is not uncommon for storage tests to be continued for up to 2 years at room temperature.

The final decision as to whether a product is suitable or not for the market-place is judgemental, based on professional assessment of the test data and past experience.

2. *Flammability testing*

Flammability potential is one of the most important properties of an aerosol product, high flammability making it potentially hazardous when shipped, stored, or used in the home. The most important tests for aerosol flammability are flame extension, flashback and flash point. Flame extension is measured by controlled ignition of an aerosol spray in a purpose-built test apparatus. The flame extension refers to the length of the ignited flame, when the aerosol valve is fully depressed.

3. *Aerosol spray characteristics*

Aerosol spray characteristics are critical in determining the performance of the finished product and are normally defined in terms of the discharge rate, cone angle and particle size. Discharge rate is measured by fully depressing the aerosol valve and expelling product for a period of 5 seconds, the weight of the aerosol can being measured before and after product expulsion. Discharge rate is calculated by dividing the mass expelled by the time, the result being expressed in grams s^{-1}.

Cone angle is not measured directly but is assessed as a function of spray pattern diameter, at a given distance from the aerosol nozzle. A piece of sensitised paper is mounted vertically, 15 cm from the nozzle of an upright aerosol can. The valve is depressed for 5 seconds and the aerosol spray allowed to impinge on the paper. The diameter of the resultant spray pattern on the paper is measured.

Accurately measuring spray particle size is somewhat more complex. To obtain accurate distributions of particle sizes, it is necessary to use specialist equipment such as the Malvern particle size analyser. This equipment uses laser diffraction techniques to obtain particle size distribution curves.

4. *Weight loss*

The weight losses from aerosols take place primarily through the valve, with smaller losses occurring at the cup/container seal and at the top, bottom and side seams of the container. Weight loss is closely correlated to the degree of gasket swelling, the greater the swelling, the greater the product permeability through the gasket and the higher the weight loss. Product/gasket and propellant/gasket compatibility tests are therefore important when selecting gasket materials for a product, although it is generally accepted that some weight loss will occur with nearly every aerosol product. Whilst it is difficult to state acceptable weight losses, hairsprays, deodorants and shave foams should lose no more than 0.5 grams per annum, at room temperature. The corresponding figure for antiperspirants is about 1.25 grams per annum.

References

1. British Aerosol Manufacturers Association, "A Guide to Safety in Aerosol Manufacture" 2nd Edn (1988).
2. Rowland FS and Molina MJ, Review of Geophysics and Space Physics, 1975, 1(1), 1.
3. British Aerosol Manufacturers Association, "Standard Test Methods" 1st Edn (1980).

PACKAGING

General considerations

Packaging can be defined as a co-ordinated system of preparing goods for transport, distribution, storage and end-use and a package provides a means of ensuring the safe delivery of a product to the ultimate consumer, in a safe condition and at minimum cost. Fundamentally, the main functions of a packaging can be summarised as, follows:

- to contain the product
- to prevent spillage or leakage of the product
- to protect the product against mechanical and environmental damage
- to identify the product manufacturer
- to display "product-use" instructions, where applicable
- to enhance product sales appeal, through design and graphics
- frequently, to act as a product dispensing aid

All the above factors should be considered when developing a package for any particular product.

The use of effective and well designed packaging, and its associated costs, is an integral part of the manufacturing operation and should be clearly recognised and understood. The development of the product packaging should be considered at the start of the product development exercise, during conception. It therefore follows that a sample of the product should be produced at an early stage in the development process, to enable investigation and evaluation of the packaging design. Packaging development will involve many departments within a company, including marketing, research and development, purchasing, storage and distribution. To enable correct and effective packaging development to occur, information is required about the type of product and method(s) of distribution, in addition to the way in which the product will be marketed.

The most important aspect in packaging design and development, is the physical form of the product itself. In the area of cosmetics and toiletries this may be a mobile liquid (e.g. perfumes), a viscous liquid (e.g. shampoo), a free-flowing powder (e.g. talc), an emulsion (e.g. lotions and creams) or a solid (e.g. soap). In addition, it is also important to know whether the material to be packaged has any particular requirements, dictated by factors such as volatility, corrosiveness, odour, density and moisture sensitivity. Lastly, the effects of environmental conditions on the product should be understood and consideration to factors such as temperature, relative humidity, and exposure to air, light and water, should be given.

Another important factor in the pack development, is the method of distribution that will finally be used. Knowledge of the type of transport to be used (i.e. road, rail, sea, air or a combination of these) and whether it is directly controlled by the manufacturing company, or hired through a third party, is important. In addition, the types of handling system used (i.e. mechanical or manual) and whether or not the temperature and relative humidity of storage areas are controlled, will also have implications on the final pack design. Consideration should be given as to whether the volume utilisation makes a significant contribution to transport costs. The cost of storing any product is normally calculated per pallet, so if the amount of product that can be stored on the pallet is maximised, costs can be saved. This optimisation can be achieved by designing packaging for maximum utilisation of pallet space.

The final factor, important in any packaging development exercise, is the way in which the product will be marketed and sold and pack design is one of the most important elements in this exercise. Pack design will vary depending on the type of consumer (i.e. age, sex, socio-economic class) that the product is targeted at, whether the product is "standard" or is intended as a gift or seasonal item and convenience in use.

The use of paper in packaging

Paper is still one of the most extensively used materials in packaging. The use of paper was first recorded in the first century by the Chinese, but it did not reach Europe until the fifteenth century. Paper was first made in England in the sixteenth century and in 1799 the first paper machine was manufactured in the UK. The process of making paper has essentially remained the same, although production technology has advanced considerably in recent years.

1. *The components of paper*

The types of paper used in packaging consist of two major components, cellulose and non-fibrous additives, the latter being used to modify the paper to give the required properties. The main source of cellulose is timber which, after debarking, contains approximately 50% cellulose, which is extracted by pulping, before being used in the paper making process. Non-fibrous additives are used to modify paper properties. These include filler/binders, binders, sizes and special additives. Filler/binders are generally inorganic pigments, such as titanium dioxide, chalk and china clay, and they are used to improve paper opacity and brightness. Binders are normally either starches or synthetic resins and are used to increase paper strength. Sizes are generally emulsions or latex materials and they are used for decreasing the absorption of inks into the paper. Special additives include optical brightening agents, for improved brightness, fungicides and manufacturing aids.

2. *The manufacture of paper*

Before paper can be fabricated, the cellulose must be extracted from the source timber by pulping, either mechanically or chemically. Mechanical pulping involves debarking the trees and abrading them against large mechanical discs, to produce cellulose pulp. The pulp produced is somewhat impure and is frequently used only for producing newspaper and some types of board. Chemical pulping is, by far, the most important method for the packaging industry. Essentially, there are three different chemical pulping methods for the extraction of cellulose, choice being dependent on the type of paper in which the extracted cellulose is to be incorporated.

The first method, *acid chemical pulping*, is used on both hardwoods and softwoods. The trees are debarked, chopped into chips and digested for several hours, under heat and pressure, using a mixture of sulphuric acid and calcium bisulphite. The lignin and carbohydrates in the timber become water soluble and the waxes are removed. This method of pulping gives a very pure grade of cellulose, known as *sulphite pulp*, which is much whiter than *sulphate pulp*, but not as strong. Sulphite pulp is used for the subsequent production of high quality paper types.

A second method of chemical pulping is *alkali pulping*, which is applicable to softwoods only. The trees are debarked, chopped into chips and digested for several hours, under heat and pressure, with a mixture of caustic soda and sodium sulphate. This method gives a pulp known as sulphate pulp, which is light brown in colour and stronger than sulphite pulp. Paper subsequently produced from sulphate pulp is known as *kraft paper*.

The third method of chemical pulping is known as *semi-chemical pulping* and is a combination of mechanical and chemical methods, the pulp obtained by mechanical grinding before being chemically treated. The chemical treatment is less extensive than that used in either acid or alkali pulping techniques and the resultant pulp is less pure. This pulp is used for the fluffing media in the manufacture of corrugated board.

After the pulp has been produced, it is bleached with sodium hypochlorite and formed into "mats" for the paper making process. In the paper making process itself, the mats are firstly broken down by beating them in water, forming cellulose fibres which are themselves further broken down into fibrils. The process of fibrillation, which gives much better fibre cohesion and strength in the finished paper, must be carefully controlled and terminated at the correct point, otherwise the finished pulp will comprise of fibres with too short a length. Careful control of the fibrillation process and resultant fibre length, gives optimum paper strength. Any non-fibrous additives are also added at the fibrillation stage.

When the fibrillation process is complete, the pulp is converted into a crude paper, commonly referred to as machine-finished paper, on a Fourdrinier paper making apparatus. This is illustrated diagrammatically in Figure 1.

At the drying cylinder stage, the water content of the paper is reduced from about 95% to between 5% and 10%. Machine-finished paper produced by this process is fairly matt in appearance and is often used for general wrapping applications.

The quality of machine-finished paper can be significantly improved by modifying the paper properties at the dry end of the production process. For example, paper with an improved glossy surface can be produced by passing machine-finished paper around a giant chromium-plated roller, to produce machine-glazed paper. Alternatively, machine-finished paper can be subjected to a high degree of friction by passage through a calendar stack, to produce a highly polished, super-calendered paper, often referred to as *glassine*.

FIGURE 1

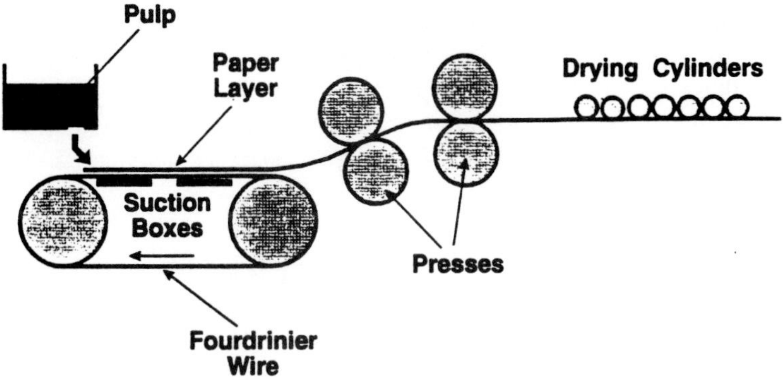

)ther paper finishes, such as waxed, varnished and pigment-coated, can also be produced
·y appropriate after-treatment of the crude machine-finished paper. The properties of the
inished paper depends not only on the after-treatment but also on the degree of fibrillation
·btained during the beating process.

. The manufacture of carton-board

'arton-board greater than about 100 microns in thickness cannot be made using a 'ourdrinier machine, as the drying time to remove the necessary water would require a 'rying stage of impractical length. For the manufacture of carton-board, a cylinder pparatus is used, as illustrated diagrammatically in Figure 2.The board substrate forms

FIGURE 2

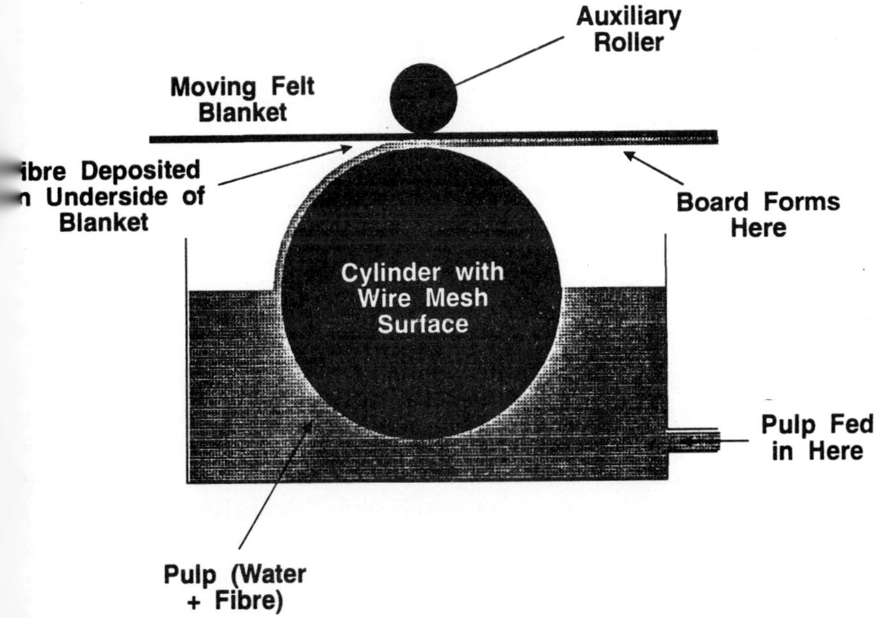

on the underside of the moving felt blanket and is then subsequently dried. Laminated boards can be manufactured using a series of cylinder machines, each machine vat being fed with a different fibre feed stock. The surface finishes obtainable for carton-board are similar to those described for paper earlier.

Amongst the various types of carton-board available, some of the most commonly found are chipboard, white-lined chipboard, duplex board and fully bleached board.

The use of glass in packaging

Glass is sometimes called the "original thermoplastic material". It was first discovered about 5000 years ago, and was used to make containers for creams and lotions in 1500 B.C. Although glass does not enjoy the same popularity that it did, it is still the material of choice for the packaging of certain types of cosmetic and toiletry products, notably perfumes and prestigious skin care products.

In terms of its usefulness as a packaging material, glass still has many, sometimes unique, advantages. Glass is inert and will therefore not react with, or produce malodours in, the product itself. It is rigid, with high compression strength, completely impermeable to gases and water vapour and will not degrade with time. Despite these properties, glass can be classified as "thermoplastic" and can therefore be easily moulded to give a wide variety of shapes and designs, with a pleasant appearance. The disadvantages of glass are that it is heavy, resulting in high handling and transport costs, exhibits poor impact strength, and, when broken, is injurious to other objects.

1. *The manufacture of glass*

The manufacture of glass is normally a continuous process. Glass normally used in packaging is known as flint glass and is composed of silica, lime and soda, with a low level of aluminium oxide. Coloured glass is produced by adding various materials at the manufacturing stage. Amber glass, for example, is produced by the addition of ferric oxide, carbon and sulphur, whilst green glass utilises ferric, manganese and chromium oxides, in various proportions.

The first stage of glass manufacture involves heating the components of the glass in furnaces, which can often accommodate up to 400 tons of glass, at temperatures of approximately 1500°C. The raw materials are loaded into the furnace with 25% cullet (broken glass of the same type as that being manufactured), which improves the melt/flow index of the glass. From the furnace, the glass, in its molten state, passes onto the gob former which forms a lump of molten glass, known as the "gob", for pressing or blowing. This process is illustrated in Figure 3.

The gob is then dropped into the parison mould, where it is either pressed or blown to form the parison, which is the "pre-moulded" glass object. The final shape required, for example a glass container, is then produced in a final blowing operation. When the bottle has been pressed or blown, it is removed from the mould and annealed at a temperature of about 450°C. This annealing process is normally carried out in a heated tunnel, known as a *Lehr tunnel*. This tunnel, up to 30 metres in length, prevents the glass from cooling too quickly, which would produce unwanted stresses in the finished glass container.

FIGURE 3

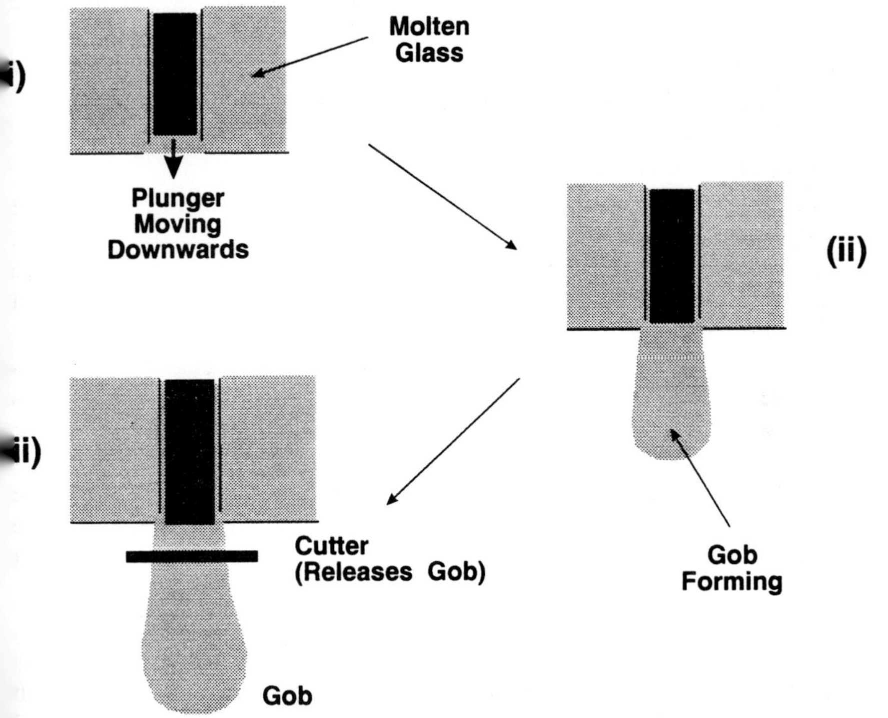

'inally, the finished container is inspected for flaws or taints in the glass and irregular pe, before being packaged for use.

'ottle design

ss bottles or containers must be carefully designed before manufacture, to ensure that / have maximum functional efficiency and aesthetic appeal. Poor design can lead to blems occurring in manufacture, subsequent processing, or even in the container's end

roblems in manufacturing often result from the fact that, during the blowing operation, s will naturally flow into a sphere, thus creating equal pressure at all points. Blowing glass into a shape which deviates from the ideal sphere will tend to cause material ribution problems, resulting in inconsistent wall thickness, particularly in corners of are-shaped containers. This type of problem can be overcome by slightly rounding the corners, enabling the glass to flow into a partial sphere. Good and poor corner design is shown in Figure 4.

FIGURE 4

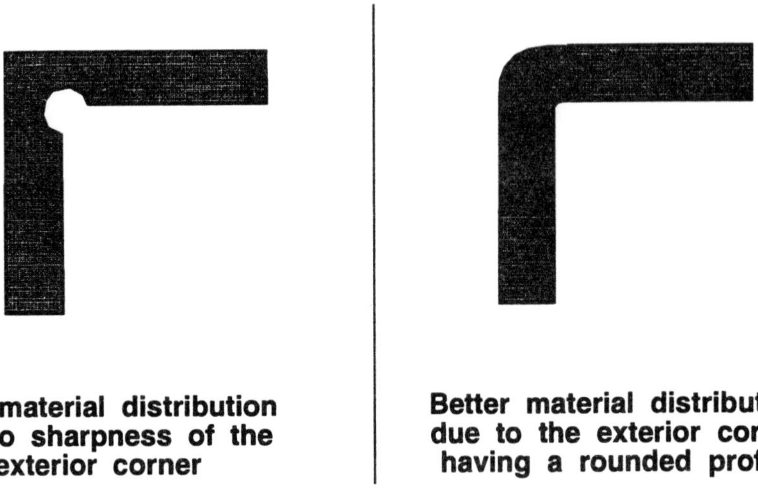

Poor material distribution due to sharpness of the exterior corner

Better material distribution due to the exterior corner having a rounded profile

Similarly, containers with an elliptical cross-section can exhibit distribution problems, in which the walls at the ends of the ellipse are thicker than the rest of the container. Another type of problem can occur in glass containers with an asymmetric design, where the mould, which is also asymmetric in shape, can be fouled by the bottle as it is removed, causing collapse of the bottle. The last manufacturing problem, allied to container design, is associated with embossed bottles. If the embossing is too pronounced, the differential stress relief in the glass produces cloudiness around the embossing. Generally, the maximum height that can be used for embossing is approximately 0.5 mm.

If a glass container is poorly designed, problems can also occur in subsequent processing. The most basic consequence of poor container design is physical instability resulting in difficult handling and the possibility of jamming on production filling lines. If the shape of the container is such that point-point or line-point contact can occur, then the likelihood of scratching, or even breakage, will increase. Finally, is the problem of bottle shape versus strength. For a constant volume, a spherical bottle is approximately ten times as strong as the equivalent bottle with a square profile.

Poor design can also result in problems for the intended end use of the container. To avoid these, the container must be designed such that there is a large enough area on which to stick a label and, in the case of paper labels, the design should have only one radius to avoid application problems or creasing. Additionally, the shape and design of the container should be such that volume utilisation during storage is as high as possible.

If decoration is required on the finished glass container, this can be achieved in the mould, as in the case of shaping, embossing or faceting, or out of the mould, as in the case of frosting which can be produced by sandblasting.

The classic method of printing on glass containers is the ceramic silk screen process, which a ceramic ink, composed of finely powdered glass, a coloured pigment and wax, printed onto the bottle, through a silk screen, and the bottle subsequently placed into a ırnace at 1000°C. The powdered glass softens and flows, thus providing a vehicle for the gment, and bonds to the main body of the glass container.

If the glass container is to be labelled, several methods are available but a flat surface invariably required. Labels are generally made of high quality laminated paper or astic, using emulsion or hot-melt adhesives.

Closures for glass containers

he functions of a closure are to contain and retain the product in the container and prevent ıy loss of volatile components from the product, that may occur. The closure must also event spoilage by foreign materials and, in some cases, may also act as a dispenser. inally, the closure can also enhance the design and appearance of the finished container.

Typically, two types of closure are used on glass containers, pre-threaded closures and oll-on closures. Pre-threaded closures are normally made of tin-plate, aluminium, or a ide variety of plastics, including phenol-formaldehyde or urea-formaldehyde resins, olyethylene, polypropylene and polystyrene. This type of closure may, or may not, have wad, depending on fabrication.

Roll-on closures are much less common and are nearly always made of aluminium loy. They are placed on the container and the thread subsequently "rolled-on" *in situ*, by otating metal wheels. This method of closure is sometimes used on tamper-proof caps.

. *Final packing of glass containers*

ilass containers are almost always packed into a high quality board carton and, if the ackaging is of very high quality, a single-faced corrugated board sleeve will be inserted etween individual units.

he use of plastics in packaging

lastics are, by far, the most commonly encountered materials in the packaging of osmetic and toiletry products. The reasons for this are many fold but key factors such as ost, production flexibility and safety in use, have played a key role in the increasing opularity of plastics over recent years. More recently, with an ever increasing emphasis n environmental preservation and biodegradability, the types of plastics used in the ackaging of cosmetics and toiletries has changed but the overall popularity of plastic emains unchallenged.

Essentially, there are two basic types of plastic material, the *thermoset resins*, such as henol-formaldehyde and urea-formaldehyde, and the *true thermoplastics*, examples of hich include polystyrene, polyethylene and polyethylene terephthalate.

. *Thermoset resins*

listorically, thermoset resins have enjoyed wide use in the packaging of cosmetics and oiletries, particularly for items such as screw-topped bottle closures. With the rapid development of different types of thermoplastics in the post-war years, thermoset resins have now been largely replaced by other materials.

Two thermoset resins, phenol-formaldehyde and urea-formaldehyde, are still of some importance.

1.1. *Phenol-formaldehyde*

Phenol-formaldehyde was first produced in 1926 and is more commonly known as "bakelite". It is produced by the reaction of phenol with formaldehyde, by one of two different manufacturing processes.

In the "one-stage" manufacturing process, phenol is reacted with formaldehyde, in the ratio of 1:1.5, under alkaline conditions. The reaction is carefully controlled and terminated at the point where a linear polymer, known as a "resol" is formed. The resol can then be used for casting and laminating purposes, by dissolving it in a solvent and catalysing under acidic conditions, to give a rigid 3D structure.

In the "two-stage" manufacturing process, phenol is reacted with formaldehyde in the ratio of 1:0.7, under acidic conditions, to produce a linear, soluble, thermoplastic resin. This resin is then dried under vacuum and mixed with pigment, filler, and hexamethylene tetramine, to produce a powder which can be used for compression moulding into desired shapes, for example a screw-thread bottle closure. During this compression moulding operation, the hexamethylene tetramine acts as a cross-linking reagent to form a rigid thermoset 3D structure.

1.2. *Urea-formaldehyde*

Urea-formaldehyde is manufactured in a two-stage process, by reacting urea with formaldehyde in the ratio 1:1.5, under alkaline conditions, to give a linear, soluble material, which is subsequently dried under vacuum and mixed with filler, pigment and an acidic curing agent. This material can then be compression-moulded in a similar way to phenol-formaldehyde.

Whilst phenol-formaldehyde and urea-formaldehyde are similar in many ways, phenol-formaldehyde is more stable to high humidity. Phenol-formaldehyde is always brown or black in colour, thus precluding its general use in cosmetics and toiletries packaging, whereas urea-formaldehyde can be fabricated in white or pastel colours.

2. *Thermoplastics*

Unlike thermoset resins, thermoplastics soften and can be reshaped when they are reheated, although, in practice, slight degradation will occur each time the plastic is reworked. Thermoplastics are widely used in the production of packaging for cosmetics and toiletry products to produce containers and closures, in addition to many types of wrapping film. The range of thermoplastics available is enormous but some of the key materials currently in use are considered below.

2.1. *Polyethylene*

Polyethylene, commonly referred to as "polythene" is one of the most widely used materials in modern packaging. It was first developed in 1933, by ICI, and became commercially available approximately ten years later. Early polyethylene was produced at high temperature and pressure, using an oxygen catalyst, yielding a material with a density of 0.915-0.940. This material was first used in packaging after the Second World War. In the mid-1950's, a new manufacturing process was developed using atmospheric pressure and a temperature of 70°C, with titanium and aluminium salt catalysts. This process was known as the *Ziegler Process* and produced high-density polyethylene, with a density range of 0.941-0.951. Shortly afterwards, a third process was developed, by the Phillips Company in the USA, using a pressure of 500 psi at a temperature of 175°C, with a chromium oxide catalyst, yielding material with a density range of 0.951-0.970.

The density of the final polyethylene directly affects other properties of the material, such as gas and water vapour permeability, softening point, rigidity and tensile strength. In selecting materials for any particular packaging application, it is therefore very important to accurately specify the density of the material required.

Polyethylene possesses many properties which make it of immense value in producing packaging components and, because of its true thermoplastic nature, can be reworked or reformed by one of many techniques, including injection moulding, extrusion, blow moulding, vacuum or pressure thermoforming and rotational casting. Polyethylene does not degrade significantly over time and all grades are chemically inert, making them unreactive with any product placed in contact with them. The one common exception is in the case of environmental stress cracking of polythene, which sometimes occurs when the material, particularly the lower density grades, is brought into contact with concentrated detergent solutions. This problem is easily overcome by selection of a high molecular weight polymer with a lower melt-flow index. Typically, material with a melt-flow index of approximately 1.0 is most suited to packaging applications. Polyethylene also has moderately low water vapour permeability and is therefore suitable for packaging aqueous-based liquid or emulsion-based toiletry products.

Some of the major disadvantages in the use of polyethylene for the fabrication of packaging components are associated with high oxygen permeability, poor resistance to some oily materials and its tendency to attract dirt and dust, because of static charge. Polyethylene is also very difficult to print, as the non-polar surface is not easily wetted by printing inks. This problem can be overcome in several ways, the most common of which is flame treatment. The container is rotated and its surface flamed for approximately 15 seconds in an oxidising flame, which oxidises the plastic. Subsequent printing onto the oxidised surface is then straightforward. Other methods used for improving the printability of polyethylene include corona discharge, primarily used for plastic films, and chemical oxidation.

Polyethylene has many applications in the packaging industry. Both high and low density polyethylene can be extruded into films or lay-flat tubing for general wrapping applications, although the optical properties of these films, particularly the high density grades, are very poor. Another application of polyethylene is in the production of shrink-wrap, which is used extensively in distribution applications. Shrink-wrap is made by extruding a flat polyethylene film onto a take-off reel, which is travelling at a faster speed than the rate of extrusion. The film is stretched before it is fully cooled and the final product is therefore in a stressed condition. Subsequent heating of the film causes stress relaxation, accompanied by film shrinkage.

One of the most common uses of polyethylene is in the production of polyethylene mouldings, such as bottles, which can be produced in both low-density and high-density grades. High-density polyethylene produces a more rigid bottle, whilst the low-density grades produce containers that are more flexible. Low-density polyethylene can also be used to make bottle inserts and flexible tubes. More recently, polyethylene has also been used in the production of plastic laminates, where its high inertness and good thermal sealability, at low temperatures, are advantageous.

Polyethylene can also be copolymerised with a variety of different plastics, including polypropylene, vinyl alcohol and vinyl acetate, to produce a number of packaging materials with various applications. Polyethylene-propylene copolymers are used for the production of caps and closures, whilst polyethylene-vinyl alcohol copolymers and polyethylene-vinyl acetate copolymers are used in film liners and packaging inserts.

2.2. Polypropylene

The most important type of polypropylene used in packaging applications is isotactic polypropylene, which has a density of approximately 0.9 and a softening point of 165°C. This material is almost totally crystalline and is less transparent than even high-density polyethylene.

Many of the fundamental properties of polypropylene are similar to those of high-density polyethylene, except that it is more rigid than the latter. Two outstanding properties of polypropylene are resistance to environmental stress cracking and good flexing properties, making it suitable for single-piece "box and lid" mouldings.

Polypropylene is used in the manufacture of films, which are clearer than their polyethylene-based equivalents, and are used extensively in packaging. Thin polypropylene films are made by firstly extruding a thicker pre-formed film on to a chilled roller (to prevent a crystalline structure forming, thus reducing opacity) and then stretching the film to a final thickness of approximately 0.25 mm. This gives an orientated polypropylene film, with good optical properties. Films of this type have a fairly high melting-point and are therefore often used in combination with polyethylene to improve their heat sealing properties.

Polypropylene is also used to manufacture moulded bottles and closures. Bottles fabricated from polypropylene are stiffer than those produced from high-density polyethylene, for an equivalent wall thickness. Polypropylene is therefore the preferred material where combinations of lightness and strength are required.

2.3. Polyvinyl chloride (PVC)

Pure polyvinyl chloride is extremely hard and unsuitable for thermoforming, because of its tendency to degrade with heating, and therefore a number of additives, including plasticisers and heat and light stabilisers, are used to improve its properties.

In cosmetics and toiletries packaging, three types of modified polyvinyl chloride are used. Plasticised polyvinyl chloride, with plasticiser levels of up to 30%, is a very flexible material, and is used for flexible tubes and refill packs. Rigid polyvinyl chloride contains up to 4% plasticiser and is used for the manufacture of thermoformed boxes and trays.

Finally, non-plasticised polyvinyl chloride is a very clear grade, which has an impact modifier added (typically a butadiene/styrene resin, at levels of up to 10%) and is used for bottle production.

Modified polyvinyl chloride has a higher water vapour permeability and lower gas permeability than polyethylene and, unlike the latter, prints easily without pre-treatment.

Over the last five years, there has been a significant reduction in the quantity of polyvinyl chloride used in the packaging of cosmetic and toiletry products. This is due mainly to environmental concerns over the non-biodegradability of the material, in combination with the problems of recyclability.

2.4. *Polyvinylidene chloride*

The principle application of polyvinylidene chloride is in the coating of various substrates, such as paper, regenerated cellulose and polypropylene, to give good barrier properties against water vapour.

2.5. *Polystyrene*

Polystyrene is a very hard, brittle material, with very poor impact strength and high gas and water vapour permeability. It is also highly prone to stress-cracking. Despite these disadvantages, polystyrene is commonly used in packaging components where its thermoforming properties, ease of printing and excellent optical properties, are of value. Polystyrene is nearly always used in modified form, either by copolymerisation with acrylonitrile to give a styrene-acrylonitrile copolymer (SAN), or by physical blending with butadiene to produce an acrylonitrile-butadiene-styrene copolymer (ABS).

Styrene-acrylonitrile, which has a much higher impact strength than polystyrene and exhibits increased resistance to stress-cracking, is used for the production of bottles and closures. Acrylonitrile-butadiene-styrene is also used for the production of bottles and closures, due to its improved impact strength, at the expense of optical clarity.

2.6. *Polyethylene terephthlate (PET)*

This material, formed by the condensation of ethylene glycol with terephthalic acid, is very strong, with excellent optical properties and moderately low gas and water vapour permeability.

Polyethylene terephthlate is frequently used for carton over-wrap, laminations and in the production of moulded containers, where optical clarity is important. Although relatively expensive, polyethylene terephthlate has enjoyed wider use in cosmetics and toiletries packaging in recent years, mainly duc to its optical clarity and environmentally acceptable properties.

Printing and printing methods

Almost all cosmetics and toiletries packaging is printed or decorated in some way. The printing and graphics on the pack identify the brand and product type and are carefully designed to maximise the visual impact of the product on-shelf. Printed material on the pack also provides instructions for use and information about the product.

Two fundamental types of printing processes are normally used for cosmetics and toiletries packaging. The first, known as direct printing, involves processes which facilitate the transfer of ink, from a plate, onto the packaging material or surface. Various direct printing techniques are used, including relief printing, planographic printing, intaglio printing and stencil printing.

The second type of process, known as indirect printing, involves the transfer of the total printed form, from a substrate onto the packaging material or surface. Methods commonly used are heat transfer and hot stamping.

1. *Relief printing processes*

There are three types of relief printing process, the letterpress process, the flexographic process and the dry offset process, all of which use a printing plate in which a surface relief is featured. In relief printing, print density cannot be changed by altering the quantity of ink applied to the surface to be printed and therefore half-tone techniques are used. Using half-tone techniques, colours are printed as a series of dots and the final print density is controlled by varying the density of dots per unit area.

1.1. *Letterpress printing*

This is a very old relief printing method, in which the design to be printed is formed on the plate cylinder, from where it is transferred to the surface to be printed. A schematic diagram of this process is illustrated in Figure 5.

FIGURE 5

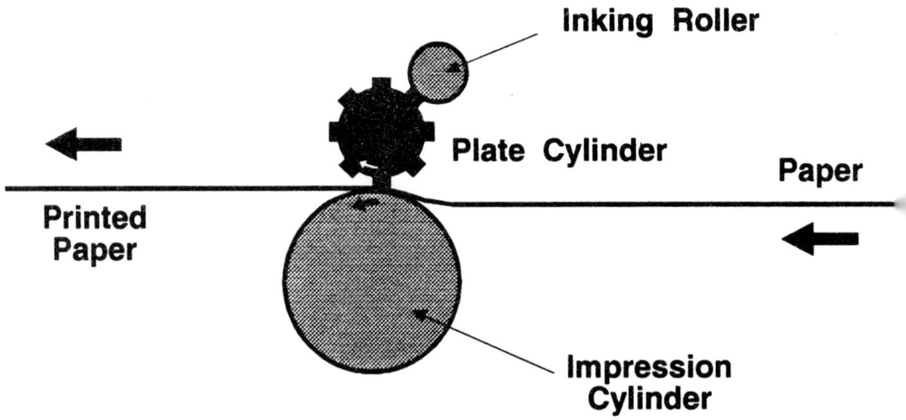

The types of ink used and printing speeds obtained, depend on whether a sheet-feed or reel-feed printing machine is used. Sheet-feed machines use oil-based inks, which dry by absorption, and printing speeds of 3000 sheets per hour are typical. Reel-feed machines use water-based inks which dry on contact with the material to be printed and typical machine speeds are in the order of 200 feet per minute.

The letterpress technique is normally used for the printing of paper-based substrates and board.

1.2. *Flexographic printing*

This is identical to the letterpress process, except that rubber characters are used on the printing plate, instead of metal ones. Half-tone printing cannot be carried out using the flexographic method. This process is typically used for printing plastic films, regenerated cellulose and aluminium foil and the inks used are based upon aniline dyes. The aniline dye is dissolved in alcohol, in the presence of a resin, and the ink dries by evaporation after printing. Machine speeds are similar to those obtainable in the letterpress process.

1.3. *Dry offset printing*

This process is similar to the letterpress process, except that an extra cylinder, known as the offset blanket, is used to transfer the ink from the plate to the surface to be printed. Print density using the dry offset method is controlled using half-tone techniques. A schematic diagram of the dry offset printing process is illustrated in Figure 6.

FIGURE 6

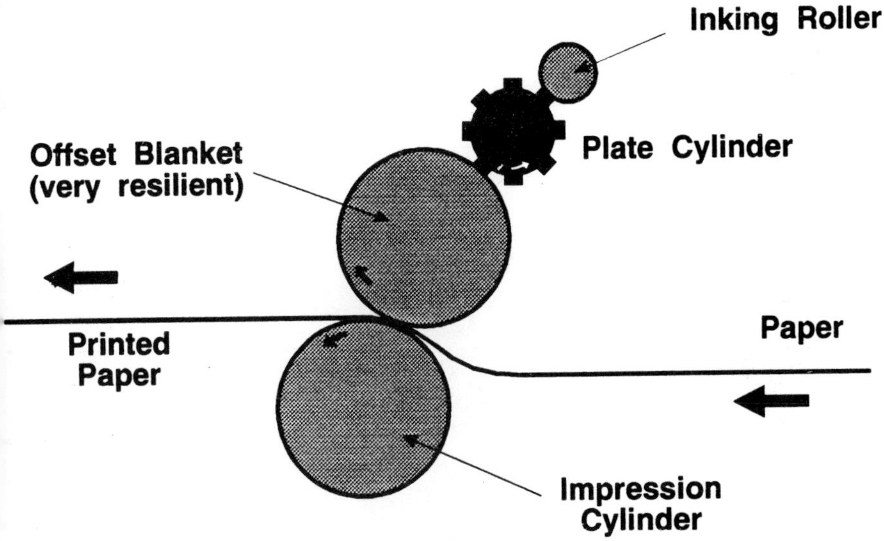

This process is used for printing paper, board, bottles and tubes. In the case of the latter, he blanket cylinder provides a more even printing pressure, resulting in a better quality print. Another advantage of the dry offset process is that more than one colour can be printed simultaneously, in one pass of the printing machine. The type of inks used in the dry offset process are similar to those used in the letterpress process.

2. *Planographic printing*

In this process, more commonly referred to as *offset lithographic printing*, the ink and the plate cylinder are actually in the same plane, as illustrated in Figure 7.

FIGURE 7

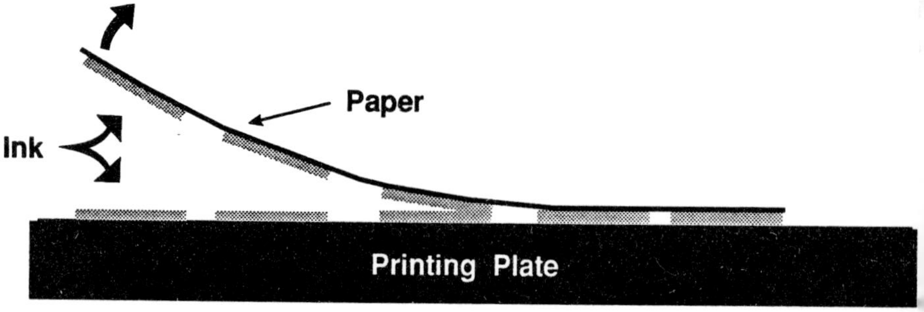

The printing plate is fabricated in such a way that the areas in which print is *not* finally required are finely etched, so that they will not accept any ink during the printing process itself. A schematic diagram of the offset lithographic printing process is illustrated in Figure 8.

FIGURE 8

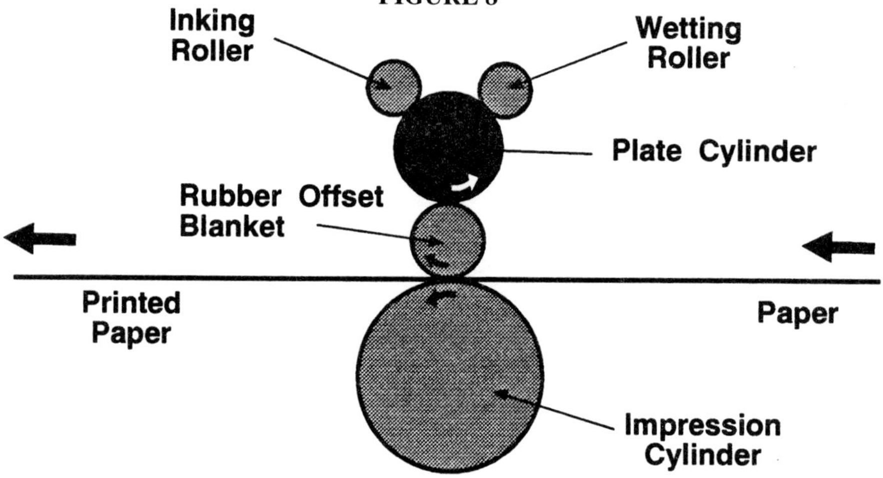

The plate cylinder is first wetted by the wetting roller(s), whereupon the areas of the plate cylinder that are etched will be preferentially wetted. Subsequently, the plate cylinder is inked, whereupon ink is preferentially accepted by the non-wetted (non-etched) areas. This process always uses a rubber offset roller which retains correct pressure between the plate cylinder and the impression cylinder, carrying the surface to be printed. This is an important aspect of the process, as the ink is laid down in a very thin film with a high pigment content. Machine speeds similar to those obtained with letterpress techniques can be obtained and print density can be controlled using half-tone methods.

The offset lithographic process is typically used for printing tin-plate, carton-board and paper labels.

3. *Intaglio printing*

Here the printing plate is etched, forming recesses into which the ink is applied, as illustrated in Figure 9.

FIGURE 9

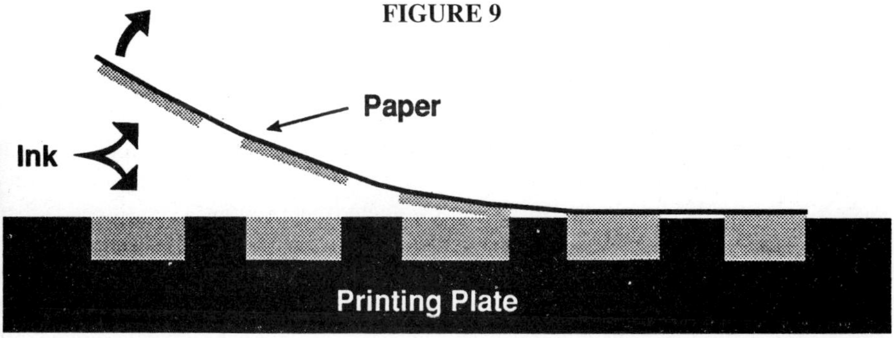

The area of the printing plate which corresponds to the area to be printed, is etched to produce minute holes in the plate surface. The only printing technique that uses the intaglio process is the *gravure printing process*, the apparatus for which is diagrammatically illustrated in Figure 10.

FIGURE 10

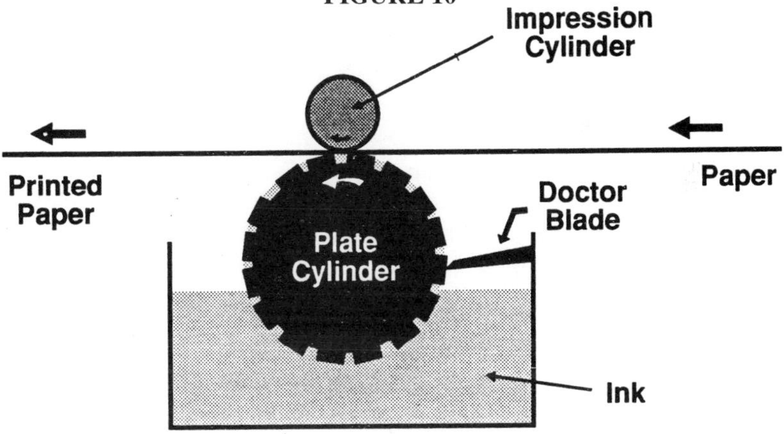

The doctor blade removes any excess ink, leaving ink in the plate cylinder recesses only. The print density can be easily controlled by modifying the depth of the etched recesses in the printing plate. Any surfaces that are to be printed using this technique must be very smooth, otherwise poor print quality will occur. The gravure method uses solvent based inks, which subsequently dry by evaporation, and machine speeds of approximately 1000 feet per minute are typical. Up to 6 colours can be printed in one pass, by using a series of cylinders, each with their own coloured ink baths, in sequence. It is also possible to perform overlap printing using this method.

One major disadvantage of gravure printing, is that the printing plates are very expensive to produce and the method only becomes commercially viable for print runs in excess of 500,000 units.

The gravure method is typically used for printing plastic film, cellulose film, paper and board.

4. *Stencil printing*

The stencil method, as its name suggests, involves printing the substrate through a stencil to give the desired design. The only printing method that utilises the stencil technique is the *silk screen printing process*. Typically the screen (the "stencil") is composed of a very finely divided mesh, stretched tightly over a wooden frame support. The screen is then dipped into a solution of photo-sensitive gelatin which fills all holes in the mesh. The screen is then exposed to an ultraviolet light source, which is passed through a mask of the image to be printed. This mask prevents the ultraviolet light from falling on the corresponding areas of the gelatin screen and when the screen is subsequently dipped into a suitable solvent, the unexposed areas dissolve away to form the "stencil" for printing. During the printing process itself, the screen is brought into contact with the surface to be printed and the ink forced through the holes in the screen, onto the object. The inks used in this method can be either solvent or oil based and, because they are deposited relatively thickly onto the surface to be printed, overlap printing or half-tone printing is not possible. The printing of two colours simultaneously is sometimes possible, using a screen divider.

The silk screen process is widely used on a variety of plastics, paper and board, and is particularly useful for small print runs and the printing of glass bottles.

5. *Heat transfer printing*

Heat transfer printing is an indirect printing technique, in which a high quality printed image is transferred, by heat, from a release paper directly onto the object to be printed. The printed object is then placed into a heated oven, to permanently fix the image. This method is relatively expensive and not very commonly used. Its major application is for the printing of plastic bottles, enabling a high quality gravure image to be put onto plastic, a task not possible using direct printing techniques.

6. *Hot stamping*

In this method, the "ink" is actually a resin and pigment mixture, which is put onto a release paper with a wax and an adhesive, as illustrated in Figure 11.

FIGURE 11

Image transfer is effected by punching down a heated die, which carries the image to be printed, onto the release film which, in turn, contacts the object to be printed. A typical heat transfer apparatus is diagrammatically illustrated in Figure 12.

FIGURE 12

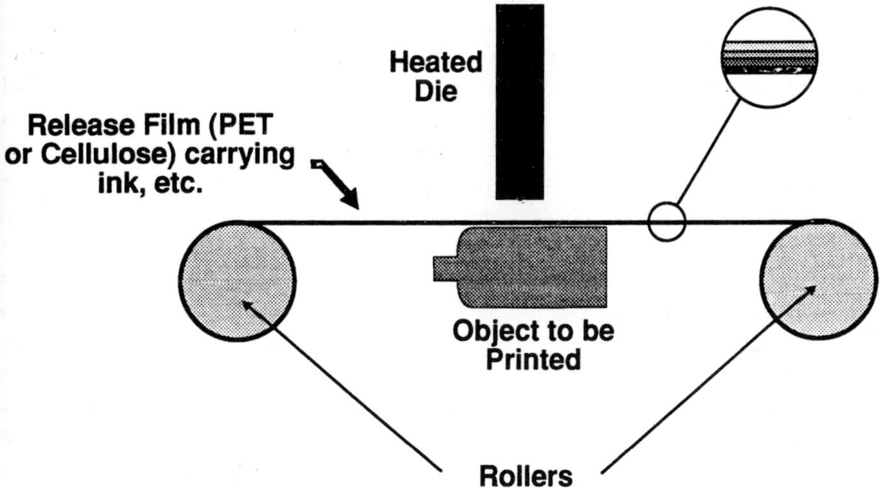

The wax melts and the image is transferred onto the surface to be printed, taking up the die shape as it does so. The adhesive serves to hold the image permanently in place.

Applications for hot stamping include the printing of cartons and gold blocking of paper and board.

Cosmetics packaging – plastic bottles

Plastic bottles are used extensively in the packaging of cosmetic and toiletry products of all types, because of the advantages that they offer over other materials. Plastic bottles are relatively inert, making them compatible with a wide range of cosmetic and toiletry products. They are light in weight but not easily broken and good impact properties are assured if the correct grades of plastic are selected. Plastics also provide enormous scope for bottle design, with a wide variety of production techniques available.

Amongst the disadvantages of plastic bottles is the generation of static on the bottle surface resulting in attraction of dust. The magnitude of this effect varies with the type of plastic but can be overcome, to some extent, by the incorporation of an anti-static agent. Plastic bottles can also exhibit panelling, a partial collapse of the bottle due to the presence of a slight vacuum inside the bottle itself. Panelling can be eliminated by correct bottle design.

1. *Plastic bottle types*

Earlier in this chapter, some of the key plastics available for the fabrication of plastic bottles were reviewed. What follows, is a brief review of the characteristics of the plastic bottles themselves.

1.1. *Polyethylene bottles*

This is, by far, the most important type of plastic bottle for the cosmetics and toiletries industry. Polyethylene bottles are flexible, depending on the density of the plastic used, exhibit good impact resistance without modification and possess good barrier properties against moisture vapour. Polyethylene bottles are susceptible to environmental stress cracking and bottle decoration presents a potential problem, which can be overcome by flaming the bottle surface. They have poor optical properties and are therefore normally pigmented to improve their appearance. Polyethylene bottles also exhibit poor barrier properties against perfume.

Flexible bottles are fabricated from low or medium density polyethylene with a typical density range of 0.925-0.940 g cm^{-3}. Rigid bottles are made from high density polyethylene, with a density range of 0.940-0.955 g cm^{-3}.

1.2. *Polyvinyl chloride (PVC) bottles*

Although this material no longer enjoys the popularity it once did, it is still used for some applications in the cosmetics and toiletries industry. Bottles can be fabricated from both plasticised and unplasticised grades, the latter being the most important.

Unplasticised bottles are rigid and possess good optical properties, providing the correct levels of impact modifier are used. They also exhibit better barrier properties to perfume than polyethylene, although permeability to water vapour is higher. Unplasticised polyvinyl chloride bottles show no tendency for stress cracking and are easily decorated. Bottles fabricated from plasticised polyvinyl chloride are less rigid, with poorer optical properties.

1.3. *Polystyrene bottles*

The use of polystyrene bottles in the cosmetics and toiletries industry is also diminishing. Whilst polystyrene is rigid and provides excellent optical properties, it has very poor impact strength in unmodified form. Polystyrene is not as inert as some other plastics and contact with oils can cause environmental stress cracking. Polystyrene bottles also have very poor barrier properties to perfume.

The most common use of polystyrene nowadays, is in the production of copolymers to produce acrylonitrile-styrene-butadiene (ABS) and styrene-acrylonitrile (SAN), both of which are use in the fabrication of rigid cosmetics containers.

1.4. *Polyethylene terephthalate (PET) bottles*

This material has become more popular over recent years, despite its relatively high cost, because of the many advantages it offers. Polyethylene terephthalate is rigid, with good impact strength, exhibits excellent optical properties and has similar moisture vapour barrier properties to polyethylene. Bottles are not prone to environmental stress cracking and are easily decorated.

2. *The manufacture of plastic bottles*

Fundamentally, there are three methods for manufacturing plastic bottles, extrusion blow moulding, injection blow moulding and, less commonly, injection stretch blow moulding.

2.1.*Extrusion blow moulding*

In this method, a tube of polymer, known as the *parison*, with a diameter of approximately the same size as the bore of the bottle neck, is extruded to a length just longer than the bottle height. A split-mould then closes around the parison and it is blown through the bottle neck, using compressed air, to fill the mould cavity.

2.2. *Injection blow moulding*

Injection blow moulding of bottles is carried out in two stages. Firstly, the parison is made in an injection moulding machine, using a core pin. The parison, with the core pin, is then transferred to a blow moulding machine and the bottle is blown into shape, through the neck. This method, which is similar to that used for making glass bottles, has the advantage of producing a bottle with a more consistent wall thickness.

2.3. *Injection stretch blow moulding*

This technique is similar to injection blow moulding, except that a physical stretching of the parison lengthways also occurs in the second stage blow moulding process. This causes the plastic to become orientated in both directions in the mould, giving a very strong, clear bottle. This method is normally used for the production of bottles fabricated from polyethylene terephthlate.

3. *Plastic bottle design*

The most basic shape of bottle design is a cylinder. Cylindrical bottles are easy to print, exhibit good pack stability over a wide range of height:diameter ratios and possess good stacking characteristics. In terms of manufacture, it is easy to obtain good resin distribution in the moulding.

Cylindrical bottles do, however, have significant disadvantages, poor size impression on-shelf and obvious exaggeration of any panelling defects, being amongst them.

The most commonly encountered bottle design in practice, is that based on the ovoid. This design possesses most of the advantages of the cylindrical design and gives a better size impression on shelf, for a given volume. Ovoid bottles also tend to conceal slight panelling defects more easily. Care must be exercised in the manufacture of ovoid bottles, as it is more difficult to ensure even resin distribution in the mould. Pack stability is also inferior to that of a cylindrical design, necessitating the use of pack orientation and stabilising guide rails on filling lines.

More rarely encountered is the rectangular bottle design. Although they are easy to print and give good size impression on-shelf, it is very difficult to obtain even resin distribution during the manufacturing process.

4. *Decoration of plastic bottles*

Plastic bottles can be decorated by printing, labelling or embossing in the mould. Printing methods commonly used include dry offset and silk screen printing, heat transfer and hot stamping. Polyethylene bottles must be pre-treated, by flaming, before printing. The application of paper or vinyl labels to plastic bottles is now very popular and pressure-sensitive adhesives are normally used to secure the label in place.

5. *Bottle closures*

Plastic bottle closures are normally either screw-type or snap-on and can be fabricated from a wide variety of materials including aluminium, polyethylene, polypropylene and some thermoset resins. Screw-type closures have less tendency to leak, particularly if wads are used, but care must be taken to tighten the closure to the correct torque, during the filling process. Snap-on closures are easier to apply but will suffer from problems of leakage if their dimensions are not accurately controlled.

6. *Quality assurance of plastic bottles*

In assuring the quality of plastic bottles, various parameters should be checked. Bottle dimensions and volume should be checked for accuracy and consistency, and the resin weight and distribution should be carefully monitored. Print quality, ink light stability and compatibility of the ink with the product, should conform to known specifications. Labels should also be examined for quality and product compatibility. Finally, the impact strength of the filled bottle should be checked using drop-tests.

Cosmetic packaging — collapsible tubes

Collapsible tubes are used for packaging pastes, creams, gels and viscous liquids and come in three types, metal tubes, plastic tubes and laminated tubes. Each of these are discussed below.

1. *Metal tubes*

Metal tubes were first introduced about 100 yeas ago and, at that time were fabricated from tin and lead. Modern metal tubes are now manufactured from aluminium, although their use in the packaging of cosmetic and toiletry products has almost been entirely superseded by the use of plastic tubes. Aluminium tubes are still commonly used in the packaging of pharmaceutical products, particularly for products of low unit quantity. Aluminium tubes are relatively cheap, impermeable to moisture and vapours and can be adapted to a wide range of products. They can also dispense any quantity of product and the tube remains collapsed after product has been expelled, reducing the possibility of subsequent drying out.

Aluminium tubes have the disadvantage that they corrode, although this problem can be overcome by lacquering the tube internally and enamelling and printing externally. They are also susceptible to mechanical damage and larger tubes tend to collapse and break open. Closures used for aluminium tubes are normally made of plastic. In practice, aluminium tubes are invariably cartoned. Cartoning reduces the chance of tube damage and facilitates distribution. The carton can also be used as a vehicle for advertising and facilitates retail display.

1.1. *The manufacture of metal tubes*

Metal tubes are commonly manufactured from aluminium slugs, with a purity of approximately 99.7%. The slug, with an external dimension equivalent to that of the tube to be made, is firstly annealed by heating, facilitating easy flow of the metal in the subsequent manufacturing process. The slug is then lubricated and put into a female die, with a corresponding lubricated metal mandrel, as illustrated in Figure 13.

FIGURE 13

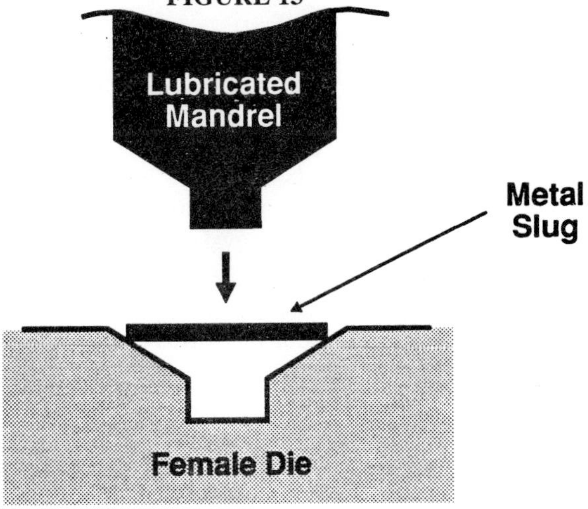

As the mandrel rapidly descends, aluminium rises up around it, giving the required tube shape. This process is known as *impact extrusion*. The tube is then cut to the required length and, if an aluminium nozzle is required, it is machined at this stage. The tube is then annealed once more to soften it, after the impact extrusion process. If the tube is to be lacquered, two coats of lacquer are sprayed on the internal surface, to prevent subsequent corrosion in contact with the product. The lacquering process should be very carefully controlled, as any pin holing will give rise to subsequent problems. The tube is then normally decorated on the outside, using a white enamel coating, before being printed using a dry offset technique. Accurate register of the print is essential, to ensure that the registration mark at the crimp end of the tube is correctly aligned. Tubes are frequently treated with a sealing compound in the crimping area, to strengthen the crimp when it is made. The tube is now ready to be dispatched for filling.

1.2. *Quality assurance of aluminium tubes*

The quality assurance measurement of aluminium tubes is a fairly involved process. Apart from the normal checks on tube dimensions and ink specification, other parameters such as the quality of the annealing, the efficiency of the closure, the uniformity of the internal lacquer and the resistance of the external decoration to cracking or flaking, should also be monitored.

2. *Plastic tubes*

Plastic tubes, commercially available since the mid-1950's, have now almost universally succeeded aluminium tubes in the packaging cosmetic and toiletry products. Unlike metal, plastic tubes will not corrode, do not require a carton for additional protection and always retain their shape. Shape retention, whilst aesthetically more pleasing throughout the usage life of the product, can result in a higher chance of product contamination or "dry-out". The reason for this is that, after use, the tube pulls air inside it as it recovers its shape. The main disadvantage of plastic tubes is their relatively high permeability to moisture vapour and perfumes.

2.1. *The manufacture of plastic tubes*

Two materials are commonly used in the manufacture of plastic tubes, low/medium density polyethylene and plasticised polyvinyl chloride, the latter becoming increasingly less popular because of environmental concerns.

Plastic tubes are manufactured using either extrusion blow moulding of an equivalent "bottle" shape, followed by subsequent removal of the base, or extrusion of the tube, onto which is welded an injection moulded nozzle and shoulder unit. Tube decoration is carried out using dry offset, hot stamping or silk screen printing methods. Tubes can also be externally lacquered, if the minimisation of moisture or perfume loss is of vital importance.

Closures for plastic tubes are similar to those used for metal tubes, except that the closure can be of full tube diameter to facilitate inverse tube storage.

2.2. *Filling and sealing of plastic tubes*

After tubes have been filled, they are sealed at the base, normally by thermal compression or ultra-sonic welding. Plastic tubes are not normally cartoned but the final package is often packed into display units, which are then packed into larger cases.

3. *Laminated tubes*

Laminated tubes were developed in the 1960's and have several advantages over normal plastic tubes. They have high barrier properties to moisture vapour and perfume and can accommodate complex decoration and graphics, with attractive designs.

Laminated tubes are manufactured in two parts. Firstly, the laminated material, which can be up to 7 or 8 layers thick, is produced on a laminator. The body of the tube is then made by taking a sheet of laminate and welding it together in a tubular configuration. Finally, an injection moulded plastic nozzle and shoulder assembly is then welded onto the tube. Closures for laminated tubes are similar to those used for plastic tubes.

After filling, laminated tubes are usually sealed using an ultrasonic welding technique. The final pack may be optionally cartoned.

Cosmetic packaging of talc

The requirements for packaging cosmetic talc are fairly specific. The pack must provide protection against moisture and perfume loss, yet should be easily re-sealable for convenience of use. Historically, talc was packaged in tin-plate containers, although these have now been completely superseded by plastic or composite containers.

1. *Plastic containers*

Plastic containers for packaging talc are normally fabricated from either medium or high density polyethylene or impact modified polystyrene. Both container types have been reviewed earlier in this chapter.

Closures are normally made of plastic, with a snap-on fit, and may be of either one-piece or two-piece design. The latter is normally referred to as a *rotex closure*.

2. *Composite container*

Composite containers for talc are much less common than plastic ones but are still used occasionally, particularly in gift-type packaging. Composite containers are made up of three components, the body, the base and the top assembly. The body is made of a flexible composite material and is wound in one of two ways, *convolute winding* or *spiral winding*. Convolute winding is most suitable for non-cylindrical containers and involves winding the body material around a pre-formed rotating mandrel. The cross-sectional profile of the container, along its length, must be constant if this method is to be used. Typical materials incorporated into the container body are paper, aluminium foil and some grades of plastic, each material being successively wound onto the mandrel during production. Spiral winding is only suitable for round containers and the body material is wound in a spiral, using similar materials to those mentioned above, the body wall being constructed to suit the particular packaging application.

The dispenser and top assembly, including the closure, is normally made of plastic with a tight push-fit into the top of the composite container. The base is applied after the product has been filled through the bottom of the container, and is normally made of plastic. The base is normally applied with a push-fit and permanently secured in place with an adhesive.

3. *Quality assurance of talc containers*

Amongst the parameters checked at the quality assurance stage are carton dimensions, carton strength and stability of the printing inks to light. Finished, filled cartons should also be checked for carton integrity, to ensure that problems in subsequent distribution and sale do not occur. Drop tests are normally performed to test carton strength and seal integrity. Vibration tests will help to ascertain the suitability of rotex closures and transit trials are normally used to determine the overall suitability of the finished pack. Storage tests, performed at different temperatures and relative humidities, will give an indication of perfume retention and flow characteristics.

Cosmetic packaging of soap

The packaging of soap presents certain problems, if a technically sound and commercially attractive packaging format is to be identified. The basic components of soap are fatty acids, fillers, perfume and water. Soap is alkaline, with a high moisture content, typically in the order of 12%, and will support mould growth. The main requirements for the packaging of soap are that it should protect the product from mechanical damage and the ingress of dirt, dust and foreign matter. When soap is manufactured, it is extruded into shape and packaged shortly afterwards, resulting in favourable conditions for mould growth. Packaging materials used should therefore inhibit mould growth and normally contain a mould growth inhibitor. Soap packaging must also be inert to the product itself.

Due to the caustic nature of soap, incorrectly selected types of paper will discolour either the product itself, or the packaging material. Care must also be taken in selecting the correct grades of printing inks, so as to avoid subsequent changes in colour, or bleeding into the soap. Finally, soap packaging should not significantly deteriorate with time. Incorrect selection of inks can cause fading and paper can discolour if it is of poor quality. High copper or iron content in the wrapping paper can act as a catalyst for discolouration of the soap and, in cases where aluminium foil is used as a packaging medium, problems with foil corrosion can occur.

1. *Soap packaging materials*

The most common type of soap packaging is a three component wrapper, consisting of a paper-based outer wrapping, a stiffener board and a glassine inner wrapper. Soap can also be packed in cartons or shrink-wrapping.

1.1. *Three-component wrapping of soap*

The most common type of outer wrapper used in soap packaging is of high quality paper, of the order of 90 g m^{-2} in weight. Paper used in the packaging of soap should not contain lignin, as this will cause it to yellow, and would normally contain optical brightening agents and be coated with an inorganic pigment or resin. This coating gives a good appearance and provides a suitable surface for gravure printing. The paper would also contain a fungicide, to prevent subsequent mould growth. The outer wrapper is normally printed using the gravure technique and any inks used must be soap, alkali and light stable. Often, an anti-scuff varnish is put over the printed surface, to protect it.

In the case of foil-laminated outer wrappers, a lower weight paper, approximately 60 g m^{-2}, is used. This paper contains a fungicide and normally has a low chlorine content, to prevent aluminium foil corrosion. The aluminium foil itself is approximately 10 microns thick and is laminated to the paper, using a sodium silicate adhesive. This gives a good, stable, water-resistant seal and also pacifies the aluminium surface, thus reducing the chance of foil corrosion. Different ink-types are used for printing aluminium foil but they must also be soap, alkali and light stable.

The use of stiffener board in soap provides some degree of mechanical protection and, at the same time, produces an elegantly shaped package for easy storage with cosmetic appeal. The stiffener board is normally made from high quality pulp and has a weight of between 150 g m^{-2} and 180 g m^{-2}. It must also be unaffected by soap and alkaline conditions and often contains a fungicide. The stiffener board must also exhibit reasonable rigidity, to effectively stiffen the soap package.

The glassine inner wrapper is an optional packaging component, which is inserted between the soap and the stiffener board as a form of protection for the soap. Glassine is made from a good quality pulp, which has been super-calendered, and is typically about 25 g m^{-2} in weight. The glassine contains a fungicide and very often an EDTA salt to chelate any copper or iron present.

1.2. Carton wrapping of soap

In cases where soap is carton wrapped, a "tuck-end" skillet carton is normally used, sometimes with a window for decoration. The carton is made from high quality, fully bleached, food grade or duplex board, and weighs between 250 g m^{-2} and 300 g m^{-2}. Cartons can be machine-coated or laminated with foil, and may incorporate a fungicide and glassine. Cartons may also be coated on the outside with anti-scuff varnish, to improve handling resilience. Any inks used in carton printing should be resistant to soap, alkali and light.

1.3. *Shrink-wrapping of soap*

The shrink-wrapping of soap involves packing 2 or 3 bars together and shrinking the package. This type of packaging is generally used for cheaper soap products.

2. *Quality control of soap wrapping materials*

All soap packaging materials should be bought to a specification, assurance of compliance being ascertained through the use of appropriately designed tests. Typical quality assurance tests for soap packaging materials would include the following:

- the weight of all packaging materials, normally expressed in g m^{-2}, should be checked
- the stability of the packaging materials to soap and alkaline conditions should be checked by placing them in contact with soap, for between 24 and 48 hours, and in contact with alkali for between 5 and 10 minutes. Any yellowing observed is unsatisfactory
- inks should be checked for compatibility in contact with soap and reverse side contact to check for bleeding
- inks should be checked for light stability. The most commonly used instrument is a "fadeometer", which uses a xenon arc lamp to generate artificial sunlight. Conditions are normally standardised using the British Standards Blue Wool Scale, stability of the inks being quoted to a particular Blue Wool Scale number
- fungicide levels in the packaging material should be checked, normally by colorimetric or ultraviolet spectroscopic methods
- the rigidity of the stiffener board should be checked using a Taber Stiffness Tester. This measures the force required to bend the board through an angle of 15°
- if a glassine inner containing EDTA is incorporated, the level of EDTA should be checked
- the "adhesion to self" of the outer soap wrappers should be checked, to identify any errors in the varnish used

Secondary packaging

Outer packaging is used on all types of primary packaging forms, to facilitate the distribution process and ensure that product reaches the retailer in good condition. Two main forms of secondary packaging are used, shrink-wrap and fibre board cases.

1. *Shrink-wrap*

Shrink-wrap packaging is normally applied to the primary package, stored in a shallow cardboard tray. A shrink-wrap tube is placed over the trayed product and the whole unit placed into a heated tunnel, to shrink the wrapping tightly over the primary packs, securing them in place. Sometimes an additional tray will be placed at the top of the package, before shrink-wrapping, for increased strength and rigidity.

The shrink-wrap material is normally polythene, of the order of 100 microns in thickness. Thicker shrink-wrap films are available for larger units. Although shrink-wrap does protect the primary packaging, it cannot offer the same degree of protection offered by a fibre board case.

Like all other packaging components, shrink-wrap should be subjected to quality assurance tests and film weight, film thickness and shrink ratio should all be monitored.

2. *Fibre board cases*

Fibre board cases have been used for approximately the last 100 years, the first corrugated board being made in about 1890 and the first solid board case made in 1902.

2.1. *Solid board cases*

These are made principally from layers of chip board, referred to as "middles", pasted together and lined on one or both sides with a kraft paper liner. The thickness of the board is normally between 0.8 mm and 2.8 mm. The manufacturing of solid board cases is similar, in principle, to the method used for the manufacture of corrugated board cases, discussed below.

2.2. *Corrugated board cases*

There are three main types of corrugated board case, single-walled, double-walled and triple-walled. The corrugated fluting comes in four different grades, the properties of which are given in Table 1.

TABLE 1

Fluting type	*Flutes m^{-1}*	*Flute Height (mm)*
Type A (Coarse)	104 - 125	4.5 - 4.7
Type B (Medium)	120 - 145	3.5 - 3.7
Type C (Fine)	150 - 185	2.1 - 2.9
Type E (Very Fine)	275 - 310	1.2 - 1.7

Flute types A, B and C are used for outer cases, whilst type E is used for display cases and large cartons, not intended for distribution. Materials used for fluting should be flexible enough to be passed through the corrugator, yet rigid enough to keep the liners apart. Fluting materials should also perform satisfactorily on the cutting and slotting machines. In practice, materials used for fluting include semi-chemical paper, kraft paper, chip board and straw-based paper.

Materials used as liners for the corrugated board must be easily printed and allow straightforward assembly, using suitable adhesives. They must also possess adequate strength for liner applications and allow easy incorporation of special additives, if required. Finally, liner materials must be available in both low chloride and low sulphate grades and should have near neutral pH. Materials normally used for the production of liners are natural kraft, bleached kraft and chipboard.

The manufacture of single-walled corrugated board cases is carried out by firstly passing a reel of fluting material over a pre-heated roller and subsequently spraying it with water, to plasticise it. The fluting material is then passed through the fluting cylinder, as illustrated diagrammatically in Figure 14.

FIGURE 14

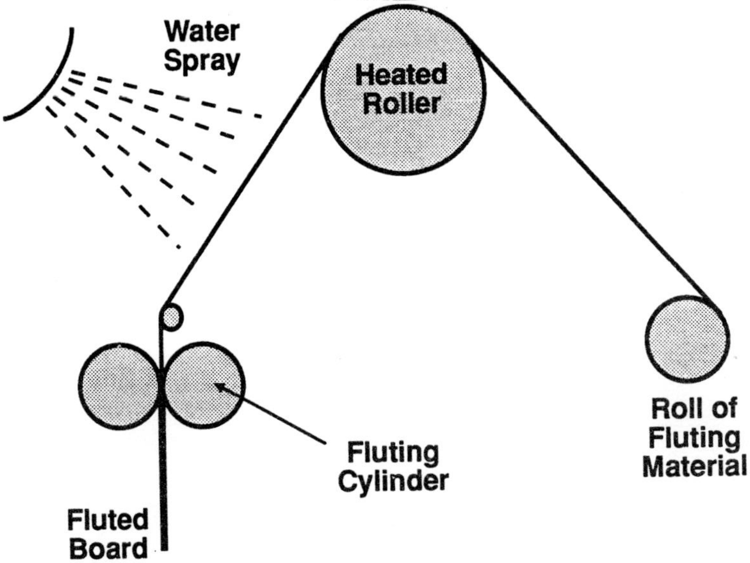

Whilst in contact with the fluting cylinder, a small amount of adhesive is applied to the tops of the fluting and a further roller then brings a pre-heated liner into contact with the sticky side of the flute. This process is called single-faced fluting. The second liner is then put on, using a machine called a double-backer, and a solid corrugated board is produced. The board is then trimmed to the correct width and cut into sheets of adequate size, for subsequent construction of the corrugated case. These pre-cut sheets are normally referred to as "blanks". The blank is then printed and passed through the slotter, where it is creased and cut. This gives a blank board which, when erected, will form a corrugated board case. The case is finished by fastening the fourth side with tape, stitching or an adhesive, before being dispatched to the customer in lay-flat form.

3. *Choice of case for distribution*

The type of case chosen for the distribution of a product will depend upon the type of transport to be used and whether or not the case will be stacked in the warehouse. Solid board cases are more expensive but provide better protection against piercing with sharp objects. Corrugated board cases have a higher compression resistance and provide better cushioning characteristics.

4. *Quality assurance of boards and cases*

Quality assurance tests for boards should include a check on the weight and burst strength, the latter being determined using a *Muller Bursting Strength Tester*. For corrugated board, all of the components should be checked and the board itself tested for crush resistance. Cases should be examined for quality of conversion (i.e. characteristics of the case when it is erected), dimensions and compression strength.

SECTION 7

Perfumery

PERFUMERY

The history of perfumery

Perfumery originated circa 3500 B.C. with the Egyptians who made pastes and creams of odorous materials. Dried rosin materials, such as benzoin, galbanum, and frankincense olbanum, existing as crystals from the dried sap of trees, were easily collected, ground and smeared over corpses to create pleasant fragrances during cremation.

The word *perfume* is derived from the Latin words "per fumen" which literally mean "through smoke", as this was how these early fragrances were perceived. Historically, a very simple method of obtaining perfume was to soak flower petals in fat, to produce a perfume saturated in fat, known as a *pomade*. This was used to anoint bodies for religious ceremonies and was probably an early form of deodorant or air freshener.

One of the most important discoveries in the advancement of early perfumery, was distillation. The Arabs discovered that if rose petals were boiled in water and the steam produced was passed over a cool surface, the fragrance was carried over into the condensate. This early perfumery product was known as attar.

In Europe, it became popular for people to make their own perfumes and an early Italian monastery managed to support itself by selling "magic water". One of the monks at this monastery, left, with details of the perfume production process, and went to Cologne in Germany. Starting production of his "magic water" there, he made the first "Eau de Cologne", a typical formulation for which is given below:

Ingredient	*Percentage (w/w)*
Oil of bergamot	30.0
Oil of lemon	30.0
Oil of sweet orange	20.0
Oil of neroli (orange blossom oil)	6.0
Oil of petitgrain (oil from orange leaves)	5.0
Oil of lavender	4.0
Oil of clary sage	1.0
Benzoin gum	2.0
Oil of rosemary	2.0

This mixture was dissolved in alcohol at a concentration of 5% (w/w) and probably represents one of the earliest created perfumes.

In about 1850, organic chemistry began opening up the route to many new compounds that could be used in the art of perfumery. Details of some of these early synthetic materials are given in Appendix I.The art of perfumery has developed rapidly and perfumes are currently used in a wide range of products in everyday use. The two most important categories of product in which perfumes are used are toiletries, (including personal care products, fine fragrances, cosmetics, bath products, deodorants, hair products, etc.), and household products (including air fresheners, laundry products, washing liquids, surface cleaners and disinfectants).

The majority of the world's perfume is used in household products, since these products are invariably sold in much larger volumes and the levels of perfume used tend to be higher than in toiletry products.

Perfumery and its function

Perfumery can be defined as the art and science of creating perfumes which, in their simplest form, can be defined as "a blend of odorous materials, providing a pleasant smell, which is used as the odorous part of a marketed product".

A typical perfume is normally a mixture of between three and several hundred ingredients, some of which may be of natural origin and some synthetic. An important property of these materials is their different volatilities, some being very volatile, others less so. Blending of materials with different volatilities in a perfume formulation, determines how the fragrance will be perceived.

In order to assess perfumes, perfumers use filter paper strips, tapered at one end. These are known as *smelling strips* or blotters. The tapered end is immersed to a depth of about 1cm in the perfume and the odour assessed by smelling with a slow drawing in of breath. Some of the ingredients will last on the paper for several minutes only, whilst others may remain for days, or even months. The most volatile ingredients are known as the "*top-notes*" and these give the initial impression of the perfume. Moderately volatile ingredients remain odorous for periods of up to several hours and are known as "*middle-notes*". The most long-lasting materials are known as "base notes" or "*end-notes*".

The perfumer must possess a good "odour-memory" to recall odours of the different ingredients. Typically a perfumer will sit at a comfortable U-shaped, laboratory bench, allowing easy access to the large range of ingredients stored in bottles around him. Other equipment used by the perfumer would include glassware for mixing and measuring, a magnetic stirrer and a balance. The perfumer's bench and the walls of the room may be formica-lined to allow easy cleaning, as maintenance of a clean atmosphere and environment are essential to prevent odour contamination. It is possible to detect the odour of some materials as low as 1ppm in the atmosphere.

The functions of a perfume

Essentially, the functions of a perfume are to provide a pleasant odour, to cover the base smell of a product, to give the product identity, to provide product concept support and sometimes to signify a product change to the end-user. Often the fragrance can help establish the image of the product. For example, if the product concept is connected with coconuts, it is important that the product reminds the consumer of coconut and the use of a coconut fragrance may be a very powerful marketing tool. The fragrance may also affect how the consumer perceives the effectiveness of the product. It has been shown that the same product can be perceived to perform differently, simply by changing the fragrance. An old product can be given a completely new brand image, simply by changing the fragrance and supporting this change with a new marketing concept.

Oflaction — the mechanism of smelling

It has been discovered that new born babies develop a sense of smell earlier than any other sense. By odour alone, a baby 6 weeks after birth, can recognise its own mother. Until this time, the baby will accept any odour as being characteristic of its mother.

The mechanism of olfaction is linked with three life-functions, sustenance, sex and sociology. The sense of smell is very powerful and most adults are not usually aware of it, until they detect something particularly unpleasant or attractive. In young children the sense of smell is very acute, although their ability to describe odours, being poor, does not generally enable them to make full use of this ability.

In order to possess an odour, a product must exert a vapour pressure, enabling molecules of the odourant to enter the nose. Most people can detect odours in materials with molecular weights of up to 250-300. It is for this reason that many flavour and fragrance materials possess a molecular weight within this range. A diagram of the adult nose and olfactory system is given below in Figure 1.

FIGURE 1

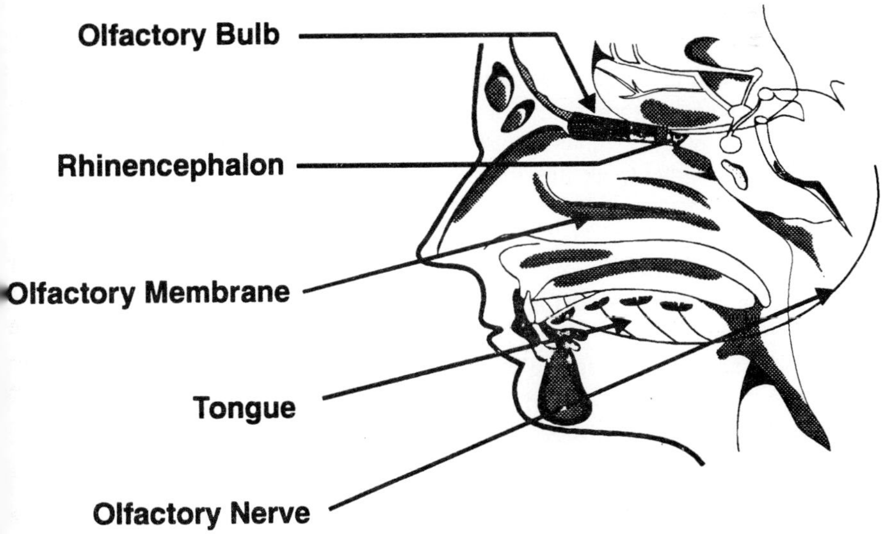

The cells that detect volatile materials are the olfactory sensory neurons and these are located in a small area (about 2.5 cm^2) known as the olfactory sensory epithelium. Volatile materials are carried in air, over these cells during the normal breathing process and are absorbed into a thin layer of mucus before illiciting an odour response. The olfactory epithelium is very similar in all higher animals and consists of three primary cell types, in a layer approximately 150-300 microns thick. The olfactory epithelium contains about 50 million olfactory receptor neurons, on either side of the nasal cavity.

The olfactory receptor cells contains the neurons which detect, encode and transmit information relating to the intensity, duration and overall quality of the stimulant molecule. This information is transferred firstly to the olfactory bulb and then to the higher cortical centre of the brain. Another type of cell that is important in the olfactory process is the sustentacular cell. These sustentacular cells separate the receptor cell dendrites from one another, isolating them by forming a hexagonal array. They are also involved in the secretion of mucopolysacharides, into the mucus layer which covers the epithelial surface.

Simplistically, different receptor neurons possess different molecular receptors, each of which detects a narrow band of odorous molecules. Man's ability to distinguish between numerous odours supports the theory of molecular receptor sites in the olfactory system.

Neurophysiological research has comprehensively demonstrated the complex nature of the membrane mechanisms and the principles of neural coding. This research suggests that the recognition molecule is a protein which interacts with the odourant molecule causing a change in membrane permeability.

1. *Odour sensitivity*

There are various conditions of the olfactory system which can affect the way in which smell is perceived. Details of some of these are given below.

1.1. *Odour fatigue*

Even a single exposure to an odourant at high concentration, will reduce the olfactory sensitivity to that molecule for a short time. This is known as self-adaptation and this phenomenon can be used by perfumers to reduce their sensitivity to the major components of a perfume blend, in order to identify minor ingredients.

1.2. *Hyposmia*

It is generally accepted that sensitivity to odours decreases with age and it has been estimated that in an adult between the ages of 20 and 70 there is a regression of one binary step in sensitivity approximately every 20 years. This would mean that a 20 year old man is, on average, 4 times more sensitive to any odour than a 60 year old. Furthermore, it has been shown that for any given age group, there is no difference between the sensitivities of smokers and non-smokers. Reduced sensitivity occurs for about 10 minutes after smoking, eating or drinking and a cold or hay fever will reduce sensitivity by about 2 binary steps.

1.3. *Anosmia*

Total loss of the sense of smell, known as general or simple anosmia, usually occurs when the olfactory or first cranial nerve is not functioning properly, as a result of a clinical or congenital condition. Although only about 0.2% of people suffer from this problem, they are incapable of perceiving any odour whatsoever. A stinging or prickling sensation may be detected after exposure to high concentrations of chemicals that can irritate the trigeminal nerve in the nasal cavity.

People suffering from *complete anosmia* also fail to show any response to these stimulants and cannot even distinguish, for example, between pure pyridine and water. As detection of flavour also depends to a great extent on the sense of smell, many people suffering from anosmia wrongly perceive that they have lost their sense of taste.

A third type of anosmia, specific anosmia, is widespread, although most sufferers are not aware of this condition. It is only when the sense of smell is specifically called upon, as in the case of perfumers for example, that sufferers realise that they have a problem. This condition seems to be inherited and has no known medical significance. The condition of specific anosomia supports the theory that the mechanism responsible for odour detection is highly selective.

1.4. *Other olfactory problems*

Acute sensitivity to certain chemicals results in a condition known as *hyperosmia*. Hyperosmia is normally associated with an accompanying clinical condition, although some women are said to develop a hypersensitivity to musk at the time of ovulation. There appears to be little evidence, however, to substantiate this claim.

Dysosmia describes a condition in which the sufferer provides inappropriate descriptions for some odours, whilst sufferers of *cacosmia* detect any odour, no matter how pleasant, as disgusting. *Phantosmia* describes a condition in which the sufferer perceives an odour, usually but not always described as unpleasant, either continuously or intermittently, even when there is no odour present.

Odour descriptions

Although objective description of an odour is virtually impossible, certain adjectives can help relate the odour to other well known odours. Some of the more commonly used adjectives are listed in Appendix II.

The classification of perfume ingredients

Essentially, there are four classes of perfumery ingredients, the terpenoids, the benzenoids, the aliphatics and, finally, other ingredients which do not fall into these three classifications.

1. *Terpenoids*

Terpenoids are naturally occurring components of living plants. A terpene hydrocarbon will have derivatives including alcohols, aldehydes and esters. Terpenoid compounds are all very closely related in structure and are based on the isoprene structural unit, shown in Figure 2 below.

FIGURE 2

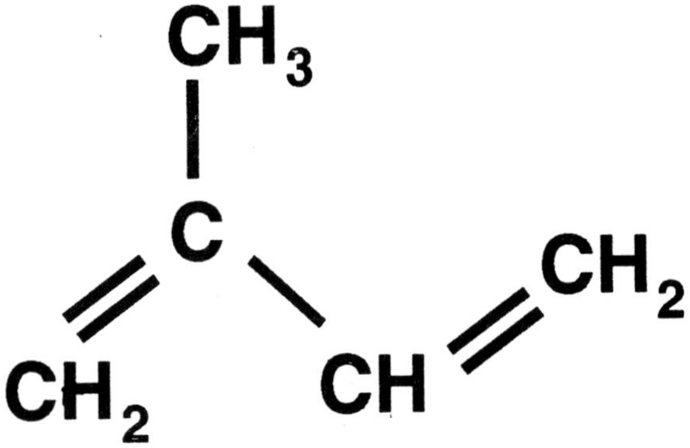

For example, two isoprene units joined together "head-to-tail" give the terpenoid compound myrcene, shown below in Figure 3.

FIGURE 3

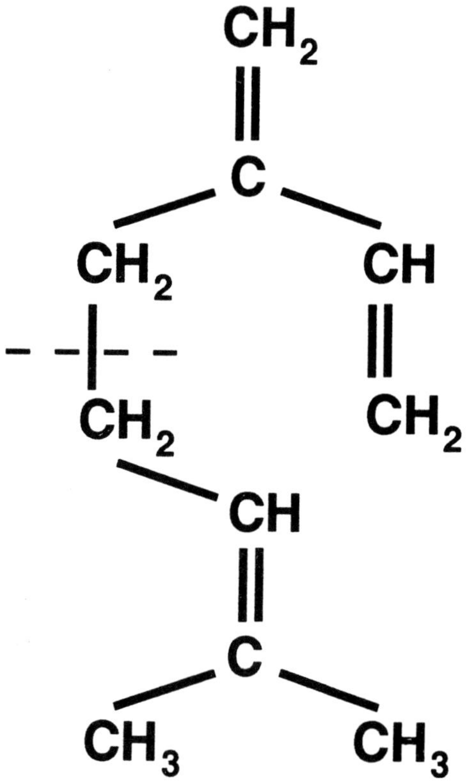

Combinations of more than 2 isoprene units exist to give the classes of terpenoid compounds listed in Appendix III. The monoterpenes and the sesquiterpenes have pleasant odours, whilst the diterpenes and triterpenes have very little odour. A list of some of the most common terpenoids, along with their respective odour descriptions, is given in Appendix IV.

2. *Benzenoids*

There are many benzenoid compounds of value to the perfumer. A list of some of the more important benzenoid compounds, along with their odour descriptors, is given in Appendix V.

3. *Aliphatic ingredients*

These are straight-chain organic chemicals, which often exhibit strong odours and can therefore be used at low concentrations. Aliphatic ingredients often form a major part of, or make a major contribution to, the total odour of a perfume compound. Many classes of aliphatic components are available, amongst which are:

3.1 *Aldehydes*

Normally saturated aldehydes, with between 7 and 12 carbons atoms are used. Unsaturated aldehydes, however, often possess a stronger and more interesting odour.

3.2 *Alcohols*

Alcohols with between 6 and 12 carbon atoms are most often used. These compounds have a bright, clean odour, similar to that of their corresponding aldehyde. Once again, the presence of unsaturation gives rise to more interesting odours.

3.3 *Esters*

Aliphatic esters are often fruity in odour and the most useful compounds are generally of molecular weights less than 200.

.4 *Ketones*

Aliphatic ketones used in perfumery are those with between 4 and 12 carbon atoms, for example methyl hexyl ketone, which is used in lavender fragrances.

.5 *Lactones*

Aliphatic lactones are also used as perfumery ingredients. Examples of these are gamma-undecalactone, which possesses an odour of peaches/apricots and gamma-nonalactone, which has an odour of coconut.

. *Miscellaneous ingredients*

Various materials, which do not fall into any of the previous three classifications, are of significant value in perfumery. A list of some of the most important materials, with details of their corresponding odours, is given in Appendix VI.

Perfumery ingredients derived from plant materials

Perfumery ingredients of natural origin come from many parts of plants, including the flowers, leaves, stems, peel, bark, fruits, wood, roots, rhizomes, grasses, seeds and gum oleoresins. Other natural ingredients are derived from animals, although these are far fewer than those derived from plant matter. Animal-derived materials will be dealt with later in this chapter. Probably the most important and widely used group of natural materials are the *essential oils*.

The term "essential oil" is credited to Philippus Aureolus Theophrastus Bambastus Von Honenheim, known commonly today as Paracelsus. Paracelsus, a 16th Century alchemist, working as a physician in Basle, believed that essential oils were the quintessence or fifth vital principal, containing the total odour and flavour of each vegetable substance from which they were extracted. In 1969, the Essential Oil Technical Committee, of the International Standards Organisation provided a definition for an essential oil, as follows, "the odoriferous product obtained by steam distillation of plant material from a specific botanical source, or by expression of the pericarp of citrus fruits and their separation by physical methods, from the aqueous phase".

That such a technical definition was proposed, indicates the complex nature of essentia oils. This definition has since been modified and now includes the following, "othe odoriferous products obtained from plant material by other distillation processes, or by fractional distillation, or which have been subject during their preparation, to ancillary processes such as filtration or centrifugal separation, rectification or treatment with absorbents, for the selective concentration or removal of particular constituents".

The Essential Oil Technical Committee defines plant material as not only the structura parts of the plant but also includes plant exudates and solvent extractions, thus furthe complicating the definition.

The function of essential oils in plants is not fully understood, although the aroma o plants is said to attract insects and lead to pollination. Essential oils commonly exhibi some degree of bacteriostatic effect and some have bactericidal action. A significan amount of research work has been carried out to demonstrate the use of essential oils i preserving food and cosmetics, against microbial spoilage. Exudates, such as gums an resins, which contain essential oils, are produced by plants to act as barriers when the become damaged, thus preventing loss of moisture and providing protection against attac by parasites.

Most components of essential oils are secondary metabolites which are produced a intermediates or by-products during the plants' primary metabolic processes. Some of th components in essential oils are said to exert physiological action in animals and man Some secondary metabolites, such as alkaloids, produce well known effects on animal and some components of nutmeg and mace have been reported to be precursors fo mescaline and adrenaline.

Whilst there have been many historical claims for the medicinal properties of essentia oils, none have been fully substantiated by correlation with chemical structure. Th practice of aromatherapy relies on the intrinsic properties of essential oils to provid beneficial effect to the patient, through the stimulus of aroma. Research work by To

(1986) substantiates the sedative and stimulant claims made for some essential oils, although it is impossible to relate these effects to any particular chemical entity.

The bacteriological and antioxidant properties exhibited by the essential oils of clove, thyme, bay, origanum, sage and rosemary, can be related to the high concentration of phenolic compounds in these materials.

1. *The history of essential oils*

Terracotta flasks reported to contain traces of essential oils have been dated back to approximately 3000 years BC. The ancient Persians were aware of the process of distillation, although it is not clear how they made use of their knowledge. Paintings found in some of the Egyptian tombs are thought to depict the anointing of bodies and statues with fragrant oils.

The Egyptians used their local cumin, marjoram, mint, rose and myrtle, all of which were found in the region of Northern Africa. They are also thought to have imported cassia, cinnamon and anise from China. The essential oils themselves were not actually produced by the Egyptians. Instead, they pressed the fragrant herbs with olive, castor and palm oils, to produce a fragrant extract. This process is known today as *enfleurage*. The Greeks and Romans imitated the Egyptians and even initiated trade in aromatic oils and ointment. Even at this stage however, the essential oils were not being physically separated from the plant source.

The first production of essential oils is recorded by Herodotus in about 450 B.C. and Pliny described the use of turpentine oil for the relief of coughing complaints in about 50 AD. Although the Arabs are thought to have invented distillation in the 9th or 10th Centuries A.D., essential oils remained available only as fatty extracts or pomades until the 13th Century, when Arnold de Villanova is thought to have introduced the practice of distillation, for their extraction. It was not until the late 16th century that the process of distillation was sufficiently understood, and widely used, to provide different types of essential oils. The first oils to be produced by this method were turpentine, rosemary, spike lavender and juniper wood. Later, the essential oils from clove, mace, nutmeg, anise and cinnamon were all produced by the distillation process.

The turning point in the understanding of the nature and importance of essential oils, is generally accepted as being linked to the publication of the "Krauterbuch" in 1551. Written by Adam Lonicir, this book stressed the medical importance of many spice and seed oils, and served to fuel the interest in the production and use of essential oils in Europe. By 1592, 61 distilled essential oils were listed in the "Dispersatorium Valeni Cordi", illustrating the pace of the development and acceptance that was occurring at that time.

During the 17th and 18th Centuries it was mainly pharmacists who were producing essential oils. French Provence was exporting oil of lavender and many other chemists were devoting attention to their study, Boerhave, Hoffmann, Glauber and Lavoisier all making contributions. The first steps in understanding the chemistry of essential oils were taken in the 19th century with De La Billardiere's analysis of turpentine in 1818 and Duma's treatise on essential oils, published in 1833. Further important contributions were made by Berthelot in 1866 and Kekule at about the same time. Gradually the use of essential oils for flavouring and perfuming products exceeded their use in medicines and by the end of the 19th Century there were extensive essential oil industries in several countries throughout the world, most notably France and North America. Today well over 3000 essential oils are known and many hundreds of these are available commercially.

2. *The production and distribution of essential oils*

Essential oil crops are grown all around the world, and cultivation of hybrid, high-yielding strains, has kept pace with improvements in agricultural methods.

The essential oil can occur in various parts of the plant, as indicated by the examples given in Appendix VII. Processing normally involves distillation and since it is not economical to transport bulky plant material this is often carried out in, or close to, the area where the plant is harvested. Mobile stills may be used although this rather primitive processing method results in contamination of the extracted oil with glycerides, tannins or minerals. Sometimes complexes may form with metal ions, typically those of iron.

Essentially, there are three basic ways of extracting perfumery ingredients from their plant matter sources, as follows.

2.1. *Distillation*

There are three commonly used distillation methods, *water distillation*, *water/steam distillation* and *steam distillation*. Before distillation is carried out, the raw plant matter may need to be pre-treated, to ensure that the best yield and quality are subsequently obtained. Pre-treatments include drying the raw plant material for a few days or, in the case of sandalwood for example, grinding the wood into fine particles so that the essential oil can easily escape. The steam and/or water used for the distillation process must have access to the raw plant material, so that the volatile portion of oil is carried over into the condensate. After distillation, the distilled oil settles on the surface of the water in the collection vessel.

It is important to note that the composition of the distilled oil is not always the same as oil found in the plant. The reason for this is that water-soluble components of the oil, for example phenyl ethyl alcohol from rose petals, are sometimes destroyed in the condensed steam, during distillation. The action of boiling water on any esters in the oil, may also lead to hydrolysis back to the original alcohol.

The sophistication of distillation apparatus used in different parts of the world is variable and may be as crude as a simple distillation device, using a metal drum. Conversely, some distillation apparatus consists of highly refined purpose-built stainless steel units, which are fully computer-controlled.

During water distillation, the raw plant material is immersed in the water within the still and heated externally, either by fire in the simple stills, or with electrical coils or steam jackets in more sophisticated apparatus. Water distillation is the simplest and cheapest method of distillation and is widely used in the less developed countries of the world. Certain materials can only be processed in this way, for example, rose petals and orange blossom. The major disadvantage of water distillation is that it is a lengthy process and because of the direct heating methods required, chemical changes can take place, which can destroy or severely spoil the oil's natural odour.

Water/steam distillation is the most common method of essential oil extraction. In this technique, the raw plant material is suspended above the bottom of the still, on a grid or in a basket, but is still immersed in water. Steam is then bubbled into the water, causing agitation of the plant material which carries the distilled oil over more quickly in to the receiver. This method has the advantage of speed and produces a higher yield of a better quality essential oil.

Steam distillation is used for materials where it is inherently more difficult to extract the essential oil, normally because the oil itself contains a higher percentage of higher boiling-point materials. The raw plant material is put into the empty distillation vessel, and super-heated steam passed through to effect distillation. This is a much quicker method than the other types of distillation but it should not be used for oils that are easily damaged or degraded by higher temperatures. Steam distillation is particularly suitable for materials such as vetyver and sandalwood but is not recommended for rose petals or orange blossom. Steam distillation is widely used for the extraction of all oils, except those with significant amounts of non-volatile or heat sensitive compounds.

A few essential oils are extracted by *dry distillation*. In this technique, the plant matter is treated in a closed vessel, in the absence of air, the oil being extracted with the smoke that is subsequently collected. Juniper berry oil and sandalwood oil have been traditionally collected in this way.

Typical yields obtained using distillation techniques are in the order of 1% - 2%, although yields range as far apart as 0.1% - 15%, depending upon the properties of the oil and the distillation method used. Essential oils are generally liquid materials, although some semi-solid types also exist.

Often, the extracted essential oil is further processed to concentrate, purify or extract particular components. *Rectification* is the name given to a process of fractional distillation, to improve the properties of an extracted oil. This process removes water, adjusts the terpene content of the oil and improves the colour. A typical example of rectification is the removal of dimethyl sulphide from peppermint oil, thus improving its properties in flavouring applications. Similarly, the cineole content of eucalyptus oil is increased by removing unwanted terpenes and residues.

The term "folding" is often applied to citrus oils. This term describes the process of concentrating the oil by the removal of unwanted components such as terpenes which have poor stability and solubility. Many folding techniques are in use today, including fractional distillation, thin-film or wipe-film evaporation and counter current-extraction. The removal of half of the volatile components present results in a double-concentration oil, normally referred to as a two-fold oil.

.2. *Solvent-extraction*

Fundamentally, there are two types of solvent-extraction processes, *volatile solvent-extraction* and *enfleurage*. Essential oils obtained by extraction processes are generally of better quality that those obtained by distillation methods, due to the absence of water and heat in the former. The main disadvantage of extraction is that unwanted materials, such as chlorophyll, are extracted from the plant matter by the extraction process and contaminate the extracted oil, resulting in poor colour. Extraction processes are mainly used for plants such as jasmine, tuberose, mimosa and hyacinth, where distillation processes cannot successfully be applied.

Volatile solvent-extraction is normally carried out in either batch processors or extraction drums. In both cases, the plant material is placed into the extraction apparatus and the solvent, frequently hexane, added. This mixture is left for several hours, after which time the solvent is run out of the extractor. The hexane solvent is then recovered under vacuum, to yield the essential oil in the form of a "concrete". Other solvents that are commonly used in addition to hexane include petroleum spirit and acetone. Volatile solvent extraction using a drum extractor is normally far more efficient than batch extraction processes and produces better yields of essential oils.

Enfleurage is a traditional method of extraction, which is rarely used today. This method involves placing the plant material on to layers of fat, typically 1 part tallow and 2 parts lard, which have been spread on to large glass plates held in a wooden frame. These glass plates are left for 24 hours, after which time the plant material is removed and replaced with a fresh stock. This process is repeated over a number of weeks, resulting in layers of fat saturated with the required essential oil. This mixture is known as a "pomade" and the essential oil is extracted by stirring the fatty mixture with ethanol. The ethanol is then distilled off under vacuum to leave an "absolute of enfleurage".

Enfleurage is a very labour-intensive and costly process, although the products obtained by this method are of very high purity. The essential oil of tuberose may be produced by enfleurage of the rose petal itself.

2.3. *Expression*

Expression is a method of extraction used for the essential oils of citrus fruits only. The technique involves squeezing and/or scraping the peel of the citrus fruits to obtain the oil Modern expressing machines use one of several methods to express the essential oil. The most common techniques involve either jabbing the fruit peel with metal spikes and washing off the essential oil with a water spray, or squeezing the fruit peel under high pressure.

Essential oils are produced in many different countries around the world. Some oils may be unique to a particular area but often they are produced in several different countries of similar climate.

The sales and distribution of the oils is usually handled in one of the following three ways. Firstly, the essential oils producers may sell direct to the customer. This is most often the case with citrus and mint oils, especially when the producer is a large commercial organisation. Secondly, specialist organisations may be used to import, export and trade the essential oils. Sometimes these organisations may also be involved in rectification and blending of oils. Lastly, flavour or fragrance companies, perhaps representing a certain producer, will simply sell the oils as raw materials or flavours.

Of the specialised companies many may use brokers, the most important being based in New York, Grasse (France) or the United Kingdom. Some countries, however, may closely control the sale and export of their essential oils, perhaps through a government agency or similar organisation. Often, oils produced in Russia, for example, are offered as payment for other goods, which may have been imported into the country. The sale of

Chinese peppermint and spearmint oils is closely monitored by various government and provincial offices.

3. *Specific essential oil types*

As previously mentioned, there are literally thousands of essential oils available and some of these are more commonly used than others, by the perfumer. Further details of some of the more important essential oil types are given below.

3.1. *The grass family*

The grass family consists of several species, the most important ones being lemongrass, citronella, palmerosa and vetivert. Typically, these grasses grow to approximately 3 feet in height.

Lemongrass contains about 70-80% of citral and occurs as two types, East Indian Lemongrass, grown mainly in India and West Indian Lemongrass, grown mainly in China, Brazil and Guatamala.

During harvesting, the grass is cut near the ground and then crudely distilled in the field, using water distillation in a simple drum distillation apparatus. The distillation vessel is heated by a fire underneath and the result oil is condensed in a second vessel, often simply stood in a river to provide cooling. Typically, a bamboo shoot is used to pass the volatiles from the distillation flask to the receiver. Distillation takes several hours, producing an average yield of between 0.2%-0.3% essential oil, which has a crude, grassy odour. Typically, batches of the crude distilled oil are blended by the broker and this blend can be used in its own right as a perfume compound for low-priced products. Originally, citral was only available from lemongrass. Nowadays the competitively priced synthetic citral is by far the most important source. Despite this, there are still over 1000 tonnes of natural citral produced every year.

There are two types of citronella, Ceylon Citronella, grown mainly in Sri Lanka, and Java Citronella, grown in Southeast Asia and South America. The Ceylon oil is less valuable than the Java oil and is used mostly in fragrances for soaps, household products and detergent powders. The Java oil is used widely throughout the perfumery industry and is also used as the starting material for the production of other fragrance ingredients.

Vetivert oil is obtained from the roots of a grass that grows in various parts of the world, including India, Reunion Island, the African Congo and Brazil. The oil, which is reddish-brown in colour and has a characteristic, long-lasting, woody odour, is obtained by steam distillation, with a yield of between 1% and 2%. The oil is complex and contains many important fragrance compounds, as well as being the starting material for the production of vetivenyl acetate.

Palmarosa oil has a rose-like, grassy, fruity note and contains about 90% geraniol. It is used in fragrances for soap and cosmetics and as the starting material in the production of high quality geranyl esters.

3.2. *Geranium oil*

Geranium oil is obtained from various species of geranium, grown in the Reunion islands, Madagascar, Morocco, China and Egypt. It has a characteristic rose-like odour and is very useful in the creation of soap fragrances. The best and most expensive geranium oil is from the Bourbon geranium, grown in Madagascar.

Distillation of the whole plant takes place as soon as possible after cutting and typically gives a yield of approximately 0.15%. Traditionally, the distillation process is carried out in a copper retort, over an open fire. Geranium oil itself has a rosy odour and it is sometimes fractionally distilled to produce an oil containing just geraniol and citronellol. This oil is known as Rhodinol ex Geranium, and exhibits a much stronger rosy odour.

3.3. *Rose oil*

Rose oil is used mainly in fine fragrances and there are two different oils that are of commercial importance. Rose absolute is produced by solvent extraction, whilst rose otto is the essential oil of rose, produced by fractional distillation. Rose oil is produced from roses grown in both Bulgaria and France.

Bulgaria grows Rosa Damascene, a pink rose with few petals, that only flowers in May. Rosa Damascene can be used to produce both rose absolute and rose otto. To obtain rose otto, distillation is used, although during distillation some of the fragrance oil is lost in solution. When the oil is first produced it is referred to as rose direct oil. The water condensate contains about 75% of the dissolved rose oil and this can be recovered by solvent extraction with petroleum spirit. Alternatively, the water can be gently distilled and fractionated, to give a redistilled rose oil. Rose otto is extremely expensive, due to the enormous number of rose petals required to produce even small amounts of the oil itself.

France grows the Rosa Centifolia rose, which is usually solvent extracted with petroleum spirit to give a concrete from which the absolute Rose de Mai is obtained by extraction with ethanol. The yield of concrete from the petals is approximately 0.25% and, of that, 40% is wax and 60% is rose absolute.

3.4. *Jasmine*

Jasmine was originally grown in Italy and France and is now also grown in Morocco, Egypt and India. The blossoms cannot be steam distilled, so solvent extraction is used to obtain the fragrance. The Jasmine concrete thus obtained contains 50% absolute and 50% wax, with a yield of approximately 3 kg of absolute per 1000 kg of flowers. Jasmine oil has a very floral odour and the absolute is one of the most valuable ingredients for use in fine fragrances.

3.5. *Citrus oils*

Citrus oils include lemon oil, orange oil, grapefruit oil, bergamot oil, neroli oil and petitgrain oil. They are obtained from the various citrus fruits, which are typically grown in Mediterranean climates in countries such as Italy, Sicily, Israel, Spain, Australia and Western America. Although citrus fruits are mainly grown for their juice, the citrus oils are a really useful by-product for the perfumer.

All of the oils are obtained by expression and are rich in monoterpene hydrocarbons. They are among the most sensitive fragrance materials in use and are particularly susceptible to oxidation. For this reason they are often protected with an antioxidant and must be stored very carefully. Citrus oils are very widely used in personal perfumes.

Lemon oil is widely used in Eaux de Colognes, because of its fresh odour. The major component of lemon oil is limonene, with about 3% citral and traces of C8 and C9 aldehydes. In the case of orange oil, most of the perfume oil used is from the sweet orange plant, the bitter orange peel being used mainly for flavours. Sweet orange oil has more than 90% limonene and although it is primarily used in flavours, it does find use in Eaux de Colognes and soap fragrances. Grapefruit is chemically similar to orange oil but it has a distinctive smell which is largely attributable to a ketone called nootkatone, which is present at levels of approximately 0.5% w/w. Grapefruit oil does not find very wide use in perfumery.

Bergamot oil is a true citrus oil, obtained by cold pressing the peel from the unripe bergamot fruit, to give an oil with a sweet, fruity odour. Bergamot is only grown in Southern Italy and, being inedible, is grown solely for the perfume oil. Unlike other citrus oils, which contain mainly limonene, bergamot oil also contains linalol and linalyl acetate, with a relatively low terpene content of between 20% and 30%. Bergamot oil is indispensable for use in high quality personal perfumes and is one of the major ingredients in both Eau de Cologne and fougère-type fragrances. Use of the untreated oil is limited by its photosensitising properties, which are attributable to the low levels of bergaptene sometimes present.

Neroli oil is the oil obtained from orange blossoms, the best quality coming from blossoms of the bitter orange. Plants were originally grown in France but are now also cultivated in Italy, Spain, Egypt and South Africa. The blossoms are harvested and immediately water distilled, giving a typical yield of approximately 0.1% oil. The blossom can also be solvent extracted with petroleum spirit, to give a neroli concrete which can be further extracted with ethanol, to give an absolute.

Petitgrain oil is a water/steam distillation product, obtained from the leaves and twigs of the orange tree. Some petitgrain oil is obtained from the bitter orange tree and this produces a special oil called petitgrain bigarade. The remaining petitgrain oil comes from the sweet orange, or the hybrid plant grown in Paraguay. Petitgrain oil produced in Paraguay has an almost floral odour, with green, citrus notes. The quality of the oil can be improved by distillation to remove the terpenes, the resultant product being known as petitgrain terpeneless, which has an orange blossom odour similar to neroli. Petitgrain oil mandarinier is a source of natural methyl anthranilate.

3.6. *The labiates*

The labiates are a botanical group of plants characterised by the pairs of leaves running up their stems and include, lavender, peppermint, spearmint, rosemary and patchouli.

Lavender oil can be obtained from three types of lavender plant, lavender itself, lavandin and spike, each of which grow at slightly different altitudes. The colours of the respective flowers differ. Lavender has deep blue flowers, spike has greyish flowers and lavandin, being a hybrid of the two, has flowers which are blue/grey in colour. Most lavender is grown in the mountain ranges of Southern France and the major components of the oil are linalol and linalyl acetate, with small amounts of camphor and cineole.

As lavender must be grown at high altitudes, it is very difficult and expensive to work. Spike, on the other hand, may be grown at much lower altitudes, although the oil is not perfumistically as good. In view of this, a hybrid lavender, known as lavandin, has been developed and the quality and composition of this oil has improved over the years. The best lavandin hybrid, lavandin super, contains about 40% - 45% linalyl acetate and 40% linalol. The odour from this oil is almost as good as that of true lavender and the oil is much cheaper to produce. For these reasons, oils from lavandin hybrids are progressively replacing the use of true lavender oils in perfumery. As a consequence, the production volumes of lavender are decreasing rapidly. Lavandin hybrid oils are extracted by steam distillation, thereby avoiding hydrolysis of the linalyl acetate present, to linalol. They can also be extracted using volatile or solvent extraction methods, producing a darker oil which is, however, of much better odour.

Spike lavender grows in Southern France and Spain and unlike lavender and lavandin has never been cultivated. Chemically, spike oil contains more camphor and cineole than lavender or lavandin and its odour possesses some of the characteristics of eucalyptus. Spike oil contains approximately 10% linalyl acetate.

Lavender and lavandin are widely used in perfumes to provide a floral, herbal, top note, forming an important part of the fougere type of fragrance. Spike oil has, in the past, been used for cheaper perfumes to give a herbal, pine-needle odour, but its use is now inhibited due to its increasing price.

Peppermint and spearmint oil are generally of use to the flavour industry and have little perfumery application.

The patchouli plant is grown in the Philippines, Indonesia and a little in India. The oil has a rich, woody, spicy, odour and is very long lasting, due to its high molecular weight composition of sesquiterpenes. This oil cannot be synthesised economically, due to its complex composition.

3.7. *Orris oil*

Orris oil is obtained from the rhizomes of the Italian Iris plant. The rhizomes are dug up, washed and stored for periods of up to 3 years, to allow the odour to develop. After this maturation stage, the rhizomes are crushed and steam distilled to give a solid known as orris concrete. The rhizomes can also be solvent extracted to give orris resinoid, a liquid.

3.8. *Clove bud oil*

Clove plants are mostly grown in Tanzania, Indonesia and Madagascar. The "cloves" are, in fact, flower buds, picked just before they open and then dried in the sun, before commonly being exported for culinary and fragrance use. Clove oil is obtained by chopping the dried cloves, whereupon they are crushed and water distilled. During distillation, the oil undergoes a chemical change and some materials such as caryophylene, not present in the buds, are formed. The principal odour of clove bud oil is warm and spicy, which is largely attributable to the high levels of eugenol present. The leaves and twigs of the clove plant can also be pruned and distilled, to give clove leaf oil, which contains about 85% eugenol. Clove leaf oil, which is much cheaper to purchase, has a drier, less sweet odour than clove bud oil. Both clove bud oil and clove leaf oil are principally used for their eugenol content. Eugenol obtained from these sources is usually converted into iso-eugenol, which has a carnation odour.

3.9. *Ylang oil*

Ylang is a tropical plant grown on the Comoro Islands, Reunion Island and a small island called Nose-Be. The plant has yellow flowers, which are picked and immediately distilled by water or steam/water distillation. Traditionally, the oil is taken off in fractions during distillation, as shown below:

1st Fraction	Ylang extra oil (best quality)
2nd Fraction	Ylang No.1
3rd Fraction	Ylang No.2
4th Fraction	Ylang No.3 (poor odour)

Ylang No.3 is obtained about 24 hours after the distillation process commences and the still operators decide at which points during the distillation the various fractions should be taken. Perfumers prefer to use the ylang extra grade, particularly when creating high quality perfume compounds. The other fractions are sold for use in cheaper perfumes. Any fractions not sold are blended and sold as ylang complete, inevitably a product of variable quality. Ylang oil has a powerful sweet, floral odour, similar to jasmine, but not as fine. Ylang oil contains methyl benzoate and methyl paracresol, materials not found in jasmine. These components mar the fineness of the ylang oil odour. Ylang is also grown in Java, where the whole plant is distilled in a crude still, heated by a log fire. This produces an oil called cananga oil, whose odour is not as fine as that of ylang, due to its cresylic nature and higher caryophylene content.

3.10. *Cedarwood oil*

Cedarwood oils are distilled from waste cedarwood off-cuts from the timber industry. There are many different types of cedar, including Texan Cedar, Indian Cedar, Virginian Cedar, Moroccan Cedar and Chinese Cedar. Each of these varieties produces a different type of cedarwood oil, which is extracted by grinding the wood to a fine sawdust and carrying out a steam distillation. The oil itself contains many sesquiterpenes and has proved very difficult to synthesise. Cedarwood oil can be fractionated to give the isolates cedrol, cedrenol and cedrene, which have much finer odours. The odour of both cedrol and cedrenol can be improved by acetylation, to give cedryl acetate and cedrenyl acetate.

3.11. *Sandalwood oil*

Some sandalwood is used for building but most is distilled for its perfume oil. Sandalwood is generally one of the purest essential oils. The best sandalwood oil is produced in Southern India and is very pure and of consistently high quality. The oil is extracted by grinding the wood into sawdust and then steam/water distilling for periods of up to 5 days. This process produces cedarwood oil with an excellent yield of 4.0% - 6.5%. The oil is composed largely of sesquiterpenes and, as most of the components have similar boiling points, it has a consistently pleasant, woody, oily odour throughout its life.

3.12. *Cistus and labdenum*

Cistus and labdenum oils are obtained from the gum of *Cistus ladaniferus*, which grows in the Mediterranean area. The twigs of the plant can actually be boiled, not distilled, in water, to give labdenum gum crude, a brown tarry material. This gum can then be further processed, to give an absolute or a concrete. Labdanum absolute is extracted with alcohol and a colourless product can be produced using hexane.

The twigs can also be solvent extracted or steam distilled, to give a concentrated odorous product. Steam distillation produces cistus oil, whilst solvent extraction with hexane yields cistus concrete, which can be further extracted with ethanol to give the absolute. Cistus products are stronger in odour and much more expensive than labdenum products. They are widely used in fragrance compositions with a balsamic, warm odour, reminiscent of ambergris. The resinoids and absolutes exhibit very good properties as fixatives.

4. *The quality and availability of essential oils*

Unfortunately, most of the commercially important essential oils are produced in areas of the world which are isolated from the main markets and many strategically important oils grow in areas susceptible to flooding, drought or political upheaval. These factors present significant problems, which affect both the price and quality of essential oils.

Since essential oils are natural products, they are produced seasonally. Often they are best purchased immediately after the season ends which, of course, results in logistical and storage problems. There may also be crop to crop seasonal variation, which may result in price changes, since high quality material naturally commands a higher price.

The quality of each crop is judged organoleptically by experts and any variation in quality can be removed by blending, to meet a desired specification. Natural products have an inherent tendency to be unstable and stability problems can occur when storing large quantities of essential oils for long periods. For this reason, essential oils are best stored in full, airtight containers, at a constant temperature of between 16°C and 18°C. If containers must be stored part-full, then the use of a nitrogen blanket in the container head space will improve the stability of the oil significantly. Citrus oils are particularly prone to oxidation and antioxidants, such as butylated hydroxytoluene, are frequently added to reduce this tendency.

Unfortunately, the low volume and high price of some essential oils can lead to the dishonest practice of adding dilutents, extenders or other adulterants, which make the oil appear of better quality than it is. False declaration of natural products and falsification of product specifications also occurs.

The availability of essential oils may vary with the weather or the political climate. Floods, drought and changes in currency exchange rates all cause prices to fluctuate, as can speculation on a crop and the actual harvest itself. Often the major international traders will visit the main growing areas and plantations of the crops they are interested in, to assure themselves that the crop is satisfactory, before agreeing a purchase price.

5. *The chemistry of essential oils*

Essential oils should not be confused with other well known oils, including vegetable and petroleum based oils. They are composed mainly of terpenes, both aliphatic and cyclic together with some oxygenated terpenoids, aromatics and heterocyclic compounds. Kekule first used the word "terpene", to describe the content of essential oils but Wallach is recognised as having pioneered their analysis.

Terpenes are derived from isoprene and each molecule is built up in multiples of isoprene units. These isoprene units are usually joined head to tail, the resulting compound having varying degrees of saturation. It is interesting to note that isoprene does not occur naturally. Any terpenes larger than the diterpenes, with a 20 carbon atom skeleton, would not usually be found in an essential oil.

Quite often, the most important component of an essential oil is an oxygenated isoprenoid, which may be an alcohol, an aldehyde or cyclic in structure. Essential oils may contain more than two hundred components and although one or two may dominate, a substance present in parts per million may be essential to the odour character. The absence of one component may change the odour profile significantly and oils extracted from the same plant grown in different parts of the world, or under different climate conditions, may have different odour profiles, due to different ratios of the same components.

An increased understanding of the chemistry of essential oils and their components, has brought about the use of increased levels of synthetic materials. Synthetics were first produced to overcome fluctuations in cost and availability of the essential oils, brought about by poor yields, drought, disease, pestilence or political unrest. Although synthesis of all components of an essential oil would not be economical, synthesis of two or three of the main components is possible and these can be blended to produce a synthetic oil, or used in isolation for their own odour value. Although the absence of some trace components may affect the aroma of a blended essential oil, many of these contribute little organoleptically.

Essential oils themselves can be considered as creative fragrances, as they contain hundreds of perfectly blended ingredients which, when added to a fragrance composition, improve the odour. Some essential oils, for example lavender and ylang, can be considered as fragrances in their own right. Before synthetics were available, all perfumes were based on essential oils. One of the oldest compounded fragrances, Eau de Cologne, is based upon the oils of lemon, bergamot, pettigrain, and lavender. Although modern fragrance compounds rely heavily on the impact and longevity that can be achieved with synthetics, many successful fragrance compounds still contain significant quantities of naturally occurring essential oils.

6. *The health and safety requirements for essential oils*

Many essential oils are approved for flavour use and therefore their physiological properties must be considered. As mentioned previously, some essential oils may contain noxious or toxic substances, which should be removed before the oil is used commercially.

Safety investigations using animal experiments have been used historically and dermal and oral toxicity results are published in monographs produced by the by the Research Institute for Fragrance Materials (RIFM). Toxic effects on man have not been documented to any great extent. Most essential oils are generally recognised as safe by the Food and Drug Administration (FDA) of the United States, and are classified as natural products. Some oils can be used for both flavours and fragrances, whilst others may only be used for one or the other application.

Recommendations on usage levels of essential oils are given by the Flavour and Extract Manufacturers Association (FEMA), the British Essence Manufacturers Association

(BEMA), the International Organisation of Flavour Industries (IOFI), the Research Institute for Fragrance Materials (RIFM), the International Fragrance Association (IFRA), the International Federation of Essential Oils and Aromas Trades (IFEAT) and the British Essential Oil Association (BEOA). The International Standards Organisation (ISO) have also issued guidelines for the safe handling of essential oils, which covers packaging, marking and labelling requirements.

7. *The future of essential oils*

In the future, it is unlikely that many new oils of commercial significance will be identified, although it is likely that novel versions of currently available products will become available.

New extraction techniques such as super-critical point fluid extraction, and new processing techniques such as spinning-band distillation, will result in the production of essential oils with enhanced characteristics. These enhanced oils will be of higher quality and exhibit more refined perfumistic value, thus enhancing their usefulness to the perfumer.

The use of the spinning-band technique, for distillation of essential oils, results in very fine separation of the components. The distillation columns employed have about 200 theoretical condenser plates and this has enabled materials present at very low levels to be extracted and subsequently added back at higher concentrations, to create versions of essential oils with new aroma profiles.

The production of essential oils through the use of modern plant-cell tissue culture techniques may soon be possible. Much research in this area is currently being conducted in both industry and academia, with promising results. By cloning high-yielding strains and growing the cells in a recirculating system, it may be possible to produce essential oils without the need to grow crops on large areas of land and without risks associated with the vagaries of climate and politics.

Other perfumery ingredients

In addition to the essential oils, there are other perfumery ingredients which have an important role to play in the art of perfumery. Some of the more important materials are detailed below.

1. *Animal derived products*

Some common animal products, such as musk, civet, ambergris and castoreum, are still occasionally used in perfumery, although generally only in special fine fragrances. These materials are usually produced as a tincture, the animal product being ground with sand in a mortar and pestle and allowed to stand in ethanol for as long as possible. The longer the mixture is allowed to stand, then the better the odour. The ethanol is not removed, since the product is produced as an alcoholic solution, with a concentration of about 20% oil. The tinctures tend to be strong smelling materials, the molecules responsible for the odour normally being of high molecular weight. Animal-derived products are most frequently used in highly substantive, long-lasting fragrances. Most them have now been duplicated synthetically.

Civet is obtained from the civet cat, which emits an odorous paste in the abdominal vicinity, when it becomes annoyed. This material is believed to act as both a sexual attractant and enemy repellent. The freshly secreted material is a liquid, with a light yellow colour, which becomes darker on exposure to light. Good civet gives a sensual, warm smell, even when used at fractions of a percent in a fragrance formulation. Dealers may dilute the civet with honey or banana puree, in order to extend it.

Ambergris is found on the coastal beaches of Ireland, New Zealand and Australia. It is produced in the stomach of the sperm whale as waxy lumps and is released as a normal excretion. The fresh material is black but turns light grey over time. Ambergris only develops its pleasant odour after exposure to light and sea water, after a period of time.

Castoreum is obtained from the Canadian Beaver, which emits an oil from a gland in the abdomen, also called the castoreum. The principle source of castoreum is the fur industry, which processes the pelts and extracts the beaver pouches. The gland is ground and gives a pleasant, slightly fruity, musk-like odour which, when mixed with hexane, gives castoreum resinoid. The resinoid can be extracted with ethanol, to give castoreum absolute.

Musk was originally obtained from the musk deer in Tibet and the Himalayas. The musk deer would rub its abdomen on tree trunks, to leave an odour for other musk deer to follow. Musk is a very highly priced material, which has now been mostly replaced by powerful synthetics.

2. *Concretes*

A concrete is the term used to describe a product which is extracted from a non-resinous raw material with a non-polar, volatile hydrocarbon solvent, such as hexane. Concretes are extracted from fresh vegetable tissue and normally contain non-volatile natural waxes.

3. *Absolutes*

An absolute is a highly concentrated, alcohol-soluble extract, which is obtained from a concrete by repeated extraction with alcohol. The alcohol extracts are bulked and the alcohol is then removed, by evaporation, to yield the absolute. Absolutes can also be obtained in a similar way from resinoids.

4. *Resinoids*

A resinoid is a perfumery material prepared from a natural resinous substance, normally a plant exudate, by extraction with a hydrocarbon solvent. Resinoids are produced from dead vegetable matter, whilst absolutes are produced from freshly harvested vegetable matter.

5. *Isolates*

An isolate is a chemical obtained from an essential oil by fractionation. An example of an isolate is geraniol, obtained from citronella oil, this material being known as geraniol ex citronella. Other isolates commonly used in perfumery include eugenol ex clove leaf oil and citral ex lemon grass. Isolates are often produced in order to give an improved odour, or simply to increase the concentration of the main fragrance component.

6. *Terpeneless and sesquiterpeneless oils*

Hydrocarbon terpenes do not contribute significantly to the odour character of an oil and their removal yields a much improved odour. In an essential oil, hydrocarbon terpenes have the highest volatility and can therefore be removed by fractionation, under partial vacuum. This method cannot be used for sesquiterpenes however and these are removed by an extraction process known as counter-current solvent extraction. This involves partitioning the oil between two immiscible liquids, whereupon the essential oil is preferentially solubilised in one layer, whilst the terpenes and sesquiterpenes remain in the other. A typical solvent system consists of a methanol/water mixture as the polar layer and pentane as the non-polar layer. Shaken together with the essential oil, and allowed to separate, the terpenes and sesquiterpenes will dissolve in the upper pentane layer, whilst the oil will dissolve in the lower methanol/water layer. The methanol/water portion is separated, the methanol distilled off and the oil, being insoluble in water, separates out.

Perfumery quality control

Quality control in the perfumery industry fulfils four main functions, as follows:

- it identifies errors in incoming raw materials
- it acts as a quality check of factory stock, ensuring that no deterioration has occurred with storage
- it examines the quality of incoming raw materials
- it examines fully compounded perfumes and fragrances, to ensure they are identical to previous batches of the same material and acceptable to the customer

1. *Raw material quality control*

If a finally formulated fragrance is to have a consistent odour from batch to batch, then it follows that all incoming raw materials must have the same odour profile as the previous batch of that material purchased. All raw materials must therefore be checked against a reference standard of target quality, which is stored in a fridge at 2-4°C. A reference standard is retained for every raw material purchased. When a new batch of raw material is delivered, samples are taken and compared with the reference standard, by the perfumer, on a smelling strip, both initially after dipping and after a few hours standing, to check the odour on "dry out". If possible, the strips should also be compared the next day, before the raw material is accepted. This test protocol examines both the identity and quality of the raw material.

A sample of the delivered raw material is also sent to the analytical laboratory for density and refractive index measurements, along with other analytical checks, such as infra-red spectroscopy and gas chromatography. The refractive index measurements give an indication of purity of the material. The raw material reference standard should be changed about every 6 months and replaced with a sample from a batch of known good quality. If a strong perfume raw material is being assessed, it is wise to dilute it to a 1% solution with an odourless solvent, such as propylene glycol or diethyl phthalate, to enable better comparison with a similarly diluted standard to be made.

Water-soluble raw materials, notably phenyl ethyl alcohol and hydroxycitronellal, can be assessed by putting a few drops into water in a beaker. If the material is pure, it will dissolve in the water with no odour. If however impurities are present, they will float on the surface of the water and exhibit an odour.

2. *Raw material adulteration*

Very sophisticated methods of adulterating perfume ingredients are used by unscrupulous brokers and adulteration of raw materials should be checked for, by the quality assurance department.

Commonly used methods of adulteration include:

- the addition of a synthetic chemical to an oil. The added material is normally one that occurs naturally in that oil and the addition of the synthetic makes an otherwise poor quality oil appear acceptable. An example of this practice would be the addition of cinnamic aldehyde to cinnamon oil, which already contains 8% of this material naturally. The only method for detecting this sort of adulteration is gas chromatography, comparing the traces with reference standards and normally acceptable samples
- the addition of a "synthetic essential oil blend" to an essential oil. This form of adulteration is impossible to detect analytically and can be detected by odour only
- adulteration arising from the production of terpeneless oils. When the terpenes are removed from an oil, they can be added to a second batch of oil to "stretch" it. This form of adulteration is only detectable by gas chromatography

Generally, the techniques of gas chromatography and optical rotation are useful in the detection of adulterated products.

3. *Finished perfume quality control*

Chemical analysis of finished perfumes is not a useful technique, due to the complexity of formulations involved. The main quality control technique used for finished perfumes is odour assessment, by the perfumer. It is important to remember that odour quality is what the customer is buying and the purchasing specification will always specify this parameter. The purchasers' quality assurance department will almost always measure some of the physical parameters of the finished fragrance compound, as they are unlikely to have expert fragrance evaluators themselves. In this case, a reference standard is retained for each perfume formulation and stored in a fridge at 2-4°C, each new batch of that formulation delivered being checked against the standard.

It is worth noting that the method of odour assessment for evaluating finished fragrance compounds is very different from that used when assessing raw material ingredients. In the case of raw material ingredients, a very brief odour assessment of both sample and standard is made. However, in the case of finished perfume quality assurance, the method employed would typically involve the following steps:

- the standard is assessed by smelling it for approximately 10 seconds, until the fragrance can no longer be smelt, that is when nose saturation has occurred. The sample is then assessed immediately by smelling it very briefly, for approximately one second. Any additional odours detected may indicate "extra ingredients" in the sample

- After a short break, the sample is assessed by smelling for approximately 10 seconds, until nose saturation occurs. The standard is then immediately assessed by smelling it very briefly. Any additional odours detected indicate ingredients "left out" of the newly delivered batch

Difficulties arise when assessing freshly made perfume batches, as they are too fresh and often smell a little harsh. Perfume compounds normally mature over the first 2–3 weeks after manufacture, whereupon the odour may improve significantly. Judgments must be made however on newly produced batches, due to customer demand for short delivery lead-times. In these cases the perfumer must consider, when comparing sample with standard, that the standard will be smoother and more mature than the freshly made batch.

The storage of perfume compounds

Ideally, perfumes should be stored in glass vessels but for practical reasons this is undesirable. Perfumes are therefore best stored in full stainless steel or lacquer-lined steel drums. The lacquer lining must be in good condition, as any contact of the perfume with the metal casing may cause discolouration and deterioration. Drums should be tightly sealed and stored indoors, in a cool place. Ideally, essential oils should be stored in a cold room at temperatures of between 2°C and 4°C. For drums that are only partly full, the head space should be purged with nitrogen, to remove oxygen and prevent deterioration. Plastic vessels are wholly unsuitable for the storage of perfumes, as they are often permeable to vapours and tend to soften when they come into contact with organic solvents.

The safety of perfume compounds

The main safety concerns with fragrance formulations involve skin contact. Historically, the perfumery industry has had few safety problems but, statistically, every perfume ingredient will cause problems for a few people. Safety considerations however are concerned with materials that affect significant numbers of people, rather than just the few isolated reactions that are inevitable with every perfume compound. It is also important to distinguish between the safety requirements for both "wash-off" and "leave-on" products. Standards of safety regarding the use of materials in formulations are regulated by the industry itself, following the guidelines and recommendations prepared and monitored by the two industry supported international bodies, RIFM and IFRA. RIFM, the Research Institute for Fragrance Materials, collates information and data about the safety aspects of each perfume ingredient. In addition to this, they test all perfume ingredients individually, with the assistance of universities and various research establishments. The results obtained, both for synthetic chemicals and essential oils, are published in the Journal of Food and Cosmetic Toxicology. The corresponding RIFM Monographs are published by a committee of experts including toxicologists, biochemists and other industry experts. To date, only very few materials have been found to give significant adverse skin reactions.

IFRA, the International Fragrance Association, tends to be a much more valuable organisation for the perfumer. IFRA study the data and reports issued by RIFM and make recommendations for the use of each perfume ingredient accordingly. These recommendations are now adhered to by all perfume houses. The practicality of IFRA's approach is endorsed by the fact that the safety guidelines for "leave-on" and "wash-off" fragrance formulations differ.

Sometimes the perfumer is able to use some otherwise irritant products, through the controlled use of a method known as *quenching*. This term is used to describe the effect that one chemical may have on another, causing the second chemical, which normally shows adverse skin reaction, to become non-irritant. For example, cinnamic aldehyde normally gives an adverse skin reaction in a significant number of people. When this material is mixed with cinnamic alcohol or eugenol however, quenching takes place and irritation does not occur. Phenyl acetaldehyde, a potential irritant, is quenched to become safe, when mixed with an equal amount of phenyl ethyl alcohol.

Creative perfumery

Simplistically, the art of creative perfumery involves the blending of many different fragrance raw materials, to produce a perfume compound which satisfies all the requirements of the creative perfume brief.

1. *Creative terminology*

Before discussing the perfume creation process further, it is important to understand some of the terms used in creative perfumery. These are as follows:

1.1. *Top note*

The top note of a perfume is the odour profile produced by its most volatile ingredients. Top notes give the perfume lift, life and impact and without top notes a perfume would appear flat and uninteresting. Typically, the top notes create the initial impression of the perfume and they contribute most significantly within the first few minutes of smelling the odour. Typical top note ingredients include citrus oils, cineole and decanol.

1.2. *Middle note*

The middle note of a perfume is the odour profile produced by the moderately volatile ingredients. Middle notes give the perfume its main character and body. Typically, middle notes will last within the first few hours of using the perfume. Typical middle notes include citronellol, phenyl ethyl alcohol and clove leaf oil.

1.3. *End note*

The end note of a perfume is the odour profile provided by the least volatile ingredients. End notes give the perfume its long-lasting, or substantive effects and also contribute towards the effects produced by the top and middle notes. The end notes are more obvious after the perfume has dried down and the odours from these materials may last days, weeks or even months.

The relationship between the top, middle and end notes of a perfume, as a function of odour impact with time, are illustrated graphically in the perfume evaporation curve shown in Figure 4.

FIGURE 4

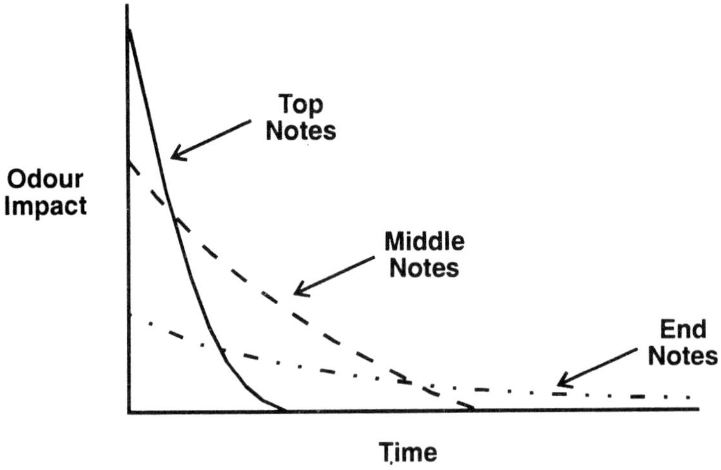

Not all perfumes require the same proportions of top, middle and end notes and the proportional ratio of these ingredients will depend upon the type of product and its intended use. Some products, for example toilet soap, require a lot of top notes but not many end notes, whereas others, for example skin creams, require a high level of end notes and very little top notes.

In a fragrance created to be used in a product with a long-lasting odour, the proportional ratio would typically be approximately 50% end notes, a high level of middle notes and a low level of top notes. Conversely, if the product was designed to have a significant immediate odour impact, for example as in an air-freshener, then a typical proportional ratio would be a low level of middle and end notes, combined with a high level of top notes.

1.4. *Basics*

When perfume ingredients are mixed, some are present at high percentages, whilst others are present at lower percentages. If this were not the case, the perfume would appear flat and uninteresting. Normally, some of the ingredients are mixed at high percentage levels and subsequently smoothed out with ingredients added at lower levels. The materials present at the higher percentages are known as basics.

1.5. *Blenders*

When smoothing out a perfume, the odour profile must be adjusted so that the odour of no single ingredient is obvious, in the final perfume. Certain chemicals are particularly useful for smoothing out others and these materials are known as blenders. The main function of blenders is to harmonise the other ingredients in the perfume.

1.6. *Fixatives*

Fixatives are materials which prolong the odour effect of a perfume, making it long-lasting and substantive. Fixatives also slow down the evaporation rate of the volatile materials in the perfume. Fixatives, which need have no odour value of their own, are normally high boiling-point materials which, when added to the more volatile ingredients in the perfume, reduce the overall vapour pressure of the mixture. Some commonly used fixatives include vetyver, dipropylene glycol and diethyl phthalate. If the fixative is a polar material, hydrogen bonding can take place and this fixes the volatile ingredients even more tightly. An example of this, is the fixation of benzyl acetate by benzyl benzoate. There is also a phenomenon known as *pseudo-fixation*, which strictly is not a fixation process at all. In this case, the end notes are formulated to give a similar odour to the top notes, thus giving the perfume a "false" long-lasting effect. For example, a rose perfume containing geraniol and citronellol can be pseudo-fixed with benzophenone and trichloromethylphenylcarbinyl acetate, which have similar rose-like odours.

1.7. *Modifiers*

Modifiers are minor ingredients, added at low levels in a perfume, to give it interest and character. The modifiers used are often completely unrelated to the base odour, for example a rose perfume may contain modifiers with a spicy odour.

1.8. *Bases*

A base is the term used to describe a small number of perfume ingredients which are mixed together to form a "part-perfume". Bases can therefore be considered as "semi-perfumes", which are used by the perfumer to save time during the creation process. A perfumer will have access to a wide variety of bases, some of which will be commercially available and many of which will have simple odour types such as rose, lilac or lavender. Bases can also be used as modifiers, when they are incorporated at lower levels.

2. *The creation process*

When the perfumer is creating a perfume formulation, he or she will normally follow a set development programme in creating the compound. Firstly, the middle notes of the fragrance are created, followed by the addition of the base notes, until the perfumer is satisfied with the odour balance and blend. Finally, the top notes are added to the perfume to give added interest and character.

In creating a perfume, the perfumer gives consideration to the technology used in the final product, before making a selection of fragrance ingredients to be used in the final perfume. There may be requirements for alkali or acid stability, or compatibility with a particular base. A perfumer creating a perfume for use in household products may have a particularly difficult task, especially given the cost restrictions that he or she has to work within. If a "natural" odour is required, the perfume may be blended to mimic a natural product and then reinforced with selected aroma chemicals. The ultimate creative challenge for any perfumer is the "fantasy" fragrance, designed to support a particular marketing brand image or concept. This form of creative perfumery often requires the use of a complex blend of ingredients, both natural and synthetic. Using premixed bases, fixatives, modifiers, blenders and a great deal of trial and error, the perfumer finally arrives at a well rounded, balanced, artistic composition, that will stand the test of time.

3. *The information required to create a perfume*

Before a perfumer can begin to create a perfume, certain information must be known about its use and application. A list of the basic information required for perfume creation is given below:

- what product type is the perfume intended for? The perfumer may need knowledge of the product formulation, in order to balance the top, middle and end notes correctly
- in which countries will the product be sold? This information is important not simply for marketing reasons but also because perfume trends and preferences differ geographically, from country to country. Any legislative requirements should also be borne in mind when considering the countries in which a product is to be sold
- the cost limits and constraints of the perfume formulation. The perfume is often the most expensive ingredient in a consumer product and generally the more expensive the perfume the more likely it is to help the end product to sell
- the marketing image of the product. This information should provide details of both product functionality and emotive benefit. The perfume type must work in synergy with the marketing concept and brand image
- the expected shelf-life of the product, both "on the shelf" and in the end-user's home
- the type of packaging to be used for the end product. The perfume that is created must work in synergy with the image created by the colour, appearance and graphics used on the pack
- the frequency of use of the end product. Products which are exposed to atmospheric oxygen regularly will require a perfume that is reasonably resistant to the effects of oxidation

4. *Developing the perfume*

Development of a fragrance is normally carried out is response to a fragrance brief, which is issued by the marketing company wishing to have the fragrance developed. It is normal for the marketing company to issue the brief to two or three different fragrance houses. In this way, the chances of receiving the best possible fragrance will be optimised and the competitive environment that results will encourage the fragrance houses to give of their best. Selection of the winning fragrance will depend upon its technical and market performance.

Technically, the fragrance should be stable in the final product over a variety of different storage conditions. Typically, fragrance stability tests are carried out 5°C, room temperature and 40°C, both in the presence and absence of light. Product intended for sale in tropical countries may also be tested at 50°C. In examining fragrance stability, it is important that the fragrance is not only chemically stable but also that performance does not significantly deteriorate over time.

Assuming that the fragrance is stable in the product, its performance is normally evaluated using some form of panel test. Typically, between 50 and 100 people use the fragranced product for a period of between 2 and 4 weeks. After this time, the participants are asked various questions about fragrance performance and preference. If time and money permit, a full scale consumer test may also be carried out to give the marketing company added confidence that the fragrance/product combination has a high chance of being commercially successful.

5. *Manufacture of the fragrance compound*

Once the client has made a decision on the winning fragrance, orders are placed with the respective fragrance house. When the perfume is manufactured, it is normally compounded to a specific procedure. Whilst it is difficult to generalise on such a procedure, any solid or crystalline materials present in the fragrance compound are normally weighed into the mixing vessel first. Next, any gums or sticky semi-solids are added, followed by any liquid ingredients that are not significantly affected by heat. The mixing vessel is then heated in a carefully controlled manner and the fragrance ingredients stirred until all the solids and gums have dissolved. The fragrance compound is then cooled to room temperature or below and the remaining heat-sensitive liquid ingredients are added. Care must be taken at the heating stage to ensure that sufficient liquid ingredients are present, so as to avoid recrystallisation or reprecipitation of any of the gums or solids, as the mixture is cooled.

The compounded formulation is then filtered and a sample is passed to the quality assurance department for release, before shipment to the customer. Application of good total quality management (TQM) techniques can ensure that fragrances are compounded correctly every time, thereby avoiding the possibility of any expensive write-offs.

Examples of simple perfume formulations

Whilst a comprehensive review of the many thousands of different types of perfume compound available is beyond the scope of this text, a few simple floral perfumes are illustrated below, to provide the reader with an insight of formulation design.

1. *Rose perfume*

The major basics used in this type of perfume are citronellal, geraniol, nerol and phenyl ethyl alcohol which, when mixed, give a weak rosy smell. Fixatives used include benzophenone, trichloromethylphenylcarbinyl acetate (TCMPCA) and phenyl salicylate, all of which are crystalline solids. A liquid fixative, phenylethylphenyl alcohol, may also be used. When mixed, the fixative ingredients have a slight rosy smell, which stretches the rose odour into the end notes. Blenders used include linalol, terpineol and cinnamic alcohol, which impart a floral odour to the perfume. The use of blenders is, however, optional.

A simple rose perfume is illustrated below:

Ingredient	*Function*	*% (w/w)*
Phenyl ethyl alcohol	Basic	41.70
Rhodinol	Basic	34.00
Linalol	Blender	2.50
Cinnamic alcohol	Blender/Fixative	5.00
Phenylethylphenyl alcohol	Fixative	1.00
TCMPCA	Fixative	6.00
Rose oxide (10% in DEP*)	Top Note	0.40
Undecylenic aldehyde (50% in DEP*)	Top Note	0.40
Citronellyl acetate	Modifier	3.00
Geranium bourbon	Modifier	1.00
Alpha-ionone	Modifier	3.00
Clove bud oil	Modifier	1.00
Lemon oil	Modifier	1.00

* Diethyl phthalate

2. *Jasmine perfume*

Jasmine absolute itself has a very rich odour and a simple jasmine perfume is normally formulated to lessen this inherent heaviness, as follows:

Ingredient	*Function*	*% (w/w)*
Benzyl acetate	Basic	35.00
Linalol	Basic	15.00
Synthetic jasmone	Basic	1.50
Phenyl ethyl alcohol	Basic	8.00
Hydroxycitronellal	Basic	10.00
Amyl cinnamic aldehyde	Basic	12.00
Benzyl salicylate	Fixative	12.50
Decanal	Modifier	1.00
Indole	Modifier	0.20
Ylang oil	Modifier	4.00
Clove oil	Modifier	0.50
Methyl anthranilate	Modifier	0.30

3. *Muget Perfume*

An example of a simple muget perfume is given below:

Ingredient	Function	% (w/w)
Hexyl cinnamic aldehyde	Basic	15.00
Rhodinol	Basic	10.00
Phenyl ethyl alcohol	Basic/Blender	12.00
Linalol	Basic/Blender	3.00
Hydroxycitronellal	Basic/Blender/Fixative	44.00
Benzyl salicylate	Fixative	11.00
Jasmine base	Modifier	1.00
Alpha-terpineol	Modifier	2.80
Citral	Modifier	0.30
Indole	Modifier	0.50
Ionone	Modifier	0.20
Phenyl acetic aldehyde	Modifier	0.20

4. *Lilac perfume*

An example of a simple lilac perfume is shown below. In this particular example, it is important that the alpha-terpineol is of very high purity, with no contamination from beta-terpineol. If this is not the case, the perfume will tend to exhibit an unpleasant pine-like odour.

Ingredient	Function	% (w/w)
Alpha-terpineol	Basic/Blender	30.00
Linalol	Basic/Blender	6.50
Phenyl ethyl alcohol	Basic/Blender	9.00
Hydroxycitronellal	Basic/Blender/Fixative	10.00
Cinnamic alcohol	Basic/Blender/Fixative	10.00
Benzyl salicylate	Fixative/Modifier	15.00
Heliotropin	Fixative/Modifier	5.00
Anisic aldehyde	Modifier	5.00
Phenyl acetaldehyde	Modifier	1.00
Dimethyl acetal	Modifier	1.00
Jasmine base	Modifier	6.30
Iso-eugenol	Modifier	1.00
Indole	Modifier	0.20

Perfumes in cosmetic and toiletry products

Fragrances are used in almost all types of cosmetic and toiletry product and are included to enhance use characteristics and convey or support the marketing concept or brand image to the end-user. The type and level of fragrance used depends very much on the nature of the toiletry product and the market that it is intended for. Some of the key product categories in the cosmetics and toiletry market are considered below.

1. *Fine fragrances*

Fine fragrances exhibit very powerful odour profiles and the perfume oils used in them are some of the most expensive in the cosmetics and toiletries industry. These perfume oils contain only the highest quality materials and costs of between £200 and £300 per kilogram is not unusual for a perfume oil. Typically, a fine fragrance may contain between 15% and 20% of the perfume oil. Most fragrance raw materials can be used in these perfume oils, providing they are soluble in ethanol. Some fragrance raw materials tend to darken when exposed to light for long periods and these should be avoided, as it gives the fine fragrance a progressively unpleasant appearance. If it is essential to use materials that exhibit this characteristic, then the darkening effect can be minimised by the addition of a low level of UV absorber to stabilise the product.

2. *Colognes*

These are similar to fine fragrances, except that they typically contain between 1% and 5% of the perfume oil. Colognes are hydro-alcoholic in nature and this restricts the perfumer, to some extent, on the types of fragrance raw material that can be used in the creation. This restriction is due to the poor water-solubility of some fragrance raw materials. Solubility problems can be overcome, to a degree, by the use of solubilisers. Cologne odour types are not as sophisticated as the fine fragrance perfumes and typically have a lighter, often citrussy, character.

After-shave products can also be considered as colognes, typically containing perfume oils at levels of around 2%. Modern after-shave products must be attractive to women and, for this reason, many of them have become more feminine in character.

3. *Toilet soaps*

The fragrancing of toilet soaps represents one of the largest uses of perfume oils in the cosmetics and toiletries industry and perfume oils are normally added at levels between 1% and 3%. Soaps are alkaline, with pH values of around 9-10, and this environment imposes severe limitations on the types of perfume ingredients that can be used. For example, any esters used in the perfume oil would very quickly be hydrolysed. Superfat soaps, which normally contain low levels of fatty acids, are often acidic in nature and this presents other compatibility problems with fragrance raw materials. For these reasons, it is important that the perfumer is fully aware of the exact composition of the soap base, before the perfume oil is created, otherwise subsequent stability problems may occur. The price of perfume oils designed for use in soap can be anything from £8-00 per kilogram for cheap household soaps, to over £30-00 per kilogram for luxury soaps.

Some perfume raw materials discolour over time, especially when they are used in soap bases. Amongst these are vanillin, vanillal, indole, methyl anthranilate, nitro-musks, eugenol and iso-eugenol. Clearly, in white or light-coloured soaps, these materials must be avoided. Some perfume raw materials are also highly coloured in their own right and their use is therefore restricted for similar reasons. Most esters hydrolyse in soap bases and some aldehydes polymerise, due to the alkaline environment in soap, thereby losing their odour.

Medicated soaps normally contain germicides or fungicides in their bases and the odour from these is much more difficult to cover with the fragrance. In these cases, the perfume is designed to act in synergy with the base odour, producing an overall product smell that endorses the medicated image.

4. *Shampoos*

Shampoos normally contain perfume oils at levels of between 0.5% and 1.0%. Unlike soap, shampoos are not functionally identical and product functions depend upon the additives used. Shampoo fragrances must therefore be individually created for a particular shampoo base and it is important to carry out full perfume stability and perfume/packaging compatibility tests, before the product is marketed.

Shampoo odour types cover a very wide spectrum but the perfume created must always support the brand image of the product, and the claims made for it. Many shampoo products are transparent and any perfume ingredients that discolour, in the presence of light, should be avoided. Care must always be taken that the perfume oil does not significantly affect the viscosity of the shampoo product. In addition, lathering properties may be adversely affected by some perfume oils and this potential incompatibility should be checked early in the development of the perfume.

5. *Creams and lotions*

Perfume oils are normally added to creams and lotions at levels of between 0.2% and 0.5%. Both creams and lotions are invariably emulsion-based systems and perfume oils will partition between the oil and water phases. The result of this partitioning, is that the same perfume may smell completely different in oil-in-water and water-in-oil emulsions. Perfume oils often have a significant effect on the rheology of emulsions, generally causing them to thin. This can give rise to significant stability problems on storage. Creams and lotions are often white in colour and perfume raw materials which are either highly coloured, or discolour with time, should be avoided. The perfume oil should always be designed such that it covers the base odour of the product, which is often fatty in nature.

6. *Aerosols*

Probably the most important aerosol products are hair sprays, deodorants and antiperspirants. Perfume oils designed for use in these products should not contain too many solid or resinous ingredients, as these may accumulate around the valve, eventually causing blockage.

Hair spray formulations often have an unpleasant odour due to the hydrocarbon propellants used, and even perfume oils added at levels of up to 0.5% may not completely cover these. Antiperspirant and deodorant products normally contain perfume oils at levels of up to 1%, although the so-called "deo-colognes" may contain levels of up to 3%, in some cases. Antiperspirant actives, typically aluminium chlorhydrate, are very acidic in nature and will react with some perfume raw materials. Clearly, these ingredients should be avoided when the perfume oil is designed for this end use. Some fragrance raw materials will contribute to underarm staining of garments, after the product has been used by the consumer. Again, such ingredients must be avoided.

7. *Talc*

Levels of perfume oil in talc are normally between 0.5% and 1.0%. The perfume is normally added to the product by a spraying process, before the powder is filled. Talc is a very difficult product to perfume, due to the very high surface area of the product, leading to a high level of contact of the perfume oil with atmospheric oxygen. Perfume raw materials used in talc perfumes must be very stable and esters and citrus oils should be avoided. Amongst the most stable perfume raw materials that can be used in the creation of a talc fragrance are sandalwood, patchouli, vetyvert, coumarin and many of the odoriferous alcohols. Aldehydes and ketones exhibit moderate stability and may also be used at fairly low levels.

8. *Permanent wave lotions*

These products are very difficult to fragrance due to the presence of ammonia and thioglycollic acid in the formulations. Ideally, the perfume should cover the base odour completely but, in practice, this is a virtually impossible task for the perfumer. The best compromise is to use a very powerful perfume, at the highest levels that safety and commercial considerations will allow. Perfume raw materials used in applications of this type may be chosen for their substantivity to hair, as this will ensure that the fragrance remains after the product has been used.

9. *Lipsticks*

"Perfumes" for lipsticks are normally created by flavourists, because taste is often more important than odour in this application. All raw materials used must be orally safe and positively listed on the Generally Regarded as Safe (GRAS) listings. Levels of flavour/ perfume oils used in lipsticks varies from about 0.2% up to 2.0%, although too high an inclusion level may well result in consumer fatigue. Ingredients used in lipstick perfumes/ flavours can also affect the colour and distribution of the pigment within the product and therefore a full stability investigation is imperative.

10. *Bath products*

These include foam baths and various types of bath oils. A wide range of perfume types are used in these products but they are often either relaxing and luxurious, or invigorating, fresh and lively. A typical inclusion level for perfume oils in bath products is between 1% and 5%.

11. *Tissue products*

Market research has clearly demonstrated that if a tissue-based product is perfumed, it will often be perceived as softer and gentler by the consumer than the corresponding unperfumed product. Perfumes for tissue products are often long-lasting, with a high percentage of end notes in the perfume oil itself. The high surface area of the tissue fibres results in a high level of contact of the perfume oil with atmospheric oxygen and therefore no easily oxidised materials, such as esters, may be used.

References

1. Allan AJ & Fowler MN, "Biologically Active Plant Secondary Plant Metabolites for the Future", Chemistry in Industry No 12, Pages 408-410, 12 June 1985.
2. Guenther E, "The Essential Oils", Volume 1, Van Nostrand & Co., New York 1947.
3. Korwek EL (1986) "FDA Regulation of Food Ingredients Produced by Biotechnology", Food Technology, 40, pages 70-74.
4. Lawrence BM, "A Review of World Production of Essential Oils (1984)", Perfumer and Flavourist, Volume 10, No. 5, October/November 1985.
5. Lewis C & Kristiansen B, "Chemicals Manufacture via Biotechnology - The Prospects for Western Europe", Chemicals & Industry, No 17, Pages 571-576, 2 September 1985.
6. MAFF (1987), "Memorandum on the Testing of Novel Foods"
7. Pearce S, "The Value of Biotechnology to the Flavour Industry", Proceedings of the September 1987 Meeting of the International Association of Plant Cell Tissue Culture, Cambridge University Press, 1988.
8. Rogers JA Jnr, "Oils, Essential", Encyclopaedia of Chemical Technology, 3rd Edn, Volume 16, Pages 307-332.
9. Theimer ET, "The Science of the Sense of Smell", Academic Press, 1982.
10. Bauer K, Garbe G & Surburg S, "Common Fragrance and Flavour Materials - Preparation, Properties and Uses", 2nd Edn, VCH 1990.

Date	Material	
1837	**BENZALDEHYDE** (First organic chemical prepared) Used in almond and marzipan flavours and fragrances	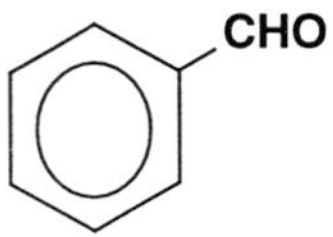
1855	**BENZYL ACETATE** Fruity smell; quite volatile Occurs naturally in Jasmin Oil	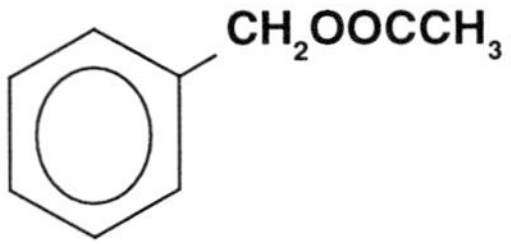
1868	**COUMARIN** White powder. Haylike. Widely used in soap fragrances	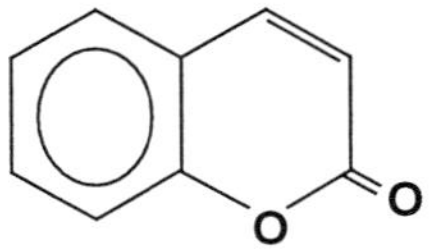
1874	**VANILLIN** Ice-cream smell	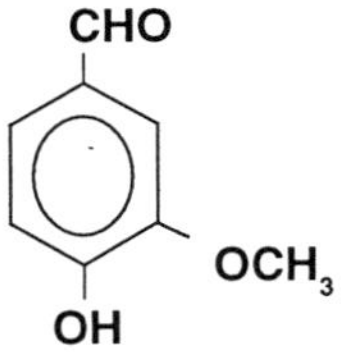
1876	**PHENYL ETHYL ALCOHOL** Smell of roses	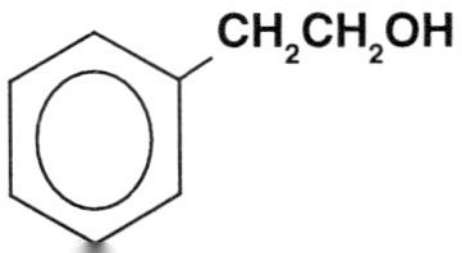

Date	Material	
1897	**INDOLE** Occurs in jasmine flowers Animal like odour turns floral at high dilution	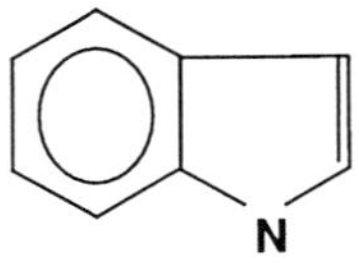
1903	**METHYL HEPTENE** Green smell; useful in a violet perfume	$C_5H_{11}-C\equiv C-COOCH_3$
1921	**LILIAL** Lily of the valley smell	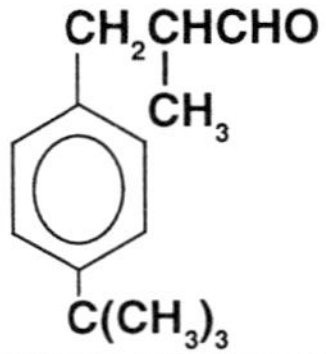
1948	**CIVETTONE** (discovered in Civet) Weak musk type of smell 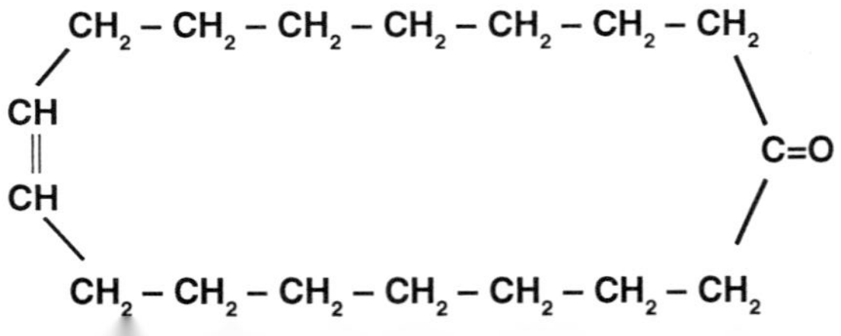	

1885	**HELIOTROPIN (PIPERONAL)** (named after heliotrope flowers) Sweet, spicy odour	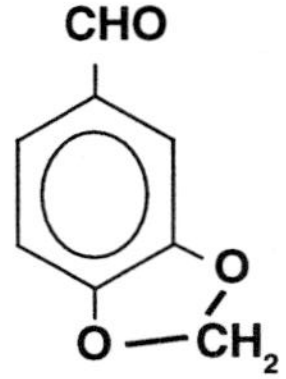	1959	**ROSE OXIDE** Found in rose petals Has Geranium like odour	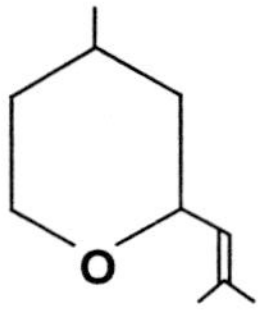
1888	**MUSK KETONE** Does not occur in nature Has a sweet, long lasting musk odour, useful as a fixative	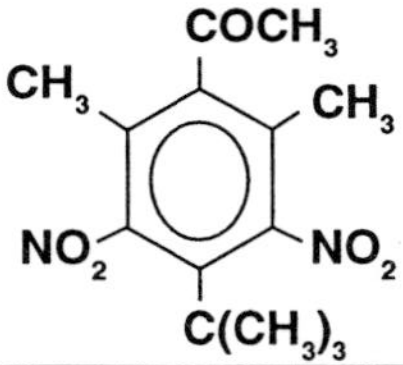	1960	**METHYL DIHYDRO JASMONATE** Found in tea. Has a fruity, jasmin like odour	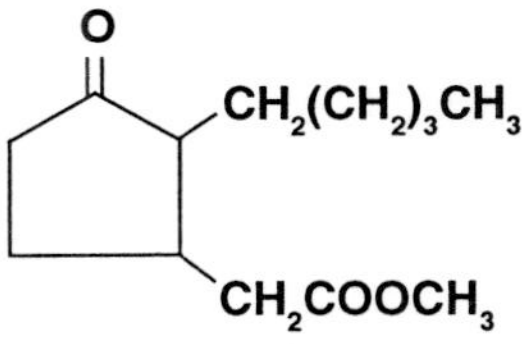
1893	**IONONE** Smell of violets Synthesised methyl derivatives of this called methyl ionones	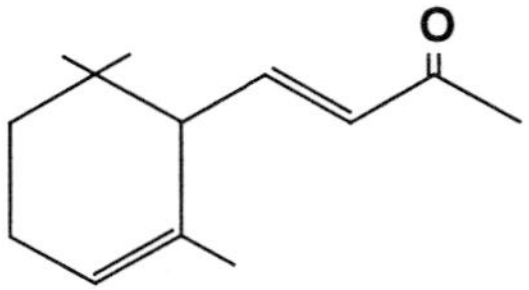	1965	**GERANONITRILE** Smells like lemon grass	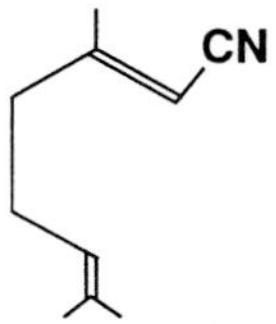
1896	**ANISIC ALDEHYDE (p-ANISALDEHYDE)** Sweet mimosa like odour	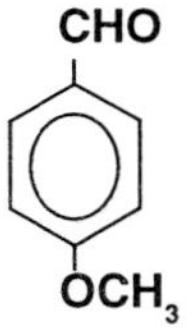	1968	**2-METHOXY-3-ISOBUTYL-PYRAZINE** Green pepper like odour	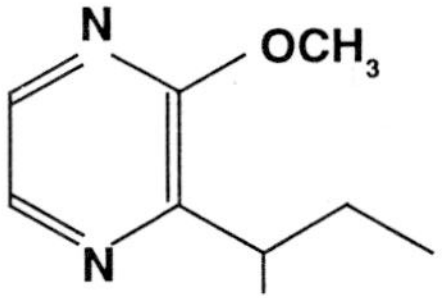

APPENDIX I

APPENDIX II

Adjective	*Odour characteristic*
Aldehydic	Fatty, sweaty or marine odour, straight chain aldehydes
Animalic	Animal-like odour, typical of civet and skatole
Balsamic	Sweet, vanilla character
Camphoraceous/Herbal	Odour of camphor, eucalyptus or sage
Citrus	Odour of oils from citrus fruit peel
Earthy	Odour of damp, humid earth, especially after rain
Floral	Flowers
Fruity	Odour of fruits
Green	Odour of freshly crushed leaves and freshly cut grass
Medicinal	Phenolic, as in disinfectants and carbolic soap.
Metallic	Typical of metal surfaces such as coins
Minty	Peppermint-like
Mossy	Characteristic of lichen, algae and fungus, normally from trees
Powdery	Sweet and powdery
Resinous	Tree resins
Spicy	Odour of cooking spices, such as cinnamon and ginger
Waxy	Reminiscent of candle wax
Woody	Odour of freshly sawn wood but also generically for cedarwood and sandalwood

APPENDIX III

Number of isoprene units	*Number of carbon atoms*	*Terpenoid class name*
2	10	Monoterpenes
3	15	Sesquiterpenes
4	20	Diterpenes
6	30	Triterpenes

ACYCLIC MONOTERPENOIDS

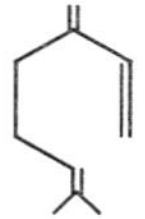

MYRCENE
Occurs in plant oils
Aggresive, fruity, floral odour

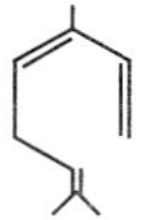

OCIMENE
Occurs in plant oils
Has a similar fragrance to myrcene

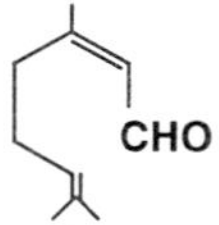

Cis-CITRAL (NERAL)
Occurs in essential oils plants

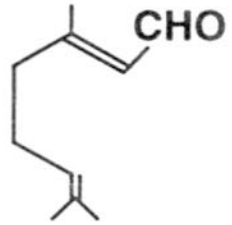

Trans-CITRAL
Citral is also a component of Litsea Cubeba.

BOTH OF THESE MATERIALS ARE MAJOR COMPONENTS OF LEMON GRASS OIL

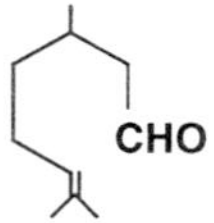

CITRONELLAL
Citrussy, grassy odour

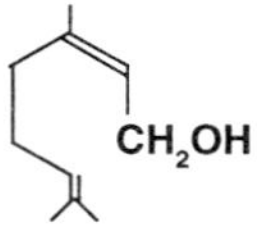

NEROL
Rose odour
Occurs in neroli oil (orange blossom oil)

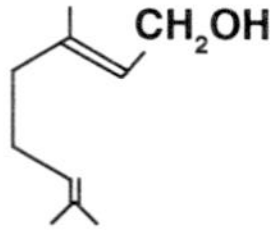

GERANIOL (Trans isomer of nerol)
Red rose odour. Occurs in rose petals, geranium oil, and in citronella oils (50% in Java oil and 35-40% in Ceylon oil).
Palmerosa oil (from grass) contains 90% geraniol

SOME NATURALLY OCCURRING SESQUITERPENOIDS

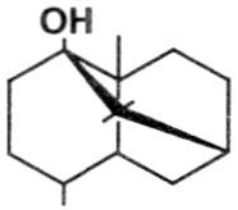

PATCHOULI ALCOHOL
Camphoraceous odour
A major constituent of patchouli oil (30-40%)

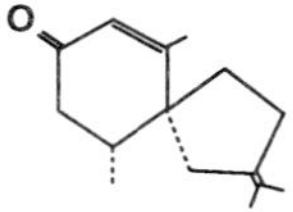

(+) β-VETIVONE
Woody, spicey odour
A major constituent of vetyver oil

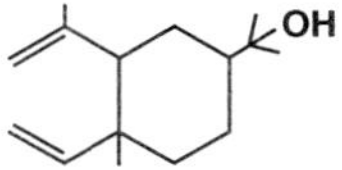

ELEMOL
Weak, citrussy odour
Occurs in citronella oil

BICYCLIC MONOTERPENOIDS

α-PINENE

β-PINENE

α-pinene and β-pinene together make up turpentine. They possess a resiny piney odour. These materials provide the starting point for the preparation of many other terpenoid compounds

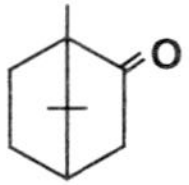

CAMPHOR
Very camphoraceous odour
Obtained from the wood of the camphor tree (China)

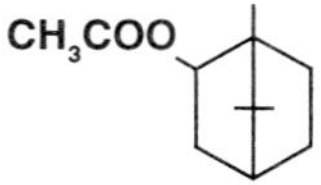

ISOBORNYL ACETATE
Pine needle odour
Occurs in pine needles but can also be synthesized

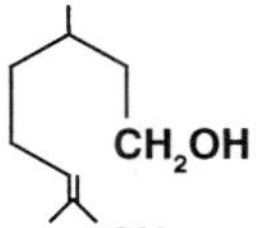

CITRONELLOL
White/yellow rose odour
Occurs in geraniums and roses

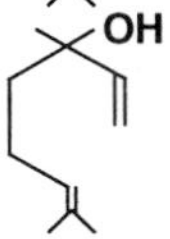

LINALOOL
Fresh floral odour. This is a tertiary alcohol and is therefore more stable than some of the other compounds. Occurs in many essential oils including lavendar, rosewood, ho-leaf (camphor tree) and petitgrain oils

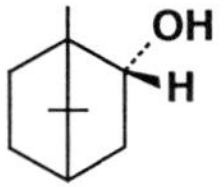

n-BORNEOL

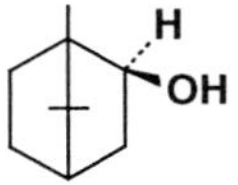

iso-BORNEOL

n-borneol and iso-borneol (isomers) both have a camphoreous odour

MONOCYCLIC TERPENOIDS

LIMONENE
Citrus odour. Major constituent of orange, grapefruit and lemon oils. Derived from orange oil it has a "D" optical rotation but from U.S. pine oil it is racemic (no optical rotation) and is kown as <u>dipentene</u>. Dipentene has an unpleasant "pine-oil" odour

TERPINOLENE
Has a lime odour when good and a "phenolic" odour when bad
Obtained from U.S. pine oil

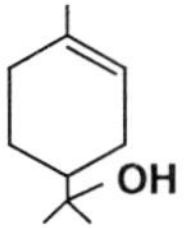

TERPINEOL
Normally has a pine odour but when <u>very pure</u> has a floral (lilac) odour. There are two isomers α and β. The α isomer produces the pleasant odour whilst the β isomer produces the pine odour

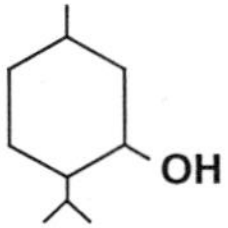

MENTHOL
Best material obtained by freezing peppermint oil to obtain menthol crystals. Can also be synthesized but this is not of such high quality due to different isomer ratios

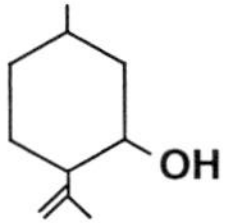

ISOPULEGOL
Minty odour
Occurs in geranium oil and imparts the minty character to this oil

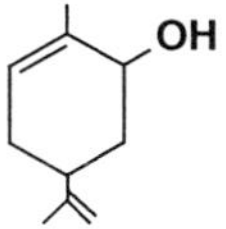

CARVEOL
"Seedy" odour
Occurs in spearmint oil and caraway seed oil

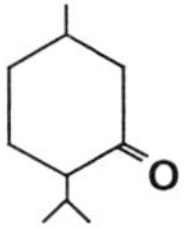

MENTHONE
Minty odour
Peppermint arvensis contains 30% menthone

I,8-CINEOLE
Fresh, camphoraceous odour
Major ingredient of eucalyptus oil and <u>eucalyptus globulus</u> (Australia) contain 80% cineole

APPENDIX IV

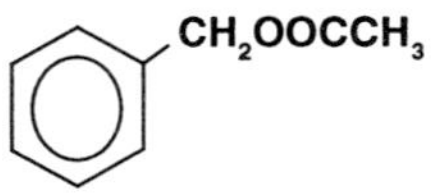	**BENZYL ACETATE** Fruity odour Found in white jasmine blossom	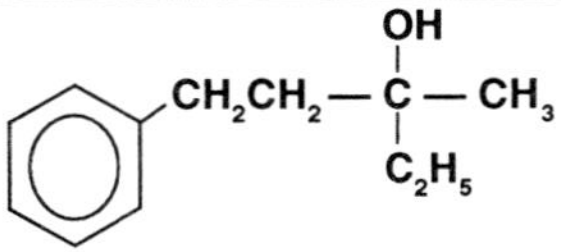	**PHENYLETHYL METHYL ETHYL CARBINOL** Delicate floral, slightly fruity odour
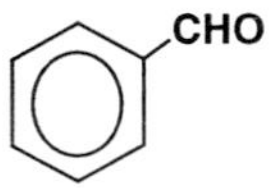	**BENZALDEHYDE** Marzipan/almonds odour	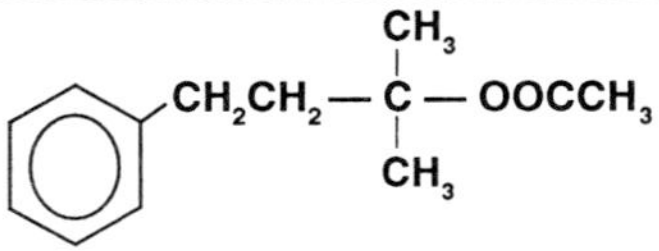	**DIMETHYL PHENYL ETHYL CARBINYL ACETATE** Rose odour
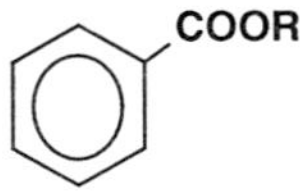	**ESTERS OF BENZOIC ACID (BENZOATES)** These are often strong smelling materials	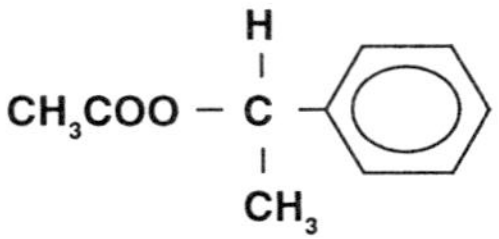	**STYRALLYL ACETATE** Bright green odour
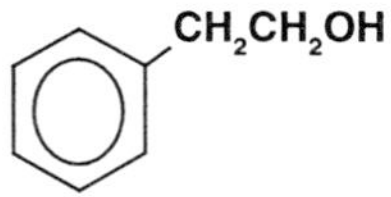	**PHENYL ETHYL ALCOHOL** Weak rose odour Occurs naturally in rose petals	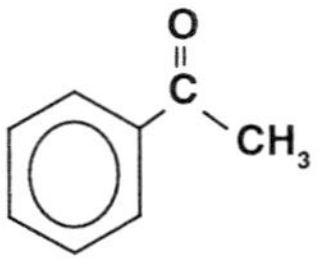	**ACETOPHENONE** Sweet orange blossom like odour
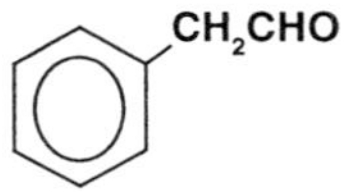	**PHENYL ACETALDEHYDE** Hyacinth odour	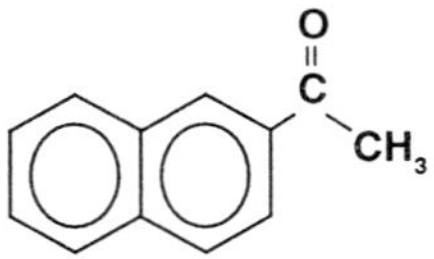	**METHYL-β-NAPHTHYL KETONE** Sweet odour Useful for orange blossom perfumes
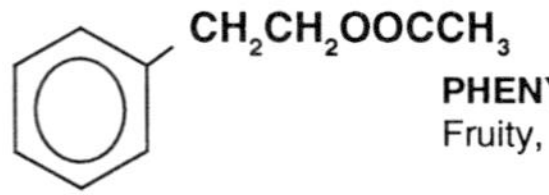	**PHENYL ETHYL ACETATE** Fruity, rose honey odour	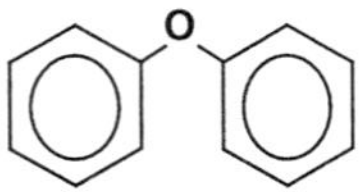	**DIPHENYL ETHER** Geranium/rose odour Otherwise known as diphenyl oxide

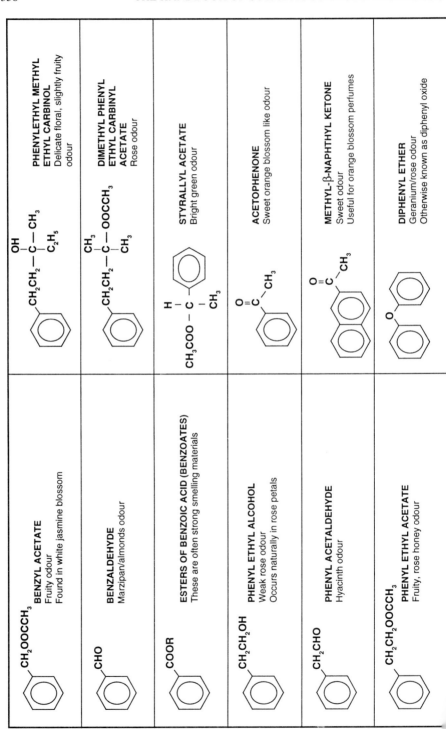

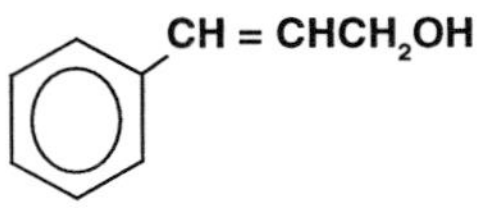	**CINNAMIC ALCOHOL** Sweet, balsamic odour	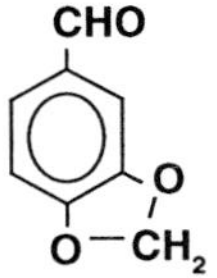	**HELIOTROPIN** Sweet, spicy odour
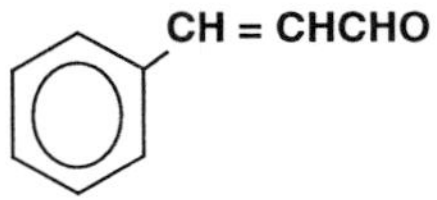	**CINNAMIC ALDEHYDE** Spicey odour Occurs in cinnamon bark as the main odour/ flavour constituent	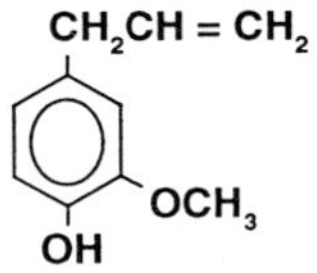	**EUGENOL** Clove odour Major constituent of clove oil and is mostly obtained from cloves
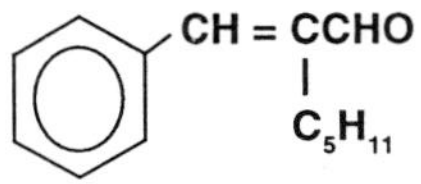	**AMYL CINNAMIC ALDEHYDE** Floral, oily, odour (Jasmine petal odour)	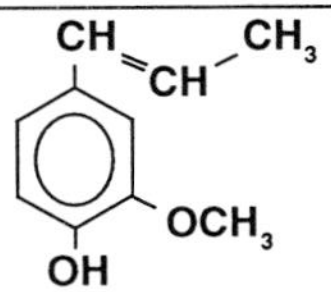	**ISOEUGENOL** Odour not as spicey as eugenol, more like carnation flowers
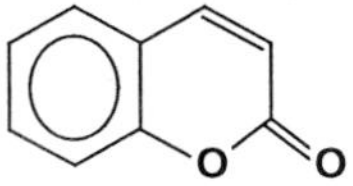	**COUMARIN** Sweet, balsamic odour of newly mown hay	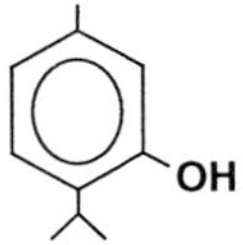	**THYMOL** Thyme odour (herbal, camphoraceous) Major component of thyme oil
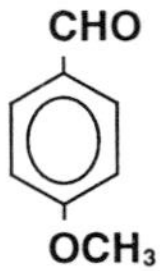	**ANISIC ALDEHYDE** Sweet, floral odour	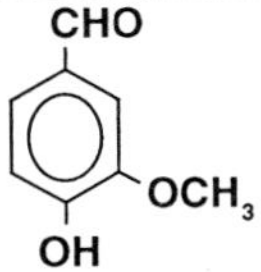	**VANILLIN** Very sweet balsamic odour Found as a white powder on vanilla pods but synthetic product is used most often

APPENDIX V

APPENDIX VI

MISCELLANEOUS PERFUMERY INGREDIENTS

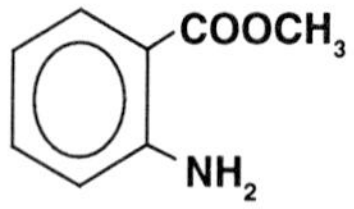

METHYL ANTHRANILATE
Occurs in a range of blossom oils and has an orange blossom odour

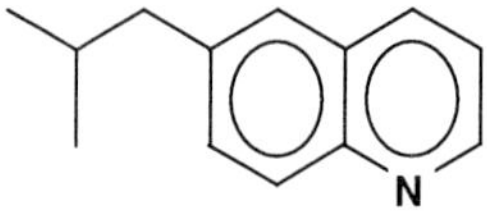

6-ISOBUTYL QUINOLINE
Leather, mossy, earthy

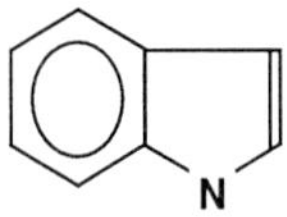

INDOLE
Animal-like, but floral on high dilution

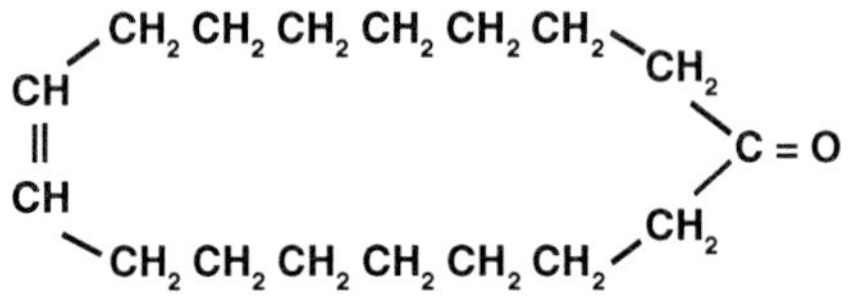

CIVETTONE
Naturally found in civet

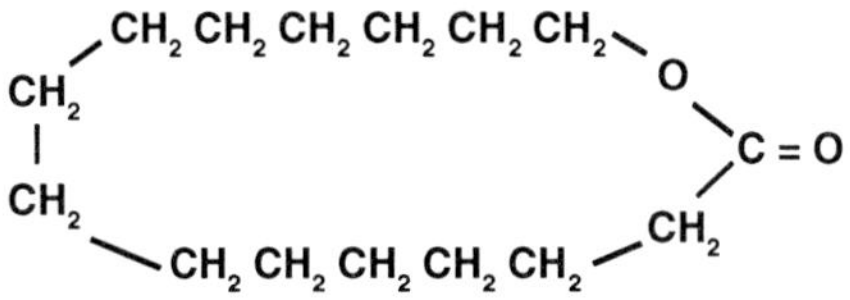

EXALTOLIDE (15-PENTADECANOLIDE)
Occurs in angelica root oil

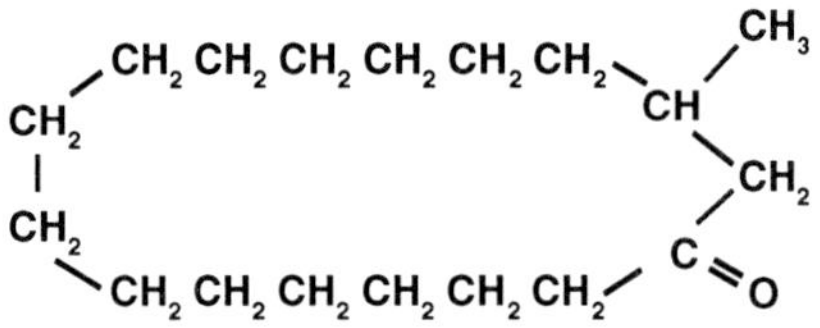

MUSCONE
Naturally found in musk deer excretion

All these have a musk odour (civettone, exaltolide, muscone)

APPENDIX VII

Plant part (source)	*Examples of essential oil*
Leaves	Patchouli, Peppermint, Rosemary, Eucalyptus
Fruit	Orange, Lemon, Anise
Bark	Cinnamon, Sassafras, Birch
Root	Ginger, Vetivert
Grass	Citronella
Wood	Amyris, Cedarwood, Sandalwood
Berries	Pimento
Seed	Caraway, Coriander, Fennel, Cardamom
Flowers	Rose, Chamomile
Twigs	Clove stem
Bud	Clove bud
Balsam	Tolu
Bulb	Garlic

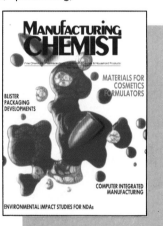

SECTION 8

Manufacturing

PRODUCTION
QUALITY CONTROL AND ASSURANCE
STABILITY TESTING
INDUSTRIAL MICROBIOLOGY, HYGIENE & PRESERVATION

PRODUCTION

Introduction

In the cosmetics and toiletries industry, the production process involves the acquisition of a pre-defined quantity of raw materials and packaging components and subsequently converting them into finished goods, for the retail or consumer marketplace. In reality, this conversion is a very complex operation, the efficiency of which depends upon many factors. The objective of any commercial company is to make profit, which can only be maximised if goods of the correct quality are manufactured in the most cost-effective way.

The manufacturing process, in its broadest sense, involves interactive input from all departments and their personnel, throughout the organisation, working effectively. The manufacturing facility, and the way in which it is designed, is a crucial factor in the manufacturing process. Where appropriate, capital should be invested in carefully selected plant, and recruitment of the best human resources will ensure optimal use of the manufacturing facilities available.

The production process

In any production process, there should be a logical production "flow" through the manufacturing facility, commencing with raw material and packaging input at one end and finished goods, of suitable quality, at the other. This process is illustrated, diagrammatically, in Figure 1.

Figure 1 does not illustrate the role of ancillary support functions in the manufacturing process. Support functions play a very important part in the manufacturing process and are described, in more detail, later in this chapter. Note from Figure 2, the large amount of storage space required for the manufacturing process. The provision for sufficient storage and warehousing is essential in the design of any factory and, frequently, the ratio of storage space to manufacturing area is between 3:1 and 5:1, for the cosmetics and toiletries industry.

Conformance to the principles of Good Manufacturing Practice (GMP) is essential at all stages of the manufacturing process and is necessary to ensure that the finished product is of the correct quality from a physical, chemical and microbiological standpoint.

FIGURE 1

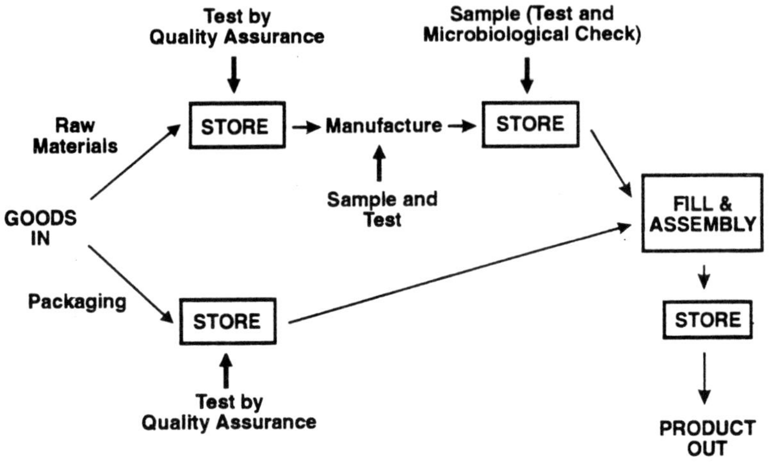

Adherence to GMP also fosters a clean and hygienic working environment which, in turn, will help to increase production efficiency and maximise work force morale.

Designing a factory layout

When designing a factory layout, it is important that the design should facilitate maximum efficiency of the production "flow". This requires careful placement of the different functions within the manufacturing complex, so that unnecessary transfer of product and personnel is avoided. The factory layout should also accommodate the support functions of the business such as, quality assurance, purchasing, planning, engineering and office functions, as well as canteen and personal hygiene facilities. Functions such as research and development, and sales and marketing may be provided for, depending upon the policies and working structure of the organisation. The floor plan illustrated in Figure 2 depicts a good basic design of a single storey factory layout, for achieving maximum efficiency and throughput.

Sometimes a factory is designed around a two storey building. In these circumstances, the manufacturing facility should always be on the upper floor, to maximise utilisation of gravity filling techniques, with the filling and assembly facility on the ground floor directly below it.

Production and interdepartmental relationships

For an organisation to run efficiently, it is important to consider the relationship between departments within the organisation. Departmental interaction is very complex, with the operation of each function having at least some bearing on the efficiency of all others within the organisation. Figure 3 illustrates the key functions within an organisation and the likely lines of interaction that take place.

FIGURE 2

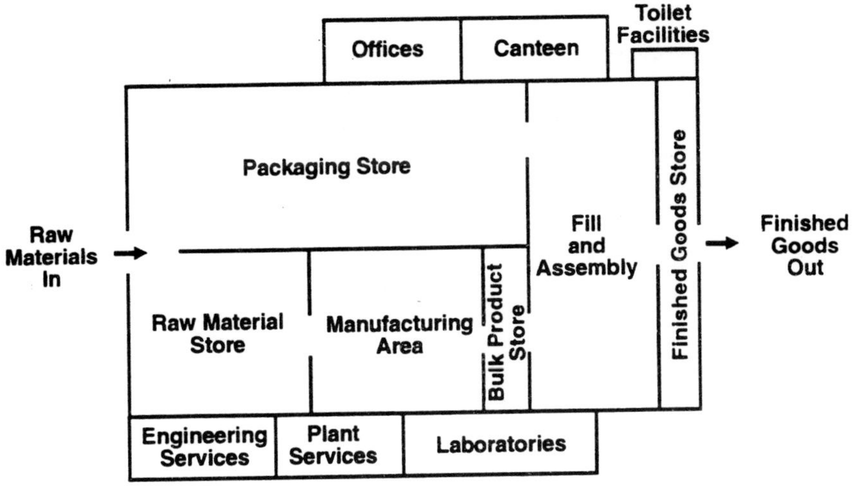

FIGURE 3

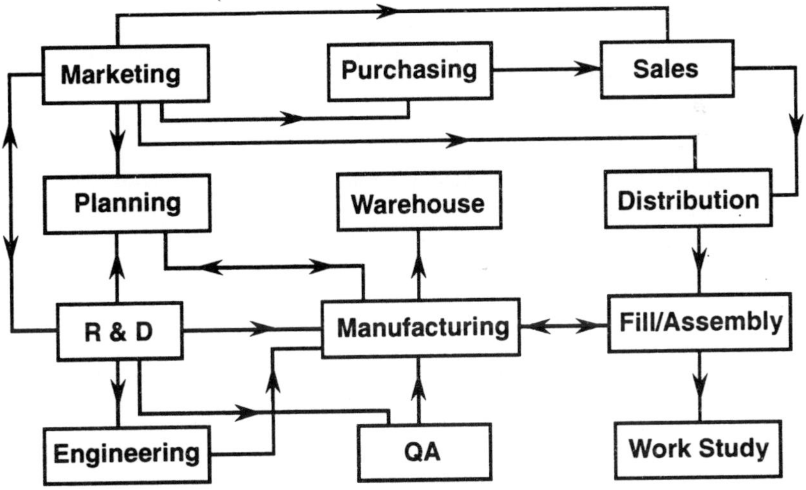

Each function has a very important part to play in overall organisational efficiency. Specific tasks associated with some of the key functions are listed below:

1. *Marketing*

The marketing function is generally responsible for the generation of new product ideas and concepts. It also provides specifications to other functions in the organisation, during the development of any new product idea. Ultimately marketing makes the business decision of whether or not to launch a new product, or idea, into the market-place.

2. *Research and development*

Research and development have the responsibility to develop new products, or improve existing ones, according to a specific marketing brief. In technologically led organisations, research and development will be responsible for the generation of new ideas, upon which marketing concepts can be based. Research and development is also responsible for the technical support and improvement of existing products, including the generation of efficacy data to support marketing claims.

3. *Purchasing*

Purchasing is responsible for the acquisition of all components (raw materials and packaging) used in the manufacturing process, to produce the finished product. Purchasing, working closely with research and development and quality control/assurance, should identify the most cost-effective sources of component supply, consistent with obtaining the quality required. They should also ensure that any supplier of components used, is commercially sound and can meet volume requirements.

4. *Planning*

Planning has responsibility to examine the total capacity of the manufacturing plant and optimise the logistics of the production process, with the resources available. As such, the planning function is often regarded as the "control centre" of the production process, with a direct impact on the utilisation of components and human resources. Further details of the planning process are given later in this chapter.

5. *Manufacturing*

Manufacturing have the responsibility of production and assembly of the finished product and, in so doing, must have good interdepartmental relationships with all other functions in the organisation. Manufacturing must be able to produce product of the correct quality, in the shortest possible time, with optimal use of materials and human resources.

6. *Engineering*

The engineering function is responsible for the identification, installation and maintenance of plant and operating services. Good interdepartmental relationships with manufacturing and research and development are essential, particularly in the areas of new plant installation and process development.

7. *Quality control and assurance*

The terms quality control and quality assurance are often used synonymously. Strictly speaking, quality control ensures the quality of the materials and components that are used in the manufacturing process, whilst quality assurance controls the quality of the finished product. Historically, no part of the quality assurance function was carried out by manufacturing personnel, due to the high risk of sacrificing product quality for production throughput. Recently, the concept of total quality management (TGM) has encouraged self-assessment of quality in manufacturing, with the quality assurance function taking an auditing role.

8. *Sales*

The sales function distributes and markets the product to a given strategy and must work closely with marketing. They have a responsibility to communicate retail response to manufactured products, back into marketing, to ensure that optimum customer satisfaction is achieved.

Planning

Planning is a very important part of production management and factory operations. Any production plan, however, cannot be strictly adhered to because of unexpected outside influences, and therefore tends to be based on expectation and experience. The planning operation can be sub-divided, in terms of time, into three categories.

1. *Long term planning*

Long term planning is associated with a knowledge of growth rates and expansion and contraction of the business, in the context of the market-place. Issues covered in long term planning, include marketing strategy, technological developments, product range and recruiting levels.

2. *Medium Term Planning*

Medium term planning normally refers to time scales of between 3 and 6 months. This involves assessing factors such as performance against budget, promotional activity, capacity of existing plant and the need for temporary labour.

3. *Short term planning*

Short term planning normally refers to the weekly production plan, concerned with factors such as labour distribution, manufacturing schedule and production capacity.

In any planning operation, it is rarely possible to achieve the theoretical 100% target. Typical achievement against a weekly plan can be represented graphically, as illustrated in Figure 4.

FIGURE 4

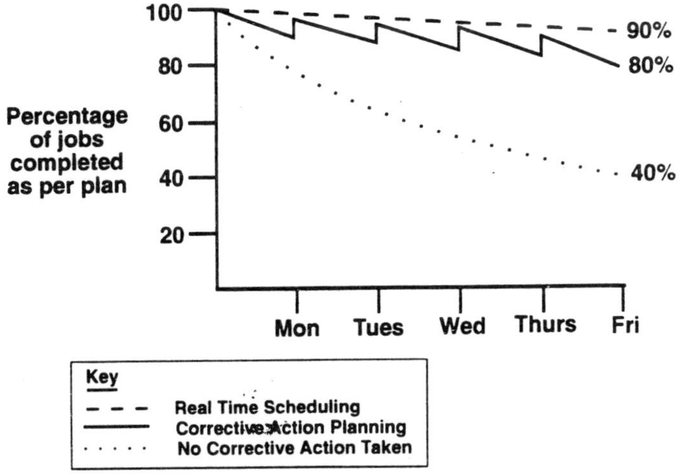

The maintenance of a 100% plan cannot normally be achieved, due to continuous "unforeseen" events occurring, such as labour shortages, component shortages and machine breakdowns. One option to counteract this effect is to "under plan" but this practice is non-productive and wholly wrong. With reference to Figure 4, it is clear that the weekly plan implemented on a Monday morning will, without any corrective action during the week to compensate for outside influences, fall to a value of about 40% by Friday afternoon. In practice, this phenomenon is corrected by constantly updating the plan to maintain a maximum target figure and, if this is done efficiently, 80% of the original plan can normally be achieved. Continuous corrective planning, referred to as *real time scheduling*, is often carried out using computer-assisted techniques where achievement of figures of the order of 90% of the original plan are not unusual. Inefficient or "bad" planning will have adverse effects on production which, in turn, will affect staff morale, thus compounding the problem.

4. *Planning methods*

Essentially, there are two different techniques used in the planning operation, depending on the time scales involved.

4.1. *Long term planning*

Long term planning involves the way in which a large number of interactive tasks can be brought into a plan, assesses the causes of failure to meet plan and determines the way in which corrective action may be applied. An example of long term planning would be the introduction of a new range of skin care products, within a time scale of 12 months. This type of plan requires a high level of interdepartmental interaction.

The technique used for this type of planning is called *critical path planning*, or critical path analysis. Using this technique, the process to be carried out is broken down into "activities", which are each assigned durations and dependences upon other activities associated with the process. The critical path can be determined giving information on the duration of each activity, the duration of the total process and dependency of activities upon each other. The critical path itself is defined as the shortest possible overall time required to complete the process, based on individual activity dependency and duration. A critical path plan can be continually reassessed, to monitor the change in the original overall duration of the exercise.

4.2. *Short term planning*

Short term planning is associated with the day to day allocation of time and labour, to various production operations. This operation is normally carried out with the aid of a planning chart, on a short term, often weekly, basis. Short term plans normally tackle problems such as effective use of time and labour and the allocation of different product types to specific areas of the manufacturing facility.

The manufacture of cosmetics and toiletries

For the most part, the manufacturing of cosmetics and toiletries involves the mixing and transfer of bulk materials, in some cases in the presence of heating and cooling. The aim of the manufacturing operation is to convert raw materials into bulk finished product, ready for filling or subsequent processing. There is a vast number of different product types and formulation designs, all of which will have specific mixing requirements. One of the main problems facing the cosmetic manufacturer is selecting the type of plant required to give maximum manufacturing flexibility, particularly in cases where the product range is large and diverse in nature.

Manufacturing involves the mixing of solid, liquid and gaseous raw materials, with each other. Perhaps the most frequently found mixing processes used in the manufacture of cosmetics and toiletries, are the mixing of miscible liquids (skin toners, alcoholic fragrances), immiscible liquids (emulsions) and liquid/solid mixing. In the decorative cosmetics industry, solid/solid mixing techniques are common, such as in the manufacture of eye shadows. Each of the mixing processes used in the manufacture of cosmetics and toiletries are described below, in more detail.

Manufacturing techniques — solid/solid mixing

Solid/solid mixing normally applies to the mixing of powders or small solid particles. Solid/solid mixing techniques can be categorised, depending upon the flow characteristics of the powder that is being mixed. Free-flowing powders, known as *segregating powders*, are rarely encountered in the cosmetics and toiletries industry. The second type of powder is the *cohesive powder*, in which individual powder particles tend to stick together, forming aggregates. The reason why cohesive powders tend to aggregate is that, because the individual particle size and weight are small, the normally relatively weak Van der Waals forces dominate the effects of particle separation, due to gravitational mass, hence aggregates are formed. In the case of segregating powders, individual particle size is too large for the Van der Waals forces to have appreciable effect. The fundamental property that dictates whether a powder is segregating or cohesive in nature, is particle size, as illustrated in Table 1.

TABLE 1

Particle size	*Nature of powder*
> 75 μ	Strongly segregatingwith no cohesive nature
50 μ-75 μ	Both segregating and cohesive nature observed
< 50 μ	Predominance of cohesive nature
< 10 μ	Strongly cohesive with no segregation observed

With reference to the particle sizes in Table 1, it is obvious that most powders encountered in the field of cosmetics will exhibit some degree of cohesive nature. A classic example is cosmetic talc, which normally has a particle size distribution in the order of 5-50 μ. In practice, this material exhibits significant cohesive nature. Another category of cosmetic materials that show strong cohesive nature are pigments, both organic and inorganic.

Cohesive powders

Any powder which is observed to exhibit appreciable flow problems can be assumed to be predominantly cohesive in nature, with very little, if any, segregating characteristics.

The forces responsible for this cohesive nature are the Van der Waals forces referred to earlier, which ultimately lead to clumping or aggregation of the powder. These forces of attraction between the particles are greatly enhanced by the presence of a liquid, particularly if it wets or spreads over the particles easily.

In order to effectively mix powders of this type, it is necessary to break the forces of cohesion, and thus the aggregates, so that the powder particles are in discreet form. If this level of mixing is not achieved, the resulting powder mixture will not be truly homogeneous.

When processing dry powders, or dry powder mixtures, it is necessary to use a ribbon blender, as illustrated in Figure 5, to carry out the mixing process.

FIGURE 5

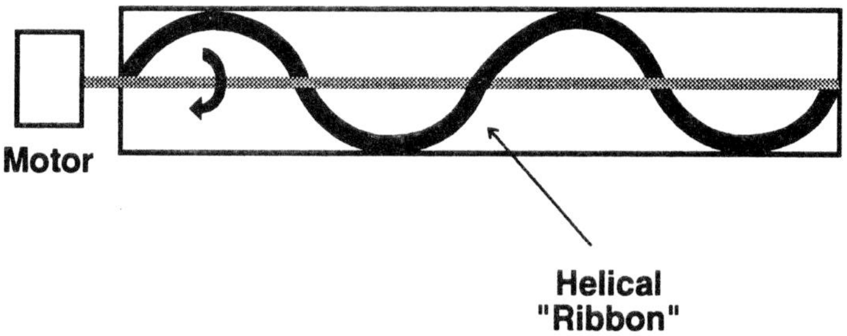

The ribbon blender is a low shear device, which breaks down the powder aggregates using a metal ribbon, rotating in the form of an Archimedian screw. The powder is continuously agitated using a tossing motion, inter-particle friction helping to break down any aggregates present. In cases where powders show stronger cohesive character, a simple ribbon blender may not be sufficient to produce an homogeneous mix, and high shear blender designs, or even milling methods, must be utilised. One of the most common high shear blenders is the plough-shear blender, which possesses a much more efficient mixing characteristic than a simple ribbon blender. This apparatus has the ability to break down aggregates, even in powders with strong cohesive character. The alternative method, using milling techniques, is particularly useful in cases where the powder mix contains hard materials. A typical process of this type might utilise a simple ribbon blender for the initial powder mixing stage, followed by mixing with a high shear milling device, such as a hammer mill.

In cases where it is necessary to mix cohesive powders in the presence of a liquid, as in the case of some decorative cosmetic applications, a much stronger mixing action must be applied, to ensure effective mixing. In the presence of a small quantity of liquid, cohesive powders exhibit much higher cohesive strength. This is due to the fact that the liquid draws the discreet powder particles together forming so-called "liquid bridges", as illustrated in Figure 6.

The increased forces of attraction between the particles derive from the forces of surface tension, at the respective liquid/solid interfaces, in combination with the negative capillary pressure within the liquid bridge itself. This increase in cohesiveness, magnifies the Van

FIGURE 6

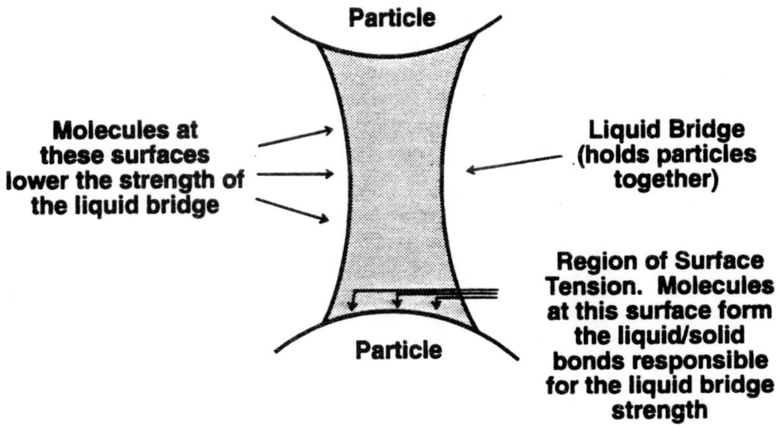

der Waals forces originally present in the dry powder, although even these forces are relatively weak when compared, for example, to a true covalent bond.

If there is a higher level of liquid in the powder mixture, then the inter-particle voids become completely wetted, as shown in Figure 7.

FIGURE 7

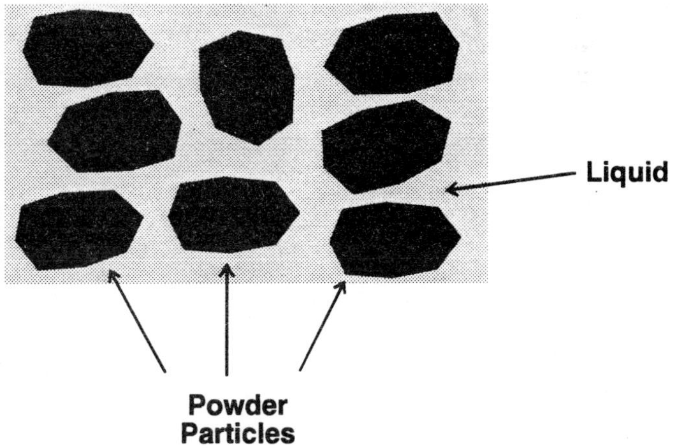

The bonding forces observed in the case of the liquid bridge, are now effective over the entire surface of the powder particles. This leads to an even higher binding strength within the aggregate itself. Providing the quantity of liquid is small, compared to the quantity the powder present, then the aggregate bonding forces are proportional to the level of liquid present. If the quantity of liquid is increased further, the aggregate may be completely enveloped by the liquid present to form a droplet, as illustrated in Figure 8. In these circumstances, the inter-particle cohesive forces are small, compared with the cohesive forces present within the droplet, as a whole.

FIGURE 8

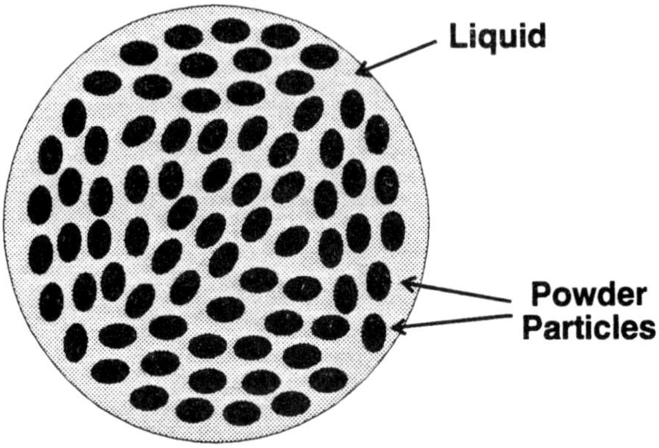

These enveloped aggregates can combine with others that are similar, resulting in the formation of even bigger aggregates, with higher cohesiveness.

Any inter-particle bonding forces present in systems of this type, will become stronger as the viscosity of the liquid is increased. This assumes, of course, that the chemical nature of both the liquid, and the surface of the particles, are such that complete wetting occurs.

In cases where it is necessary to mix powder products, containing small quantities of liquid, the original method of choice was a simple blending operation, using a ribbon blender, followed by a milling process. More recently the plough shear type blender has been utilised, as mixing can be carried out in a single step.

Figure 9 illustrates the theoretical tensile strengths of various types of cohesive bond, in comparison with each other, as a function of powder particle size.

Other types of inter-particle attractive force, present in powder mixtures, include electrostatic attraction and form-closed bonding. The latter describes the tendency for particles to aggregate due to shape.

Manufacturing techniques — solid/liquid mixing

The simplest type of solid-liquid mixing, is the dissolution of large water-soluble crystals or water-soluble dyes, in water. If the liquid, which is very often water in the case of many cosmetic or toiletry products, is of low enough viscosity, then virtually any type of mixing will suffice, the best type being a propeller stirrer.

Solid materials which are either insoluble or partially soluble, or materials which require wetting before dissolution, present more difficult problems. Typical examples are powdered cellulosic gums, which must be thoroughly wetted, before the gum structure is allowed to build, or pigments which must be pre-dispersed in oil, before use in lipstick manufacture. In these cases, it is necessary to disperse the solid efficiently, into the liquid. There are two critical requirements, if an effective dispersion is to be obtained. Firstly, the

FIGURE 9

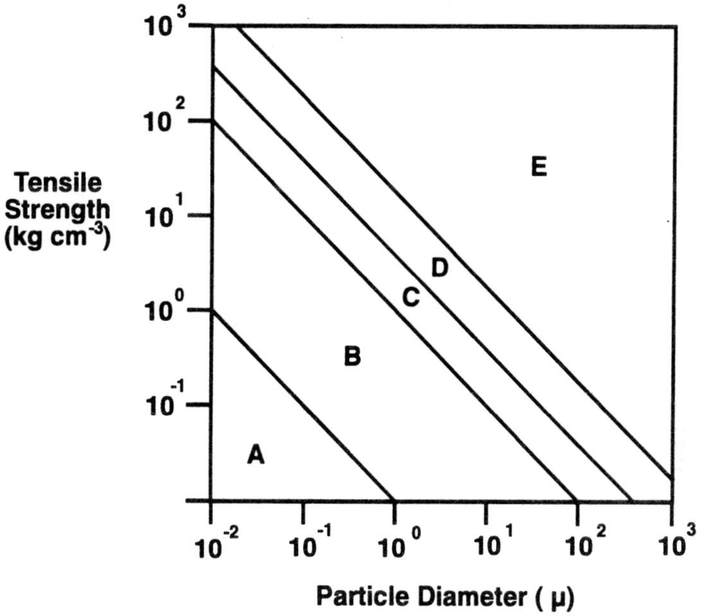

Key

A = Van der Waals forces
B = Van der Waals forces with adsorbed moisture
C = Presence of liquid bridges
D = Inter-particle voids filled with liquid
E = Non-mobile (viscous) binders and solid binders

solid material to be dispersed must be completely wetted out by the liquid and, secondly, it is vital that all the solid aggregates are broken down, in so doing. During subsequent processing, aggregates must, of course, not be allowed to reform. In the case of gums, it is necessary to produce a clear dispersion in a medium, usually water, which increases in viscosity as dispersion occurs.

If a solid is brought into contact with a liquid surface, the process of *adhesion* occurs, which involves the elimination of the airspace between the solid and the liquid. If the solid is of irregular form, is not easily wetted by the liquid, or contains entrapped air itself, then dispersion problems will immediately occur. The only solution in these cases is to use a liquid which more readily wets the solid to be dispersed, or to use a much more efficient mixing process, ensuring that dispersion is effected more quickly.

As the solid is added to the liquid, the process of immersion takes place. Provided that the liquid has a low surface tension and the interfacial tension at the solid/liquid boundary is low, thus promoting easy wetting of the solid, then this process can be carried out relatively easily. After immersion, the final stage is that of spreading, in which the liquid must make intimate contact with the individual solid particles.

Problems with both immersion and spreading can occur in some liquid/solid systems. For example, when attempting to disperse a cellulosic gum into water, agglomerates introduced into the liquid will wet on the surface, but gum particles inside the agglomerate will be prevented from dispersing. The extent of this problem will be dictated by two factors, how easily the solid is wetted by the liquid and whether or not the agglomerate will form a gel-like "skin" on its exterior, when immersion into the liquid takes place. If the latter phenomenon occurs, gel lumps will occur in the system. Very often, gel lumps of this type cannot be broken down, once formed, and the mix must be rejected. A similar phenomenon can occur when attempting to disperse pigments into oils, although the problem of hydration, forming a gel-like structure, does not occur in this case. When dispersing a solid into a liquid, it is therefore very important to apply the correct degree of mixing action to the system, so that the solid is completely dispersed, in discreet particle form, into the liquid.

Some of the most difficult dispersion problems occur when attempting to disperse a hydrophobic solid into a hydrophilic liquid and, in order to produce an effective dispersion, a high degree of shear must be used. This task can be made somewhat easier by pre-dispersing the solid in a liquid which wets the solid easily, but will not hydrate or dissolve it. A typical example would be the pre-dispersion of a cellulosic gum into propylene glycol, before introducing the pre-mix into water, thus allowing the gum to hydrate. If using a pre-dispersion step of this type, it is important to note that hydrophilic solids are normally pre-dispersed in hydrophilic liquids and hydrophobic solids should be pre-dispersed in hydrophobic liquids. The application of shear to the dispersion is normally carried out using a high shear mixing device, the most common of these being the rotor-stator mixer. This will break-up the agglomerates, allowing wetting of the discreet particle to take place.

Once the solid has been dispersed into the liquid, the dispersion is normally kept in a stable state using conventional mixing systems that create adequate *axial flow*, up and down the mixing vessel. The use of this type of mixing is covered in more detail, later in this chapter.

Liquid flow and Reynolds number

When a liquid flows inside a pipe, it experiences either *laminar flow* or *turbulence*. Laminar flow is said to occur when fluid particles move along in straight lines, parallel to the direction of flow, as illustrated in Figure 10.

In the case of laminar flow, mass transfer, or mixing, can only take place due to molecular diffusion of adjacent layers of liquid, within the pipe. This is a highly inefficient method of liquid mixing, involving appreciable time periods for any significant mixing to be observed. As the speed of flow in the pipe is increased, however, interaction of adjacent liquid layers occurs, giving rise to a shearing effect which causes the onset of the formation

FIGURE 10

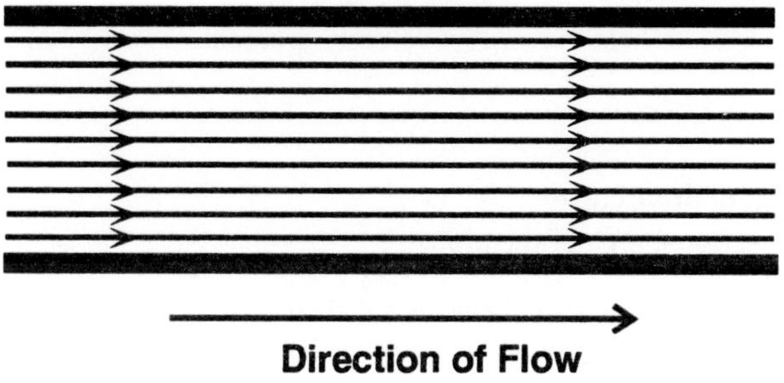

of turbulent currents, known as *eddy currents*. Thus, as the velocity of liquid flow increases, laminar flow gives rise to turbulent flow, the extent of the turbulence being variable and dependent on several factors. It is turbulence which provides the basis for all practical, effective, liquid mixing processes, providing a very efficient method for mixing liquid products. A valuable aid in determining the characteristics of liquid flow in a mixing vessel is *Reynolds number*, a dimensionless value, given by the following equation:

$$R_e = \frac{D^2 N \rho}{\eta}$$

where

R_e = Reynolds number
D = diameter of the impeller
N = tip speed of the impeller (rpm)
ρ = density of the liquid
η = viscosity of the liquid

This calculation of Reynolds' number is only valid for liquids with Newtonian flow, rarely found in practice. In order to understand how this equation needs to be modified for practical use, it is necessary to revert to the case of liquid flow through a pipe, this time experiencing turbulent flow, as illustrated in Figure 11.

FIGURE 11

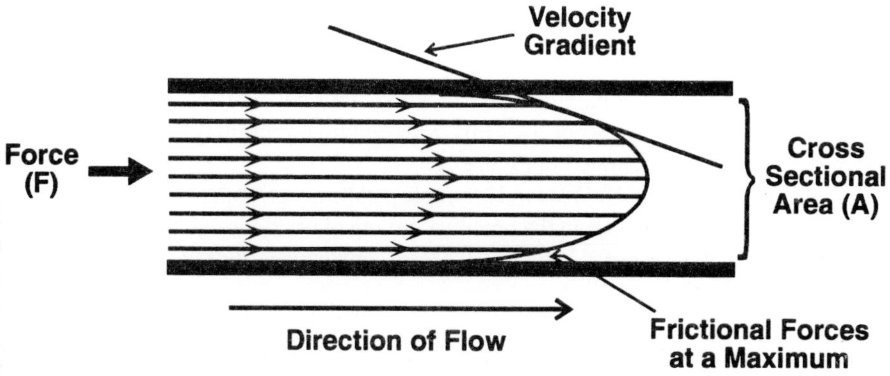

The force, F, is given by the following relationship:

$$F = \eta A \text{ x velocity gradient}$$

where

F = force
η = coefficient of viscosity
A = cross-sectional area of the pipe

The term F/A is normally referred to as the shear stress and velocity gradient as the rate of shear. The above relationship however, is only strictly true for Newtonian liquids and is therefore of limited value. For the majority of products, exhibiting non-Newtonian behaviour, the equation is modified thus:

$$F = (\eta_{app})^n A \text{ x velocity gradient}$$

where

η_{app} = apparent viscosity
n = integer (normally between 0 and 1)

Substitution of this information, back into the Reynolds number calculation, gives the following relationship:

$$R_c = \frac{D^2 N \rho}{\eta_{app}}$$

Onset of turbulence is generally considered to occur at values for the Reynolds number, R_e, of approximately 2 x 10^3. An R_e value of greater than about 10^4 indicates the existence of fully developed turbulence. In practice, this formula is often used to determine the required impeller speed and dimensions, to perform a specific mixing task.

Liquid mixing equipment

When mixing fluids on a production scale, the type of flow patterns that exist in a mixing vessel are important, if effective mixing is to take place. The basic requirement for mixing is the generation of turbulent flow.

Three different types of liquid flow pattern can occur in a mixing vessel, as described below.

1. *Tangential flow*

In this case, the liquid moves parallel to the direction of the impeller, with little movement perpendicular to the blades, except at the tips, where eddy currents may be generated. This type of flow pattern is illustrated in Figure 12.

Tangential flow is normally observed in simple paddle mixers, operating at low speeds. This type of flow is of little use in the practical mixing of cosmetic and toiletry products, except perhaps in the case of mixing low viscosity, miscible liquids.

2. *Radial flow*

When radial flow patterns are created, liquid is discharged outwards from the impeller by

FIGURE 12

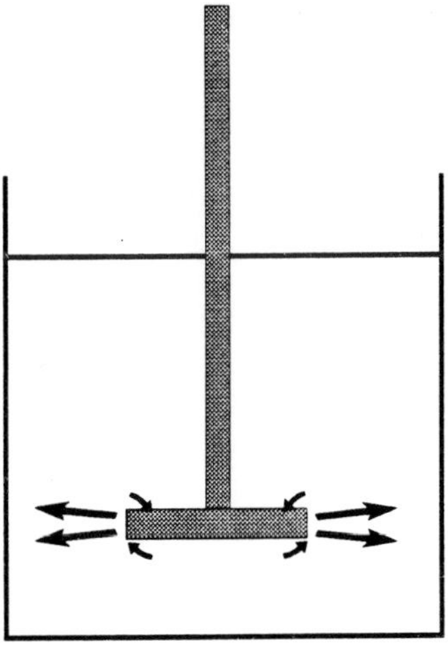

centrifugal force. If the moving liquid strikes the vessel wall, its flow pattern splits into two, circulating back towards the impeller, where it is entrained once more. Radial flow gives some degree of turbulence by producing some axial flow, the degree of which is determined by the impeller speed. A typical radial flow pattern is illustrated in Figure 13.

FIGURE 13

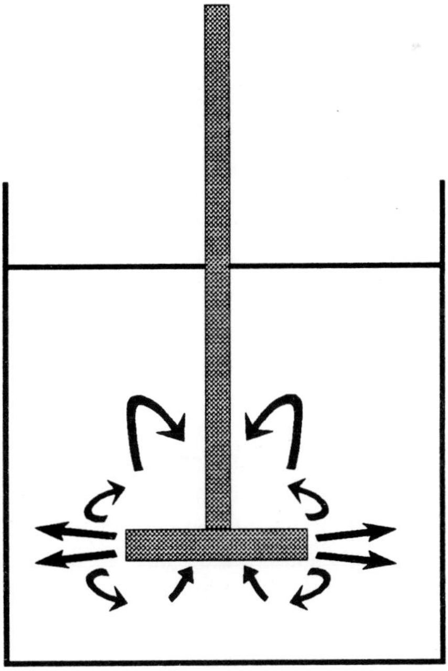

3. *Axial flow*

Axial flow is, by far, the most important type of flow pattern and describes the movement of fluids up and down the mixing vessel, parallel to the axis of rotation. Usually the impeller blades are pitched so that liquid is discharged axially and the direction of flow may bc from top to bottom of the vessel, or vice versa. Axial flow is the most difficult type of flow pattern to maintain and the presence of *baffles* in the vessel can significantly assist in this process. A typical axial flow pattern is illustrated in Figure 14.

FIGURE 14

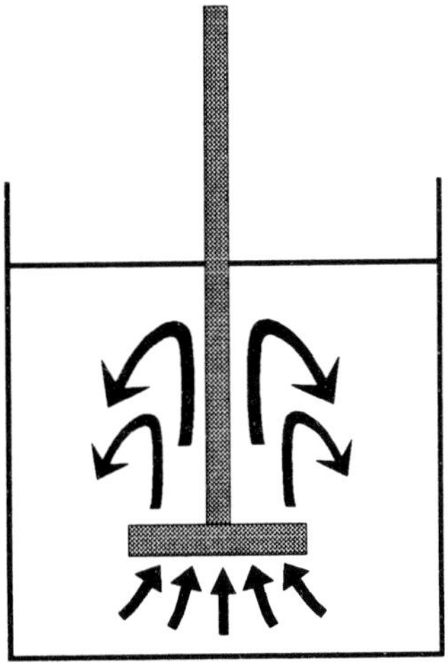

Axial flow exhibits the highest degree of turbulence of the flow patterns described and is therefore the optimum choice for effective mixing of cosmetic and toiletry products. It is also the most difficult to obtain in practice.

In reality, any mixing system will create elements of tangential, radial and axial flow, each to a varying degree. The absence of good axial flow characteristics in a mixing vessel gives rise to layering, particularly in cases where the liquids being mixed are not easily dissolved in each other. The type of flow pattern that occurs in any manufacturing configuration will depend largely on the size and shape of the vessel itself, in addition to the speed, shape and type of impeller or mixing device employed.

4. *Vessel shape and design*

When designing a mixing vessel, it is important to remember that the requirement for optimum mixing is the creation of turbulence, in as much of the vessels volume as possible. Any parts of the vessel where little or no turbulence occurs, will create a dead spot, in which liquids or, more commonly added solids, may accumulate and never be incorporated into the mix proper. In general, vessel designs should possess no sharp corners or irregular shapes, as these are the most likely cause of dead spots. Vessel shapes should therefore be smooth and rounded wherever possible and almost all vessels used in the manufacture of cosmetic and toiletries are based on a cylindrical shape, with either a hemispherical or dished bottom.

With the exception of the simple mixing of low viscosity miscible liquids, it is also very important to be able to create axial flow patterns within the vessel, throughout as much of its depth as possible. From the point of view of space utilisation in the factory, it is sensible to incorporate a vessel design which is as tall and narrow as possible, for a given volume. The taller the vessel, however, the less likely it is that axial flow will be easily created throughout its depth, unless a multiple impeller system is used. The problem of creating axial flow increases with an increase in the viscosity of the liquid(s) mixed and consideration of the type of materials to be processed is an important factor in vessel design. Generally, for low/medium viscosity liquids, vessel design can be tall and thin with typical height:diameter ratios of 4:1 and 6:1. For higher viscosity liquids the tank design should be more squat, to ensure that adequate axial flow occurs, and corresponding ratios of 2:1 or 3:1 are not uncommon. Tank diameter is also an important consideration in vessel design. The larger the tank diameter, the bigger the impeller required to produce effective mixing.

The presence of internal baffles in the vessel will markedly increase turbulence, due to the disturbance of existing flow patterns. The wider the baffle, the higher the degree of flow pattern disturbance and the higher the level of turbulence created. For vessel baffling to have any effect on the turbulence created, the onset of turbulence must have already occurred.

5. *Impeller design and placement*

Impellers are used in almost all mixing processes involving low or medium viscosity liquids. The ideal impeller should be able to effectively produce turbulence and axial flow in the mixing vessel, with the minimum of power input, in order to obtain the most cost-effective route. There are many different types of impeller design available, all in a wide variety of sizes and configurations. The simplest and cheapest type of impeller is the simple *paddle stirrer*. These devices create some turbulence but the flow patterns in the mixing vessel are largely tangential in character. Paddle mixers are therefore only normally used for mixing very low viscosity liquids and are operated at slow speeds for maximum efficiency.

Another type of impeller design is the *marine propeller* which, as its name suggests, is modelled on a ships propeller. Marine propellers typically have three or four blades with a variable pitch along their length. This type of impeller can be successfully used for the mixing of low/medium viscosity liquids, where they are used at operating speeds of between 200 and 2000 revolutions per minute.

The last, and most frequently encountered type of impeller, is referred to as the *turbine*, which can be obtained in many different sizes and configurations, each able to deal with different range of liquid viscosities and mixing requirements. One of the most common turbine designs is the *fixed pitch axial flow impeller*, which generates a reasonable level of axial flow in the mixing vessel, even at moderate speeds. This type of impeller also creates a significant level of radial flow, which can be used to give further turbulence by the use of internal baffles in the vessel.

As a mixing vessel becomes larger, it becomes more difficult to achieve sufficient axial flow to produce good mixing characteristics and, irrespective of impeller design, placement in the mixing vessel is most important, particularly in cases where significant amounts of axial flow are produced. Figure 15 illustrates an upright mounted impeller, centrally placed in a mixing vessel.

FIGURE 15

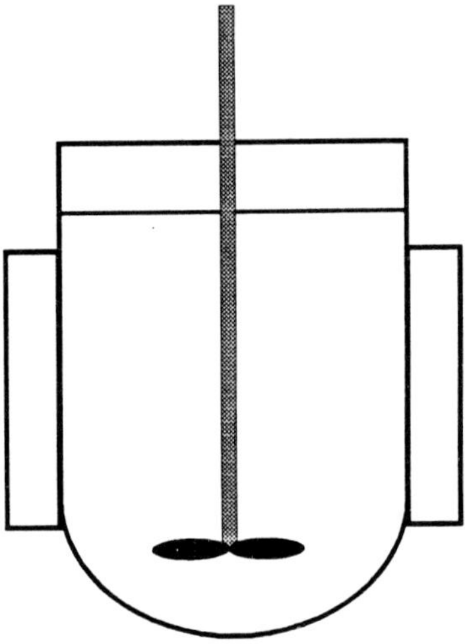

Fundamentally this is a poor design as the mixing action will result in depression of the liquid surface, thereby creating a vortex at higher mixing speeds. In any mixing operation, the creation of a vortex should be avoided because of the likely entrapment of air in the

FIGURE 16

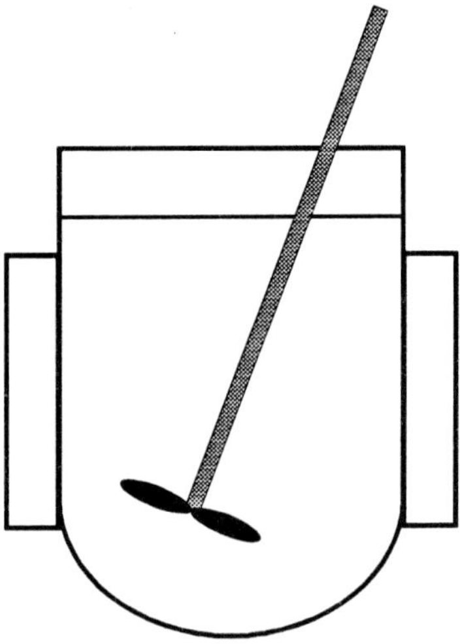

mix. Additionally, vortex creation represents wastage of energy and low capability for the creation of turbulence. The likelihood of vortex creation can be slightly reduced by moving the axis of the impeller away from the centre of the vessel but the most satisfactory solution is to mount the axis of the mixing device at an angle, as illustrated in figure 16.

The creation of a vortex can also be reduced by the use of internal baffles in the mixing vessel.

Manufacturing techniques — liquid/liquid mixing

A wide variety of liquid mixing operations are used in the manufacture of cosmetics and toiletries, ranging from simple mixing of two miscible liquids, to complex emulsification processes. Each particular task requires a different approach and, often, a different type of plant design, to be carried out efficiently.

1. *Mixing miscible liquids*

The mixing of miscible liquids is probably the easiest mixing operation in cosmetic and toiletry manufacture, only requiring agitation to cause sufficient turbulence in the vessel.

2. *Mixing immiscible liquids*

The mixing of immiscible liquids refers to the creation of emulsions, a very important manufacturing process in the cosmetics and toiletries industry. Simplistically, emulsions contain an oil phase, a water phase and an emulsifier to stabilise the system. The properties of a typical oil-in-water emulsion are dependent on the viscosity of internal and external phases and the particle size and distribution of the internal phase, in addition to parameters such as pH and specific gravity.

If, after manufacture, the particle size and distribution is variable, then the emulsion will be inherently unstable. For maximum stability the particle size should be small and

FIGURE 17

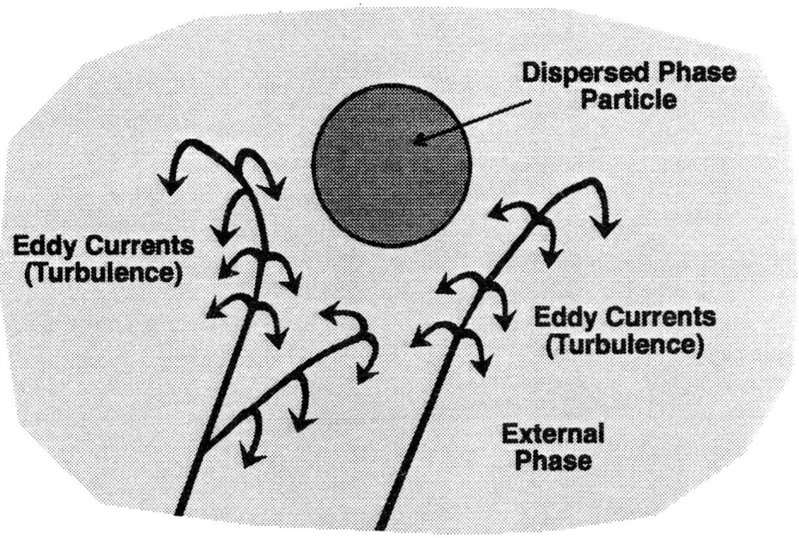

distribution should be as uniform as possible. Any large dispersed phase particles present in an emulsion system, after manufacture, will catalyse instability. Apart from the type and quantity of emulsifier present, the main determinant of emulsion particle size and distribution is the manufacturing process and the mixing methods used within it. The reduction of particle size during the manufacturing process is a function of the degree of turbulence created in the emulsion. A dispersed phase emulsion particle or droplet is held together by two mechanisms, the surface tension of the droplet itself and the magnitude of its viscous inertia, which is a measure of the cohesiveness of the dispersed phase liquid molecules. In order to further break up this droplet, it is necessary to lower the interfacial surface tension, by the addition of a suitable emulsifier, and create turbulence, as illustrated in Figure 17.

The size and force of the eddy currents, which create the turbulence, is dependent upon the energy input of the mixing system. In order to reduce the dispersed phase particle size, it is often necessary therefore, to apply a high degree of shear in the mixing process, such as that produced by a rotor-stator mixer. In some cases, medium or even low shear conditions are sufficient to cause dispersed phase particle size reduction, particularly if the emulsifier used is highly efficient.

For ideal mixing of emulsions on a production scale, the same degree of shear or turbulence, relative to batch size, as that used in the laboratory should be used. In practice, this is very difficult as the size and power of a high shear mixer required to produce laboratory-scale degrees of turbulence and shear in a production-sized vessel, would be impractical. In order to overcome this situation practically, the production of emulsions is normally carried out in two different stages. The first stage, normally referred to as the phasing operation, involves mixing the oil phase and the water phase, in the presence of the emulsifier, under relatively low shear conditions. This will produce a coarse emulsion, often with a relatively large dispersed phase particle size and poor uniformity of distribution. In the second stage, this coarse emulsion is passed through a high shear mixing device, such as a rotor-stator mixer. This will provide the required reduction in emulsion particle size and create a much more uniform particle distribution. This latter stage is normally referred to as the *homogenisation* process.

In order to ensure uniformity of emulsification throughout the production batch, it is important that all of the product is homogenised. If the high shear rotor-stator mixer is placed inside the main mixing vessel, it is critically important to ensure that good bulk movement of material takes place within the vessel. If this is not the case, some of the batch may not pass through the high shear mixing head, leading to potential problems with the properties and stability of the finished product. Under these circumstances, the only way to maximise the chances of subjecting the entire batch to homogenisation, is to extend the mixing time, with the resultant negative implications in terms of manufacturing cost and throughput.

A more elegant solution is to move the high shear mixing device outside of the main mixing vessel and pump the entire batch through it. This is known as *in-line homogenisation* and is diagrammatically illustrated in Figure 18.

Using this technique, assurance is given that the entire batch has been homogenised in the shortest possible time. The extent of homogenisation can be controlled by the design of the high shear mixer and the contact dwell time of the product with the mixing head.

FIGURE 18

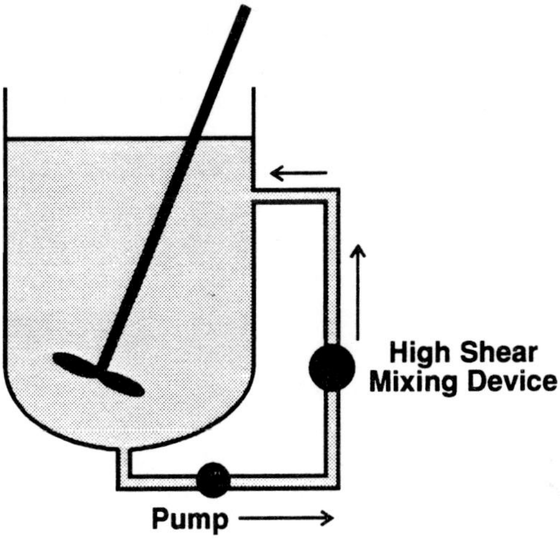

This latter parameter is, in turn, controlled by varying the pumping rate through the high shear mixing device.

In some cases cosmetic emulsions can be produced by the *phase inversion* technique, in which uniform particle distribution and small particle size can be obtained, without the use of a high shear mixing device.

Many cosmetic emulsions exhibit pseudoplastic behaviour, in which a reduction in apparent viscosity occurs, as the applied shear is increased. The viscosity of the system normally recovers with time. Some emulsions, if sheared at ambient temperatures, may not recover viscosity, although raising the temperature of the product back to the temperature it was manufactured at, and then allowing it to cool again, normally effects complete recovery. Very few cosmetic emulsions exhibit dilatent, or shear thickening, behaviour. Those that do, often have a high solids content, whereupon the solid particles become "less lubricated" upon the application of shear, resulting in apparent viscosity increase.

3. *Heating and cooling of emulsions*

It is frequently necessary to use heat during the manufacture of emulsions, because the oil phase often contains materials that are solids at room temperature. Provided that an emulsion is manufactured correctly, its viscosity will be determined by particle size distribution and the rate of cooling of the emulsion after manufacture. Apart from this, however, there is no practical reason why manufacturing processes for emulsions should use heat except, perhaps, that it is slightly easier to ensure microbiological integrity when heat is used. If the materials in both phases of the emulsion are liquids at room temperature, then there is no reason why the emulsion should not be manufactured using low temperature processing.

If it is necessary to manufacture an emulsion using heat, the vessel design and construction play a very important role in producing a successful product. Heating and cooling is normally applied using steam and cold water respectively, each of which is passed, in turn, through a jacket surrounding the outside of the manufacturing vessel. The application of heat, and subsequent cooling, during the manufacturing process, must be carefully controlled, as both will affect the viscosity and properties of the finished product. If the vessel is of simple configuration, containing just a top mounted turbine mixer, then other problems may occur during emulsion manufacture. Given that the rate of cooling will affect the finished properties of the emulsion, the product nearest the vessel walls may exhibit a different emulsion structure to that in the centre of the vessel, the comparative rates of cooling being significantly different. Product may also adhere to the vessel inside walls during the heating process, resulting in burning of the product adjacent to the wall surface. The resultant layer of product will also reduce the efficiency of heat transfer during subsequent heating and cooling processes, thereby affecting processing times.

In order to overcome these problems, a specific design of manufacturing vessel must be used.

Given knowledge of the likely viscosity range of finished products, the turbine mixer is designed such that good axial flow patterns are produced in the vessel. This provides a fundamental mixing action and ensures bulk transfer from the top to the bottom of the vessel. If the manufacture of very high viscosity emulsion products is required, the top mounted turbine mixer is normally replaced by a contra-rotating stirrer, driven around the

FIGURE 19

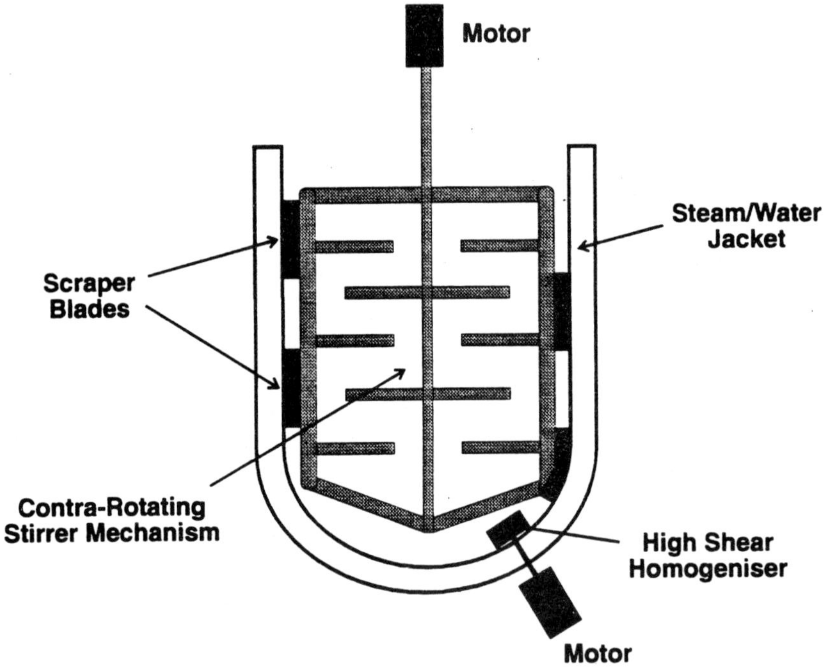

central axis of the vessel, as illustrated in Figure 19. The side-sweep anchor stirrer assists in producing an effective mixing action within the vessel and ensures that material adjacent to the side walls of the vessel is continuously moved away, thus preventing variability in the finished product and assisting in the heat transfer process. The anchor stirrer is also fitted with scraper blades, which scrape the inside of the vessel walls, as it rotates, thus preventing burning on of the product. The scraper blades are normally held in contact with the vessel wall by a spring-loaded device or simply by the positive pressure created as the blade itself moves through the product.

Homogenisation is provided by an in-line or in-vessel high shear mixer. In the case of the former, product may be pumped either back into the manufacturing vessel or directly into a holding tank for subsequent filling. In the case of the latter, placement is normally at the base of the vessel, to prevent the possibility of introducing air into the product. If placement at the vessel base is not practically possible, then a top-entry high-shear mixing device can be used but this should also be placed in such a position as to minimise air entrainment.

High shear mixing systems

Although the creation of turbulence is a critical requirement in a high proportion of mixing processes for cosmetics and toiletries, for certain applications it is desirable to generate a very intense degree of shear stress within the mixing environment. A typical example would be the homogenisation of emulsion products.

The most frequently encountered high shear mixing device, in the cosmetics and toiletries industry, is the *rotor-stator* mixer. This principle involves the placement of a high speed rotor, typically rotating at speeds between 3000 and 6000 revolutions per minute, in very close proximity to a fixed stator. Product can then be forced through the very narrow gap between the rotor and the stator, where it experiences a very high degree of shear, as illustrated in Figure 20.

FIGURE 20

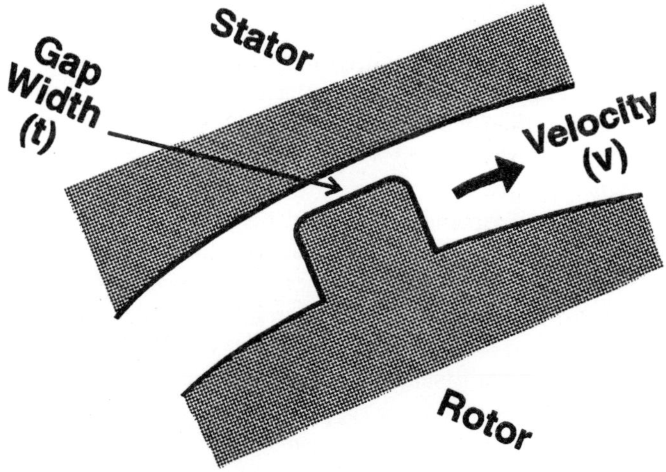

The shear rate is determined by the speed of the rotor and the width of the gap between the rotor and the stator, and is given by the following equation:

$$S = \frac{v}{t}$$

where

S	=	shear rate
v	=	velocity of the rotor
t	=	magnitude of the gap between the rotor and the stator

Therefore the shear rate, or degree of shear, experienced by the product, increases with an increase in rotor speed and a decrease in the gap between the rotor and the stator. In practice, the shear rate applied to the product also increases with the viscosity of the product itself.

Other types of high shear mixing device are also used in the cosmetics and toiletries industry, although less frequently. These include valve homogenisers, ultrasonic homogenisers and various types of colloid mill.

Other manufacturing devices

In addition to the mixing equipment discussed so far, there are many other devices used to support the various manufacturing processes. Some of these are briefly discussed below.

1. *Vacuum vessels*

Products likely to entrap air, for example thick gels, emulsions and toothpaste, are sometimes manufactured under vacuum, using a vacuum mixing vessel. These are of conventional design, except that they are fitted with a vacuum system to generate a vacuum within the vessel itself. Raw materials are introduced into the vessel, through an entry port, by virtue of the negative pressure within the vessel.

2. *Pumps*

Pumps are essential in virtually every manufacturing operation, either for transfer of raw materials into a manufacturing system, for the manufacturing process itself, or simply for transferring finished product from one location to another. Many different types of pump are used in the cosmetics and toiletries industry but the most common are the rotary *tri-lobe pump* and the *diaphragm pump*.

The operation of the tri-lobe pump is based on the rotational movement of two interlocking tri-lobe rotors, placed inside a pump chamber, with a very small gap between the lobes and the chamber itself. Pumping action takes place through the positive pressure produced by the rotating tri-lobes within the pump chamber. The tri-lobe pump is suitable for the transfer of a wide variety of materials, even those of very high viscosity. Tri-lobe pumps are, however, not particularly suitable for very thin liquids.

The diaphragm pump operates, as its name suggests, through an oscillating diaphragm, which creates a negative pressure in the pump chamber, followed by the creation of a subsequent positive pressure, on the return stroke of the diaphragm. When the negative pressure is created in the pump chamber, product enters at the chamber inlet, through a one-way valve. On the return stroke of the diaphragm, the product is forced out of the chamber outlet, through a second one-way valve, thus producing a pumping action. Diaphragm pumps can be use for a wide variety of applications, including the transfer of corrosive materials, although they are not particularly suitable for pumping products with very high viscosity.

3. *Heat exchangers*

Apart from the normal method of heating and cooling products, using vessel jackets, heat exchangers can also be used. Whilst there are many types of heat exchanger available, including *plate heat exchangers*, *counter current heat exchangers* and *thin film heat exchangers*, they all operate on a similar principal. The product to be heated, or cooled, is passed through a chamber constructed of very thin metal walls, with high thermal conductivity. Liquid, or steam, is passed around the outside of the chamber and thermal transfer across the chamber wall rapidly heats or cools the product, as required. Heat exchangers are normally installed in-line and often provide the most cost-effective and energy-effective method for heating and cooling products. In cases where there is a danger of product burning to the chamber wall, *swept surface heat exchangers* are normally used. These are equipped with some sort of scraping device, which continuously moves the product passing through the heat exchanger, away from the chamber wall.

Filling techniques

A wide variety of filling machines are available, ranging from simple hand-operated systems, to filling machines with highly complex automatic mechanisms.

1. *Liquid filling*

The term " liquid filling" normally applies to products that flow readily under gravity, typical examples including lotions, shampoos and aftershaves. There are two fundamental techniques used to fill liquids, volumetric filling and vacuum filling.

1.1. *Volumetric filling*

The principle of volumetric filling is best illustrated by considering the simplest type of volumetric filler, as shown in Figure 21.

As the cam rotates clockwise, the piston moves back, creating a vacuum in the chamber, thus causing product to flow from the hopper through the t-valve and into the chamber. The t-valve then rotates anti-clockwise by 90° and, as the cam continues to rotate, the piston commences its return stroke, thus pushing the product from the chamber out through the filling nozzle. Using this method, the volume of fill can be infinitely adjusted by adjusting the position of the connecting rod and a constant volume is delivered, irrespective of the product form and container size. A typical cycle time for a small volumetric filling machine would be 3-4 seconds , corresponding to 6000-9000 units filled in an 8 hour day. Larger volumetric filling machines will run at much higher speeds, up to 100 cycles per minute.

The advantages of the volumetric filling method is that it is very accurate, with good

FIGURE 21

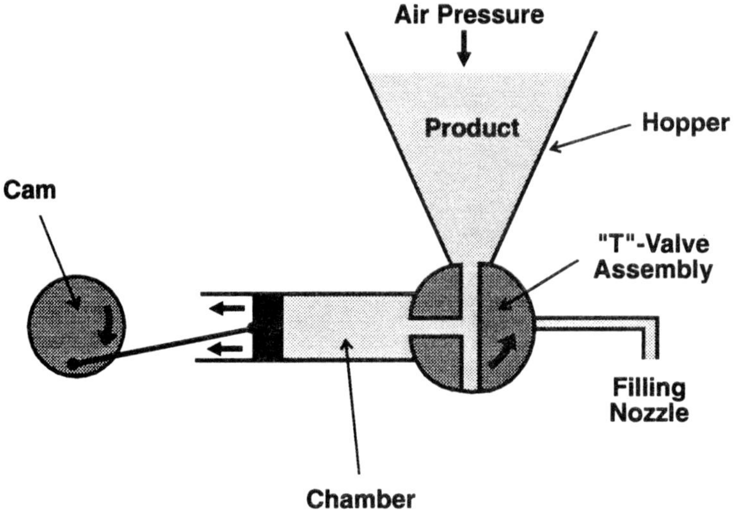

reproducibility. This technique is ideal for filling higher viscosity products, such as creams, and is also excellent for products which have a tendency to foam, as the filling process is carried out under pressure. The volumetric method is also very flexible, as the hopper into which product is placed can be of any size.

Disadvantages of volumetric filling is that it is not very efficient for low viscosity liquids, such as aftershaves, and the apparent volume can change, if the volume of the packaging container is not maintained consistently.

1.2. Vacuum filling

The effectiveness of vacuum filling relies upon the design of the filling nozzle. A typical vacuum filling apparatus is illustrated in Figure 22.

When the container is in position, the vacuum sucks air out of it, creating a partial vacuum. The product is drawn down the outside tube, and into the container, until it touches the bottom of the nozzle assembly, whereby the product starts to rise up the vacuum tube, whereupon the vacuum is released, or the nozzle withdrawn. The advantage of this filling method is that it always fills containers to the same constant level.

One of the disadvantages of vacuum filling, is that the container must be able to withstand a vacuum, without collapse, and must therefore be fabricated from glass, rigid polystyrene or a similar material. Vacuum filling is not very suitable for products that tend to foam and can also give problems with declared volumes or weights, if the volume consistency of the product containers is not carefully controlled.

Irrespective of whether a volumetric or vacuum filling technique is used, a higher degree of accuracy can be obtained by the use of a fluid logic circuit. A detector in the filling nozzle senses the level of the liquid being filled into the container and, when the

FIGURE 22

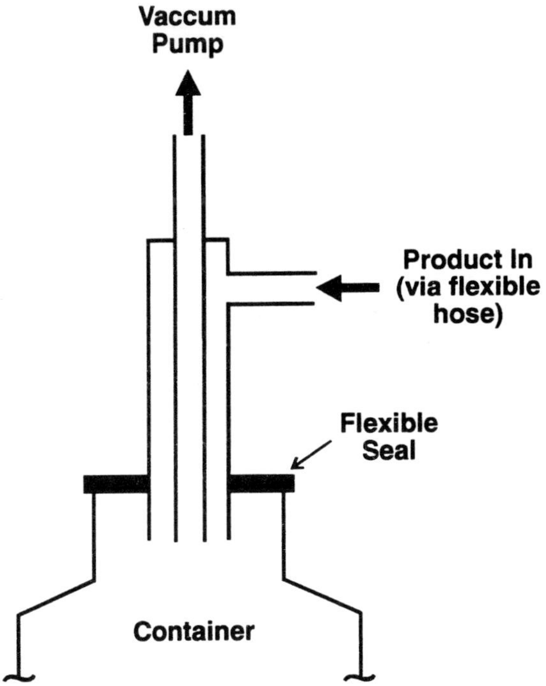

required fill has been achieved, transmits a signal to the filling valve, closing it. This technique is very useful for accurate filling of liquids, with moderately high viscosities. It is not particularly suitable for thin liquids or liquids that tend to foam. Where it is necessary to fill liquids that foam, this can be minimised by the use of a specially configured filling nozzle, which dispenses the product on to the side walls of the product container, thus minimising aeration.

2. *Filling flammable liquids*

The filling of flammable liquids does not differ greatly, in principle, from the filling of ordinary liquids. There are, however, a number of additional precautions which must be taken, to ensure a safe working environment. Flame-proof motors should be used on all equipment, to prevent accidental spark generation. The filling area should be flame-proofed and, where possible, isolated from the rest of the factory. All apparatus and equipment must be earthed, to prevent sparking caused by the generation of static electricity and the use of nylon protective clothing for operators should be avoided. The filling area should be safely ventilated and strict fire precautions adhered to.

The most commonly encountered flammable liquids in the cosmetics and toiletries industry are solvents, such as ethyl acetate, used in the manufacture of nail varnishes, and hydrocarbon aerosol propellants. Flammable raw material storage areas should be flame-proofed and isolated from the main factory.

3. *The design of filling lines*

Despite the many different types of filling line that exist, fundamentally they all fall into two basic categories, *in-line or "batch" filling* and *rotary or "continuous" filling*. The particular technique chosen for any given task, is dependent upon such factors as the quantity of product to be filled, the filling rate required and cost constraints.

3.1. *In-line filling techniques*

In-line filling uses a conveyor system, which feeds the containers to be filled, in batches, to the filling machine. The number of containers in each batch normally corresponds to the number of filling nozzles on the filling machine. Small in-line filling machines typically fill containers in batches of four, although batches of eight or sixteen are not uncommon in larger units.

Typically, machines of this type can fill approximately 10,000 units in an 8 hour day.

This is the simplest form of filling line, with the minimum number of change parts, and down-time when changing products is relatively short. In-line filling is also very flexible, allowing a wide variety of containers to be filled. The major disadvantage of in-line filling is that it is relatively slow, when compared with some other techniques.

3.2. *Rotary filling techniques*

All rotary filling techniques are based on the use of a rotary carousel, diagrammatically represented in Figure 23. The simplest form of rotary fillers have only one filling nozzle and this type are often fed manually. More complex rotary filling systems have a filling nozzle over each container holder, the filling operation for each container being carried out whilst the table, and nozzles, are in motion.

The latter type of rotary filler is much faster and container feed is carried out automatically, using an unscrambling table. Filled product is removed from the carousel, using an automatic ejection system. The number of filling heads on a rotary filler varies but commonly 8, 16, 24 or 32 head machines are used. 64 head machines are also available but these are very expensive and only used for very large capacity filling requirements.

Rotary filling systems often occupy less floor space than in-line fillers and their rate of fill is much faster. The disadvantages of rotary filling systems are that they are difficult to set-up. Changing products is time consuming, due to the number of change parts that have to be fitted and, because of this, rotary filling techniques are only economically viable for large scale filling operations of greater than 100,000 units.

The choice of filling method, in-line or rotary, is normally based on cost and required throughput. For smaller capacity tasks, in-line filling is the method of choice but for larger filling operations, particularly where there is not a great variety of products to be filled, rotary filling is much preferred.

4. *Ancillary operations*

Besides the actual filling operation, there are a number of other tasks that are completed on the line. These include capping, labelling and batch identification, in addition to the optional activities of cartoning or shrink-wrapping.

FIGURE 23

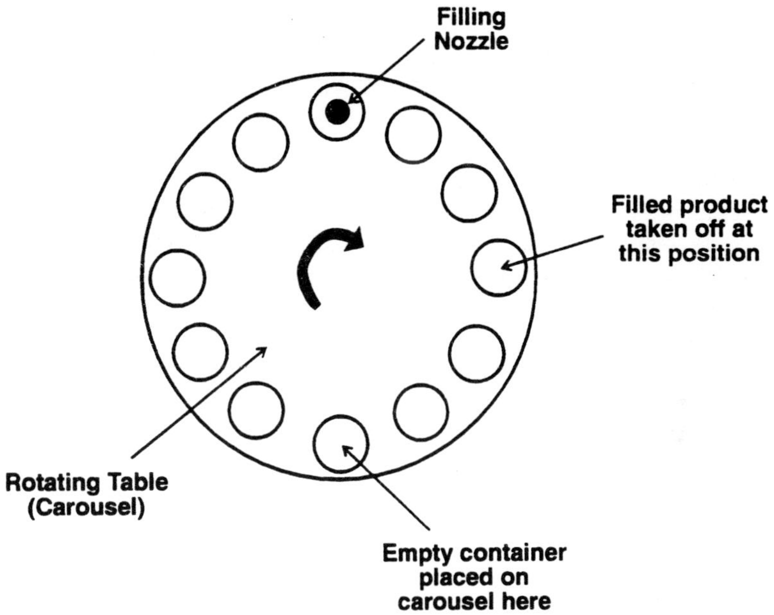

4.1. *Capping*

Automated capping operations are often designed around a machine using a rotary carousel, although in-line capping machines are also available. The cap is dropped from the storage hopper, down an automatic chute, on to the container. In the case of push-fit caps, a plunger pushes the cap on to the container. Screw-fit caps are normally applied using a rotating wheel screw. It is essential that screw-caps are applied with the correct degree of tightness, or torque, otherwise the product may evaporate or leak out of the container. Incorrect torque can also lead to a slackening of the cap in subsequent warehousing or distribution.

4.2. *Labelling*

Whilst silk screen printing is the cheapest form of container labelling, paper or vinyl labels are often used to enhance the appearance of the finished product. Labels are normally applied using an automatic labelling machine and are either self-adhesive, or require the application of adhesive before they are placed on to the container. Self-adhesive labels are most convenient and frequently used, the label being removed from its backing paper by the labelling machine, before being applied to the container. Application of adhesive requires a more complex operation but fast throughput can still be achieved.

4.3. *Batch identification*

Batch identification, or date coding, is essential, in order to facilitate traceability in the event of a subsequent product complaint. Coding can be carried out in a number of ways, ranging from a simple date stamp to a highly sophisticated ink-jet coding system.

4.4. *Shrink-wrapping*

Shrink-wrapping involves wrapping the product in a heat sensitive plastic film and applying heat to shrink it tightly onto the product. Automatic shrink-wrapping machines incorporate heated tunnels, through which the wrapped product is passed to give the finished packaging, ready for warehousing and distribution. Both individual and boxed products can be shrink-wrapped, although care must be taken if passing heat sensitive products through the tunnel. Shrink-wrapping is more commonly used nowadays, particularly in the case of products distributed to supermarkets or cash and carry.

4.5. *Cartoning or final packaging*

Cartons can either be erected by hand, or with automatic carton erectors which fold and assemble the carton and, if required, place the product and any associated literature in it. Products are finally packed into cardboard outers, or shrink-wrapped, before being loaded on to pallets for storage and distribution.

Specialist filling techniques

Although the vast majority of filling operations are carried out using techniques very similar to those described above, there are a number of filling tasks that require special consideration.

1. *Tube Filling*

Nowadays almost all tube containers, plastic or metal, are supplied with an open-ended base, through which product is filled prior to sealing. The filling and sealing operations are normally carried out on a rotary carousel and the method of sealing depends upon the type of tube being filled. During the filling operation, it is important to avoid contamination of the area to be sealed with product, otherwise the seal will not form properly.

Plastic tubes, the most frequently encountered type in the cosmetics and toiletries industry, are sealed thermally. An infra-red heater softens the bottom of the tube, making it soft and pliable, and sealing jaws close on the softened tube and seal it. The seal is then force-cooled to harden it and the excess plastic trimmed off, before the filled product is ejected from the carousel for packing. In order to ensure that sealing takes place with correct orientation, with respect to pack graphics, the tube carries a register mark which is sensed by an optical sensor on the carousel. Each tube is then automatically rotated before sealing, to ensure correct orientation.

One of the main problems that occurs in tube filling is *tailing* of the product, which describes the tendency of the product to "string" at the filling nozzle. The extent of tailing is dependent on the rheology of the product and the temperature at which it is filled and if care is not exercised, can cause poor sealing, due to contamination of the sealing area. The batch coding of plastic tubes is normally carried out during the sealing process, the thermal sealing jaws carrying the appropriate coding embossed on their internal faces.

Metal tubes are filled and sealed using similar principles, except that the tubes are double-folded and crimped to seal them. Metal tubes are rarely used in the cosmetics and toiletries industry, although they remain popular for some types of pharmaceutical product.

The newest form of tube packaging is the laminated tube, which is becoming increasingly popular for cosmetic and toiletry applications, particularly in the filling of toothpaste. Heat-sealing is sometimes possible with certain types of laminated tube but they are more commonly sealed using ultrasonic techniques.

2. *Aerosol filling*

Principles for the filling of aerosols are similar to those described earlier, except that the contents of the pack are filled under pressure. Filling operations can either be in-line or rotary and the filling process takes place in two stages, crimping of the valve cup and subsequent pressurisation of the aerosol can. In a typical pressure-filling operation, the product is put into the aerosol container and the aerosol head space purged with propellant. The valve cup is then placed into position and crimped. Finally, the pack is pressurised, normally by filling through the aerosol valve.

3. *Eye-shadow powder pressing*

The forces which bind powders, such as eye-shadows, together, operate at very short distances and the powder must be compressed during the filling operation. This operation is carried out using a powder press, the powder being pressed into a container, known as a godet. Firstly, the powder is loaded into the godet and compressed with a punch at pressures of up to 1500 psi. During the filling operation, a nylon ribbon is placed between the punch and the powder surface, to prevent sticking, and leave the surface of the filled powder with an attractive finish. Care must be taken to ensure that all the air is expelled during the pressing operation, otherwise the pressed eye-shadow is likely to break up into layers, particularly in cases where pearled pigments have been used.

In the laboratory, powder pressing is carried out using a hand-press and a die, to hold the godet. Air is squeezed out of the sides of the punch as it descends. On a production scale, eye-shadow powder pressing is usually carried out on a rotating carousel, each part of the filling process taking place at a particular position on the carousel, as it rotates. A plan view of a simple eye shadow pressing carousel is illustrated in Figure 24.

The godet is delivered to the carousel and lifted into place at position 1. The carousel rotates and the godet is filled with loose powder, using a mass-flow hopper, at position 3. The mass-flow hopper is equipped with a rotating archimedian screw, known as an *auger stirrer*. This ensures free-flow of the powder into the godet. The degree of fill is controlled by adjusting the speed and time of rotation of the auger stirrer. After filling, and before the powder is pressed, there is a chance that the powder will become disturbed, leading to poor distribution in the godet. This occurs because the filling action, combined with the rotation of the carousel, causes a depletion of powder at the front edge of the godet. When the powder is subsequently pressed, the front edge of the pressed eye-shadow will be very weak and exhibit a tendency to break. Pressing takes place at position 5 using either a spring-loaded or, more frequently, hydraulic punch. As the pressing operation occurs, a

FIGURE 24

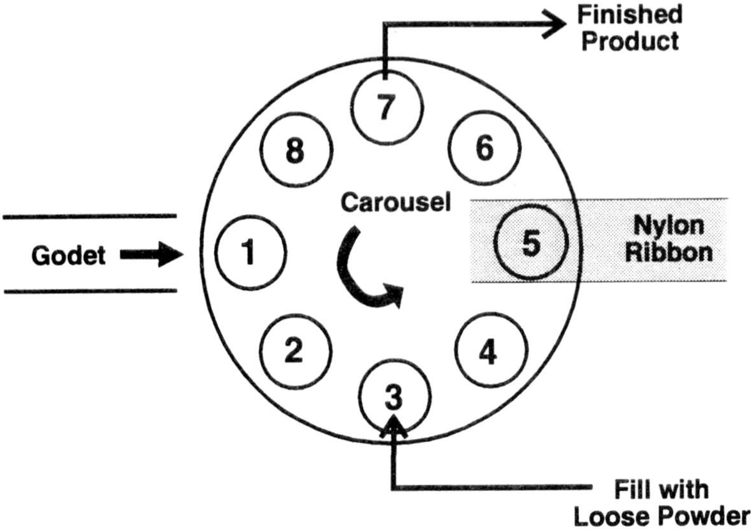

new section of nylon separating ribbon is fed under the punch, for each godet pressed. Finally, at position 7, the godet is pressed out of its slot with a rubber plunger and removed for final packaging.

This process is a continuous operation, with theoretical production rates of up to 14,000 units per day. In practice, however, typical production rates are somewhat less than this. Throughput can be increased by using a double pressing head, although control of finished product quality is much more difficult using this technique.

Filling line management and automation

When considering the manning or labour requirements for a filling line, it is important to use sufficient human resources, so as not to reduce the speed of the filling operation. It is the filling machinery, and not manning levels, that should determine the speed of filling, allowing more accurate capacity estimates to be made, thereby optimising production planning.

The operation of a filling line, depending on the degree of automation installed, can be a very labour intensive process. Management decisions have to be taken regarding the level of capital investment made in automatic filling equipment, versus the cost of labour and high manning levels required, in low automation systems.

The decision to automate a filling operation must be carefully examined with respect to labour costs, capital investment, pay-back and volume capacity. Automation will normally allow a filling line to be run at higher speeds, although automatic filling machines may have to be adjusted to slightly below their maximum rate, to achieve the minimum filling cost per unit. A disadvantage of automation is that it is less flexible and cannot easily accommodate minor changes in the filling process. In this respect, the use of higher manning levels, instead of automation, provides greater flexibility.

QUALITY CONTROL AND ASSURANCE

Introduction

The quality function in manufacturing industry has roots that go back many decades. Its history can be traced to the development of mass production, when the traditional skills of the craftsman were replaced by machines. The craftsman performed personal quality checking, as a part of his normal routine during the manufacture of whatever artifact he was making at the time. The advent of mechanisation and mass production, led to a situation where the craftsman was no longer a necessary part of the manufacturing operation, since most aspects of his role were taken over by the machine. Mass production also produced product at a much faster rate than the craftsman could ever do and, as a result, traditional techniques of inspection throughout the manufacturing operation, were no longer applicable.

It was soon recognised that quality of production had to be checked and this led to some of the early ideas of inspection. The quality control department, as we know it, was born. The war years brought further challenges to industry and to those who were responsible for procuring armaments. In times of war, the production of arms has to be rapidly accelerated, not always allowing the time to check output as thoroughly as may be done in peacetime. However, a gun that will not fire, because it is defective, is obviously pretty useless. More to the point, this creates a life-threatening situation, one that cannot and must not be tolerated. So what could the armed forces do to minimise the chances of such occurrences? The pioneering work of people such as Shewart, in the use of statistics and their application to manufacturing processes, led to the development of control charts, which could be used to monitor the quality of manufacturing while the process was underway.

Another problem requiring resolution, was how to check a consignment without having to test every single item within it. This was answered by the introduction of sampling plans, which were published by the Defence Ministries in the USA, as Military Standard 105-D and in the UK as Defence Standard 131A. This has now been superseded in the UK by BS 6001, although the latter is loosely based on Defence Standard 131A.

After the Second World War, different influences were to impact upon the industrial development of the West and the East. In the West, the major thrust was to produce the volume of goods required by soaring consumer demand, to the point where output became the dominating factor. In the East, on the other hand, and specifically in Japan, the lesson was soon learnt that this approach could only lead to one thing, declining quality standards and falling demand.

Japan launched a major quality initiative at government level and this was enthusiastically adopted by many of the leading industrialists. This activity was fuelled and directed, somewhat ironically, by the teachings of two Americans, Juran and Deming. Both men had already tried to make their fellow countrymen aware of quality but they were largely ignored. In Japan it was different and, only a few years later, the West started to see the rapidly improving quality of Japanese goods. At the heart of these changes was a well-structured quality control system. Moreover, the Japanese were to take quality control to new dimensions, involving the whole work force in the achievement of company quality and cost goals.

The development of the quality department within industry has naturally changed, along with other influences. Quality control became quality assurance, which became, in some companies, quality audit. Some experimented with quality circles. Now we have total quality, quality improvement, quality systems, BS 5750/ISO 9000, and so on.

The array of approaches to quality can be confusing. However they all have one thing in common, the desire to manufacture product of regular quality which satisfies the customer.

Quality

The dictionary definition of the word quality would reveal a number of perfectly acceptable phrases such as "goodness" or "degree of excellence". However these are limited in their application to the more scientific uses of the word and particularly in association with the assessment of manufactured goods in industries such as cosmetics and toiletries. To understand its meaning a little more, it is necessary to delve deeper, by considering two aspects of quality when applied to products, *quality of design* and *quality of conformance.*

Quality of design, by definition, is the quality that is built into a product at the design and development stage. It is the thought that is put into the original brief for a cosmetic product, before research and development start to develop a formulation. It is the materials that research and development use during the development process. It is the experimental design used by the research and development scientist, as the formulation is constructed, taking into consideration all the knowledge that already exists, regarding the type of formulation that is being developed. It is the assessment programme that the final formulation is subjected to, before the product is allowed to be marketed. It is also the design of the machinery that the product is going to be made on, once it is in production, and the development and evaluation of the packaging that is going to be used.

Quality is always associated with a cost and the relationship between cost and quality of design is illustrated in Figure 1 on page 399.

FIGURE 1

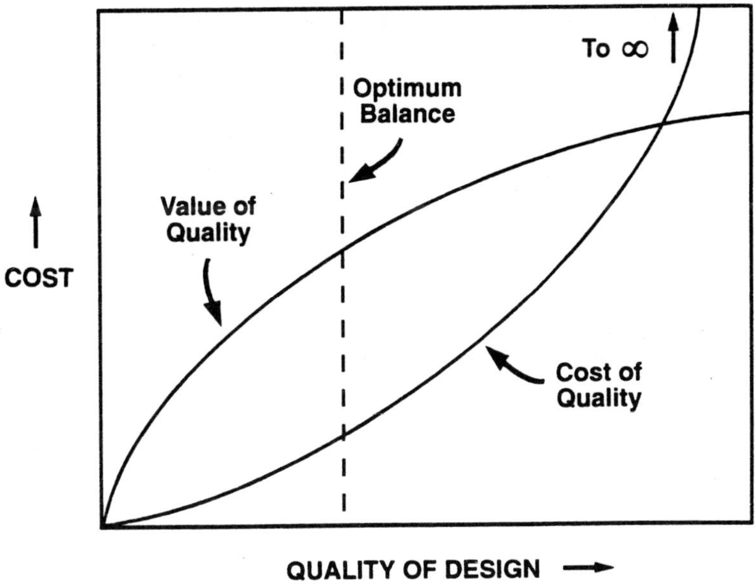

This relationship helps interpret the cost of quality of design.

In the case of cosmetic products, there is very often a high perceived value and it becomes feasible to spend relatively more on their design, particularly in the packaging that is used and, perhaps, more use of esoteric raw materials.

However, the majority of toiletries and cosmetics bought by the average consumer are mass market, relatively low-cost products. In these cases, the cost of the design becomes more critical and must be considered very carefully. Even if production costs are reduced, the consumer may still be perfectly happy with the resulting product because it satisfies his or her expectations and has good quality of conformance.

Whatever the product type, it is important to obtain the correct balance between the cost put into the design and the value obtained from it. This is depicted by the vertical line on the graph in Figure 1, where the value and cost are at their widest separation.

Quality of conformance, on the other hand, comes later in the process and refers to the conformance of the product to the design, once it is in full production. It is the way in which the production department interpret the formulation and process detail, provided by the research and development department. It is the way in which full scale machinery operates, compared with laboratory scale. It is the degree of control which an operator has over the process and how that control is used. It is also the way in which machinery is maintained, allowing it to consistently function correctly, and the variability in any of the raw materials which are provided to the production lines.

The relationship between cost and the quality of conformance is illustrated in Figure 2.

It is useful here to look at this relationship in terms of the loss due to defective product, that is product that has to be scrapped, reworked or sold at a low price, because it is faulty. Again, the objective is to obtain the right balance between the amount spent on quality

FIGURE 2

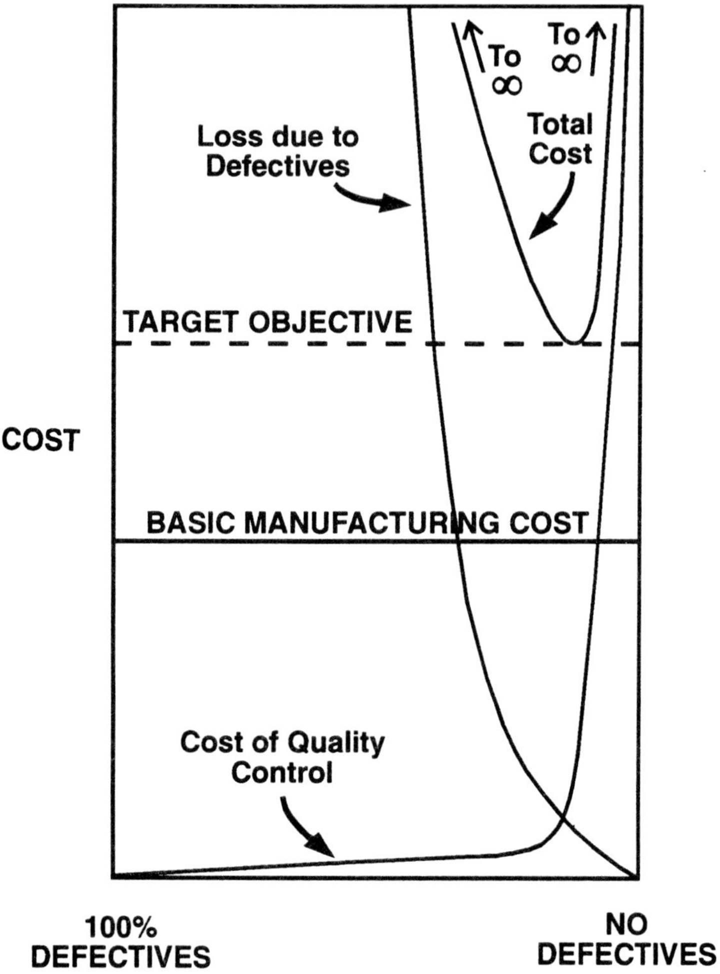

control and the losses which might otherwise be incurred. In this case that balance is depicted in the total cost graph and the target is the minimum point on that curve.

1. *Quality control*

Quality control is, by definition, the control of quality in a manufacturing operation. In the early days of quality control, one of the guiding principles was that a separate operating unit should be established, to measure and report on quality of product. The name of this department was, and often still is, the quality control department. This is really a misnomer, as an independent department such as this is unable to control quality and invariably can only monitor and advise. The only people who can actually control quality are those involved directly in manufacturing the product. This is the combination of all those working in the production environment, including production operators, maintenance engineers and service operators, not forgetting supervisors and managers of the department who have a vital role in engendering the right attitudes and motivation amongst their staff.

2. *Quality assurance*

Quality assurance (QA) is the more correct terminology for the function of monitoring and reporting on quality on behalf of the organisation. The QA department operates by means of a number of techniques, which are designed to give a database of information about the quality aspects of the product in question. This database will provide the information required for the quality assurance department to ensure that the quality of the product is in accordance with the requirements of the organisation.

3. *Total quality management*

The latest philosophies of quality management go way beyond the more traditional views and radically places more responsibility on each individual involved in the whole process of making a product, throughout the whole company. This is the basis of total quality management (TQM). Some companies have been so successful in developing these systems, that the quality department virtually ceases to exist because all the faults are eliminated at source, or corrected as they occur by line operators. It is beyond the scope of this text to comprehensively cover this topic but the essential key principles of TQM are reviewed below.

Quality can be defined as continually satisfying customer requirements and total quality can then be defined as achieving quality at lowest cost. Total quality management can then be defined as achieving total quality by harnessing commitment throughout the organisation. So, how is TQM introduced and what are the requirements for it to succeed?

There is no doubt that for any new technique to succeed in an organisation, it must be management-led, with a high degree of commitment and support. This must extend from board of directors down through senior management, and to all levels within the organisation. However this, in itself, is not enough. There are too many examples of companies which have ploughed enormous resources into introducing TQM, with everyone being committed to its success, only to see it fail after a relatively short time. Often, the reason for this is that although management understand and are committed to its success, the work force do not have, or management do not provide, sufficient time for it to succeed.

TQM must be introduced on a company-wide basis and it must not be seen as a "quick-fix" for the ills of the organisation. Those companies that can demonstrate success, have often been working with TQM for many years. It is a long term commitment. Very often the culture of the company has to change, to ensure that all employees throughout the organisation recognise the role they have to play in the achievement of total quality, whether they are in a direct or an indirect service function.

The method of achieving total quality is prevention, rather than detection, of defective products or service. The standard to which everyone must work is "get it right first time", with an attainment target of zero defects. Quality performance must be measured by using the cost of failure and TQM is a continuous process of improvement.

The quality assurance system

An essential component of any attempt to manufacture products to a consistent quality standard, is the implementation of a well-structured, comprehensive quality system. In modern parlance, this could be achieved by seeking accreditation to BS 5750.

Whichever route of quality attainment is chosen, there are some basic elements which must be put into place, as listed below:

- specifications
- raw materials and packaging control
- process batch control
- finished product control

All of these, without exception, demand precise, clearly defined and agreed procedures and standards to be set, in conjunction with suppliers, internal departments and customers.

1. *Specifications*

Specifications are used to define the requirements for a raw material, process or product. They are essential in assessing whether correct results are being achieved, at any particular stage in the manufacturing operation. A few key parameters that can be used to ensure that specifications achieve the desired results, are listed below:

- wherever possible give the specification a unique code number, so that there can be no confusion as to what it refers to. This is preferably the computer stock code number for the material or the product, so that the specification can be used as a source document for the other systems operating within the company
- link the code number to an unambiguous name. In the case of a raw material, use the trade name or the full chemical name. Extend this name through, in full, to all other specifications which may be linked with it. In the case of a product, use the product name on the product label, or that used in the company catalogue or order form
- always provide clear and easily understood requirements, so that there is no margin for error in the interpretation of the instructions or targets. These requirements must, wherever possible, be measurable, with a target value and a specified range of results, which the material, process or product can tolerate. Where the parameter cannot be measured, it must be specified with reference to an agreed standard
- ensure that the requirements are necessary and do not specify things which will only waste time and resources in checking for them
- ensure that the requirements set are realistic and achievable; if the target has not yet been defined, or the process not fully developed, targets should not be specified until sufficient data is available
- ensure that all specifications are dated and that previous issues are cancelled to avoid users referring to outdated requirements
- before issuing any specification, ensure that it is agreed with suppliers, customers and all appropriate personnel/departments within the organisation

There are further requirements that are specific to the three types of specification needed, to provide the starting point of a good QA system, as follows:

1.1. *Raw material specifications*

These should include a list of the following parameters:

- description and appearance
- identification tests, such as infra-red or chemical analysis
- physical tests
- chemical tests
- storage conditions. This is particularly important for those materials that are sensitive to heat, cold, light or other environmental factors
- safety requirements, such as handling precautions and action to be taken in the event of a spillage

Packaging material specifications are an extension of the raw material specification system and should include detailed dimensional requirements of the packaging components, with reference, where appropriate, to approved drawings. They should also specify details of approved materials for all the components, including approved grades for specific material types.

1.2. *Process specifications*

These should include the following parameters:

- batch size(s). If more than one batch size can be manufactured, ensure that clear definition is made between the two
- full formulation, in a "recipe" format, to include raw material code number, raw material name, percentage (optional) and weight of raw material per batch (essential)
- safety requirements to be observed, highlighting any hazardous materials and process steps, with precautions to be taken and actions required in the event of a spillage

1.3. *Product specifications*

These should include the following parameters:

- description and appearance
- physical tests such as colour, feel, odour, specific gravity, pH, viscosity
- chemical tests such as active matter and preservative assay
- microbiological requirements
- packaging requirements on the finished product such as visual appearance, cap torque, label position, date coding and fill levels

2. *Raw material and packaging control*

Under this heading, the following items should be considered.

2.1. *Raw materials*

A system is required to ensure that raw materials are controlled properly, throughout the process. The purpose of the system is to prevent the use of untested or rejected raw materials in manufacturing, to provide traceability, to maintain records which can be referred to later if required, to assist in monitoring supplier performance and to help prevent the manufacture of defective product.

An outline of a raw material control system is given below. The system must be designed to reflect the requirements of the company and the type of operation being carried out.

- ensure that there is a comprehensive raw material specification system in place. There should also be a full set of raw material standard samples, for reference purposes, together with reference IR, UV, GC and LC traces, where appropriate. These standards must be regularly reviewed and updated, at least on an annual basis. The frequency of review will vary, depending on the nature and stability of the material
- all raw materials should be received into a quarantine area, which may be the unloading bay itself. The important criterion is that the QA department has the opportunity to sample the material, before it is stored away. When it is put into store, it must be either a physically separate raw material quarantine store or, if the stores are controlled by a computer system, the status of the material must be controlled by QA
- identify the material as being under test, either physically with a "hold" or "on test" label, or by flagging its status in the computer
- a log of all raw materials received should be maintained, noting the suppliers' batch reference, together with any other appropriate internal data, such as stock code number. A QA control number may also be assigned in order to maintain better traceability. Computerisation of this information is preferred wherever possible
- a sample should be taken for laboratory testing. The sample should be clearly marked to ensure that there is no possible mix-up with other materials. At least one sample for every suppliers' batch that is delivered, should be taken. For deliveries consisting of multiple containers, it will be necessary to sample more than one, in order to ensure that a representative sample of the delivery is obtained
- tests should be performed as detailed in the agreed raw material specification and results recorded. Results should be compared against specification limits and the disposition of the material decided upon. The decision taken should be recorded on the raw material record sheet
- a sample of all the raw material batches tested should be retained, for later reference
- action should be taken, according to the decision reached. If the material is accepted, then its release should be progressed, either by removing the "hold" label and replacing with an "accept" label, or by releasing it via the computerised control. It is important that the only department with access to the computer system and responsibility for releasing materials, is the QA department
- if the material is rejected, its status should be indicated with a red "reject" label and it should be moved immediately to a reject material store. The appropriate department (usually the purchasing department) should be notified of this action immediately, allowing follow-up procedures to be implemented

2.2 *Packaging components*

Although the principle behind controlling incoming packaging components is the same as that for the control of chemical raw materials, the techniques required are somewhat different. The prime reason for a different approach , is the numbers of items in a single lot, or a delivery. In the case of raw materials, a delivery will normally include only a few drums or a few dozen sacks. With a delivery of packaging components, the quantities involved can be several thousands, or even tens of thousands, at any one time. It is obviously insufficient just to take a few samples from a single box.

This is, perhaps, the first point at which statistical techniques can be used. The requirement is for statistical sampling and, more particularly, sampling plans such as those laid down in BS6001. Further information on the use of sampling plans is provided later in this chapter, under the heading "Statistical quality control".

Sampling plans provide a technique for assistance in the assessment of the quality of a delivery of packaging components. In other respects, the same types of procedures can be used, as described for raw materials. Specifications must be available, standards must have been agreed and a proper procedure for quarantining and subsequent approval, or rejection, must exist.

2.3 *Supplier certification*

This can be implemented to make the control system easier to operate. Supplier certification works on the principle that suppliers are producing to an agreed specification and are already carrying out all the required tests on the material. They are therefore able to certificate all the test results to a receiving organisation, confirming that all the requirements have been met. As a result, duplication of testing can be easily avoided, saving considerable cost, time and effort. Supplier certification is very often an essential first step towards a "just-in-time" (JIT) approach to inventory control.

Before implementing full certification, it may be necessary to carry out a detailed audit on the supplier. This is usually done by both QA and the purchasing department, as it is essential to assure that the supplier can be qualified both on commercial and technical grounds. The audit will include a review of the management structure, the QA organisation, QA systems and procedures and documentation.

3. *Batch control*

The next stage in the quality assurance system, is batch control. The main requirements of this part of the system are to extend the traceability chain on from the raw material control system, to help monitor process performance and, as always, to help prevent the manufacture of defective product, by maintaining control over its quality at all process stages.

The following steps outline the batch control system:

- QA prepare a "batch sheet", also known as a batch card or batch record. This is based on the process specification and should contain a detailed ingredients list, process details and the process control checks which will determine the finished batch release requirements. It will act as a set of instructions for the operator, in preparing the batch of product

- each batch of product should be assigned with a unique batch number, to identify it throughout the process
- when raw materials are dispensed, the operator must fill in relevant details, such as the raw material control number or the supplier's batch number, into the appropriate spaces on the batch sheet. Confirmation that all ingredients have been correctly weighed, should also be provided. The properly completed batch sheet should then accompany the raw materials to the mixing area, where it will be used to assist in the control of the mixing process
- critical process parameters, such as temperature and mixing times, should be recorded in the appropriate places on the batch sheet
- when the process has been completed, and the mixing operator has determined that the product properties are as specified, a sample of the batch must then be taken and submitted to QA, accompanied by the batch sheet, for final confirmatory testing
- in a similar manner to that described for raw material control, the finished batch must be held until full QA clearance has been given. It is important not to release the product for filling, until full QA clearance has been obtained. To do so, adds tremendous additional cost to the product, incurred as a result of the filling process
- QA must ensure that correct actions are taken, depending upon the test results obtained, in a similar fashion to that described previously for raw materials. They should also file the batch sheet, so that future reference may be made to it when required

4. *Finished product control*

Control of finished product completes the traceability chain and provides a permanent record of outgoing product quality from the filling lines. Monitoring filling line performance, is the final opportunity to assure that product quality is as it should be, preventing defective product being shipped to the customer.

Line inspection should be carried out on a routine basis, at least three or four times a shift, and preferable on an hourly basis, in order to monitor the quality of the day's production. The parameters against which the product is being inspected must be clearly specified and agreed throughout the organisation of the company.

The method of checking and inspection will then need to be determined. It may be considered sufficient just to take regular samples of, say, a box of product and inspect each individual unit. If defective product is found, then one can reject everything since the previous inspection. However, this approach is unhelpful, providing little or no information on the percentage of defective product detected. Conversely, a 100% inspection could be carried out but this, of course, would be very expensive. In reality, the solution to this dilemma lies in the use of statistical techniques, which are dealt with later in this chapter under the heading "Statistical quality control".

There are also legal requirements that implicate on the production and QA departments, to ensure that the product complies with local regulatory requirements, particularly weights and measures controls and consumer-protection laws. Although these laws are similar throughout the EEC, it is essential that a organisation supplying international markets complies with the appropriate regulatory controls.

All manufacturers have a legal requirement to ensure that they comply with the Weights and Measures Act, 1985, and its associated Statutory Instruments. The Act should be read in conjunction with the Code of Practical Guidance for Packers and Importers, Issue No.1, otherwise known as the "Packers Code".

The responsibility for monitoring this compliance invariably lies with the quality assurance department, although the prime responsibility for actual compliance rests with the manufacturing department.

Under the Weights and Measures Act, cosmetics and toiletries for retail sale are controlled by average fill, at the point of manufacture. As such, it is incumbent upon the manufacturer to sample, measure and maintain adequate records, to demonstrate that the goods produced comply with the legal requirement.

The legal requirements are enforced in the United Kingdom by the Trading Standards Office. The Trading Standards Officer, or Weights and Measures Inspector, may visit a manufacturer and take whatever samples deemed necessary, without prior notice. Under normal circumstances, the frequency of visits is once or twice per year.

A series of legal duties must be carried out by the manufacturer, or importer, of cosmetics or toiletries. These are as follows:

- a duty to ensure that product complies with the "three rules for packers". These are:
  - Rule 1 - The actual contents of the packages shall be not less, on average, than the nominal quantity.
  - Rule 2 - Not more than 2% of the packages shall have negative errors larger than the tolerable negative error (TNE), as specified for that nominal quantity. TNE values for various pack quantities are given below in Table 1.
  - Rule 3 - No package may have a negative error larger than twice the specified TNE.

TABLE 1 – TOLERABLE NEGATIVE ERRORS FOR VARIOUS PACK SIZES

Nominal Quantity (Q_n) in grams or millilitres	*Tolerable Negative Error (TNE)*	
	as % of Q_n	*g or ml*
from 5 to 50	9	–
from 50 to 100	–	4.5
from 100 to 200	4.5	–
from 200 to 300	–	9
from 300 to 500	3	–
from 500 to 1000	–	15
from 1000 to 10000	1.5	–
from 10000 to 15000	–	150
above 15000	1	–

- a duty to mark the container clearly and legibly, with the nominal quantity and the name and address of the packer or importer. The "e" mark may also be used to facilitate free trade within the EEC
- a duty to check and record the weights and volumes of packages, at the time of production. Generally, this requires that a sample be taken every hour and weighed and recorded in such a way that production which does not comply with the three rules for packers can be detected. This can be achieved by carrying out a sampling check, as described in the Packers Code, but is most easily done with a computerised scale, which automatically records and calculates the necessary statistics
- a duty to assist the Weights and Measures Inspector in carrying out his task. This includes allowing full access to all documentation and providing facilities for the inspector to carry out his tests

Statistical quality control

The use of statistics in quality control is well documented. Although a comprehensive review of the techniques available is beyond the scope of this text, some of the key principles are detailed below.

1. *Sampling plans*

As previously mentioned, sampling plans are available, as an official publication, in a number of countries. In the UK, at the time of writing, the standard reference is the British Standard, BS6001. This publication clearly sets out the background to, and application of, the sampling plans contained therein.

The use of statistical sampling plans in quality assurance, allows the collection of statistically valid information of the quality of, for example, a delivery of packaging components, or a batch of filled product taken from a filling line. The main problems in sampling in such circumstances, are how many samples to take and, having inspected them, how to interpret the information obtained. The plans defined in BS6001 provide clear guidance on achieving these objectives.

BS6001 covers, specifically, inspection by attributes. Attributes are those qualities of an item which cannot be measured. Each item being inspected is therefore classed as either "accept" or "reject". It is therefore important, prior to carrying out any inspection, to define the acceptance or rejection criteria for the item to be inspected. These criteria are the quality level and types of defective which might be found.

1.1. *Acceptable quality level*

An acceptable quality level, or AQL, must first be set. AQL is defined as the maximum number of defective units, per hundred units, which can be considered satisfactory for the process being carried out, or the product being manufactured. When setting the AQL, it should be borne in mind that the tighter the AQL, the more costly the inspection exercise will be.

1.2. *Types of defective*

It is important to define the different types of defect - these can be defined as critical, major or minor. BS 6001 defines critical defects, for example, as those which would be life-threatening. This is not really appropriate for the cosmetics industry and it might be better to define a critical defect as one which renders the product unusable or unsafe. A major defect could then be defined as one which detracts markedly from the appearance or performance of the product and a minor defect one which is noticeable, but does not markedly detract from the appearance or performance of the product.

The same definitions can also be applied to the sampling of packaging components, except that the effects of any defective material on the production line, as well as the consumer, must be considered.

A typical selection of critical, major and minor defect types, is given in Table 2.

TABLE 2 – EXAMPLES OF CRITICAL, MAJOR AND MINOR DEFECT TYPES

Critical Defects	*Major Defects*	*Minor Defects*
Safety instructions absent from the label Hole in bottle wall Lipstick mechanism will not operate Pressed powder tablet crumbles Stripped thread inside cap No date code on pack	Large section of informational print missing from label Moulding flash remaining on bottle Gouge out of the side of lipstick bullet Particles of undispersed pigment visible on tablet Deep scratches from the capping machine, on the outside of the cap Incorrect date code	Small defects on individual letters of print Surface moulding defect e.g. "orange peel" Slightly uneven flaming of lipstick bullet Slight unevenness of surface embossing Scuff marks on outside of cap Smudged but legible date code

1.3. *Setting AQL's*

Having considered the defects and classified them as critical, major and minor, it is then necessary to set AQLs for each class of defect. This will depend on a number of factors but, as a rough guide for an economical plan, an AQL of 0.1 or 0.25 could be set for critical defects, an AQL of 1 or 2.5 for major defects and an AQL of 6.5 or 10 for minor defects.

This does not mean, of course, that all outgoing production will contain 0.1% unusable product, rather a maximum acceptable level has been set, against which production quality performance can be monitored.

1.4. *Selecting the correct plan*

Having established the AQL parameters, the tables in BS 6001 can be used to decide on the correct sample size. The vast majority of sampling will be carried out using single sampling plans for normal inspection. With reference to the tables presented in BS 6001, the two most useful are Table I (sample size code letters) and Table II-A (single sampling plans for normal inspection master table). The information in these two tables gives sufficient guidance for most routine sampling situations. Extracts from these tables are reproduced below, in Tables 3 and 4 respectively.

TABLE 3 – TABLE OF SAMPLE SIZE CODE LETTERS

Lot or Batch Size			*General Inspection Levels*		
			I	*II*	*III*
2	to	8	A	A	B
9	to	15	A	B	C
16	to	25	B	C	D
26	to	50	C	D	E
51	to	90	C	E	F
91	to	150	D	F	G
151	to	280	E	G	H
281	to	500	F	H	J
501	to	1200	G	J	K
1201	to	3200	H	K	L
3201	to	10000	J	L	M
10001	to	35000	K	M	N
35001	to	150000	L	N	P
150001	to	500000	M	P	Q
500000	and	over	N	Q	R

A sampling routine can be summarised as follows:

- the lot, or batch size, to be sampled, is determined. For example, suppose that a delivery of caps consists of 100 boxes, each containing 1200 caps. This gives a total lot size of 120,000
- referring to Table 3 above, the row corresponding to the lot, or batch size, is selected and the appropriate sample size code letter for the desired sampling plan is identified. As a rule, particularly for occasional or ad hoc sampling, general inspection level II (normal inspection) is used. In this example, for a lot size of 120,000, sample

TABLE 4

Sample size code letter	*Sample size*	AQL																					
		0.10		0.15		0.25		0.40		0.65		1.0		1.5		2.5		4.0		6.5		10	
		Ac	Re	Ac	Re	Ac	Re	Ac	Re	Ac	Re	Ac	Re	Ac	Re	Ac	Re	Ac	Re	Ac	Re	Ac	Re
G	32							0	1					1	2	2	3	3	4	5	6	7	8
H	50					0	1					1	2	2	3	3	4	5	6	7	8	10	11
J	80			0	1					1	2	2	3	3	4	5	6	7	8	10	11	14	15
K	125	0	1					1	2	2	3	3	4	5	6	7	8	10	11	14	15	21	22
L	200					1	2	2	3	3	4	5	6	7	8	10	11	14	15	21	22		
M	315			1	2	2	3	3	4	5	6	7	8	10	11	14	15	21	22				
N	500	1	2	2	3	3	4	5	6	7	8	10	11	14	15	21	22						
P	800	2	3	3	4	5	6	7	8	10	11	14	15	21	22								

REFERENCE TABLE OF SINGLE NORMAL SAMPLING PLANS

plan N would therefore be used. The plans also allow for the introduction of tightened inspection, level III, when quality has been consistently poor or reduced inspection, level I, when quality has been consistently good

- with reference to Table 4, the row corresponding to the sample size code letter, identified with the help of Table 3, is selected. The figure in the second column gives the sample size to be taken. In this example, the sample size required for Plan N is 500
- having determined the correct sample size, the sample is drawn as randomly as is practical and inspected against the agreed criteria. The number of defective units is recorded
- the number of defectives found is compared with the figures quoted under the appropriate AQL column in Table 4. If the number of defectives is equal to, or less than, the "Ac" figure, then the batch may be accepted. If the number is equal to or greater than the "Re" figure then the batch should be rejected. In this example, using sample plan N, and a sample size of 500, the criteria for critical faults, at an AQL of 0.10, are to accept the lot if only 0 or 1 defective items are found but to reject if 2 or more defective items are found

The great advantage of a single sampling plan, is that it gives an unambiguous answer, the lot being either accepted or rejected.

1.5. *Other plans*

BS 6001 also defines double, multiple and sequential sampling plans. Double and multiple sampling plans allow smaller samples to be taken, but there is an increased risk that the results of the inspection will not provide a definitive decision. There are a significant number of "grey areas" in these plans, requiring a second sample or a number of additional samples to be taken, before a decision can be reached. These plans can lead to significant additional work, even though they appear, superficially, to use smaller sample sizes and require less time. For general sampling inspection purposes, the single sampling plan described above will satisfy most requirements and should be used as the general starting point when adopting a BS 6001 quality control system.

Process control and process capability

All processes have an inherent variability, which depends upon a number of different factors. The QA department needs to be able to monitor these variations and interpret the information collected. By so doing, the QA inspector is better able to detect, or more preferably prevent, an out-of-specification event arising.

To understand the different techniques of inspection, it is first necessary to examine the different types of information which can be collected. Fundamentally, these divide into two classes, variables and attributes. Variables are those qualities which can be measured quantitatively, for example length, volume, viscosity and pH. Attributes are those qualities which cannot be measured quantitatively, but are determined by a "pass" or "fail" assignment, for example the presence of a label on a bottle, or the existence of a legible date code.

Each of these require different statistical methods for assessing how well the process is being controlled, as detailed below.

1. *Control by variables*

Any process has a degree of variability and if any parameter of a product is measured, for example the weights of injection mouldings, then the results may be recorded and plotted to produce a *frequency distribution*. The simplest way of handling such data, is by plotting a "tally chart" and this can show, very clearly, the way in which a process is running. Figure 3 shows a typical collection of results from a QA inspection exercise, accompanied by a corresponding tally chart, drawn up using these results.

FIGURE 3

10.77	10.76	10.73	10.75	10.78
10.76	10.79	10.75	10.75	10.76
10.77	10.78	10.76	10.78	10.75
10.80	10.77	10.74	10.79	10.74
10.72	10.75	10.82	10.76	10.73
10.76	10.77	10.79	10.76	10.77
10.75	10.76	10.77	10.78	10.75
10.81	10.74	10.81	10.74	10.73
10.75	10.78	10.76	10.80	10.77
10.74	10.79	10.78	10.77	10.80

MEASUREMENT DATA

DATA AND CORRESPONDING TALLY CHART

	No. of pieces of each size									
Size	1	2	3	4	5	6	7	8	9	
mm										
10.67										
10.68										
10.69										Lower Limit
10.70										
10.71										
10.72	X									
10.73	X	X	X							
10.74	X	X	X	X	X					
10.75	X	X	X	X	X	X	X	X		
10.76	X	X	X	X	X	X	X	X	X	
10.77	X	X	X	X	X	X	X	X		
10.78	X	X	X	X	X	X				
10.79	X	X	X	X						
10.80	X	X	X							Upper Limit
10.81	X	X								
10.82	X									
10.83										
10.84										

TALLY CHART

It is immediately obvious from this particular tally chart, that the process is running towards the upper end of the specification, with the result that the occasional reading is actually above the upper specification limit. This indicates that the production equipment needs to be adjusted, to bring the range of results into the middle of the specification.

The tally chart could equally well be plotted as a smooth curve, to represent the distribution of measurements. However this graph, in isolation, does not provide sufficient information on the running of the process and also requires that many samples be taken over a longer period of time.

In order to make better use of inspection time, it is necessary to use probability and statistics, to help reduce sample sizes and maximise the amount of information which can be obtained.

Fortunately, most processes which are running correctly produce what is known as a *normal distribution*. The two important statistical parameters which can be used to describe a set of data, are the average or *mean* and the *standard deviation*.

Whilst the mean is fairly easy to understand, the standard deviation is mathematically more complex and the reader may wish to refer to a mathematical text for further explanation. In the context of process control, the standard deviation can be best described as a measure of the variability of the process.

The normal distribution has some very precise characteristics, as follows:

- approximately 68.2% of all observations will fall within ± 1 standard deviations of the mean
- approximately 95.4% of all observations fall within ± 2 standard deviations of the mean
- approximately 99.8% of all observations fall within ± 3 standard deviations of the mean

Figure 4 illustrates a typical normal distribution, showing the weight distribution of a filling machine, which has been set at 20 grams nominal.

FIGURE 4

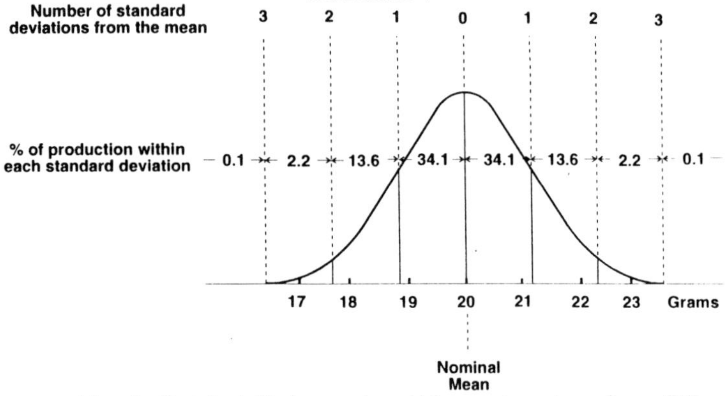

When a machine is first installed, or when it is newly set-up for a filling run, it is therefore possible to carry out a process capability study, in order to determine the characteristics of that machine. As described earlier, a tally chart could be used to provide some information. However, it is more useful to determine the mean and standard deviation of the process, which can then be used to set the specification limits at, for example, ± 4 standard deviations. This then provides a control specification, against which all future production can be measured.

2. *Process control*

Having set the mean, and determined the standard deviation, these parameters can be used as the basis for establishing a process control system, by the use of control charts. Process control techniques are used to monitor how a process is performing, usually by observing how the mean or standard deviation changes with time. This becomes particularly important where the mean moves towards one specification limit, or the variability of the results increases. Either circumstance can lead to out-of-specification product, resulting in rejection and increased levels of scrap, or rework. Process control techniques are therefore designed as preventative measures, to prevent deviation from specification, by revealing the trends in the process and allowing corrective action to be taken, before the process drifts out of specification.

Graphical examples of such process changes are shown in Figures 5 and 6.

FIGURE 5

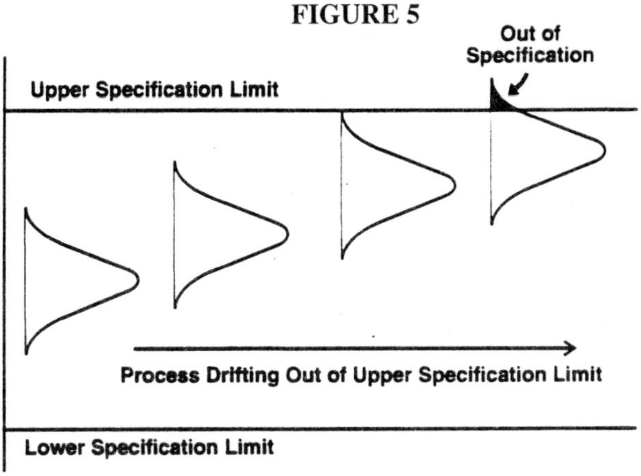

FIGURE 6

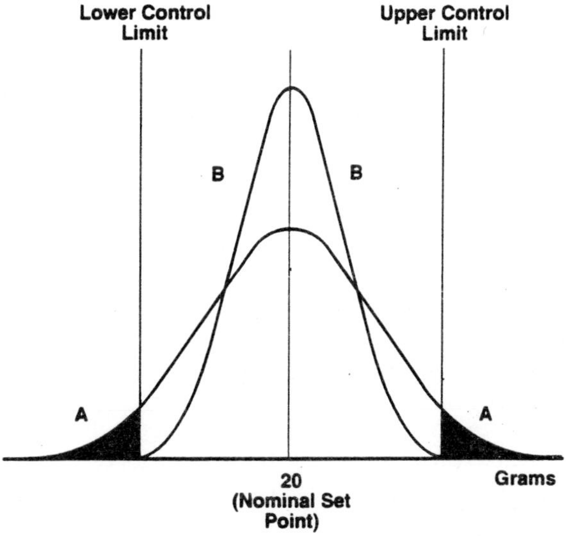

Process change from Condition "B" to Condition "A" represents process drifting out of specification

3. *Process control charts - variables*

Plotting individual results is not a very sensitive technique and a method is required to improve the sensitivity. This is achieved by plotting the results in groups of data, the basis of the Shewart control chart. Typically, the chart is prepared by plotting the results obtained from groups of 5 samples, although control charts can be generated for a number of different group sizes. Figure 7 demonstrates the advantage of the control chart technique quite clearly.

FIGURE 7

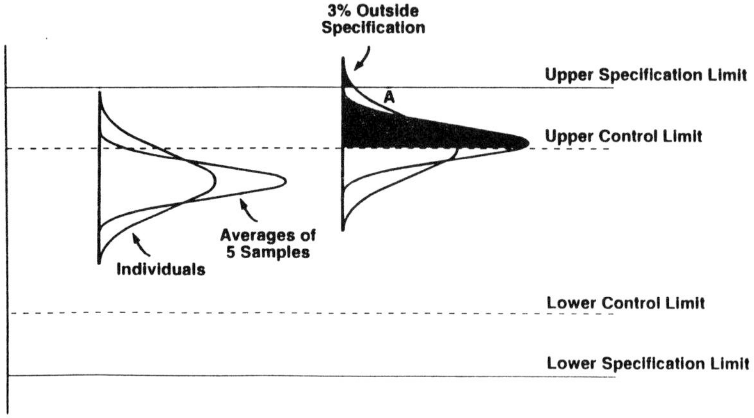

When the distribution curve of the individual results begins to move outside the specification slightly, for example about 3% as shown in curve A, the plot of the averages of groups of five moves 77% outside the control limit.

Control limits can be calculated by using the data obtained in a process capability study and plotting the data in groups of five. When sufficient data is available from the process or machine, that is between ten and twenty samples of five, then the control limits can be calculated using established techniques. This operation should only be attempted when the process is running in a steady state.

A full control chart will actually contain two sections, an *average chart* and a *range chart*. Each chart has its own inner and outer control limits, often more meaningfully referred to as the warning and action limits. An example of a full control chart, with all limits in place, is given in Figure 8.

The procedure outlined below, illustrates how a process control chart may be set up, using an example based on taking samples of five items. Other sample sizes may be used but the conversion factors to calculate the control limits will be different.

- samples of five items are regularly taken, for example every 30 minutes, from the production line
- each sample is measured, for example the weight of a filled bottle, and the average of each group of five results calculated. The range for each group of five samples, that is the difference between the highest and lowest result, is also recorded

FIGURE 8

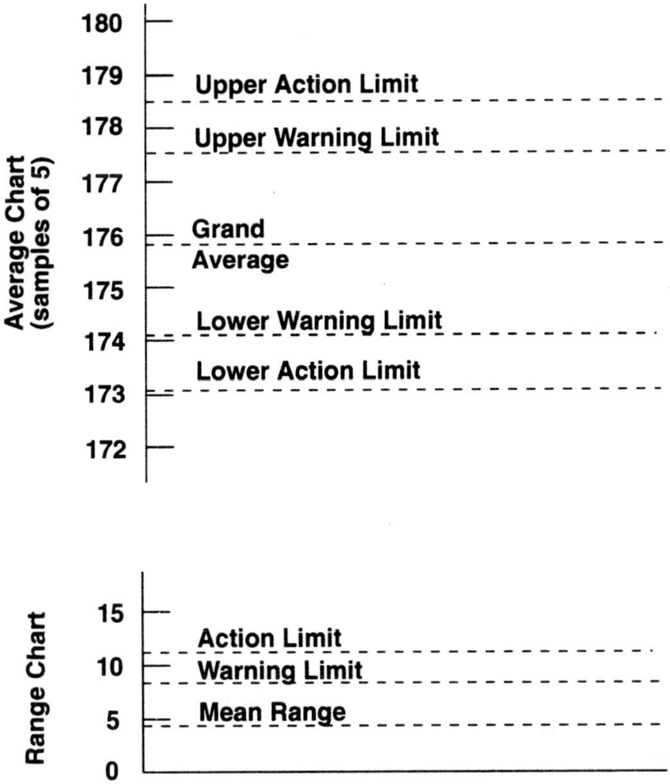

Average and Range Control Charts, with Limits in Place

- sampling is continued in groups of five, until at least ten, and preferably twenty, results have been obtained and recorded (a total of 50 and 100 individual samples respectively)
- all the sample averages are themselves averaged, to give the grand average, and this value is plotted on the control chart as a horizontal line, as shown in Figure 8, where the grand average is approximately 175.8
- the mean range is calculated by averaging all the ranges, and this value is then plotted on the range chart. In the quoted example, this is approximately 4.5 in Figure 8
- the warning limits for the average chart are calculated by multiplying the mean range by 0.38, the correct constant for samples of 5. This figure is then used to plot the warning limits, by drawing horizontal lines above and below the grand average

at that distance away from it. In the example quoted, the value is ± 1.7, as shown in Figure 8

- the action limits for the average chart are calculated by multiplying the mean range by 0.59, the correct constant for samples of 5. The control limits are then plotted in the same way on the control chart, ± 2.7 in Figure 8, for the example quoted
- the warning limit for the range chart is then calculated by multiplying the mean range by 1.81, the correct constant for samples of 5. A line is then plotted at the resulting figure, approximately 8.0 in Figure 8 for the example quoted, above the zero range
- the action limit for the range chart is then calculated by multiplying the mean range by 2.34, the correct constant for samples of 5. The resulting figure is then plotted above the zero range. In the quoted exampl,e this is approximately 10.8, in Figure 8

The completed control chart is now ready for use. Samples continue to be taken in groups of five and the average and range values are plotted consecutively on the control chart.

The average chart is used to monitor and control the process, to the centre of the specification. In this way, if the distribution starts to drift towards one of the specification limits, it can be seen at an early stage, and action can be taken to correct the situation before defective product is made. If the process drifts towards the warning limit, no action is necessary. Even if it drifts beyond the warning limit, no immediate action is necessary. By definition, it is a warning that the process might be going out of control. However, if two or more consecutive samples fall outside the warning limit, this confirms that the machine or process should be adjusted, to prevent an out-of-specification situation occurring.

If the process drifts beyond the action limit, then an immediate resample should be taken to confirm the data obtained. The machine or process should be immediately adjusted to bring it back within specification and all production since the last sample must be impounded for further sampling or 100% inspection.

The range chart is used to monitor and control the variability of the process, for example when one or more heads on a multi-head filling machine go out of adjustment. It is effectively a monitor of the standard deviation and it is possible to calculate the standard deviation of the process, from the range chart.

The control limits on the range chart are used in exactly the same way as described for the average chart.

4. *Process control charts - attributes*

Control charts can also be constructed for use in line inspection but, because of the nature of inspecting by attributes, which results in a "pass" or "fail" result, larger sample sizes are required than for inspection by variables, for which precise measurements are made. In general, a sufficiently large sample size must be taken, to give a small number of defectives to plot. For example, in looking for 1% defectives, as in the case of controlling major faults to an AQL of 1%, it would be necessary to take samples of about 200 which would be expected, on average, to give 2 defectives per inspection.

Control limits can also be set for an attribute control chart, although the method of calculation is different than that for variables. The procedure is as follows:

- the appropriate sample size for the agreed AQL is inspected and the number of defective items plotted on the control chart, as in the example shown in Figure 9, below.
- when between ten and twenty samples have been taken, the average number of defectives per sample, "c", taken is calculated. This number is then plotted on the control chart
- for a sample size "n", and an average number of defectives per sample "c", the standard deviation, L, can be estimated by using the following equation:

$$L = \sqrt{\left[c \left[1 - \frac{c}{n}\right]\right]}$$

- the warning control limits are plotted at ±2L from the average and the action control limits are plotted at ±3L from the average, as shown in Figure 9

FIGURE 9

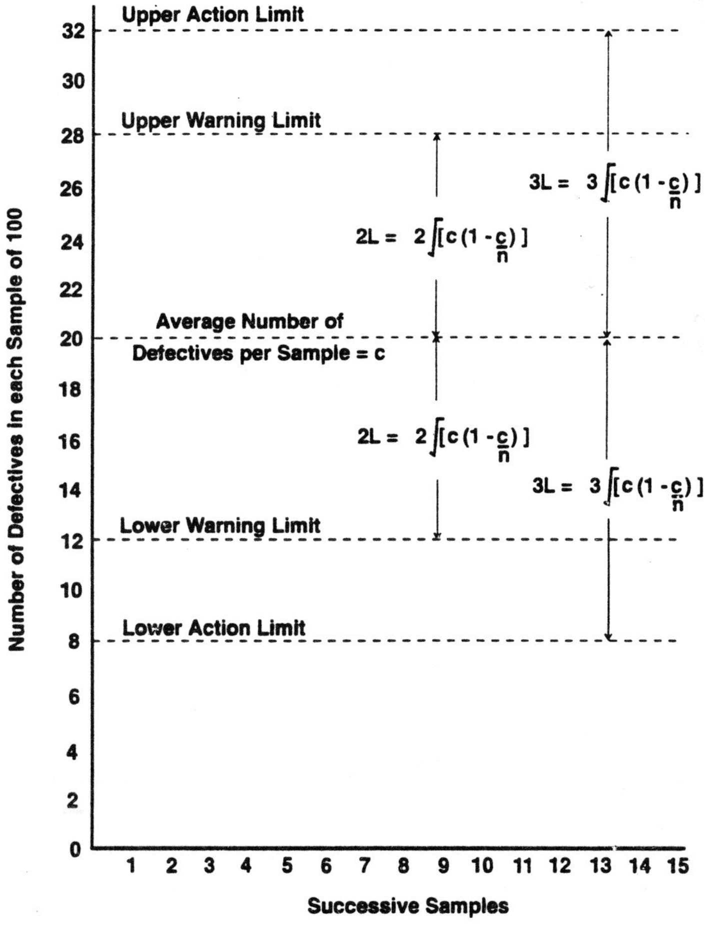

The control chart can then be used in a similar way to that described for variables. The disadvantage of this technique is that during the course of a day, many hundreds, or even thousands, of samples may need to be taken.

An alternative technique for inspecting attributes is based on *sequential sampling plans*. This is still a control chart method but the type of chart is somewhat different. The advantage of the sequential sampling technique, is that it allows an earlier decision to be taken with regard to accepting or rejecting a batch of production. At the same time, it provides the statistical data necessary to reassure management that the ongoing quality of production is at the agreed level.

There are rather more mathematics required in order to set up this control chart and the reader is referred to the detailed explanation in "Facts from Figures" by M J Moroney. An example is given in this reference, from which the chart shown in Figure 10 was produced.

Four parameters need to be defined in order to set up this control chart. These are:

- good quality, which is the fraction defective below which there is a high probability of accepting
- bad quality, which is the fraction defective above which there is a high probability of rejecting
- producer's risk, which is the chance of a good quality batch being rejected
- consumer's risk, which is the chance of a bad quality batch being accepted

In the case of the chart calculated for Figure 10, good quality is defined as 1% defective and the Producer's risk as a probability of 0.10, in other words a 10% chance of rejecting a batch containing 1% defective. Bad quality is defined as 5% defective and the Consumer's risk as a probability of 0.10, in other words a 10% chance of accepting a batch containing 5% defective.

Once the chart has been set up, samples can be taken from the line, inspected, and the number of defective items plotted on the chart. In the case of the chart in Figure 10, it may well be possible to make a decision to reject after the first sample, whereas it will not be possible to make a decision to accept the batch, until approximately 60 samples have been inspected, with no defectives being found.

There is also quite a large "grey area" with this plan, in between the two control lines, which may cause uncertainty about the quality of the batch. This can be managed, when required, by stopping the plan and reverting to a full BS 6001 sample, as described previously.

BS 5750 — Notes on interpretation

As discussed earlier it is important to have a quality system that ensures that products are supplied to the customer, as promised. The following notes should assist in the interpretation of BS 5750, which would provide the minimum requirements of a quality system, that will effectively ensure the supply of products to the agreed specification.

British Standard (BS) 5750 Part 2, 1987 is a version of the original BS 5750 Part 2 issued in 1979, slightly modified to make it identical to the International Standard (ISO 9002) and the European Standard (EN 29002), in order that registration under BS 5750 will also confirm that the company operates a quality system, to the approved international

FIGURE 10

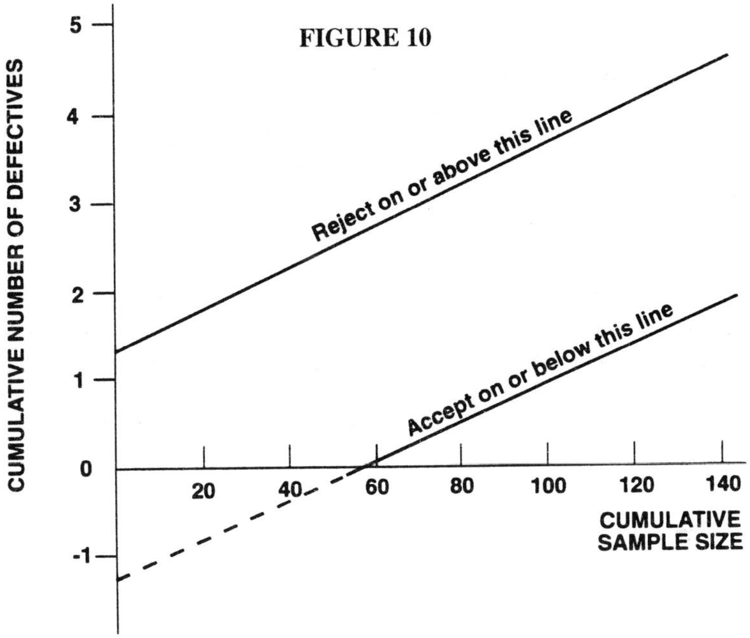

· European Standards. Anyone wishing to put together such a system should read the andard and refer to BS 5750: Part 0: Section 0.2, 1987 (ISO 9004-1987). This does not over the requirements of health and safety or the environment, and these requirements ıould be considered, as indicated in BS 7750, the new environmental standard, when ssembling operating procedures. In recognition of the widespread use of computers, ectronic equivalents of all documents are acceptable.

For ease of reference, the headings and sub-headings of BS 5750, Part 2, have been used this chapter.

Each quality system will vary and must be tailored to meet individual company equirements, depending upon the size of the organisation, the processes employed, the pe of products and chosen form of distribution.

Other documents likely to be of use are as follows:

- BS 4778 : Part 1 : 1987 Quality Vocabulary. International Terms
- BS 4778 : Part 2 : 1979 Quality Vocabulary. National Terms.
- BS 5295 : 1989 Environmental Cleanliness in Enclosed Spaces
- BS 5750 : Part 0 : Section 0.1 Guide to selection and use
- BS 5750 : Part 0 : Section 0.2 Guide to Quality Management and Quality Systems Elements
- BS 5750 : Part 2 : Quality System specification for production and installation
- BS 5750 : Part 4 : Guide to the use of BS 5750 : Part 1 : Specification for design and development, production, installation and servicing. Part 2 : Specification for production and installation. Part 3 : Specification for final inspection and test
- BS 5781 : 1988 Measurement and Calibration systems
- BS 7229 : 1989 Quality System Auditing

Where new product development is carried out, it may be useful to consider this as a design activity and apply ISO 9001\EN29001\BS 5750 : Part 1.

1. *Scope and field of application*

1.1. *Scope*

The activities that have to be covered by the approval should be briefly described, with specific coverage of each particular operation, process or product type. Individual products cannot be covered. An appropriate statement may be "The development, production, packaging and distribution of cosmetic and toiletry products, as regulated in the European Community, under the Cosmetic Directive 76/768/EEC (as amended) and in the UK, under Statutory Instrument 1984 No. 1260, Consumer Protection, Cosmetic Products (Safety) Regulations 1984 and amendments". The addition of a definition of cosmetic products, under Article 1 of the EEC Directive, should be considered.

1.2. *Field of application*

The field of application should be stated, for example, as distribution through retail trade or contract manufacture. BS 5750 can be used to demonstrate a manufacturer's capability to control the processes that determine acceptability of the final products supplied under contract. If a manufacturer supplies branded products to his own design, for general distribution and sale, it is implied that he will design and develop his product to an in-house specification, which is subject to his control system and forms the basis of a purchase contract.

2. *References*

The reader should refer to ISO 8402, Quality Vocabulary and ISO 9000, Quality Management and Quality Assurance Standards - Guidelines for Selection and Use.

3. *Definitions*

The reader should refer to ISO 8402 (1986) for a list of definitions. Some terms are defined in the glossary of this handbook.

4. *Quality systems requirements*

4.1. *Management responsibility*

4.1.1. *Quality policy*

There should be a Quality Policy document that is signed by the chief executive or managing director. This should clearly set out the management objectives, policy and commitment to the quality system. It should also include a statement on environmental and safety issues. The Quality Policy should be displayed on company premises and should be made available to anyone that wishes to see a copy. An example of a typical Quality Policy is given in Appendix I.

4.1.2. *Organisation*

4.1.2.1. *Responsibility and authority*

This should clearly indicate the positions of authority and responsibility that need to be allocated, to ensure an effective quality system. It is necessary to define these responsibilities and authorities within the organisation and clearly state them. At the same time, it is important that they are fully understood, accepted and rigidly adhered to. It should be made clear that production or sales personnel have no authority to apply pressure to quality assurance, in order to obtain release of a product or batch of material.

4.1.2.2. *Verification resources and personnel*

All staff carrying out verification work must be suitably qualified, trained and provided with adequate and appropriate resources. Verification should apply for inspection, testing, and monitoring of processes and their output, as well as the final auditing of the system. Verification need not be carried out entirely by staff reporting directly to the quality manager. Normally, auditing should be carried out by personnel independent to those having direct responsibility for the work being performed. However, a proficiently trained auditor should exhibit complete independence, even of his or her own operation.

4.1.2.3. *Management representative*

A quality manager must be appointed by the chief executive and given the responsibility and authority to ensure that quality procedures (BS 5750) are implemented and maintained. It is recommended that the appointed quality manager should report directly to the senior executive on site (ideally the Managing Director), or to the senior manager responsible for quality (for example the Director of Quality Assurance), but not to any manager who has responsibilities that become subject to the reviews, audits and follow-up actions that are part of the quality manager's responsibility. Under no circumstances, should the quality manager report to a manager who has responsibility for production or sales.

4.1.3. *Management review*

Regular management reviews of the quality system should take place. This ensures that continued improvement of the system occurs and that best use of the information gathered from internal quality audits, and any other available data relating to quality performance, is made. Examples of such data would include material rejections and supplier performance; packaging line performance records; records of batch control; customer complaints; microbiological results and summaries and records relating to non-conforming products. Ideally, the data can be shown as historical trends, or in charts using statistical techniques, thus facilitating easier understanding.

4.2. *Quality system*

Since a documented system is specified, it is recommended that two manuals are used. The first is a quality manual, showing in broad detail the policy, responsibility and organisational structures. This can be non-confidential and shown to suppliers and customers, as necessary. The second manual should be a quality procedures manual. This should provide greater detail than the quality manual and will refer to more detailed procedures that will be found at the point of use, for example, specific instructions for cleaning equipment or operating filling machinery, laboratory test methods and supplier records. This manual should be considered as highly confidential.

In view of the complexity of the quality system, it is acceptable for the manual to describe the organisation section by section, making reference to the available separate documentation, including specifications, procedures, methods and working instructions, as detailed in Appendix II.

It should be remembered that the essential requirement is for the system to ensure that the end-product conforms to its specification. Product quality can be directly affected by legislation and various codes of practice, so the system must contain procedures for monitoring, maintaining, updating and documenting these. An effective quality system is not fixed or rigid, it should be dynamic and allow for feedback and improvements, in the light of experience.

4.3. *Contract review*

Any order accepted from a customer constitutes a contract and should have a written procedure to ensure the production and delivery requirements are met accordingly. This documentation should be subject to regular review by management, to make sure that the terms and conditions can be met, that the order is clear and unambiguous and that any problems, concessions or complaints, are formally resolved and documented. Sometimes customers need to be informed of changes made to the processes, or raw materials, used. The system should ensure that this occurs, when required.

4.4. *Document control*

Document control applies to all documents that implicate on the quality of the finished product, including all specifications, written procedures, operating instructions and approved lists of suppliers and sub-contractors. The range of controlled documentation may be extensive and some examples are given Appendix II.

4.4.1. *Document approval and issue*

It is important that a designated person is given the responsibility for the preparation, issue and review of each document. The designated person will then become responsible for authorising changes, amendments and reissues. Normally, this person would be the quality manager.

Authorisation of a document must serve as an assurance that the document has been prepared satisfactorily, and vetted. Documents must be accurate, up to date and available for everyone who needs to use them. When issuing revised documents an acknowledgement system should be used, thereby maintaining a register of controlled documents which should be reviewed periodically. Any obsolete documents should be removed and destroyed.

4.4.2. *Document changes or modifications*

All changes and modifications to controlled documents should be formally authorised.

Ideally, the change should be identified on the amended version and the document withdrawn for amendment, rather than issuing an amendment instruction.

4.5.4. *Purchasing*

4.5.1. *General*

Purchasing includes, for example, raw materials and packaging, sub-contracting any part of the manufacturing process and any other services affecting quality, such as analytical services, microbiological test services, transport, distribution, cleaning and maintenance. All quality requirements should be defined and the requirement for all materials and services to conform to the desired specification, should be ensured.

For the purpose of registration to BS 5750, the inspecting body may require independent assessment of the sub-contractor's quality systems. Third party certification may be useful, if a significant part of the operation is undertaken by the sub-contractor or contract packer.

It should be noted that the term "sub-contractor" may include everything from raw material suppliers, to contract packers or suppliers of services. A contract packer is a a sub-contractor who undertakes part, or the whole, of a manufacturing process.

4.5.2. *Assessment of sub-contractors*

A system for assessment of all suppliers and contractors should be operated and a list of those approved kept for review.

Selection of suppliers and contractors will depend on the operation but consideration should be given to the establishment of a system of vendor appraisal. Vendor appraisal should include a site visit, the completion of a quality systems questionnaire and the maintenance of records detailing past performance.

Supplier appraisal should be undertaken, in the first instance, with the aid of a questionnaire. Questions should be designed such that they are appropriate to the business and supply of the raw material, or service, in question. The questionnaire should be completed and signed by a responsible person in the vendor's organisation, rather than the sales person involved. When visiting a new supplier, a team approach should be considered, combining the technical and purchasing requirements, using trained auditors where possible. It should be remembered that the prime purpose of the visit is to assess the suppliers' systems and facilities, confirming ability to meet the agreed specifications and requirements.

In the case of delivery of bulk materials, it is important to ensure that individual batches are identified, tested and certified, at the time of delivery. Agreement on the conditions of acceptance or rejection should be made with the vendor, prior to the first delivery, including any requirements for labelling.

In the case of packaging materials, on-line control systems and achievable standards for critical, major and minor defects should be agreed. It is also essential to agree on the vendor's tests and purchaser's acceptance criteria at this stage, including sampling requirements and arrangements for reviews of results and performance.

In the case of contract manufacture or packing, it is particularly important to closely examine hygiene practices, microbiological control and safety. Product liability issues are extremely important and attention to good manufacturing practice should be insisted upon. Finished product sampling test methods and warehouse dispatch systems should also be examined.

It may not always be possible, or desirable, to visit a vendor or supplier. In such cases, a commercial specification of the supplied material, or service, should be used, followed by detailed testing of deliveries. A strict quarantine and rejection procedure should be implemented.

4.5.3. *Purchasing data*

When issuing formal purchase orders or contracts, the full specification, or other appropriate documents for the product or service, should be referred to. The temptation to use phrases of the type "as previously supplied" should be avoided, as these can be easily misinterpreted. Before any order is released, the responsible person should ensure that it is correct and any problems occurring at this stage should be reviewed with the supplier, before authorisation is given.

4.5.4. *Verification of purchased products*

Evidence of conformance to the required specification should not depend upon inspection performed solely by the supplier, and should be complemented by inspection upon receipt of goods.

Verification of purchased products must depend on the assessment of the suppliers' own verification system, and on quality performance. For some products, flexibility can be introduced by specifying a system based on acceptable quality limits, such as that detailed in BS 6000/6001.

The decision to operate using supplier acceptance and verification, must be considered carefully. Whilst there are obvious advantages and savings to be made, it is almost always necessary to carry out some testing on receipt of the goods, or product. In these cases, statistical techniques can be used to examine the suppliers' past record, before a final agreement is made.

It is important to remember that acceptance of the quality assurance system of the supplier, does not absolve that supplier's responsibility to provide acceptable product quality and it should be made clear that this procedure does not preclude rejection.

Under BS 5750, verification by the purchaser cannot be accepted as effective contro of quality by the supplier.

4.6. *Purchaser supplied product*

If a manufacturer arranges for a contract packer to be supplied with materials, then he mus ensure that they have been verified within the guidelines provided by his own quality system. The contract packer should then ensure that all the materials received are acceptable, for the purpose specified in the contract.

4.7. *Material identification and traceability*

Material identification and traceability is designed to ensure that the finished produc contains only materials specified in the product specification and provides a means of tracing defective material in the end product.

Traceability at all levels is a fundamental aspect of any good quality management system and is essential if a recall operation is to be effective. The cosmetic and toiletry industry, like the pharmaceutical industry, requires a system for product recall, if it becomes necessary.

It is mandatory to have a batch coding method for cosmetic products, thereby enabling the identification of faulty or defective products and facilitating their easy recall from the market place. The batch number should enable manufacturing records to be traced which, in turn, should enable the test results for all the ingredients in the product to be located.

All specifications for finished products should clearly identify the components in the formulation, ensuring that they comply with the legislative and mandatory requirements, by providing numbers or codes to identify every batch. Each raw material or component should be given unique identifying numbers and kept in quarantine until approved by the quality assurance department.

4.8. *Process control*

4.8.1. *General*

Control of processing and packaging is the responsibility of production personnel, who should be provided with the equipment, documentation and training, to enable them to carry out all necessary functions to produce an acceptable product. Appropriate processing equipment, which should be capable of producing the required standards, at speeds intended for the normal production operation, should be available.

Detailed documentation should be referred to in the quality procedures manual, including plant operating instructions, codes of manufacturing practice, safety, and housekeeping.

4.8.2. *Special processes*

Special procedures, such as the pre-treatment of raw materials by irradiation, should be documented. Details of the verification methods for the proof of acceptability of the raw material, after processing, should also be provided.

4.9. *Inspection and Testing*

This refers to all inspection and test methods used in-house and referred to in the quality procedures manual. It includes everything from receipt of raw materials, to the dispatch and delivery of the final packed product. All test methods must be documented and if a test method differs from a standard or specified method, then it must be validated.

Special reference needs to be made to microbiological testing because of its importance within the industry. The programme of testing should include susceptible raw materials, work in progress, finished product stored in bulk and packed products before issue.

Organoleptic assessment is a specialised skill and requires training. It is based on appearance, odour, taste and texture and is used for the examination of flavours, fragrances, colours and textured materials, such as brushes. When carrying out an organoleptic or sensory evaluation it is important to ensure that everyone employed in the activity is carefully selected and appropriately qualified, trained and tested. Particular attention should be paid to the possibility of colour blindness and anosmia.

If panels are used to assess organoleptic properties, they should consist of appropriately trained staff, whose assessment abilities are re-evaluated periodically. Testing should be carried out under appropriate conditions, such as an odour-free "clean room" for fragrance evaluation and in good lighting of the correct colour temperature for colour assessment. It is also important to ensure that the correct reference standards are used, for comparison with test samples. Such reference samples should be correctly stored, so as to minimise their degradation over time. For example, fragrance references should be kept in a cool, dark place, ideally in a refrigerator. A formal procedure for testing, such as the triangle test, should be used.

4.9.1. *Receiving inspection and testing*

A system of inspection and test should exist, to ensure that all materials are tested, or otherwise verified, and positively released, before being used. Only in special limited circumstances, should concessions be allowed for the use of untested material. All untested materials should be stored in quarantine, away from other approved materials, until verified as acceptable, and released.

4.9.2. *In-process inspection and testing*

Tests carried out by process operators, to control the product during processing, should be included here, along with tests used to ensure that the final product meets specification, before release. Statistical techniques should be employed wherever possible. Products and part-processed materials awaiting the results of test, must be held in quarantine pending release.

4.9.3. *Final inspection and testing*

When the final product has met all the specification requirements, authorisation for its release and dispatch should only be given after completion of all appropriate documentation.

4.9.4. *Inspection and test records*

A system providing records and results of all tests should exist, as corroboration of the conformance to specification. This system should also ensure that certificates of analysis, and retained samples of the batch to which they refer, are kept for a pre-determined period of time.

4.10. *Inspection, measuring and test equipment*

BS 5750 is quite clear on this point and lists the requirements for the selection, maintenance, control and calibration of all equipment to be used for quality measurements. Such requirements apply equally to equipment used in both the laboratory and the factory and all equipment should be clearly labelled with details of the relevant testing and a service record. Equipment not used for testing or measurement, within the scope of the quality system, should be clearly labelled accordingly.

When selecting or purchasing test equipment, it is important to specify the required capability and accuracy. Any special requirements for the test environment should be carefully considered and personnel carrying out the test procedures should be adequately trained in the relevant techniques. Accurate laboratory equipment, such as analytical balances, must be checked against standards, periodically.

4.11. *Inspection and test status*

The status of all materials and finished products should be clearly labelled, at all times. It is necessary to make clear which personnel possess the authority to allow progression from quarantined to approved status. Approved personnel would normally include the quality manager or his deputy, both of whom can be overruled by the chief executive or managing director of the organisation.

The CTPA recommend a system based on classification into three basic levels, as follows:

Classification	*Status*
Hold	- awaiting sampling - awaiting tests/results - awaiting authorisation
Accepted	- verification given - approved for use - test completed, issue authorised
Rejected	- non-conforming - for disposal - to be reworked - not to be used

4.12. *Control of non-conforming product*

Non-conforming product must be identified with a "hold" or "rejected" label, according to status 4.11, and clearly segregated in a quarantine area, to prevent onward transfer in the process. The documented system must ensure that action leading to correction or disposal, is controlled.

4.12.1. *Non-conformity review and disposition*

In the event of any non-conformity, it is essential to ensure that no safety hazard is introduced into the end product. This consideration must be made in the context of any decision relating to concession or correction.

In practice, typical actions taken on review would include correction by reworking and release under a concession.

4.13. *Corrective action*

Corrections should be made to the quality system as appropriate, normally as the result of a follow-up action after management reviews, quality audits, any report of non-conforming product, or evidence of other failure in the system. Particular attention should be given to customer complaints, and procedures for monitoring, investigating and implementing follow-up action, to such complaints, should be established. Corrective actions can be summarised as follows:

- the cause of non-conforming product should be investigated and any corrective action needed to prevent a recurrence should be identified
- all processes, work operations, concessions, quality records, service reports and customer complaints should be analysed thoroughly and periodically, to detect and eliminate potential causes of non-conforming product
- controls should be applied to ensure that effective corrective actions are taken and that the resulting changes in procedure are recorded

4.14. *Handling, storage, packaging and delivery*

Precautions must be taken against damage by handling, or deterioration. The documented system of warehouse quality should include provision for the issue of goods, on a first in, first out, basis. Checks on the general condition of stock held in bulk and checks on the condition of product after prolonged storage periods, should be made. Periodic audits on stock rotation should be carried out. Checks for security, environmental protection and vermin/pest control in the warehouse, should also be undertaken. Examination for evidence of environmental damage and contamination in transit should be made.

When contractors are employed for transport and distribution, they should be monitored to ensure reliability and satisfactory performance.

4.15. *Quality records*

The requirements of BS 5750, relating to quality records, are clearly stated and should demonstrate the achievements and effective operation of the quality system. (See BS 5750 Part 0: Section 0.2: 17.3)

These records should include certification and test results provided by suppliers and contractors, incoming materials test results, process control and verification test results, on-line test results, final inspection test results and complaint records. Reports and reference information relating to the quality system, including audit reports, calibration data, costs and regular review documents, should also be included. Records and samples should be held for a defined period of time, depending on regulations and customer requirements.

The legal connotations provide some uniformity within the industry, although retention times will vary from company to company, depending on the advice of legal advisers, bearing in mind the type of product and its shelf-life.

4.16 *Internal quality audits*

The purpose of internal quality audits, is to ensure that the quality system is being operated correctly, in accordance with specified procedures. The audit should review the quality system, making proposals for improvements where necessary.

Audits must be carried out, according to a set procedure, at regular intervals and must be conducted by trained auditors, who are independent of the function being audited.

Guidelines to the audit plan, activities to be carried out and actions following-up audits are given in BS 5750 Part 0: Section 0.2: 5.4.

4.17. *Quality training*

The need for quality training, extending across the company at all levels, should be addressed. Guidelines on this are given in BS 5750: Part 0. Section 0.2: 18.1.

In the cosmetic and toiletries industry, it is of particular importance that training in hygiene and cleanliness is given to all personnel engaged in the manufacturing operation.

4.18. *Statistical techniques*

Guidelines on statistical applications and techniques are given in BS 5750 Part 0: Section 0.2:20. It should be stressed that applications related to quality, may be found at all stages in the operation.

Of particular note are incoming material inspection, sampling and acceptance testing (possibly based on BS 6001), assessment of process or machine capability, weight control, analysis of trends, performance, defects and complaints.

Statistical techniques constitute an important part of the training needs of personnel engaged to perform quality duties.

References

Caplen, RH – "A Practical Approach to Quality Control", Business Books Ltd.
Juran, JM – "Quality Control Handbook", McGraw Hill.
Moroney, MJ – "Facts from Figures", Pelican Books.
Crosby, PB – "Quality Without Tears", Plume Books.
Crosby, PB – "Quality is Free", McGraw Hill.
Price, F – "Right First Time", Wildwood House Ltd.
BSI – "BSI Handbook 22: 1983 Quality Assurance", BSI.
BSI – "BSI Handbook 24: Quality Control", BSI
HMSO – "Code of Practical Guidance for Packers and Importers", Issue No 1, HMSO.

APPENDIX I

AN EXAMPLE OF A QUALITY POLICY STATEMENT

1. The Company has a commitment to producing products that will meet the needs of its markets and the expectations of its customers
2. The Company will ensure that its strategies do not endanger the Quality Policy
3. The Company will make itself aware of its legal obligations in the control of quality and will ensure that its products meet those needs. It will play a constructive part in framing regulations that are of benefit to its customers
4. The Company will ensure that its quality standards satisfy any advertising claims made for its products
5. The Company will pay attention to customer complaints and take action to minimise their incidence
6. Any contract packer or manufacturer given the responsibility of producing goods for the Company will be required to maintain the Company's quality standards. Formal auditing procedures will be applied to processes and the goods issued
7. The Company will actively co-operate with other companies in theindustry, through the medium of the Trade Organisation, to help set standards, formulate reviews, or enter into discussion on quality matters of interest to the Company or its customers
8. The Company will ensure that all its employees understand their quality responsibilities and are given appropriate authority, facilities, education and training for performing their duties in respect of quality
9. The Company will regularly review its effectiveness and audit its performance in achieving quality objectives
10. The Company will constantly endeavour to enhance quality and reduce quality costs
11. The Company will ensure that all its employees are made aware of its Quality Policy and of the commitment to quality by senior management
12. The Company will ensure that all its obligations regarding environmental legislation are met
13. The Company will ensure that it maintains and implements a safety policy, as part of the quality system. The Company will provide all necessary training to assist its employees in meeting their obligations under safety legislation

Signature: ....

CHIEF EXECUTIVE

APPENDIX II

A SUMMARY OF DOCUMENTATION GENERAL REQUIREMENTS FOR A MEDIUM/LARGE QUALITY SYSTEM

	Specifications Systems and Procedures	*Inspection and Tests*	*Reviews*	*Special Procedures and Records*
GENERAL QUALITY	Quality manual Quality policy Organisation chart Responsibilities and authorities Pack/brand/ product specifications Quality assurance specifications		Quality audits Management reviews	Register of controlled documents Job specification Staff training schedules and records Instrument calibration and servicing schedules Procedure for monitoring legislation and Codes of Practice
PURCHASING/ SUB-CONTRACTING	Contracts Supplier/ contractor assessment Raw material specifications Contract packing specifications		Contract reviews Supplier performance reviews	Approved supplier list Procedure for supplier notification of changes

APPENDIX II *(continued)*

	Specifications Systems and Procedures	*Inspection and Tests*	*Reviews*	*Special Procedures and Records*
RECEIPT OF MATERIALS	Sampling plans and procedures Acceptance procedures	Analytical test methods Packaging material test methods	Materials rejection summaries	Certificates of conformance
PROCESSING	Plant operating instructions Manufacturing instructions Hygiene and sanitization procedures	Control test and release methods and standards Micro-biological control tests and procedures	Process performance reviews Micro-biological performance reviews	
PACKING	Line operating instructions Line control test standards	Final test and release procedures	Line quality performance reviews Reviews of non-conforming products	Weight control procedures
POST-PRODUCTION	Recall procedures	Warehouse procedures and test schedules	Warehouse reviews Delivery contractor reviews Complaints summaries	

STABILITY TESTING

Introduction

Any cosmetic or toiletry product destined for the consumer market-place should be stable and, typically, a product will not be bought by the consumer until weeks, months, or sometimes years, after it has been manufactured. Additionally, any product sold through a retail chain is likely to be on the shelf, ready for purchase, for variable lengths of time and should therefore have a known shelf-life.

The stability profile of any product has implications for its safety, efficacy and aesthetic properties. Whilst there is no legal requirement for the stability of a cosmetic or toiletry product to be quantified, the Consumer Products (Safety) Regulations of 1989 requires a product to be safe for its intended use, throughout its usable life, currently regarded as 30 months in the UK. If the product does not conform to this criterion, then it is a legal requirement to label it with an expiry date. Products which are not stable may become unsafe, thereby violating the legal requirement.

There are also implications for product stability when referring to the Consumer Protection Act, although this only states that a product must be fit for the purpose for which it was bought. An unstable product may not adequately perform the task for which it was bought, a good example of this being an unstable anti-dandruff shampoo, in which the active ingredient settles to the bottom of the bottle on standing.

Notwithstanding the safety and efficacy of the product, it is unlikely that a consumer will buy a product which does not possess its intended appearance, odour and feel. If a product is unstable, there is a very high chance that these properties will be adversely affected, thus reducing the chance of purchase.

Product stability

When examining a product for stability, there are many parameters which must be taken into account. The stability of a particular product can only be measured by consideration of its critical properties, which vary enormously, depending on product type. Some of the more common characteristics which should be considered are listed on page 436.

1. *Physico-chemical parameters*
   - appearance
   - colour
   - odour
   - pH
   - viscosity
   - specific gravity
   - signs of phase separation
   - pearl stability
   - light stability

2. *Product/package compatibility*
   - incompatibility of product and pack
   - weight loss
   - moisture loss
   - perfume loss
   - pack degradation/softening/cracking
   - pack operation (pumps/aerosols)
   - can corrosion (aerosols)
   - propellant loss (aerosols)
   - valve blockage (aerosols)

3. *Microbiological stability*
   - preservative stability

Product stability testing

There are many different ways of testing product stability. Some stability tests can be carried out in a very short space of time, whilst others must be carried out over days, months, or even years.

Perhaps the most important type of stability test, for any finished product, is storage testing, in which an attempt is made to understand how the product will behave, if stored over a period of time, after the date of manufacture. It is this type of stability test that provides information about the likely stability of the product in the market-place. The steps for initiating and carrying out a product stability storage test must be carefully planned and the following five step approach is helpful in obtaining the best possible testing protocol.

Step 1 - Attribute checklist

The first step in setting up any storage testing stability programme, is to decide which attributes of a product are to be measured. Clearly, it is possible to measure a very large number of attributes for any product but, from a practical viewpoint, it is only valuable to measure those that have implications on product safety, efficacy and aesthetics, in the market-place. When drawing up the attribute checklist, the following factors should be considered.

- product function - what is the product designed to do and how will the end user assess its performance?
- critical attributes - what attributes determine the quality and consistency of the product?
- product appearance - what attributes are likely to affect the appearance of the product?
- packaging format - what type of packaging is used?

When the attribute checklist has been drawn up, the test protocol can then be established.

Step 2 - storage conditions

The objective of this type of study is to measure the stability of a product, as a function of time. One option is to take a freshly manufactured product and measure any change in chosen attributes over a 30 month period, in order to assess the product's stability profile. Technically this option is sound but from a commercial perspective is it clearly unworkable, in view of the time taken to complete the test. In reality, a predictive method is used, in which the product is subjected to extreme conditions of storage, in an attempt to predict the effects of long term storage in a moderate environment. This technique is known as *accelerated storage testing* and relies upon the fact that products stored at high temperatures will reveal any potential instability in a shorter space of time, than if it were stored at ambient temperatures. When carrying out this test, it is normally accepted that any change in the product will occur twice as fast, for every 10°C increase in temperature. In addition to the requirements for accelerated storage testing, other storage conditions, which the product may experience at some time during its life, are also used. Storage conditions are normally chosen from the following options:

- freeze/thaw[#]
- storage at 5°C[#]
- storage at room temperature (20°C or 25°C), in darkness[#]
- storage at room temperature, under ultraviolet light[1]
- storage at room temperature, under conditions of known humidity
- storage at 40°C, in darkness[#]
- storage at 40°C, under conditions of known humidity
- storage at 50°C[2]

[#] indicates minimum requirement
[1] used for products packed in transparent containers
[2] used for product to be sold in hot climates

The storage conditions selected, beyond the minimum requirement indicated above, are determined by consideration of the following factors:

- will the product be sold in countries with a hot climate?
- will the product be sold in clear containers or opaque packs?
- will the product be subjected to freezing at any time during its life span?

Care should be exercised if carrying out accelerated storage testing at temperatures of 50°C or above, as certain changes in a product can take place at this temperature, which would never take place at lower temperatures, irrespective of storage time. A 50°C storage condition is normally only used for assessing the stability of products during their development phase, or in cases where the product will be sold in extremely hot climates.

Step 3 - define testing schedule

The frequency and duration of product testing will depend upon many factors, many of which are mentioned above. A typical schedule for stability storage testing is detailed below:

- Initial#
- 1 month#
- 2 months
- 3 months#
- 6 months
- 1 year#
- 18 months
- 2 years
- 30 months#
- 3 years

\# indicates minimum requirement

The duration of testing required, before a product can be released for general sale, will depend upon the following factors:

- is the product totally new, or a result of a minor change to an existing formulation, for which good stability data exists?
- what market is the product intended for?

Whilst it is very difficult to specify requirements for the release of each type of product, it must be remembered that the longer the duration of testing, the higher the degree of certainty that the product will be stable in the market-place. Guidelines for the minimum requirement of duration of testing are indicated below:

Minor product changes or changes in raw material source	2 months (moderate risk) 3 months (minimum risk)
Significant formulation changes or manufacturing process change	3 months (moderate risk) 6 months (minimum risk)
New formulation	3 months (moderate risk) 6 months (minimum risk)

Step 4 - product testing

Having established the test protocol, conditions of storage, duration of testing and criteria for formulation release, then the testing programme can commence. The test methods used are normally industry standards, examples of which are detailed in Table 1.

TABLE 1

Product attribute	*Test method*
Appearance	Visual
Colour	Visual or spectrophotometric
Odour	Sensory
pH	pH meter
Viscosity	Brookfield viscometer
Specific gravity	Densitometer
Light stability	UV light cabinet
Actives	Instrumental analysis

Step 5 - assessment

Perhaps the most difficult part of storage stability testing, is the decision as to whether any change that has been observed in the product, over the period of testing, is significant enough to reject the product as a commercial proposition. Again, guidelines are difficult to give and assessment must be made on the basis of professional judgement and experience.

Obviously any significant change in the desired specification, for a given attribute, is cause for formulation rejection. Minor changes in attributes may not, however, prohibit subsequent marketing of the product. In making the judgement on suitability for the market-place, the following should be borne in mind:

- consider the function of the product under test
- consider the implications of any changes on the likely safety, efficacy or aesthetics of the product under test
- look for any significant <u>trends</u> in attribute changes

If any attribute changes by more than ±20% of its original specified value, then product rejection should be considered.

INDUSTRIAL MICROBIOLOGY, HYGIENE AND PRESERVATION

The microbial quality of all cosmetic and toiletry products is of extreme importance. Contamination by micro-organisms can affect not only product functionality and stability but, most importantly, the safety of the product in use. Opportunities for product contamination are numerous, during the manufacturing process, and this chapter provides a review of where potential hazards may lie. Finally, a discussion of effective product preservation is presented.

What are micro-organisms?

Micro-organisms, unlike visible contaminants, are viable and can exist within, or upon the surface of, any material where conditions for their growth are favourable. The three main categories of microbes which affect cosmetic and toiletry products are, *bacteria*, *yeasts* and *fungi*. These are autonomous entities, which can exist individually as a stable form and are capable of reproduction.

1. *Bacteria*

Bacteria are single-celled organisms. Most have a characteristic cell wall, although some have no wall at all. In size, individual bacterial cells range from approximately 0.5 μm to 10 μm. Bacterial cells can exhibit one of four basic shapes, spherical *(coccus)*, rod *(bacillus)*, spiral *(spirillum)* and comma *(vibrio)*. Cells may occur singly, in pairs, as a cluster, as chains, or in various arrangements such as 'V'-shaped. Some bacteria possess various appendages, which enhance their mobility.

2. *Fungi*

Fungi can be either unicellular or multicellular organisms, with sizes ranging from approximately 1 μm, to several centimetres. Their size enables further classification of fungi; mushrooms and puff-balls are *macrofungi* whereas *Penicillia* and *Aspergilli* are *microfungi*. Fungi are typically aerobic organisms but some are facultative or obligate anaerobes. Most fungi reproduce asexually, often by the formation of conidia. In sexual reproduction, spore bearing fruiting bodies are formed, many of which are visible. Fungi utilise substrates ranging from simple sugars to cellulose, hydrocarbons, liginin, pectin, etc, and are widespread in nature. Certain fungi are essential to man in antibiotic, alcohol and fermented food production, whereas pathogenic types can cause disease in man, animals and plants. *Aflatoxins*, a group of *mycotoxins* produced by strains of *Aspergillus flavus* are regarded as highly lethal, interfering with the immune system and inhibiting DNA, RNA and protein synthesis.

3. *Yeasts*

Yeasts are a category of fungi. A typical yeast is a unicellular saprotroph which can metabolise carbohydrates by fermentation. Larger than bacteria, reproduction occurs by means of budding or cell division by fission. Yeasts can be either beneficial, or extremely dangerous, to man. For example, the common yeast *Saccharomyces cervisiae* has important industrial applications, whilst *Candida albicans* is classified as a pathogen.

Irrespective of organism type, bacteria, fungi or yeast, there are certain requirements that must be fulfilled before microbial growth, and proliferation, can occur. The general requirements for microbial growth are listed below:

- water
- sources of carbon and nitrogen
- minerals and trace elements
- oxygen (for aerobic organisms only)
- correct temperature profile (warm conditions)
- suitable pH

Extremes of temperature, pH and osmotic pressure can be harmful to most micro-organisms.

Microbial content

Microbial content refers to the total sum of bacteria, yeasts and fungi present in a batch of bulk product. For example, if 10^7 microbes are present in a 100 kilogram batch of a toiletry product, this is equivalent to a *total viable colony count (TVC)* of 100 colony forming units (CFU), per gram of that product. The most common sources of microbial contamination are listed below:

- insufficiently pure water used for manufacture
- contaminated raw materials
- non-sterile production and filling equipment
- non-sterile bulk storage containers
- poor operator hygiene
- airborne contamination
- contaminated primary packaging
- changes in cleaning and disinfection procedures
- poor assessment of hygienic requirements

Raw material microbial quality

Poor microbial quality of raw materials, accounts for a high proportion of cosmetic and toiletry contamination incidents. Good raw material quality, and effective microbial quality control, therefore greatly assists the minimalisation of potential contamination in the manufacturing environment, and the finished product itself. Whilst most raw materials used in the cosmetic and toiletry industry are subject to some degree of microbial contamination, the most susceptible categories are:

- natural ingredients (botanical extracts and concentrates)
- carbohydrates and glycosides
- higher molecular weight alcohols
- surfactants/detergents
- fatty acids and esters
- proteins and protein derivatives
- talc, kaolin and cornstarch
- water

It therefore follows that the microbial quality of any raw material incorporated in the manufacturing process of a cosmetic or toiletry product, should be adequately controlled before use. It is practically impossible to eliminate all microbial contamination from a raw material, so an acceptable quality level must be defined. What can be deemed acceptable is largely at the discretion of the individual manufacturing organisation, but existing guidelines suggest maximum microbial limits of not more that 1000 cfu per gram, for use in adult products and not more than 100 cfu per gram, in products intended for use on babies and young children.

Perhaps the most susceptible raw material, to poor microbial quality, is water. Although two thirds of the world's surface is covered by water, less than 1% is surface and ground water, which is of potential use to man. Of this, a significant quantity is toxic or of doubtful quality. Apart from the need to remove particulate matter and neutralise toxins, the grade of water chosen should be of high standard.

The microbial quality of water is of paramount importance to the cosmetic and toiletry industry, as it is a major constituent in the majority of products manufactured. Water also plays an important part within the factory, in washing and cooling processes. Raw mains water, softened water, deionized water, and water produced by reverse osmosis or distillation, are all used in different types of manufacturing procedure.

Distilled water, although free from contamination as it leaves the still, can quickly become contaminated with micro-organisms. These organisms will grow and proliferate quickly, in water held in storage vessels or trapped in pipe work. Often, a significant proportion of the microbial contaminants of water are gram negative bacteria, with other species such as *Micrococcus* spp., *Cytophaga* spp. and various yeasts and fungi being present. Bacteria indigenous to waters are nutritionally undemanding and often have low optimum growth temperatures. They include *Pseudomonas* spp., *Alcaligenes* spp., *Flavobacterium* spp., *Chromobacter* spp., and *Serratia* spp.

Water used in the manufacture of cosmetic and toiletry products must be of known good microbiological quality and many different methods of purification are used in industry. Amongst these are chemical treatment, filtration, UV light and ozone treatment. The effectiveness of any of these methods, relies heavily upon correct application and monitoring of the technique chosen.

Assuming the raw materials used in the manufacture of a cosmetic or toiletry product are carefully controlled, it is essential that subsequent storage of the raw material is performed in the correct manner, otherwise adventitious contamination may occur. The conditions for correct storage of raw materials are listed below:

- a dedicated storage area for raw materials should be clearly identified
- store should be maintained in a clean and hygienic condition
- all materials should be stored off of the floor, in a dry environment
- storage area should be maintained at constant temperature and humidity
- raw materials should never be stored in close proximity to packing materials, finished products and work in progress.
- storage tanks for liquid raw materials, including water, must have well fitting lids with capped inlets and outlets
- raw material lot/batch number rotation should be used, to cut down storage times
- raw material containers must be returned to the storage area after use and never left in the manufacturing area
- raw materials should be tested at regular intervals (every 3-6 months), during the period of storage, to check their microbial quality

Hygiene in manufacturing and filling areas

The fabrication (construction of ceilings, walls and floors) and work surfaces, in both manufacturing and filling areas, should be from materials with impervious surfaces that can easily be cleaned and sanitised, if required. Any exposed drainage systems present on the factory floor, including channel drainage, should be kept clean and, where possible, sealed. Dust traps, such as high shelving or window-ledges, should be avoided where possible or, where present, should be cleaned regularly.

The manufacturing and filling areas should be kept clean and tidy, with particular attention being paid to the rapid removal of unused raw materials, cartons and other waste. Any waste receptacles that must be used in the manufacturing and filling areas, should be kept covered until disposal can take place.

Manufacturing equipment used in the production of cosmetics and toiletries must be of the correct grade of stainless steel, either type 304 or, preferably, type 316. Associated pipe work should be designed so as to avoid sharp bends and dead-legs. Pipe work joins and valves should be designed to allow easy cleaning and exposed bolt heads should be avoided where possible, as they may provide a site for accumulation of product or raw material residues. Any vessel outlets or tapping points should be designed such that they slope downwards, to avoid accumulation of product residues.

Protective clothing, including effective headware, should be worn by all personnel entering the production and filling areas, irrespective of their discipline or function.

1. *The manufacturing process*

It is important before manufacturing any cosmetic and toiletry product, to observe certain precautions and carry out specific procedures, to ensure that the requirements of Good Manufacturing Practice (GMP) are met. This involves the correct preparation and cleaning of manufacturing equipment and the use of properly trained personnel, within the factory environment.

1.1. *The cleansing of manufacturing equipment*

The eradication of pathogens and product-tolerant micro-organisms is essential, prior to manufacture. Common practices such as forcing steam around a manufacturing system and associated pipe work are often insufficient if proper sanitisation is to be achieved. All equipment must be cleaned thoroughly, using the correct detergent and methodology, before subsequent use in the manufacturing process. The choice of detergent or cleansing agent is most important and the following critical aspects should be considered very carefully, before selection is made:

- the type of soil to be removed
- the type of surface(s) to be cleaned
- the hardness of the water used in the cleansing process
- the particular germ killing properties required
- safety considerations
- cost

There are four different aspects to any equipment cleansing process, the contribution of mechanical action, the chemical cleansing process itself, the temperature at which cleansing is carried out and the contact time. Increasing one of these parameters, will often allow the reduction of another. For example, if the temperature in a machine washing process is increased, then the extent to which mechanical cleansing must take place may be reduced.

The principal methods of cleansing, employed in the cosmetics and toiletries industry, are listed below:

- soaking
- spraying
- cleaning in place (CIP)
- foaming
- gelling
- high pressure rinsing

Every manufacturing unit should practice appropriate cleaning techniques regularly and carry out a comprehensive cleaning programme, at least once a week. Soil and product debris on the production area floor should never be simply flushed away but should be removed and disposed of in the correct manner. Following removal, the floor surfaces can then be cleaned using a suitable technique. Cleaning and sanitising operations must be taken seriously and the manufacturer's instructions for the disinfection procedure should be strictly adhered to. Any sanitisation techniques used, on an ongoing basis, should be comprehensively validated and the appropriate disinfectant concentrations and contact times should not be altered.

Disinfectants should never be used as a substitute for poor cleansing processes. Theoretically at least, sanitisation may involve removal of pathogenic organisms, possibly in very large numbers, and the application of a disinfectant to previously improperly cleaned surfaces, will almost certainly reduce the effectiveness of the sanitisation process itself. Any sanitising techniques used should effectively address the elimination of "in-house" micro-organisms, which may be peculiar to one factory, or even to one area within a factory. If elimination of such organisms is attempted with the same sanitising regimen routinely, there is a likelihood that a gradual build up of organism resistance to that particular disinfectant will take place. This risk can be minimised by alternating the sanitising regimen, at regular intervals.

A summary of the sanitisation techniques commonly employed in the cosmetics and toiletries industry, is illustrated diagrammatically in Figure 1.

FIGURE 1

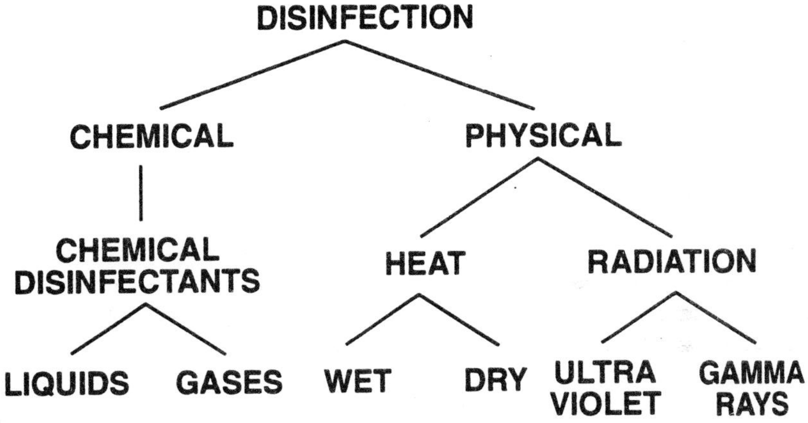

The execution of all sanitising procedures should be carried out by competent production operatives, who have been given a fundamental training programme in Good Manufacturing Practice (GMP). To ensure that a microbially acceptable manufacturing environment is obtained, where risk of product contamination is kept to an absolute minimum, a rigid sanitisation programme should be in operation throughout the manufacturing area and generally high standards of factory hygiene should be implemented.

The cleansing and sanitisation of all manufacturing equipment, including mixing vessels, holding tanks, stainless steel weighing bins and mixing scoops, must be carefully controlled and have demonstrated proven efficacy. Any procedures used should be effective against commonly encountered micro-organisms and against any specific factory isolates. Whilst it is difficult to provide a universally acceptable cleansing and sanitisation regimen for all circumstances, the protocol illustrated in Figure 2 should act as a guide for general use.

Providing it is commercially viable, it is advisable to use freshly sanitised equipment

FIGURE 2

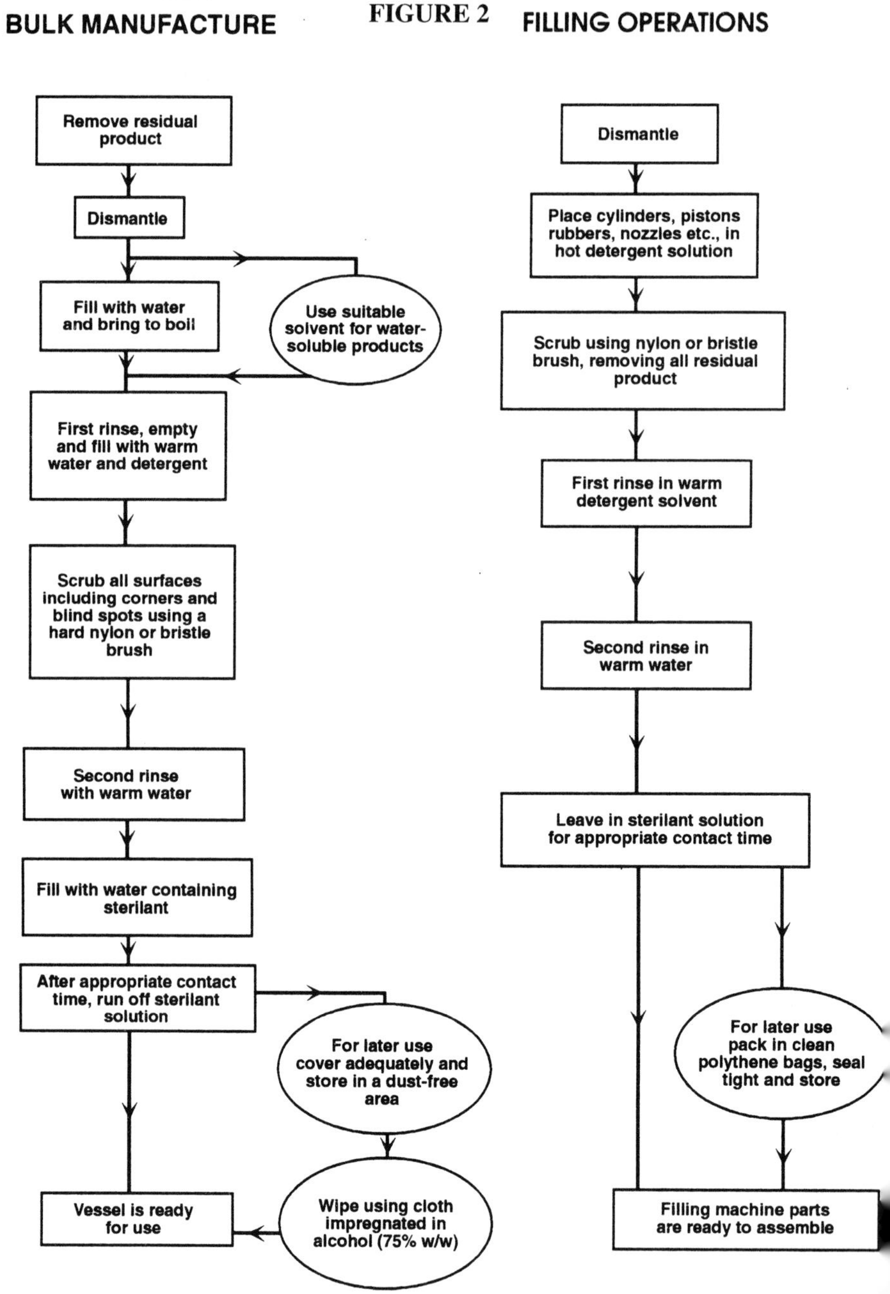

every day, even if the same product is being manufactured on a continuous basis. At the very least, thorough cleaning and sanitisation should take place every time a product change-over takes place.

1.2. *Factory personnel*

Personnel adequately trained in the principles of Good Manufacturing Practice and personal hygiene, are rapidly becoming an essential requirement for any company involved in the manufacture of cosmetic and toiletry products. In the absence of good personal hygiene, potentially harmful micro-organisms could be transferred to finished products, by direct contact with factory personnel. The health of factory operatives should not be overlooked and personnel should be free from communicable diseases and should have no open lesions on exposed body surfaces. The deployment of any person suffering from acne or other skin disorders should be carefully considered, due to the increased chance of product contamination from the increased levels of skin microflora.

Protective overalls, headwear and gloves must be provided and worn at all times when working in a designated production area. Headwear, in particular, should be designed to provide total hair coverage and, particularly for microbially sensitive products, facial hair should be covered as well. Access to properly maintained hand washing facilities, equipped with anti-bacterial liquid soap dispensers, should be provided within close proximity, but not within, the manufacturing area. Purpose-designed areas for changing from day wear into factory wear should be provided and no factory garments should be worn in any area designated for the consumption of food and beverages.

Personnel personal hygiene procedures should be carefully designed and simple to implement, with relevant training being given where necessary. Standards of personal hygiene may be monitored regularly by swabbing or rinse tests, if appropriate.

1.3. *Packaging components and product applicators*

Products manufactured under controlled conditions, require protection from extraneous or adventitious microbial contamination throughout their shelf-life which, for some products, may be an indefinite period. This protection may be described as the "physical preservation" component of overall product preservation. The cap should provide an integral seal for the main body of the pack, thereby giving protection against any airborne contamination, and appropriate capping techniques should be specified and implemented. Packaging components such as jars, bottles, tubes, sachets and godets should not be ignored as possible sources of microbial contamination. Brushes used in eye-shadows, powder-puffs and other loose powder products represent a high risk for undesirable microbial contamination, particularly if the bristles used are of natural origin.

Brushes made from animal hair should undergo microscopic examination for the presence of lice, lice eggs and other zoonotic parasites, before they are released for use. Mascara brushes, powder-puffs, sponges, pads and wipes should bear no micro-organisms and should be pre-treated by impregnation with suitable anti-microbial agents, where appropriate, to ensure lasting anti-microbial activity. Woven baskets, such as those used in the packaging of gift cosmetics and toiletries, can also present a significant microbial hazard and can often be carriers of parasites and fungi. It is therefore essential to ensure that suppliers of such packaging materials have very rigid standards of microbial quality control.

Packaging such as soap wraps, soap boxes and labels, should all possess sufficient anti-fungal activity, especially if the products are to be marketed in tropical countries.

Suppliers of all packaging components, and associated applicators, should be requested to deliver their goods in properly designed, protective secondary packaging and all components destined to come into direct contact with human skin should be delivered in sealed plastic bags.

Microbial quality control

Having implemented procedures to ensure conformance to Good Manufacturing Practice and good factory hygiene, it is important to monitor their effectiveness carefully. This involves the identification and quantification of the following:

- the numbers of micro-organisms present
- the nature (pathogenic or otherwise) of any micro-organisms present
- confirmation of static or nil microbial activity

The isolation and identification of micro-organisms, and the determination of microbial counts, involves skilled laboratory techniques. High counts of micro-organisms in a finished product and the presence of a certain species of *Bacilli*, *Pseudomonas*, *Staphylococci*, *Clostridia* and *Candida*, can represent a significant public health hazard. Resuscitation and recovery techniques, used for the retrieval of micro-organisms from finished products, need to be evaluated regularly, thereby eliminating the possibility of obtaining false negative results. The adaptive capacity of the *Pseudomonas* and others is well documented in the cosmetic and pharmaceutical industries and these bacteria can sometimes exist and survive in large numbers, even in otherwise well-preserved products. Testing procedures therefore need to be reexamined constantly, to ensure that they meet with the requirements for the detection and quantification of any biochemically altered bacteria.

A vigilant microbiologist should test at least the first ten batches of any new product very thoroughly, validating microbiological test methods in the process, before finally selecting an appropriate technique that is suitable for a given formulation.

Quality control of the final product alone, can not be relied upon to ensure that the whole batch is of good microbiological quality, whereas quality assurance (the monitoring of total product quality at different stages of the manufacturing process) will inevitably lead to expected quality. Quality assurance, therefore, is aimed at testing and monitoring all susceptible raw materials, packaging components and the manufacturing environment (including floors, walls and air) and equipment, as well as controlling potential contaminants from personnel. Quality control test results, obtained by the microbiological testing of finished products, are only a measure of Good Manufacturing Practice.

Detection of high microbial counts from a total viable colony count test, or the detection of pathogenic micro-organisms in a finished product, should be treated as serious matter. The source and cause of contamination should be investigated and eradicated, before any further manufacturing operation takes place. Failure to observe this precaution will result in further contamination appearing in other manufactured products, within a very short space of time. Realistically, it is likely that some form of product contamination will be detected in the manufacturing area, given enough time, but effective microbiological quality assurance provides the best defence against such an occurrence. Detection of frequent, or low-level, microbial contamination is often due to an isolated cause within an otherwise well-controlled manufacturing area. Commonly encountered isolated causes include the use of an inadequate sterilant, poor cleaning prior to santisation, disinfectant incompatibility, insufficient sanitisation contact time, or resistant contaminants.

Preservation and preservatives

Preservation is an anti-microbial process, used to control the prevention of microbial spoilage of cosmetics and toiletry products. Effective preservation, which ensures the safety and stability of formulations, may be implemented either with, or without, the use of added chemicals or preservatives. Furthermore, preservation should enable a formulation to cope adequately with all micro-organisms likely to enter the product, during repeated use. Higher preservative levels should never be used to eradicate contaminants introduced during the manufacturing process. Such problems should be addressed by the introduction of GMP, rather than increasing preservative levels.

Product formulation should aim to create conditions within the product, that are unfavourable for allowing the survival and growth of micro-organisms. This is normally achieved by the use of effective preservatives. The choice of preservatives is very wide, depending upon the circumstances encountered, and the formulator is advised to refer to the EEC Cosmetics Directive, which lists the positively approved and the provisionally approved preservatives, intended for use in cosmetic and toiletry products. A criteria-dependent guide to preservative selection is provided in Appendix I of this chapter.

1. *The requirements of a preservative*

No preservative fulfils all the requirements of an ideal system and, in practice, mixtures of preservatives are often used to effectively preserve products. If an ideal preservative did exist, it would need to fulfil the following requirements.

1.1. *Toxicity profile*

Freedom from all toxic effects, irritation and sensitisation on the skin and mucous membranes at the concentrations required for effective preservation. In practice, the most effective preservatives are often the most toxic and typical examples such as phenyl mercuric nitrate and thiomersal have very limited use.

1.2. *Activity profile*

A broad spectrum of activity against all gram positive and gram negative bacteria, yeasts, fungi and mixtures thereof. In addition, a high level of activity at low concentrations is highly desirable, particularly in the context of toxicity and cost considerations.

1.3. *Solubility profile*

Since micro-organisms multiply in the aqueous phase, it is important that an active preservative concentration remains in the water phase of a product, and good preservative water-solubility is therefore important. When dealing with emulsion systems, where both water and oil phases are present, the partition coefficient of the preservative is important. For simple systems, without emulsifying agents, the concentration of preservative in the aqueous phase is given by the following equation:

$$C_w = \frac{C(\phi + 1)}{(K_w^o \phi + 1)}$$

where

C	=	total preservative concentration
C_w	=	concentration of preservative in the aqueous phase
ϕ	=	oil:water phase ratio
K_w^o	=	oil:water partition coefficient

Ideally, a preservative should exhibit a high water solubility and low oil solubility, therefore giving a low oil:water partition coefficient (K_w^o). If K_w^o <1.0, then the aqueous concentration is increased by increasing the proportion of oil. If K_w^o >1.0, then the aqueous concentration is decreased by increasing the proportion of oil. The partition coefficient itself, varies with pH and the nature of the oil under consideration.

1.4. *Wide pH activity*

Since micro-organisms may grow between pH values of 2 and 11, the ideal preservative should exhibit activity over this range.

1.5. *Physical properties*

An ideal preservative should be colourless and impart no odour or taste to the finished product.

1.6. *Non-volatility*

Preservatives should be non-volatile, thereby preventing loss of efficacy if the finished product is subjected to elevated temperatures, either during manufacture, or in use.

1.7. *Compatibility*

A fundamental requirement of a preservative, is that it should efficacious in the product and therefore good compatibility with other raw materials in the product is essential. Of particular concern is compatibility with the three main types of surfactant, as indicated below:

Anionics: Many anionic surfactants exhibit mild anti-microbial activity at higher concentrations but tend to support microbial growth of gram negative bacteria and fungi, at lower concentrations. Anionic surfactants can sometimes interfere with preservative action.

Cationics: Cationic surfactants often possess good anti-microbial activity and provide added protection, when used in combination with preservative materials. Cationics rarely interfere with preservative action.

Nonionics: Nonionic surfactants exhibit no anti-microbial action and provide a good substrate for microbial growth. Nonionics often interfere with the efficacy of a preservative or preservative system and good compatibility is essential.

A summary of the compatibilities of the different types of surfactants, on specific preservatives, is detailed in Appendix I. Other than surfactant materials, compatibility in the presence of some metallic salts, particularly those of aluminium, zinc and iron, should be carefully considered.

1.8. *Heat stability*

Ideally, preservatives should be heat stable, giving maximum flexibility for use, in various types of manufacturing process.

1.9. *Long-term efficacy*

Preservatives should retain their efficacy over prolonged storage, thereby protecting the product throughout its useful life. Additionally, the preservative should not break down over time, thereby avoiding the risk of producing by-products injurious to the product or the health of the user.

1.10. *Cost effectiveness*

A preservative should be cost-effective, thereby facilitating its use on a commercially viable basis.

In reality, most commonly used preservatives interact, to some extent, with product ingredients, as well as providing anti-microbial efficacy. These interactions may affect overall anti-microbial efficacy of the preservative, with only a small proportion of the total preservative being available for the inactivation of microbial contamination.

2. *Factors affecting preservative efficacy*

There are many factors which may affect the efficacy of a preservative in a finished product. Some of the more important considerations are detailed below.

2.1. *Effect of pH*

The pH of a product should be stable over a long period of time. The activity of any ionizable preservatives, typically organic acids, depends on the proportion of acid present in the undissociated form which, in turn, depends on the pH of the system and the dissociation constant of the acid concerned. The percentage of undissociated acid, for various acidic preservatives, as a function of pH, is given in Table 1.

TABLE 1

Ionizable preservative	*Percentage of preservative undissociated*				
	pH 3	*pH 4*	*pH 5*	*pH 6*	*pH 7*
Benzoic acid	94.0	60.0	13.0	1.5	0.15
Dehydroacetic acid	100.0	95.0	65.0	15.8	1.90
p-chloroacetic acid	91.0	52.0	9.7	1.06	0.107
Salicylic acid	49.0	8.6	0.94	0.094	0.0094
Sorbic acid	98.0	86.0	37.0	6.0	0.60

The esters of p-hydroxybenzoic acid, with their non-ionizable, esterified, carboxyl groups and poorly ionized hydroxyl groups, have meagre activity at neutral pH. In the case of chlorhexidene and quaternary ammonium compounds, activity probably resides mainly with the associated cation. The activity of these materials will therefore be increased at neutral to alkaline pH values.

2.2. *Effect of preservative concentration*

Some ingredients used in the aqueous phase of a formulation make the product virtually self-preserving, whereas others may be so susceptible to microbial attack, that the aqueous phase provides a source of nutrients for micro-organisms. The concentration of preservative required in the formulation, therefore depends upon the overall susceptibility of the aqueous phase to microbial attack. Obviously, in the cases where susceptibility is high, a more powerful type of preservative, often used at higher concentrations, is required. In general, a preservative should be effective at concentrations between 0.05% to 0.20% w/w.

Where a degree of self-preservation exists, additional preservation should not be excluded, as the self-preservation capability of the product may be limited. Products which typically exhibit some form of self-preservation properties include permanent-wave lotions, neutralising lotions, depilatories, alcoholic perfumes and products containing high concentrations of humectants such as sorbitol or glycerol. As discussed earlier, the availability of a preservative to inhibit microbial growth, is more important than the overall preservative concentration itself. Preservative efficacy is often enhanced by the use of synergistic combinations of two or more preservatives. Such combinations often exhibit the following characteristics:

- a wide spectrum of anti-microbial activity
- enable the use of lower concentrations of each of the preservatives, thus reducing the risk of irritancy and solubility problems
- the mode of preservative action on microbial cells is normally by absorption, diffusion through the cell wall, or reaction with the cell cytoplasm. The use of preservative combinations often provides more than one mode of action, thereby reducing the likelihood of microbial tolerance or resistance
- the total anti-microbial activity of a preservative combination, is often greater than the additive effects of the individual preservatives

2.3. *Effect of the partition coefficient*

Preservative molecules will distribute themselves between the aqueous phase, oil phase, surfactant micelles by solubilisation, and solutes by competitive displacement of water. Thus, the available preservative concentration in each of these phases will be reduced. A formula for calculating the available preservative concentration in a simple oil/water system has already been reviewed.

2.4. *Effect of micro-organisms present*

Most organic preservatives, present at below normal use concentrations, are readily metabolised by many bacteria and fungi and may even serve as growth substrates. Perhaps the only exceptions are the quaternary ammonium compounds, which are only degraded very slowly. For example, Pseudomonads and related bacteria have been known to utilise p-hydroxybenzoic acid esters, even at normal use concentrations, as substrates for microbial growth. If interaction with micro-organisms occurs, the preservative is effectively removed from the formulation and the degree of preservation capacity is reduced. In cases where the level of interactive micro-organisms is high, for example the unintentional introduction of a high bioburden during product manufacture, the net preservative efficacy in the finished product may be negligible.

2.5. *Effect of solid particles*

Water insoluble solids such as talc, kaolin, titanium dioxide, chalk, zinc oxide, synthetic pigments and natural materials such as olive stone and apricot kernels, all present surfaces on to which adsorption of preservatives may occur. The effects of particulate solids on various commonly used preservatives, and the resultant loss of preservative activity due to absorption, was studied by McCarthy in 1969. Clarke and Armstrong studied how the regulation of pH can alter the extent of adsorption of benzoic acid on to kaolin. The order of addition of ingredients, during product manufacture, can also influence preservative adsorption. For example, if the preservative is dissolved in a slurry carrying the solids, then greater adsorption will occur than if the preservative is added after the solid surfaces have become coated with another major ingredient, such as an oil or surfactant.

To determine the amount of preservative lost to solid particle surfaces in a product is quite complex. In such cases, factors to be considered include the magnitude and type of surface electrical charges at the product pH, the total surface area presented to the aqueous phase and any ion exchange mechanisms that may be in operation.

2.6. *Effect of temperature*

Preservative activity may be affected by temperature in two different ways. Firstly an increase or decrease in temperature can significantly affect the killing power of the preservative system, depending on the type of preservative and micro-organisms present. For example, a drop in temperature from 30°C to 20°C, can result in a fivefold reduction in the killing rate of *E.coli* by phenol and forty-five-fold reduction in the killing rate by ethanol on the same types of bacteria. *E. Coli* exposed to 0.1% w/v chlorocresol is completely killed at 30°C in 10 minutes, whereas at 20°C it would require 90 minutes.

Secondly, the effectiveness of preservatives will be affected if temperatures during manufacture are not carefully controlled. Many preservatives are unstable at higher temperatures and those that are susceptible should be added towards the end of a manufacturing process, when the temperature is below about 50°C. In every case, the preservative suppliers' recommendations on temperature stability should be followed carefully, in order to avoid the inadvertent reduction of anti-microbial activity.

2.7. *Effect of natural oils*

Natural oils such as wheat-germ oil, olive oil and coconut oil, will affect the preservative partition coefficient, allowing less preservative to migrate into the aqueous phase of the product, thereby increasing the likelihood of product spoilage. Substitution of the natural oil with a mineral oil will minimise this effect and therefore favours preservative performance. It is worth noting that some natural oils (e.g. citronella oil, clove oil) exhibit anti-microbial properties in their own right, thereby increasing the preservative capacity of products in which they are used.

2.8. *Effect of product container*

Routine opening and closing of containers and loose fitting lids, will readily deplete volatile preservatives such as alcohols. Phenols and other preservative types may also be lost by permeation into the container walls. Adsorption of preservatives on to the walls of plastic or glass containers can significantly reduce the available preservative level, thereby adversely affecting preservative capacity. For example, quaternary ammonium compounds, benzoic acid, sorbic acid, salicylic acid and parabens are all adsorbed, to some extent, by nylon, polyvinyl chloride and polythene. Thus, the type of container used can play a significant role in determining the preservative capacity of a finished product. The only way of measuring the effect of the packaging, is to carry out a microbial challenge test on product which has been stored in the final containers.

Microbiological challenge testing

A very critical test for all cosmetic and toiletry products is the challenge test, which is used to assess the efficacy of preservative systems. Successful completion of a microbial challenge test, gives reasonable assurance that the product is adequately preserved to cope with all practical circumstances. The basic microbial challenge method is outlined in the British Pharmacopoeia (1988), although extensions of this method, with much more stringent requirements, such as multiple-challenges at periodic intervals, are frequently used by independent manufacturers. The different types of challenge testing methodologies currently in use are listed below:

1. British Pharmacopoeia (1988), Efficacy of Anti-microbial Preservatives in Pharmaceutical Products.
2. United States Pharmacopoeia XXI (1985) Antimicrobial Preservatives - Effectiveness.
3. European Pharmacopoeia Methodology.
4. Cosmetic Toiletries and Fragrance Association Methodology (1979).
5. Mixed innoculum, repeat challenge test.

The main differences in the official guidelines of the three main test methods, BP, USP and Ph. Eur, listed above, lie in the age of the inoculum and counting methods used. These differences are summarised in Table 2.

TABLE 2

Methodology	*Age of innoculum*	*Counting method (cfu)*
BP	Fresh	Pour plate, spread plate or membrane filtration
USP	Fresh/stored	Pour plate
Ph. Eur	Fresh/stored	Pour plate, spread plate or membrane filtration

A summary of the acceptance criteria, laid down in the official guidelines for challenge tests, are shown in Table 3 below.

TABLE 3 – CHALLANGE TESTING ACCEPTANCE CRITERIA - CHANGE IN MICROBIAL POPULATION (LOG REDUCTION)

Method		*Period after initial challange*				
		48 hrs	*7 days*	*14 days*	*21 days*	*28 days*
BP	Bacteria	3	ND	ND	ND	ND
	Fungi			2	NI	NI
USP	Bacteria			3	NI	NI
	Fungi			1	NI	NI
CTFA*	Bacteria			3	NI	NI
	Fungi			1	NI	NI

* Re-challenge at 28, 56 and 84 days
ND not detected in 1 gram
NI no increase

1. *Challenge test methodology*

Challenge test methodology varies to some extent, depending upon the method used, but the following procedure gives an outline of the basic steps taken in each case.

1.1. *Selection of test organisms*

The types of organism used will depend upon the method employed and individual requirements. Test organisms are usually selected from the following list:

Bacteria: *Staphylococcus aureus* (NCTC 10788)
Pseudomonas aeruginosa
Escherichia coli
Bacillus subtilis
Streptococcus faecalis (ATCC 7080)
Salmonella cholerasuis (ATCC 10708)
Proteus vulgaris (ATCC 9921)

Yeasts: *Candida albicans* (ATCC 10231)
Saccharomyces cerevisiae (ATCC 560) - used when the alcohol content is 5% or more)

Fungi: *Aspergillus niger* (IMI 149007)
Aspergillus oryzae (ATCC 10196)
Trichophyton mentagrophytes (ATCC 8125)
Penicillium expansum (ATCC 1117)
Cephalosporium spp.

In addition, "in house" micro-organisms or mixed organisms cultures may also be used, including those isolated from any customer complaints received.

1.2. *Inoculation*

The product, containing the preservative system under test, is then inoculated with known levels of the chosen organisms. Inoculation is normally carried out at a level of 1 ml of cell suspension per 100 ml/g finished product. The cell suspension itself contains between 1 x 10^6 and 2 x 10^6 bacteria per gram and between 0.5 x 10^6 and 1 x 10^6 yeast/fungi per gram. An identical inoculum is made on unpreserved product, which acts as a negative control.

1.3. *Sampling*

The product is sampled at various intervals after the commencement of the test and the levels of microbiological contamination determined. Sampling intervals vary, depending upon methodology, but typically samples are taken at 24 hours, 48 hours, 7 days, 14 days, 21 days, 28 days, and, in some cases, 42 days, 56 days and 63 days. Test samples may be re-challenged at 28 days and 56 days, if multiple-challenge methodologies are used.

1.4. *Inactivation and plating out*

When each sample is removed, as detailed in step 1.3, the residual preservative must be inactivated, before plating out to obtain a microbial count. Inactivation is normally carried out using a nonionic surfactant, such POE (20) sorbitan monooleate. Subsequent plating out is usually carried out using an agar growth medium.

1.5. *Judgement criteria*

A preservative system is deemed to have passed the challenge test if the reduction of viable micro-organisms reaches a certain level after a given period of time. The criteria for successful completion of the challenge test differs, depending upon methodology used, as indicated in Table 3.

Ideally, challenge testing should be carried out not only on finished products, but also on the product during the various stages of its development. An initial screening should be performed at very early stages in the development, when raw materials are finalised, to confirm that there are no ingredient incompatibilities present. A second evaluation may be made at laboratory preparation stage, when a new formulation is usually assessed for potential toxicity. A third evaluation should be made, in parallel with long term stability tests, when the final packaging has been selected. This will enable early detection of any product/packaging incompatibilities. A fourth evaluation should then be carried out on the first factory-produced batch, to ensure that the manufacturing procedures used have not adversely affected the preservative system.

Occasionally, it becomes necessary to re-evaluate the preservative efficacy of a product, some time after it has been developed and launched. For example, if new raw materials or even new sources of raw materials are substituted into the product formulation, then the microbial challenge test should be repeated. Optionally, preservative challenge testing may also be carried out on samples of product returned as a result of consumer complaints. This type of testing may indicate whether a product has been misused by the consumer and will also help to determine the preservative capacity of the system against customer-introduced micro-organisms.

A well preserved product will normally reduce a microbial challenge by at least 99.99% in a relatively short period, ideally within 48 to 72 hours. A preservative system giving no detectable counts after 7 days from commencement of the challenge test is normally considered to be acceptable.

2. *Rapid screening methods*

Sometimes it is desirable to make a prediction of preservative efficacy, for example in initial screening, without carrying out a full preservative challenge test. In these cases the linear regression method, which determines the D-value for each organism under test, is useful. The D-value is defined as the decimal reduction time of an organism, as calculated from a plot of the log of the number of surviving organisms per gram, as a function of the time after inoculation of the product. When used properly, this method will provide accurate information about a preservative system within hours or days, rather than weeks.

D-value determinations, however, do not represent an accelerated mode of challenge testing and normal challenge tests, as described above, should still be carried out before a finished product can be deemed to be commercially acceptable.

Further reading

Wilkinson, JB and Moore, RJ – "Harrys Cosmeticology", 7th edition, 1982.

Ramp, JA and Wilkowski, RJ – "Developments in Industrial Microbiology", Volume 16, 1975.

Kabara, JJ – "Cosmetic and Drug Preservation", 1984.

Lawrence, CA and Block, SS – "Disinfection, Sterilisation and Preservation", 1968.

Hugo, WB and Russell, AD – "Pharmaceutical Microbiology", 4th edition, 1989.

PRODUCT EVALUATION

Product evaluation is a useful tool for the cosmetic scientist and provides a means of assessing the quality or attributes of a product. This assessment must be quantified in some way and very often numerical scales are used for this purpose. Historically, some very spurious and exaggerated claims have been made about cosmetic and toiletry products, and the benefits that they can provide to the end-user, but current legislation and advertising standards have significantly reduced such malpractice.

The basis of good product evaluation techniques, is scientific investigation and application of scientific principles. These investigations, however, must never be carried out in the absence of a common sense approach and an ever present awareness of what the evaluation exercise is designed to achieve.

The purpose of product evaluation

There are many different reasons why a product evaluation exercise may be carried out. Some of the key areas in which product evaluation may help, are listed below.

1. *New product development*

This is the most obvious use for product evaluation and throughout new product development the formulator will constantly refer to various evaluation techniques, in order to optimise product attributes. Methods used range from simple objective assessments in the early stages of development, right through to a full consumer research study, prior to product launch.

2. *Raw material resourcing*

Sometimes, very often because of commercial considerations, it is necessary to replace or substitute one or more raw materials in a formulation. In these cases, it is essential that the consumer's overall perception of the product performance does not change. Product evaluation can be employed to quantitatively assess whether raw material substitution will affect product performance, using techniques such as paired comparison tests and triangle tests.

3. *Quality control*

Although product evaluation is used less commonly in quality control, it can be employed to ensure that perceived performance does not change over a period of time. Evaluation methods can also be used to assist in the assessment of customer complaints, where the returned product can be compared against a known standard of the correct quality.

4. *Claim support*

In order to make claims about any product, it is now necessary to be able to produce claim support data, if requested. A wide variety of product evaluation techniques are used to generate claim support information, ranging from laboratory-based instrumental methods to quantitative panel testing.

5. *Competitive product assessment*

Before developing any new product, it is essential that the formulator is given an action standard, against which the product must perform. This is very often the brand-leading product in the appropriate market sector and techniques such as paired comparison testing or panel assessments will help the formulator to ensure that the performance objectives are met.

Types of product evaluation

There are two main classifications of product evaluation, *objective evaluation* and *subjective evaluation*. Objective techniques involve the generation of quantifiable data, normally by trained personnel, using parameters that are reproducible and devoid of any personal bias or judgement. Objective evaluation can be made using either instrumental or sensory techniques.

Subjective evaluation, on the other hand, uses techniques that are based on personal judgements, often by subjects that are completely untrained in assessment techniques. Subjective methods are invariably sensory in nature, using the basic human senses of sight, touch, smell, taste and hearing. As such, the data obtained from subjective evaluation methods, is based on personal opinion. Subjective evaluation is sometimes referred to as *hedonic evaluation.*

The inter-relationship between objective and subjective assessment, is best illustrated by the use of an example. Consider an exercise designed to assess the mint flavour in a mouthwash product, using the three basic methods, objective instrumental evaluation, objective sensory evaluation and subjective sensory evaluation. In this example, it is assumed that five different samples of mouthwash are under test, the only differences between them being the level of mint flavour, which is added at 0.5%, 1.0%, 1.5%, 2.0% and 2.5%. The types of question used, and the data obtained, for each of these methods can now be considered.

In the case of objective instrumental evaluation the data obtained would be quantified as a function of the actual level of mint flavour in the mouthwash. For example, one of the flavour ingredients may exhibit a characteristic absorption peak in the ultraviolet wavelength, the intensity of which is directly proportional to its concentration. Typically, a calibration curve could be constructed as shown in Figure 1, from which the level of flavour present in any mouthwash sample could be accurately determined.

FIGURE 1

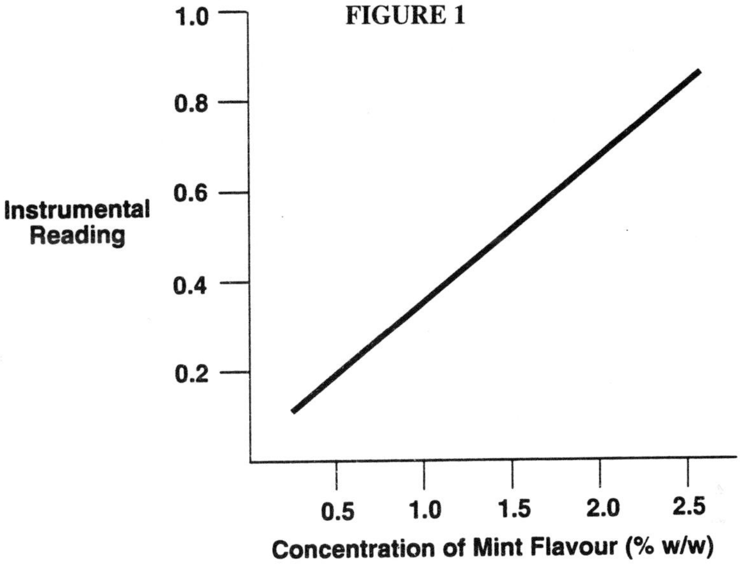

The data obtained from this method is entirely reproducible.

Using objective sensory evaluation, the concentration of flavour is determined by the sensory property of taste and the assessors used would be from a highly-trained expert taste panel. The question posed to the panel would be of the type "how strong is the flavour in this mouthwash?", using a rating scale of 1 to 5. Even with expert assessors, the assessment of flavour at either very low or very high concentrations becomes difficult to determine, as the discriminatory power of the taste sensation is significantly reduced. The type of curve obtained, using this method, is illustrated in Figure 2.

FIGURE 2

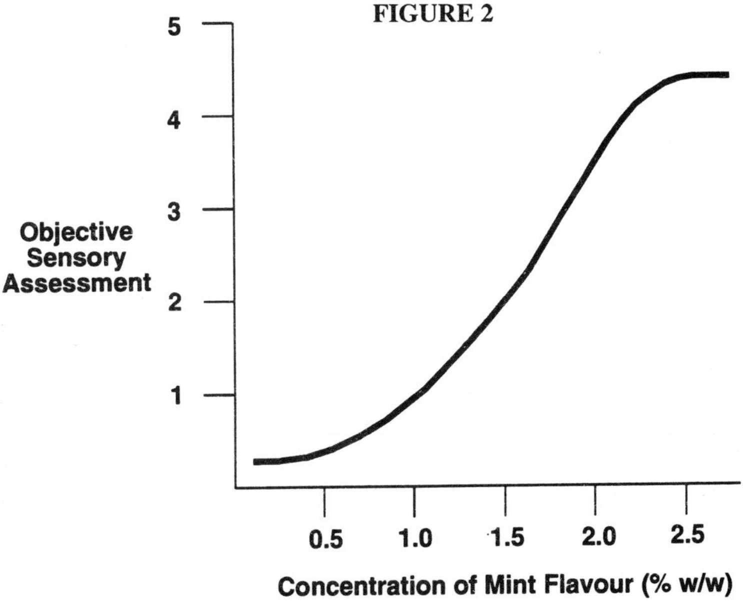

The data obtained from this method is only as reproducible as the discriminatory taste skills of the panelists concerned. The utilisation of these skills is discussed, in more detail, later in this chapter.

Subjective sensory evaluation, unlike the previous methods, is not concerned with the actual level of flavour present in the mouthwash, as the purpose is to determine what level of flavour is preferred. This method is normally carried out with untrained assessors and is reliant upon opinion. Typically, the question posed would be of the type "what do you think about the strength of flavour in this mouthwash?" and could be rated using phrases of the following type:

- not nearly strong enough
- not quite strong enough
- just right
- slightly too strong
- much too strong

Typically, the type of curve obtained by this method would be similar to that illustrated in Figure 3.

FIGURE 3

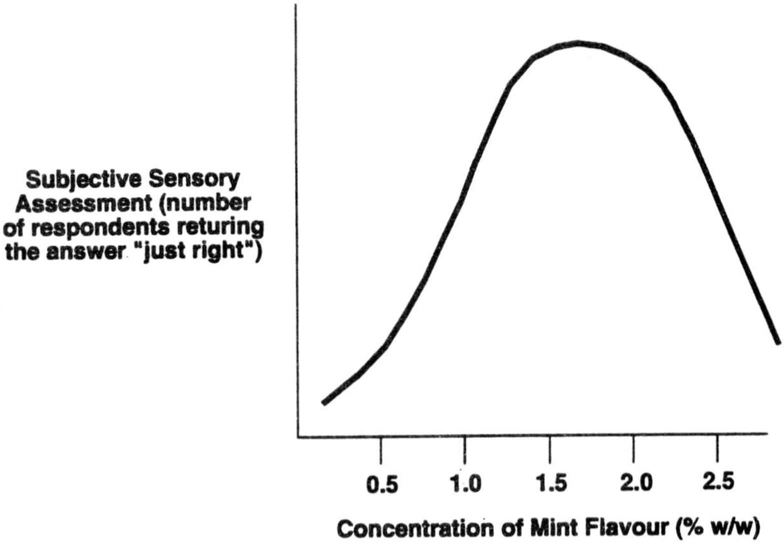

Data from this method is not always reproducible and will be entirely dependent upon the opinion or preferences of the assessors used.

Objective sensory evaluation

Objective sensory evaluation uses the physical senses of touch, sight, smell, taste and hearing, as instruments for measurement. In the evaluation of cosmetic and toiletry products, the most commonly used senses are sight, touch and smell, although taste is also used in the evaluation of oral care products.

The sight sense is used to gauge one of the key attributes of any product, appearance. Appearance generates an immediate impact and is an important factor in the early sensory assessment of a product. This assessment can be used to measure the effectiveness of the packaging, the appearance of the product itself, or it may even be applied to the product in use.

The touch sense is important in measuring the feel of both the packaging and the product but, in practice, is more commonly used to assess the feel of the hair or skin, after a product has been used.

The sense of smell is highly developed and a highly trained assessor may be able to recognise up to 2000-3000 different odours. The main difficulties in using the sense of smell for objective product evaluation, are the achievement of true objectivity (i.e. absence of bias from personal preference) and odour fatigue, which can occur very rapidly in some people. Nevertheless, the sense of smell is successfully used in the assessment of product odour, both during and after use.

The taste sense is normally used only for oral care products and is perhaps one of the less sophisticated human senses. Fundamentally, the tongue can only detect four different tastes, as illustrated in Figure 4.

FIGURE 4

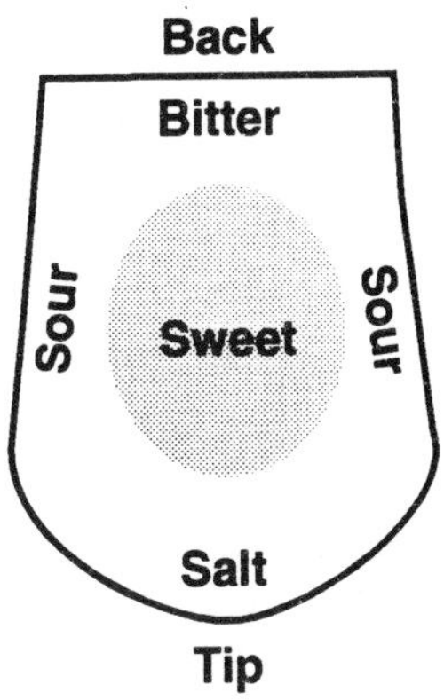

All other differences in taste are caused by the interaction of taste and smell, which is integrated in the brain to produce an overall perception.

Test methods used in product evaluation

Any product evaluation exercise should be based on sound scientific principles, if it is to have any subsequent value. Irrespective of the actual method used, the approach to the evaluation study should be based on the following steps:

Step 1: Define the test objective

The parameter(s) which are to be assessed or measured should be clearly identified, before the test begins, along with the conditions under which the test will be run. These objectives should then be written down, to form the basis of the test protocol.

Step 2: Consider/outline the method of test

Whilst several different test methods may be applicable, it is important to define which one will best fulfil the objectives of the study. In making this decision, consideration should be given to the sensitivity of the test, the complexity of the test and the time available before an answer is required. All of these factors must be balanced against the amount of funding available, with which to carry out the evaluation.

Step 3: Define test protocol

It is important to define the test protocol accurately and precisely, before the test begins. The protocol should be free from any ambiguity and should be written in a step wise fashion, such that a properly trained assessor can perform it exactly.

Step 4: Carry out test

The test must be carried out exactly in accordance with the protocol and the results recorded accurately at the time of testing.

Step 5: Analyse/validate results

Results obtained should be carefully analysed and the correct statistical method(s) applied to give a measure of their reliability. The statistical methods employed should be relevant to the protocol used and validation must relate back to the test objective.

General test methodologies

Some of the more commonly employed test methodologies used in product evaluation, are discussed below.

1. *Paired comparison testing*

This technique, as its name suggests, compares the performance or properties of one sample, directly against another sample of similar type. This is best illustrated by an example comparing two different skin creams, formulation A and formulation B, with the objective of assessing whether formulation B leaves a smoother feel on the skin than formulation A. This example also assumes that both time and financial constraints exclude the possibility of carrying out a consumer test.

Using the guidelines discussed earlier, the approach to this evaluation exercise would be as follows:

Define the test objective:	To determine which of the two skin cream formulations leave the skin feeling smoother, under conditions of normal use.
Outline test method:	Consideration in this case must be given to the simplicity of the test and the requirement to compare two products directly, using sensory methods. The test protocol should also be able to provide a measure of test reliability. The paired comparison test is the ideal choice in this case, as the protocol is quick and simple, and a direct comparison of two products is required. At least 15 assessors are required to provide meaningful test results.

Define test protocol:

a) obtain 15 volunteers
b) wash inner left forearm with dilute liquid soap and leave to dry for 5 minutes
c) apply both skin creams to adjacent sites on inner forearm, rub in and leave for 2 minutes
d) assess the smoothness of the skin for each product
e) rank skin creams for ability to leave skin feeling smooth after use
f) repeat test for all volunteers

Carry out testing:

Obtain test results and tabulate them. Suppose that 2/15 volunteers ranked formulation A as leaving the skin feeling smoother and 13/15 ranked formulation B as leaving the skin feeling smoother.

Validate results:

The binomial test would be applied in this case. This statistical method predicts the chances of obtaining 2 from 15 responses if the products are really of equal benefit in leaving the skin feeling smooth. Using binomial tables, the test statistic = 0.004.

The conclusion that could be drawn from this paired comparison evaluation, is that there is strong evidence, greater than 99.5%, that formulation B leaves the skin feeling smoother than formulation A, under conditions of normal use (i.e. there is less than a 0.5% chance of this result being wrong).

2. *Difference testing*

There are many types of difference testing used in product evaluation. Some of the most commonly used methods are discussed below.

2.1. *Triangle test*

This is one of the most important types of difference test and involves the use of three samples, two of which are identical and provide the test control, the other being the sample under test. The assessors are then asked to select one sample from the three, that is different to the other two.

The protocol of the triangle test should be carefully designed, so as to avoid bias. It is essential that all three samples are presented to the assessor in an identical format, and coded in such a way as to prevent bias. This can be achieved by making the coding invisible to the assessor in some way, or by using a code which does not invoke bias. Perhaps the best way of coding the samples is with a randomly chosen four-figure number. Care must also be taken that secondary differences between the test and control samples do not introduce a bias in the selection process. For example, in trying to detect difference in odour, slight difference in sample colour may bias selection.

A suitable panel of assessors should then be selected. Either trained or non-trained assessors can be used, depending upon the degree of discrimination required. At least 20 assessors are recommended using this method, to guarantee statistically significant results. Assessors are than asked to select one sample from the three which is different to the other two. The phraseology of the question is important and can either be directed (i.e. "which sample has a different odour?") or non-directed (i.e. "which sample is different?"). The results obtained should then be tabulated in the following way:

Correct selections	*Incorrect selections*	*Total*
x	y	T (= x+y)

This result can then be compared with the result that could be expected using random choice thus:

Correct selections	*Incorrect selections*	*Total*
T/3	2T/3	T

The observed result and the expected result are then compared using the chi-squared test, thus:

$$\text{Chi-squared} = \Sigma \left[\frac{\Sigma(\text{observed frequency} - \text{expected frequency})^2}{\text{expected frequency}} \right]$$

$$= \frac{(x - T/3)^2}{T/3} + \frac{(y - 2T/3)^2}{2T/3}$$

Comparison of the chi-squared value obtained, with chi-squared probability tables, indicates the level of significance of the result of the test and hence its reliability.

The triangle test is a particularly good method for detecting small differences between products. It also has the added advantage of being easy and quick to perform and produces results that can be reliably validated, using an established statistical technique.

2.2. *Dual standard test*

This test involves two control products and two test products. Firstly, the assessor is shown one control product and then one test product. The assessor is then given the two remaining samples and asked to assess which is the same as the first control sample shown and which is the same as the first test sample shown. Only two results are possible using this methodology, the second control being matched with the first, or the second control being matched with the first test sample. The statistical technique applied to the data obtained is the chi-squared test.

2.3. *Duo/trio test*

This method uses three samples, two of which are controls and the other the test sample. One of the control samples is identified as such and the assessor is then asked to pick which of the unidentified samples is different from the control. The statistical method applied to the data obtained is the chi-squared test, with an expected frequency of 0.5.

2.4. *"A not A" test*

Assessors are shown one control product and one test product. They are then given a set of experimental samples and asked to classify each one as a control product or a test product. The statistical technique applied to the data obtained is the chi-squared test.

3. *Rating scales*

Rating scales are often used in product evaluation, although they do not represent a test methodology in their own right. There are many different types of rating scale which can be used, depending upon the test methodology and the type of property being assessed. The most commonly used types of rating scale are as follows.

3.1. *Numerical scales*

Simple numerical scales are often used in product evaluation testing, to give a quantitative measure of a sensory assessment. Typically, scales run from 1 to 5 or 1 to 10, with the magnitude of the attribute being assessed, being directly proportional to the numerical value on the scale.

3.2. *Verbal scales*

Verbal scales use words, or phrases, to describe the performance of a product, or the magnitude of one or more of its attributes. An example of a five-point verbal scale for the objective assessment of softness, is given below:

- very soft
- soft
- moderately soft
- slightly soft
- not at all soft

Difficulties arise using this type of rating scale, because the measured attributes are often difficult to quantify and non-linear in nature. Verbal scales can also be used for subjective evaluation. In this case, the structure would be thus:

- much too soft
- slightly too soft
- just the right amount of softness
- not quite soft enough
- not nearly soft enough

3.3. *Graphical scales*

Graphical scales provide a method where the respondent assesses performance by marking a pre-defined scale. The rating scale may be segmented, depending upon the attribute and application being studied. An example, used for determining the strength of a fragrance in a toiletry product, is given below:

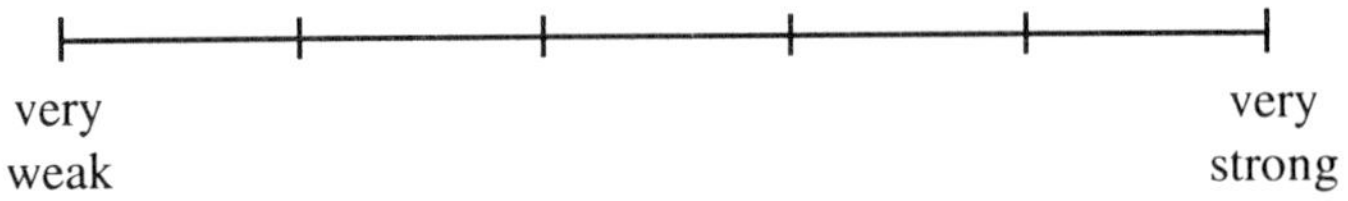

The assessor is then asked to make a single mark on the scale, which corresponds to his or her assessment of the product attribute.

3.4. *Scale of standards*

This technique involves comparison of a test sample, with a range of control samples. The respondents are than asked to match the sample under test, with one of the control samples.

Panelists for product evaluation

Product evaluation exercises can be carried out using either a trained panel, or members of the general public, and the choice of respondent depends upon the nature of the test, be it objective or subjective, and the type of data required. The use of trained panelists is often preferred over members of the general public, for a variety of reasons. Firstly, the details of the panelists are known and their reliability in different types of panel test is normally well documented. Secondly, trained panelists are invariably more discriminating and are able to "force" a decision, in circumstances where members of the general public will remain indecisive.

The gathering and training of a group of people to act as expert panelists, is a time-consuming and expensive exercise. However, if selection is made carefully and candidates trained properly, validity of the results obtained will more than offset the time and expense incurred. Subjects selected for training as expert panelists should possess the following attributes:

1. Panelists must be in good health and it is often advisable to request that a general medical examination be carried out wherever possible
2. Panelists must be able to make judgements reproducibly, within pre-defined limits of human error, when taking part in any evaluation exercise. This skill can be assessed periodically, by carrying out a difference test and comparing the results obtained with those from the same test, repeated at a later date
3. Panelists must be objective and not influenced by personal likes or dislikes. This skill can be assessed by carrying out a standard test and then repeating it using a distracting secondary influence
4. Panelists must be genuinely motivated to participate in the evaluation exercises. Stimulation can be provided with incentives and feedback of research data
5. Panelists must be able to demonstrate good communication skills and be articulate

in the use of verbal and written language. This requirement can be assessed by determining how many descriptors a person will use, to describe a given property

Specific product evaluation techniques

Having discussed some general methodologies for product evaluation, a more detailed discussion on the evaluation of certain types of toiletry and personal care product is given below. A comprehensive review of the statistical analysis used to evaluate the data in each case, is beyond the scope of this text. However, guidance is provided on the type of statistical method that may be employed in each case.

1. *Evaluation of hair care products*

Hair care products may be evaluated either *in vivo* or *in vitro* and both can be usefully employed in the development of a new hair care product. The most commonly used *in vitro* techniquc is to carry out testing on hair swatches, sometimes also referred to as hair tresses, which are bundles of human hair anchored together at one end, normally with wax or adhesive. When using this method, the swatch should be carefully specified to reduce the number of uncontrolled variables in the testing protocol. In particular the length, weight and origin of the hair should be defined and a history of any pre-treatment, such as bleaching, should be known.

Identical swatches are normally treated with the product under test and a negative (untreated) or positive (action standard) control and comparisons are made subjectively using difference testing or rating methods. Often instrumental assessments are made on treated swatches, or single hair-fibres from treated hair swatches, and properties such as shine, tensile strength, fly-away and stiffness are determined. Methods for quantifying these attributes are discussed elsewhere in this text. When carrying out any measurement on hair swatches, it is very important to carefully control the temperature and humidity of the test environment, as both can significantly affect the results obtained. Typically measurements are made between temperatures of 20°C and 25°C and relative humidities between 65% and 75%.

One particular method using hair swatches, the curl retention test, is often used to evaluate the performance of hair styling products, such as hair sprays and styling mousses. This method involves distorting or curling the hair swatch, in a controlled manner, and treating it with the styling product to be evaluated. Again either a negative or positive control may be utilised, although testing is normally carried out against untreated hair, which has been curled in a similar way. After treatment, the retention of the curl or "style" is measured as a function of time, for both the test product and the control. The curl retention, at time t, for control and test swatches may then be calculated using the formula below:

$$CR_t = \frac{L_u - L_t}{L_u - L_s} \times 100$$

where

CR_t = percentage curl retention at time t

L_u = length of original hair swatch before "styling"
L_t = length of "styled" hair swatch at time t
L_s = original length of "styled" hair swatch

The data obtained from this type of test is normally analysed using the well known t-test methodology.

1.1 *Salon testing*

The method commonly used for the *in vivo* assessment of human hair, apart from the obvious market research evaluation, is salon testing. For this type of evaluation, a purpose-built salon is normally constructed, rather than relying on a co-operative local hair-dressing salon. The former of these options provides a much better test environment, eliminating any unnecessary variables from the test. It also allows the easy administration of the test subjects. Salon testing is normally carried out by a specialist assessor. This may be either a professional hairdresser, who has received additional training in the skills of sensory assessment, or a professional cosmetologist with knowledge of hairdressing techniques. A combination of skills is essential if products are to be accurately assessed and subjects are to leave the salon with the desire to return at a later date!

The salon should be equipped to full professional standards and be fitted with purpose-designed hair washing/styling chairs and a large mirror. An adequate selection of hairstyling equipment such as combs, brushes and hair-drying apparatus is essential. Provision of towels for drying the hair and face and protective garments for subjects clothing should also be made.

Test subjects are normally taken from in-house staff, or members of the public who live locally. The requirements for subject selection depend upon the type of product being tested; suffice to say that all should be in good health with no known allergies to the types of hair product being tested. Exclusion of any subject taking medication, or exhibiting symptoms of clinical/ dermatological scalp disorders, should be made. Other than these requirements, the prime criterion for subject selection is hair type and condition. It is useful, if the subject base is large, to keep a regularly updated database on the type, colour and condition of each subjects hair and whether or not they normally use any bleaching, perming or colouring aids.

Salon testing may be used to evaluate a wide variety of products, including shampoos, conditioners, styling products and even perming and colouring products, providing enough willing subjects can be found. For many people the promise of a "free" hair treatment is sufficient to ensure participation in testing, although a small financial incentive may assist in many cases. Irrespective of the willingness of the subject, it is important that he or she fully understands any risks involved, however small, prior to any testing. This understanding must be documentally recorded, using informed consent.

Testing is carried out by the trained assessor using a variety of techniques but perhaps one of the most commonly used is the *half-head test*. The aim of this test is to evaluate the test product, at the same time as the positive or negative control. This method reduces the likelihood of any judgement errors, which may occur if sequential techniques were used. In half-head testing, the hair is parted longitudinally from the front to the back of the head and the test product is applied to one half of the head, whilst the control is applied to the other. The evaluation should obviously be carried out with both test and control product in unmarked containers, so that their identity is not known by the assessor. The sides of the head to which test and control products are applied should be randomised, to exclude any left/right bias which may occur in the assessors judgement.

Assessment is normally carried out using some form of rating scale and at least 20 subjects should be used to provide a statistically reliable database. Many different attributes may be assessed using salon techniques but more commonly properties such as combing force (wet and dry), shine, fly-away/static, body, softness and manageability are measured. The data obtained from salon testing may be statistically analysed in a number of ways, although techniques using various forms of analysis of variance (ANOVA) are most often used, as they can accommodate any left/right bias on the part of the assessor.

1.2 *Anti-dandruff shampoo testing*

Dandruff can be regarded as a clinical disorder of the scalp, in which clumps of individual skin scales aggregate to form corneocytes. These corneocytes are shed from the scalp, producing the visual effect known as dandruff. Symptoms experienced by the dandruff sufferer may also include a mild irritation or itching of the scalp. The condition is always accompanied by the presence of the micro-organism *Malassezia ovale*, previously known as *Pityrosporum ovale*, which is considered, by many, to be the causative factor. Shampoos designed to address this disorder contain a number of different anti-dandruff agents and are probably the most clinically active of all cosmetic hair care products. Evaluation of their ability to clean the hair and leave it in good condition is just as important as that carried out for true cosmetic shampoos and the *in vivo* and *in vitro* tests described above are equally applicable. Clearly though, evaluation must focus on their ability to alleviate or eliminate dandruff, particularly if competitive product claims are to be made.

Indeed, in the USA, anti-dandruff products are classified as drugs and any clinical evaluation protocol may need to be reviewed by an ethical committee, before any testing can take place.

Many techniques have been used to quantify the level of dandruff found on the scalp, although still the most preferred is that based on direct sensory assessment. The assessor in this type of test should therefore be a clinician or dermatologist, or someone specifically trained in the quantitative assessment of dandruff severity.

The selection of subjects for this type of test should be carefully made and an obvious requirement is that they should all be dandruff sufferers. Other than that, all test subjects should be in good health and should not be taking any medication prescribed by their general practitioner. The use of informed consent is essential before any testing begins.

Many different protocols have been used to evaluate the efficacy of anti-dandruff preparations in this way, although they are all similar in principle. It is very likely that any person suffering form dandruff, particularly in severe cases, will already be using an anti-dandruff preparation. Clearly, this would invalidate any test protocol used, and all subjects should be normalised before testing proper can commence. Accordingly, test subjects are given a placebo shampoo up to 2 weeks before the test starts, so that the effects of any previous treatment do not affect the data obtained. Care should be taken when selecting a placebo product, as some shampoo ingredients can exhibit a mild anti-dandruff effect in their own right. This can easily be checked by determining the effect of the product on *Malassezia ovale*, using conventional *in vitro* microbiological methods. Any activity against this organism, means that the placebo should be rejected as such.

Following the period when all test subjects use the placebo shampoo, the level of dandruff in each of the test subjects is then determined. This involves clinical examination of the dandruff condition and assigning a rating according to its severity. In order to ease this process, the subjects scalp is arbitrarily divided into a number of sections. Each section of the scalp is then assessed for severity of dandruff. Severity is quantified using rating scales, normally 1 (least severe) to 5 (chronic severity). A total dandruff score is then obtained, by adding the severity scores for each section of the scalp together, thus giving a total head score for that subject.

The subject base is then divided into two halves and one half is given the anti-dandruff product to be assessed. The other half is given the control product, which may be either a placebo (negative control), in the case where quantification of anti-dandruff reduction is required, or a competitive anti-dandruff product (positive control), where competitive superiority is to be claimed. Test subjects use the products for periods of up to 3 months, during which time they return to the salon for assessment of dandruff severity every two weeks.

At the end of the test period, the results are normally plotted graphically and data is statistically analysed, using a two-way analysis of variance technique.

In view of the difficulty of establishing an in-house anti-dandruff assessment facility, there are a number of established clinical research houses that specialise in these techniques. In addition to defining a precise protocol for evaluation, they will also advise on the number of subjects required for statistical significance and be able to provide an accurate costing for the whole evaluation exercise, prior to commencement.

2. *Evaluation of skin care products*

The range of methods that have been used for the evaluation of skin care products is very wide indeed but, with the exception of methods associated with clinical disorders and sunscreen efficacy, most focus on measuring either the sensory characteristics or moisture content/dryness of the skin.

The feel of the skin, used to determine the efficacy of emollient skin care products, is often determined by sensory assessment, although instrumental methods are available to measure the smoothness of the skin's surface. Perhaps the simplest technique is the use of macro- and micro-photography of the skin, whereupon smoothness is subsequently assessed visually.

A more sophisticated method for measuring the smoothness of the skin, in one particular area, is *surface profilometry*. In this technique, an impression of the skin's surface is firstly made using a room temperature curing silicone elastomer. The profilometry measurements are then made, by drawing a stylus, linked to a transducer, across the impression surface. The surface profile causes both lateral and vertical movements of the stylus, which are subsequently converted into electrical signals by the transducer. The resultant electronic signals are then converted into a visual impression of the skin's surface, on a visual display unit.

Perhaps the property of most interest to the cosmetic scientist is moisturisation of the skin, as this is commonly the benefit that a cosmetic cream or lotion promises to deliver, to the end user. Moisturisers are designed to combat the effects of dry skin, a condition that occurs through excessive transepidermal water loss (TEWL) from the skin's surface. Moisturisation is normally provide by the application of either an occlusive material to the skin's surface, thus reducing the TEWL, or by the application of a humectant, which draws moisture back into the skin, from the surrounding environment. Many methods have been used to quantify the moisturisation effects of skin care products and some of these are reviewed below. With any of these methods it is important to control factors such as temperature, relative humidity and the amount of product applied to the skin, as each of these can significantly affect the results obtained.

2.1. *Evaporimetry*

This technique involves the measurement of TEWL above the stratum corneum and is perhaps currently the most widely accepted measure of moisturisation benefit. The evaporimeter probe is touched against the skin's surface, to obtain the measurement of the TEWL. Inside the probe are two highly sensitive moisture detectors, mounted vertically above each other. The difference between the moisture content of the air at the first and second probes, is dependent upon the moisture concentration gradient above the skin's surface, itself a function of the TEWL.

Measurements of TEWL are made before and after the application of the skin care product under test, whereby moisturisation efficacy can be subsequently determined. Whilst this method is fairly accurate, it must be carried out with care in a draught-free environment, to avoid air disturbance which would affect the results obtained.

2.2. *Flexometry*

In this technique, the elasticity of the stratum corneum is determined by distortion of the skin's surface, in a controlled manner, and subsequent measurement of the time taken for the skin to recover from the deformation. Typically, a suction probe is attached to the skin's surface, using double-sided sticky tape, and deformation of the skin is produced through the use of a vacuum suction device in the probe head.

The extent of deformation of the skin is measured electronically and recorded on a chart recorder. The vacuum is then released and the time for elastic recovery of the stratum corneum to occur, is measured. Effective moisturisers normally increase skin elasticity and correlations between the deformation/recovery of the skin and moisturisation efficacy can be made. Whilst this method can yield meaningful results, the elastic response of the stratum corneum may be affected by the elastic characteristics of the underlying dermis and this may result in some inaccurate conclusions being drawn.

2.3. *Torsional measurements*

This method is similar, in principle, to flexometry and relies upon the fact that the moisture content of the skin is closely correlated to its elastic properties. In this case, a flat disc, connected to a torsiometer, is attached to the skin and a rotational torsion force is applied to the skin's surface. The skin's resistance to torque is determined electronically and is calculated as a function of stratum corneum elasticity. Correlations of the measurements obtained can be made with skin moisture content.

2.4. *Electrical impedance measurements*

The flow of electrical current through the skin's surface, is dependent upon the water content of the stratum corneum. Measurement of the impedance of small electrical currents, passed through the skin's surface, therefore provides a measure of the moisture content of the skin. Although this method can be usefully employed to determine the efficacy of skin moisturising products, other factors which independently affect the electrical properties of the skin, may represent a source of inaccuracy.

2.5. *Total-reflectance infra-red spectroscopy*

In this method, the skin surface is brought into contact with a total-reflectance cell, containing a measurement crystal made of zirconium nitrate. The infra-red radiation is totally reflected at the surface of the stratum corneum and the resultant spectrum analysed quantitatively, to estimate the water content of the skin.

3. *Evaluation of deodorants*

The primary function of a deodorant is to mask or neutralise the body malodour that arises through the microbiological breakdown of apocrine sweat on the skin's surface. Both *in vitro* and *in vivo* methods are used to evaluate the efficacy of deodorant products, although the *in vivo* method of direct sensory assessment is perhaps preferred.

The use of *in vitro* methods relies upon assessment of anti-microbial activity, using conventional microbiological techniques. Although such methods will provide accurate information on anti-microbial activity *per se*, overall effectiveness as a body deodorant cannot always be assumed. Most microbiological assessments aim to measure activity against organisms commonly found in the axilla, such as *Staphylococcus aureus*.

The most reliable way of evaluating the efficacy of deodorant products is still by direct sniffing of the axilla, in the form of a clinical trial. Many different protocols have been developed for this type of evaluation but all are very similar in principle. The requirements for test subject selection are similar to those described earlier in this chapter. Clearly, test subjects must be selected from people with a significant level of axillary odour, when abstaining from the use of deodorant and antiperspirant underarm products. The conditions and requirements of the test should be fully understood and accepted by any subject selected for the test, before the test begins. This can be acknowledged as part of the informed consent document. Specific requirements placed on test subjects, throughout the period of testing, are that they should not use any other deodorant, antiperspirant or highly scented product, they should refrain from eating highly spiced foods, particularly those containing garlic and they should also minimise their level of participation in physically active sports, which may affect the pattern of production of body malodour.

The odour assessment must be carried out by panelists specifically trained for the task. Apart from the obvious social issues connected with direct axilla sniffing, it is important that the assessors can accurately assess levels of malodour, if the results are to be meaningful. Selection and training of assessors is typically carried out by evaluating their ability to accurately rate different levels of isovaleric acid, a material with a very similar odour profile to that of body malodour.

In order to eliminate the effects of any deodorant products normally used by the test subjects, a normalisation or conditioning period is required. Subjects are given a low-fragranced, placebo soap bar, which must be used for washing 1-2 weeks before the test proper commences. During this period, subjects are permitted to use no other cleansing products and must refrain from the use of any deodorant or antiperspirant product.

Following the conditioning period, the subjects are given two products, one test product and one control product, to use throughout the duration of the test. The control product may be either a placebo (negative control) or a competitive product (positive control). Each test subject is instructed to use the test product on one axilla, either left or right, and the control on the other. Emphasis must be heavily placed on the avoidance of sample cross-over during the test period and, in the case of deodorant cleansing products, two flannels must be provided, one for each axilla.

On the first day of the test period, subjects are asked to perform a controlled washing, and odour assessment normally commences on the morning of day 2, to define the baseline odour scores. Odour is assessed by close direct sniffing of the axilla and an odour rating score is assigned, depending on the magnitude of the malodour. Numerical rating scales, normally running from 0 (no malodour) to 10 (strong malodour), are used to quantify the level of malodour and if the score is too low, a test subject may subsequently be excluded from the test. Subjects remaining are then asked to apply/use the two products that they have been given, for the remainder of the test period. The test period itself may range from two days to two weeks, depending upon the type of product being evaluated and the data that is required. Direct sensory assessment may be carried out twice per day, once in the early morning and once later in the afternoon, and the product may be applied to the axilla after the first morning assessment.

One of the main problems with direct sensory assessment of axillary odour, is the personal contact which is necessary between the test subject and the assessor, during the assessment process. In order to overcome this difficulty, an indirect odour assessment protocol can be used. In this case, soft absorbent pads are taped under each test subject's axilla and the malodorous substances from the axilla are absorbed into the pads. The pads can then be subsequently assessed by the assessors and the level of malodour quantified in a similar manner to that described earlier. The advantage of this technique is that subjects and assessors never meet, so avoiding embarrassment and possible assessor bias.

Several statistical techniques may be used to analyse the data obtained from deodorant evaluation studies. The t-test, analysis of variance and the Wilcoxon Signed Rank techniques, have all been used.

4. *The evaluation of antiperspirants*

In contrast to deodorants, the primary function of an antiperspirant product is to reduce the amount of sweating. Often confusion arises between antiperspirant and deodorant products, as the former can also act as effective deodorants, in addition to reducing the level of body sweat. An evaluation of antiperspirants should focus on efficacy of the primary function, rather than ability to reduce body odour.

As in the case of deodorant products, antiperspirants can be evaluated using both *in vitro* and *in vivo* techniques, although the *in vivo* clinical trial methodology has become the widely accepted industry standard. A variety of *in vitro* methods have been developed around the reaction of water with starch iodide, the resultant liberation of iodine producing a dark blue/black colouration. Typically, starch-iodide paper has been taped to the axilla and the quantification of sweat (water) made by optical analysis of the colour intensity produced. Other evaluation methods have sought to quantify the level of sweat produced by measuring the amount of water vapour present in the underarm area.

In the USA, an antiperspirant product which is advertised as such, is classified as OTC drug status and it is perhaps because of this fact that the commonly accepted standard method for antiperspirant evaluation is the *in vivo* clinical trial. There are many different protocols used in this type of test. Common to them all, is that the assessment of sweat reduction is made by taping absorbent pads to the axillae of the test subjects and measuring the amounts of sweat produced gravimetrically.

Selection of subjects for the test should be based on the types of criteria discussed earlier in this chapter. In this case, subjects should refrain from the use of any antiperspirant products, two weeks before the commencement of the test, to eliminate any "carry over" which may adversely affect the results obtained. A minimum of 20-30 subjects is required for each evaluation, so that the results obtained are statistically significant. Following the period of normalisation, on day 1 of the test, a pre-weighed absorbent pad is taped to the axillae of each test subject. Most gravimetric assessments of sweat reduction utilise a hot room environment, to stimulate and accelerate the normal sweating process, thereby reducing the test period significantly. The subject is placed into a specially controlled environment, normally kept at 100°F and 35% RH, with the absorbent pads attached to each axilla. It is important, whilst in the hot room environment, that the subject is kept as relaxed and as comfortable as possible. The reason for this is that the amount of sweat produced by an individual can be significantly affected by stress and anxiety, thereby biasing the results. After an initial period in the hot room environment, typically 40 minutes, the pads under both axillae of each subject are replaced with a second set of pre-weighed pads. The old pads are discarded, as it is well known that sweat production is rather variable over the initial period in the hot room. After a further 20 minutes, the second set of pads are collected and re-weighed, allowing the amount of sweat produced in the axillae of each of the test subjects to be calculated. The results obtained establishes the baseline sweating ratio.

Each subject is then given unmarked samples of the product to be tested, and a placebo, and is instructed to apply one to the left axilla and the other to the right axilla. The left/right assignment for the test and placebo products is reversed for half of the test subjects, to remove the effects of any left-right sweating bias. Application of the products is normally carried out twice a day, for a period of up to 4 days after the commencement of the test. On day 5, the subjects apply the test and control products in the normal way and, after a 30 minute conditioning period, the hot room sweat collection protocol described above is repeated.

Once the results have been collected, the percentage reduction in sweating (SR) can be calculated. Calculations are normally based on the ratios of sweat produced by the left and right axilla of each subject, pre- and post-treatment, as defined by the following formula:

$$SR = 1 - \frac{\text{(Post-treatment Sweat Ratio)}}{\text{(Pre-treatment Sweat Ratio)}} \times 100$$

The statistical method commonly employed to test the significance of the results obtained, is the Wilcoxon Signed Rank test.

References

1. "Principles of Product Evaluation: Objective Sensory Methods", IFSCC Monograph No 1, Micelle Press.

CONSUMER RESEARCH

Psychology associated with consumer research

A key aspect of any discussion on consumer research, is the psychology connected with it. Psychology became a true science during the last century, when leading exponents such as Pavlov carried out a great deal of work on animal and human senses. During the period 1920-1930, much of this work was developed further, with a more humanistic approach, to lay down the psychological foundations associated with the human senses and judgment processes, which are so important today in the field of consumer research.

1. *The psychology and perception of the physical world*

The five human senses, hearing, sight, smell, taste and touch, contribute to our perception of the physical world. For example, physical measurements such as length, mass and temperature. are contrivances of man, used to describe the world as he sees it through his senses. Only very simple phenomena can be represented in this way.

Perceptions related to physically measured phenomena, for example the "hotness" of temperature, can be represented graphically by "S" curves. At the extremes of the range, as the physical property becomes increasingly large or small, the human ability to sense differences in that property diminishes. In the middle of the range, the human perception of the property is more precise. The characteristic S-curve, associated with this observation, is illustrated in Figure 1.

This S-curve is important when considering methods of assessing the physical parameters of cosmetic and toiletry products such as viscosity, colour and fragrance. The relationship between response (perception) and physical stimulus (property), can be quantified using *Webers Law*, represented graphically in Figure 2.This relationship gives rise to the following equation.

$$d¥ = \frac{k \quad dØ}{Ø}$$

where

¥ = response (perception)
Ø = physical stimulus (property)
k = constant

FIGURE 1

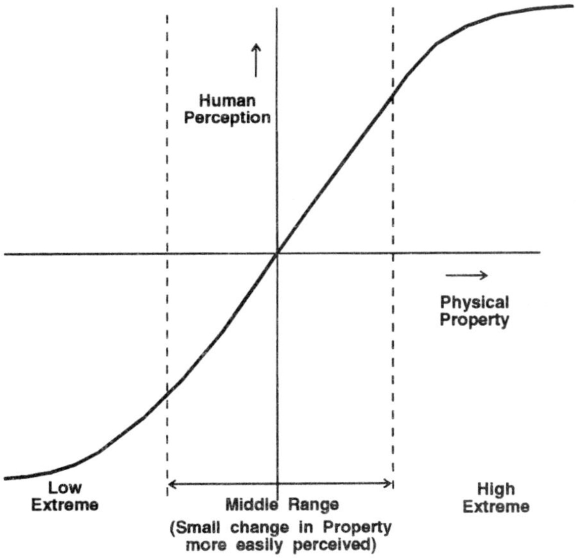

FIGURE 2

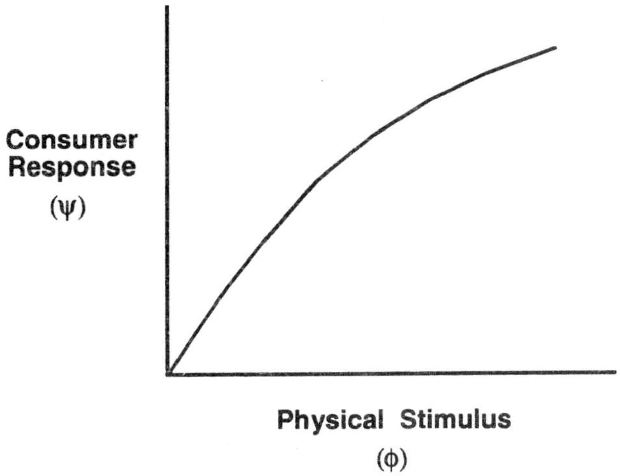

Other physical properties, for example underarm wetness, have a threshold of perception, below which that particular property cannot be perceived. When carrying out any consumer research associated with properties of this type, caution must be exercised in the research design.

The perception of physical phenomena is best described by use of the term *stimulus field*, which is defined by the proportional mix of the elements within it. For example, in

FIGURE 3

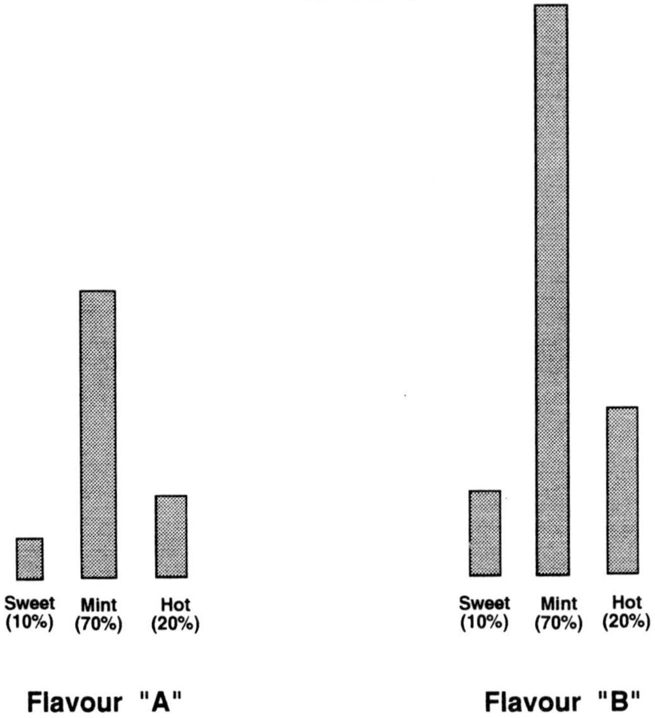

FIGURE 4

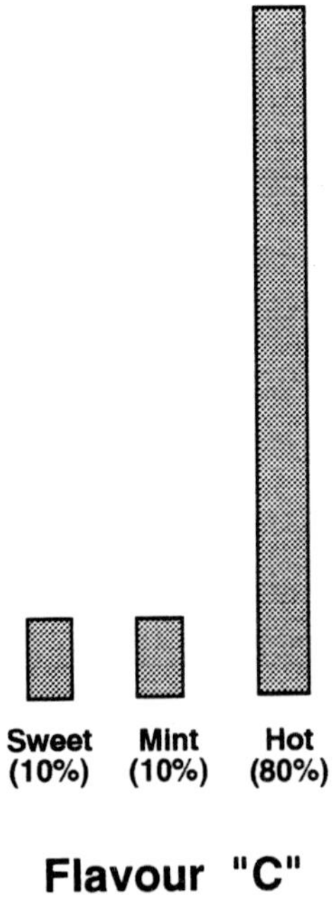

the case of a toothpaste flavour, there may be a mixture of tastes, such as hotness, sweetness and mintiness. The mix of these elements defines the stimulus field and also specifies its character. The most dominant element of the stimulus field, is the one that is most obvious to the assessor; minor elements of the stimulus field may, in some cases, not even be detected.

The concept of field stimulus is best illustrated by a practical example, in this case toothpastes flavours. Consider two toothpaste flavours, each of which have three elements of taste, sweetness, mintiness and hotness. The compositions of the two flavours are illustrated graphically in Figure 3.

The proportion of elements in flavour A and flavour B are the same but in flavour B all the elements have twice the strength. When assessed, flavour B would be perceived as being twice as strong as flavour A, although the character of the two flavours would be identical. This is because character is only dependent on the relative proportions of the elements present and not their absolute quantities. Now consider a third flavour, flavour C, the composition of which is graphically illustrated in Figure 4.

Now, the character of flavour C is different from the character of flavour A and flavour B, because the proportion of the elements is different. The dominant element in any of these flavours, is the element which is first detected when the flavour is assessed. Returning to flavour A, the elements can be characterised further, on a hierarchical basis, as illustrated in Figure 5.

FIGURE 5

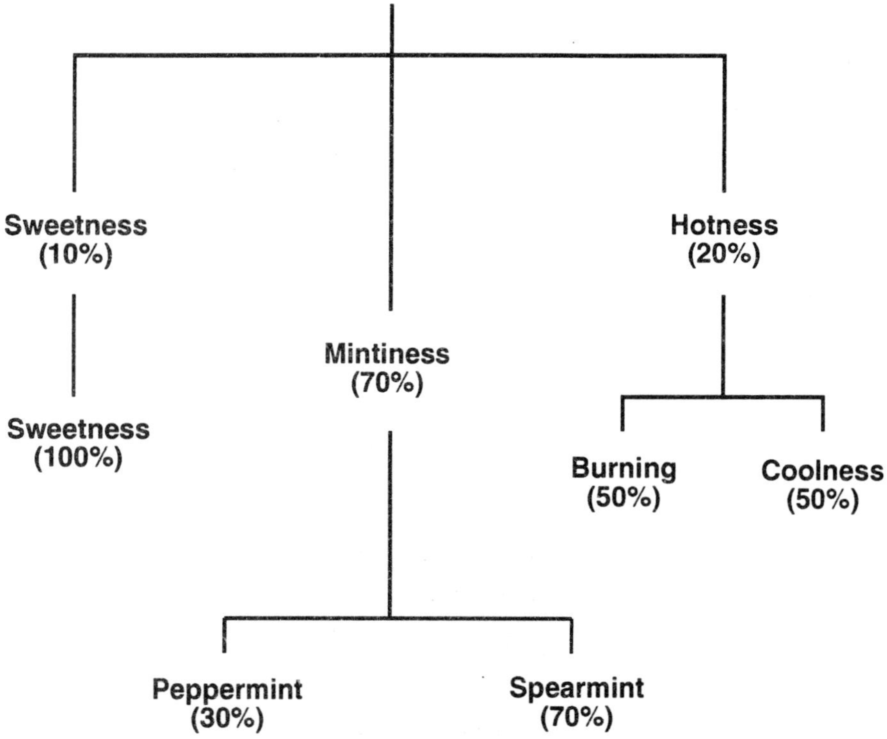

FIGURE 6

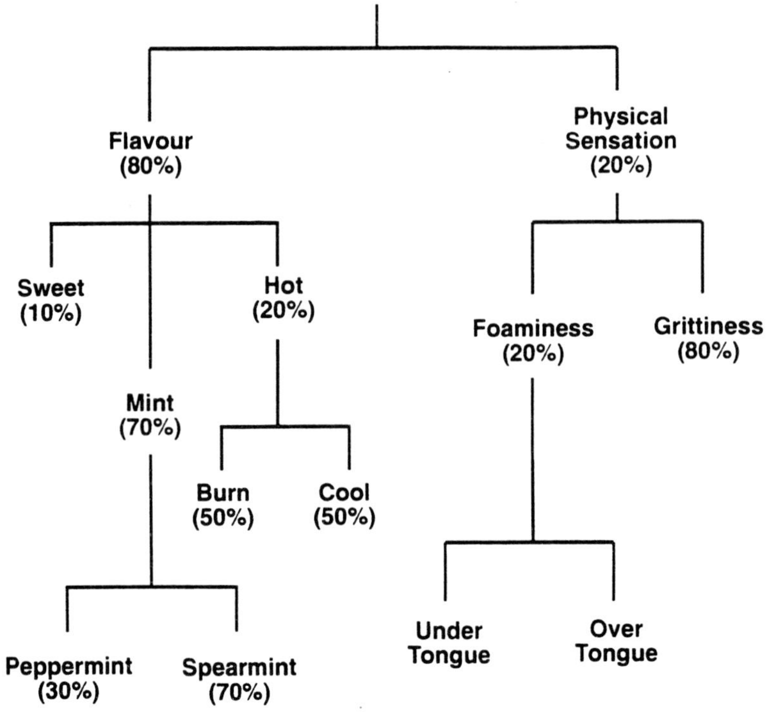

In a flavour of this type, the ratio of burning:coolness is dictated by the raw material menthol, which has a fixed ratio of these elements, and thus is uncontrollable. The ratio of peppermint:spearmint however, can be controlled, by varying the flavour blend of these two elements in the formulation. The human perceptual system can focus on certain areas of the stimulus field and, in so doing, may ignore others. In reality, a toothpaste has a stimulus field which is more complex than that associated with flavour alone. There are many physical sensations, such as grittiness of the abrasive, or mouth feel of the product, which also form part of the very complex stimulus field. This is represented graphically in Figure 6.

Human perception will always follow the dominant path, which will often reduce the perception of another element in the stimulus field. For example, grittiness in the above stimulus field may be reduced in two ways. The first, most obvious way, is to reduce the grittiness of the formulation itself. Perceived grittiness may also be reduced, however, by increasing the dominance of the flavour in the stimulus field. Although the physical level of grittiness remains the same, its perception has been reduced by increasing the dominance of another element in the field. This is known as altering the perceptual balance. By changing the perceptual balance, it is possible to draw attention away from a negative aspect of a product, by increasing the magnitude of a positive one.

Another aspect which is important in the perception of any product, is the conditions under which that product is used. For example, a shampoo formulation used by two different people, one using hot water and one using warm water, may be perceived entirely differently, as the fragrance, for example, is likely to be more performant when the shampoo is used with hotter water. Thus, the stimulus field of a product may well vary with method of use and the latter should be specifically defined, when examining the structure of any stimulus field.

Finally, when designing any piece of market research, it is important to use respondents that will be able to discriminate between products, for the attribute of interest. For example, in the market research of a functional product, such as an under arm antiperspirant, it is important to select respondents that would normally benefit from the functional advantage, in order to make valid assessments between different products. In this case, respondents with heavy under arm perspiration would be selected, otherwise the product may not appear to offer any functional benefit.

2. *The principle of conception*

Conceptual thought, in a similar way to perception, is hierarchically organised and varies greatly, depending upon the individual. In many cases, the conceptual thoughts between individual people are similar. For example, if 10 people were asked to list words describing a household item, such as a chair, each list would be very similar and use non-emotional phrases. If, however, the same 10 people, 5 of whom were smokers and 5 of whom were non-smokers, were asked to list words associated with cigarettes, then the lists would be very different. The smokers would list words with a mostly pleasant connotation, such as cooling, soothing and relaxing, whilst the non-smokers would list words with a negative connotation, such as bad-smell, smoky and unhealthy. In this case, the conceptual thoughts of the smokers and non-smokers, towards cigarettes, are very different. It is important to note however, that conceptual thoughts are highly dependent upon the context of the situation in which the assessment is made.

The conception of each individual is dependent upon the way in which they think about the world. For each individual, these thought patterns, or *constructs* as they are more properly known, are organised and related to one another hierarchically, and contribute to that individual's personality. Individuals differ in the number and type of constructs they have, but there is a great deal of overlap between the constructs of different people. Constructs may, or may not, carry good or bad connotations, the degree of each varying with different people. In any construct hierarchy, the constructs placed towards the top of the hierarchy are known as *super-ordinate constructs*, whilst those placed towards the bottom are referred to as *subsumed constructs*.

Constructs are hypotheses, induced by the mind, to explain the world. Consequently, new constructs are constantly being formed, whilst other are being rejected. Constructs are laid down in a chronological way and new constructs are only accepted if they fit within the pattern of the old constructs, or operate in harmony with them. An individual's constructs come from three different sources. The first source is learning, or what people are told. These constructs are often very factual in nature, for example "London is the capital of England". They can also be emotional as in the case of "it is wrong to kill animals". The second source of constructs is through interaction with other people and, in this case, all participants involved have an equal contribution to make. Thirdly, constructs are developed through individual thinking, where each individual generates his or her personal constructs, about a particular thing. Thus, thoughts come in three ways, what people are told to think, how people think things through with others and what people think by themselves. Everybody's personal construct system is unique, because of the development of constructs through isolated thought, and the unique way in which each individual interacts with others. In the latter case, each persons construct system can lead to the rejection of others, who have constructs incompatible with their own, and acceptance of people who have constructs similar to their own.

3. *The relationship between perception and conception*

The nature of each individuals conception, and the way in which they perceive different things, are obviously connected. Perception is the way in which things are perceived or understood, whilst conception depends upon personal constructs, although it does have an influence of evaluative ability. Therefore, each mental thought process can be divided into three components, perception, conception and evaluation. The latter is associated with judgements, such as "good" or "bad". The relationship between each of these components can be illustrated using the concept of a personality circle, as shown in Figure 7.

FIGURE 7

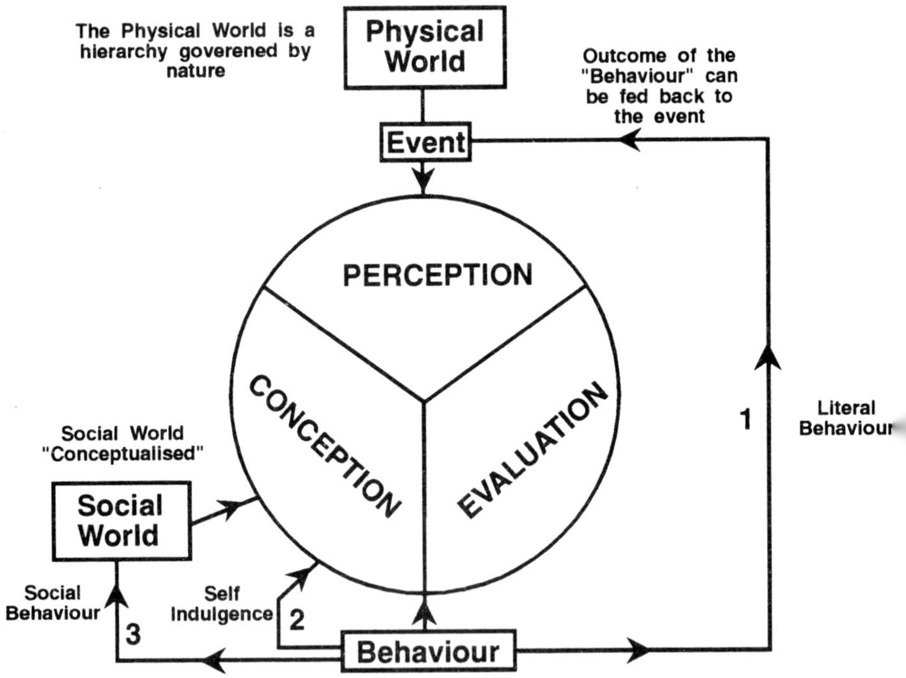

A individual will perceive a small number of things which he or she personally judges to be of importance and these may vary with time, even for the same individual. In the context of this perception, a behaviour response will be produced, as shown in Figure 7. This behaviour will be directed, with different emphasis, either at the physical source or event of the perception, as indicated by path 1 in Figure 7, or at some aspect of an individuals conception of it, as indicated by paths 2 and 3. Behaviour may be directed, with different emphasis, at any of the following areas:

- the physical event
- an individuals personal thinking or self-indulgence
- the image an individual has of themselves in the social world

The use of any product, including a cosmetic or toiletry, constitutes a particular "behaviour" and therefore such products possess an intrinsic combination of these three possibilities.

The way in which that product is marketed, referred to as the *brand position*, is associated with the particular mix of these elements. For example, the belief in a product, in social and self-indulgent terms, although primarily created by advertising and pack design, is supported by the sensual experience of using the product, which is under control of the product formulator.

Thus, when carrying out market research, the most important factor to focus on is conception, and market research should always be asking individuals what they personally think about the products they are assessing.

Marketing and the role of brands

Fundamentally, products are marketed and sold in two distinct ways, *commodity markets* and *branded markets*. A commodity is a product of which the origin is unknown, as in the case of potatoes, for example. Commodities tend to sell on price, due to the fact that retailers seek to gain a competitive advantage by buying their commodities as cheaply as possible. The pressure on suppliers, brought about by the reduced price, may finally lead to a cut in quality. A commodity market therefore tends to prohibit technical advance, be dictated by price and often contain products of poor quality. Commodity markets are also limited in size, by virtue of the fact that the products exist in peoples minds, in a unidimensional way.

Brand markets, by contrast, offer products or brands which offer additional rewards for the benefit of the consumer. Generally, the products are of higher quality and there is a value associated with them, for which the consumer is prepared to pay. This added value may be a physical difference, as in the case of a technically superior product, or it may be a symbolic difference, linked to the feeling of an "identity" that has been generated in the mind. Brands always have to be advertised to create this added value, or make a person aware of it.

The existence of brands leads to an expansion of the market, as new marketing opportunities are identified. Each brand is differentiated and characterised by its *brand position*, which is the basic offering of the brand and its appeal to different people in the market-place. The groups of people to whom brands may appeal are known as the *target markets*, which are differentiated by the basic attitude of the people in them. Unfortunately, technological advances also lead to so called "me-too" products, which do not differ in brand position from the brands they are mimicking. Thus "me-too" products do not give rise to market expansion but lead to splitting, or fragmentation, of the market. Some companies, however, recognise growing markets and often produce "me-too" brands, which are essentially positioned in the same way as the original brand in the market-place. These only give rise to short-term success, due to effective brand devaluation. The only long term successful products, are those displaying genuine product improvements. In general, brands give rise to a higher level of satisfaction in the market-place, more spend in the market category and the production of better products. The trends that may take place, when a branded product is introduced into a commodity market, is illustrated in Figure 8.

FIGURE 8

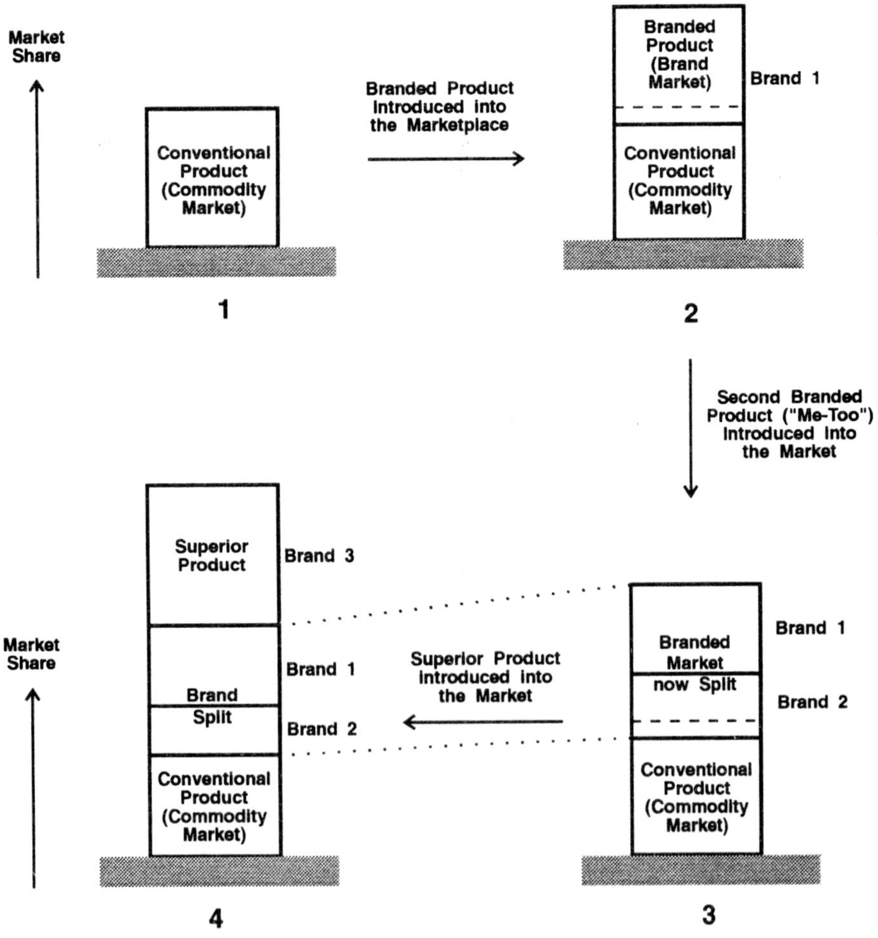

1. *The elements of a brand mix*

The basic elements of a brand mix are listed below:

- the product
- brand position and advertising
- brand name
- pack design and functionality
- price

Each of these have an impact on brand performance. For example, the physical characteristics of the product must not have any defects and, in addition, should communicate the message of the brand position, through the use of product appearance, functionality, viscosity and fragrance. The brand name can be very powerful in generating brand interest in the market-place and should preferably invoke connotations that are relevant to the brand position. The pack design is also a very important component of the brand mix. The pack graphics must clearly link the brand name, lettering design and pack colour together, in order to support and communicate the brand position to the consumer. Of course, none of these considerations must detract from the functionality of the pack, which is critical if the product is to be successful.

The *marketing mix* for a brand, defines how money is spent or allocated across the various elements within the brand mix, within a given marketing budget. It is also important that the elements of the marketing mix are complementary and reinforce one another, so that the brand positioning is communicated with maximum effectiveness.

2. *Advertising*

Advertisements must create brand awareness and generate a specific conception of a product, amongst a defined target market. This can be achieved through a variety of media, such as television, press, magazines and posters, which are collectively known as a *media mix*. Decisions have to be taken on the most suitable type of media to be used, and selection is made against the cost of reaching the target market and the context that the medium puts the advertisement into. Advertising must be based on a brand strategy. The advertisement must illustrate the basic product concept, normally through the use of a creative device that communicates the various elements of the brand. The *execution* of the advertisement describes the actual artwork or film used. A series of executions is called an *advertising campaign*.

Another important aspect of advertising, is trade advertising. This is designed to entice the retailer to purchase the product from the manufacturer or supplier.

Consumer research methodology

The design of any piece of consumer research should be very carefully planned, if meaningful results are to be obtained. Consumer research can be carried out for a number of reasons, as detailed in the list below:

- market information retrieval. Research is designed to investigate factors such as market share, which indicates the proportion of a given market that a particular product has, distribution, which analyses the category and number of outlets in which a product can be found and pricing levels

- market structure research. Research is designed to retrieve information about the size and nature of the market, in terms of consumer demographics, different types of product, pack size and the type and channels of trade outlet
- consumer usage and attitude (U and A) studies. Research is designed to analyse the types of consumer that buy different products and their behaviour and motivation for doing so
- concept research. The purpose of concept research is to identify the conception of a product, who or what is thought to be associated with it, and its symbolic overtones
- product research. Research is design to assess the consumers' view or opinions of a product, and the concept associated with it. This type of research is often carried out in a paired comparison with another product
- advertising research. This type of research is designed to analyse whether a particular form of advertising is effective in the market-place. Factors such as communication of the brand positioning, brand awareness and advertising recall, are investigated

1. *Methods of data collection*

Data collection is based on two main sources of information, distinguished as *primary* and *secondary* sources. Primary data may be collected by a number of different methods. Qualitative research involves the collection of primary data by talking to a small number of people, using in-depth interview techniques. Groups used in this type of research are normally made up of between 6 and 10 people. Quantitative research, by contrast, involves a larger number of people, responding to a questionnaire, and applying statistical validation to the data to obtain quantitative information. Primary data can also be obtained by the observation of people and classifying them and their behaviour.

A fourth method of collecting primary data is through auditing, which is used to obtain information about market share, price and distribution. Auditing can be carried out in one of two ways. The first method involves the collection of audit data obtained by the recruitment of a number of retailers throughout the country and subsequently monitoring sales, distribution and on-shelf price. The second method of auditing involves the collection of panel data. In this method, a panel of people are recruited and asked to record their purchases in a diary, on a regular basis. The collected data is processed to give information about purchase and usage behaviour.

The collection of secondary data involves desk research, gathering previously published research data and company records. This is a sensible and inexpensive preliminary step to take before conducting any consumer field study, and may negate the need to conduct any further research at all.

2. *Designing ad-hoc research programs*

Before designing any piece of research, it is important to establish whether there is a real need for it. In designing the research, it is necessary to define the problem and identify which aspects of it are amenable to market research techniques. Formulating clear objectives for the research at the outset, and specifying the necessary actions to follow it through, is essential. The research must be designed so that it that will enable the collection of data that is relevant and applicable to solving, or helping to solve, the defined problem. In drawing up the research design, the following areas must be addressed:

- define the sample. This is the type, or category, of respondent to be included in the research programme
- define the number of respondents. This is absolutely essential, particularly in the design of quantitative research, in order to ensure statistical validity
- define the questions to be asked. These should be based on the problem(s) to be solved and are often determined against the hypotheses to be tested

The research is then commissioned to an appropriate research agency, who will execute the research. The questions arc put into the field, the data collected and, if appropriate, put onto a computerised database. The research data is then tested against the hypotheses. Finally, the work is presented, in the form of a report, and its implications discussed.

3. *Qualitative research*

Qualitative research deals with small numbers, using in-depth interviewing techniques, in one of two ways. The first method uses "face-to-face" interviews, involving the researcher and a single respondent. Interviews may last for periods of up to 2 hours and will be carried out by a researcher specifically trained in interview technique. Face-to-face interviews can reveal a high level of qualitative detail but are very expensive to commission.

A second technique, which is much more common and cost-effective, is the use of group discussions. A forum atmosphere is created, usually with a group of 6-10 people, dependent upon the research topic and sample definition. The group is provided with various stimuli, designed to initiate and propagate discussion, by a moderator, for a session of between 1 and 3 hours duration. In all qualitative research, the proceedings are usually tape-recorded for subsequent analysis and, in some cases, may be recorded on video.

Qualitative research is normally used to investigate underlying attitudes, often about things which people are not conscious of. It is used mainly in new product development and advertising research, when the objective is to explore consumer feelings and attitudes, rather than collect hard data. Qualitative research requires an interviewing technique which is supportive and encourages people to talk. Appropriate stimulus material to provoke comment is supplied and an interviewing sequence, which helps people to articulate and develop their thoughts, is used.

Certain projective techniques may be used to stimulate responses, as listed below:

- sorting products into groups and explaining why
- drawing pictures of things and explaining why
- "filling in" cartoon images with "speech-balloons"
- picking the "odd one out" of three products and explaining the difference(s)
- by "role playing" the product

The interviewing sequence would normally start with the collection of simple facts about behaviour and then move on to attitudes, when people in the group are feeling more relaxed. The analysis of data involves listening to a tape recording of the proceedings and assimilating the different opinions within a psychological framework.

4. *Quantitative research*

There are three main approaches to obtaining quantitative research data, as listed below:

- ad-hoc research. This is research that is specially designed, with a particular company and purpose in mind
- syndicated research. This is continuous research that is designed by a research agency and sold to competitive companies. It is normally carried out using a sample of people, or a pre-recruited panel
- omnibus research. This involves research techniques in which private questions may be placed into a questionnaire that is fielded regularly

Quantitative research surveys are carried out in one of two ways, either by the use of an interviewer administered questionnaire, by telephone or "face-to-face" interview with the respondent, or by using a self-completion questionnaire, which is usually dispatched through the post.

In any quantitative research exercise, the different types of question that can be asked, and the way in which they can be asked through various types of questionnaire design, must be carefully considered. Question types available fall into one of two categories, pre-coded questions or open-ended questions.

Pre-coded questions are structured within the research design and may be one of the following types:

- scaled questions. These are questions whose responses are assigned to a scale, usually consisting of 5 or 7 points. A practical example would be the question "How would you rate the fragrance of this product?". Participants would then be asked to assign a value to the response from 5 (extremely good) through to 1 (not at all good). One of the main advantages of scaled questions, is that they can be computed numerically and given a mean score
- multiple response questions. These are questions used to examine people's opinions, such as "Here are a number of things that people have said about this product, which do you think applies?"
- factual questions. These are questions with a definite factual answer, such as "How many times do you shower every week?"

Open-ended questions are unstructured, using phrases such as "tell me everything you liked about the product . . . ". The respondents are then allowed to state their opinion freely and answers are recorded verbatim.

PSYCHOLOGY

Introduction

The psychology of cosmetic and fragranced products is associated with how these products affect us psychologically and socially, their benefits, when we use them, and what motivates our selection and usage.

The use of cosmetics helps determine the social image we project and our "appearance" affects the way our own personality is perceived by others. The beneficial effects obtained from cosmetic products also affects our perceptions of self image, self esteem, self confidence and the reshaping of our moods, in different situations. Inter-personal attractions, such as liking, affection and sexual attraction, are also important areas of benefit. These benefits can all be psychologically therapeutic.

Psychological benefits of cosmetics and fragrance

In research on the psychology of cosmetics and fragrances, cosmetic care, make up, skin care, hair care and fragrances have all been shown to fulfil some very important functions in relation to the "well-being" of a person. People using these products can feel better psychologically and socially, as demonstrated by their social confidence and the quality of their social encounters with other people.

It has been shown that facial make up enables a person to be seen as more attractive in appearance and to project a more positive personality in their social world. Specifically a woman is perceived as being more feminine, mature, clean, pleasant and physically attractive. They are is also seen as being more confident, secure, sociable, interesting, popular, organised and poised. From this research[1], emerged the notion of the "what has been made beautiful is good" stereotype.

In a similar fashion, the principle of "what has been made beautiful is good" should operate for fragrances too. Benefits of being seen as more attractive, in looks and personality, is attributed to the users of perfume, who have been seen to have made the effort to be beautiful, or to have taken care of themselves, by using fragranced products.

It has also been shown[2] that there are benefits to be gained from fragrances added to other products, such as a hand care lotion. A well chosen fragrance has been shown to favourably affect how the product is evaluated and positively influence perception of the user, when the product has been applied.

The benefits of cosmetics and fragrances also encompass the area of self perception. One important aspect of how a person perceives herself, is feeling or mood. Cosmetics and fragrances can both be used to help a person feel better. This has been shown with facial makeup, for example, by Graham & Kligman[3], where professional make-overs were found to provide a number of benefits in terms of self perception, appearance, self image and social confidence. It would be anticipated that similar kinds of benefits would be experienced with fragrance application but further research is needed to explicitly demonstrate this fact.

If a person wants to enhance a mood she is already experiencing, change the way she is feeling, or project a particular mood for a special occasion, then it is important to understand what type of mood is projected by a particular product. It is known, for example, that certain colours of eye and lip make-up[4] and certain perfumes[5] project their own individual moods, which may be linked with a particular kind of situation that they are appropriate for. The consumer can then decide for herself the mood she wishes to project and select the make up and/or fragrance to project it, thus enhancing the feeling of that mood within herself.

A person who feels introverted and not very confident socially, may benefit from using vibrant make-up and perfumed products, which would help her to feel more confident and outgoing. Such products would complement and balance the person's personality better than "quiet" colours and unobtrusive perfume. Likewise, an already very extroverted personality could benefit from using products that somewhat "tone down" how she feels, moderating the image she projects, rather than enhancing it.

One major area of psychological benefit in using cosmetics and fragrances is, of course, that of interpersonal attraction. One of the major social functions of fragrances and deodorising products, for example, is to conceal unpleasant odours from different parts of the body and give the impression of being clean and attractive.

A study by Graham and Furnham[6] in 1981, found that all commonly used categories of decorative cosmetic products, including perfume, used by girls in the 14 - 18 age group, were rated as more attractive when used in night-time social situations, compared with day-time situations. Perfume was rated as the most attractive product used. The highest rating of all was given by men to the perfume used by the girls. Nowadays, of course, there is more of a tendency to use all-day products. More recently there has been a trend towards the use of romantic moods to promote and advertise women's fragrances and men's colognes, where attraction between the sexes is very much emphasised as a theme.

The psychology of hair and hair care

In the psychology of hair and hair care[7], the social importance and psychological implications of different hairstyles, colours and lengths, can readily be observed. The benefits to be gained from taking good care of the hair, such as a higher level of self perception and the projection of a better image, are clearly demonstrated.

In the case of hair appearance, there appears to be traditional stereotypes associated with different styles, lengths and colourings. The stereotype, however, may or may not reflect accurately what the person is like in reality. For a long time, men were thought to prefer women with long hair, which is typically associated with connotations of being more feminine, more romantic and sexually appealing. What is considered to be the ideal hair colour changes, depending on fashion and social context. The "dumb blonde" stereotype, for example, depicts a pretty but brainless creature. On the other hand, do "gentlemen really prefer blondes?". Studies in the 1970's have shown that men actually do prefer lighter hair colouration in females. Females, on the other hand, tend to prefer darker colouration in men, for both eye and hair colouring. For women, dark hair has traditionally been associated with being more intelligent, more complex and more emotional; red hair with being quick tempered or passionate. In women, grey hair is associated with ageing but for men the connotation is that of being mature and distinguished.

These stereotyped ideas come from art, films and literature, and remain with us over the years as they become reinforced. For example, if a woman has long hair, men will be more likely to behave romantically towards her, as if she is more attractive and feminine. Today, with the large range of hair colourants and other products available, it is quick and easy to change hairstyle and hair colour, in order to change the personality stereotype.

In recent years, some examples of very extreme hair styling, colouring and make-up have been observed, in the youth subcultures of punk rock and new wave. Nowadays, because it is easy to change hair colour, many variations of hair types and colourings are evident. Each style and colour projects a different image, depending upon the person with whom it is associated. As a result, some of the traditional stereotypes discussed above are rapidly becoming less important.

Anti-ageing and camouflage cosmetics

The ageing of the skin, and what makes people appear younger than they actually are, is of great concern to many cultures throughout the world, particularly those in the West[8]. Prominent figures and famous people are frequently in the public eye these days, particularly if they project an image that is more attractive and younger looking than that typically associated with someone of their age. In one study [9], elderly US women who had aged well for their years, were compared with a group of women of the same age who had aged badly in terms of skin appearance, as judged by the presence of wrinkles, pigmentation disorders, and so on. Questionnaire responses on various dimensions of behaviour were compared for the two groups and it was found that those who had aged well, and possessed a more attractive appearance, saw themselves more positively in terms of appearance, self image, and mental and physical well-being. This study highlights the fact that not only physical damage occurs from taking insufficient care of the skin. A person who has aged prematurely also experiences a higher risk of suffering social and psychological disadvantages, in having a skin appearance that looks relatively unattractive.

Research has revealed many examples of elderly people who do not want too much make-up applied to their faces, for fear that they may appear "over made-up" or "artificial" when the cosmetic make-over has been completed. They also expressed concerns about using certain eye products because of allergic reactions. Interestingly, all of these people were found to benefit in some way from the experience of a make-over, particularly in terms of gaining a more positive mental attitude.

Cosmetic camouflaging of skin defects is also an important area. Ordinary beautifying cosmetics, such as facial foundation, can effectively cover age spots and other blemishes, such as bruising, broken capillaries and minor wrinkles. For more severe skin defects, heavy camouflage cosmetics can be used. By camouflaging, a make-over can be used to distract the attention away from the skin's defects and unattractive areas, thus highlighting the more attractive parts of the face. This is an important method of enhancing somebody's self perception, making them feel more youthful, optimistic and active. A kind and caring approach to cosmetic care, and a concern for the individual's psychological needs, enables this treatment to work effectively and uplift an individual's self image.

Another important aspect of creating an attractive image, and using physical appearance as a method providing "well-being", is self acceptance[10]. Everybody has features or physical defects that they would like to hide, or at least minimise. Such defects make people feel self-conscious and less comfortable with themselves, because they don't "feel" very attractive. Apart from wrinkles and other normal signs of the ageing process, moles, birthmarks, scars and general disfigurements all affect the psychological state of self acceptance. Skillful use of cosmetic products gives people a chance to feel more attractive because it helps them to focus on the more positive aspects of themselves. In this way, cosmetics can be used to enhance standards of health and beauty by building confidence, reshaping moods and generally improving image and appearance. These factors help people to feel better overall because their positive features are being highlighted, both physically and mentally. Enhancing outward attractiveness automatically enhances self-perception of well-being and, in this way, the entire outlook and quality of life can be changed.

It is quite likely[11] that the 1990's will bring more symptoms of nervous disorders and skin problems are likely to become more common, as a result of the increase in stress that arises from faster life styles, increased unemployment and economic difficulties. Problems are likely to include stress-related skin disorders, in addition to the more conventional signs of premature ageing, such as wrinkles, age spots, dryness and other ageing-related skin conditions. There will surely be a need for more products to deal with these problems. In addition to the skin camouflaging products discussed earlier, there will be a greater emphasis placed on basic skin care and skin protection products. More importantly, these treatments will need to be combined with appropriate psychological care, to meet the needs of each individual.

Image making

There are many ways in which the choice of cosmetics, hair care, fragrances and fashion can be linked together, in order to convey a particular "look" which is appropriate for any given situation.

In many work environments, for example, it is important that dress conveys a professional image. Research by John Molloy[12], author of "The Woman's Dress for Success" published in 1977, revealed that the ideal kind of hairstyle for maximisation of professional credibility was found to be one of medium length and not too elaborate. A more elaborate look, perhaps accompanied by long polished nails and an over-apportionment of jewellery, was found to give the impression that too much time was being spent on caring for appearance, and not enough on the work itself. More elaborate and feminine styles were thought suitable only for more social or romantic occasions. By contemporary standards, even the more elaborate styles of that time would look conservative in the modern workplace, given the recent increase in popularity of more casual looking hairstyles. Factors such as length of the hair do not matter so much today because contemporary attitudes are much more liberal. However, being well-groomed and taking care of one's appearance is still important in creating a good overall image.

If used in the correct way, fashion, colour cosmetics, hairstyle and fragrances can all convey an air of modern-day professionalism. In employment recruitment particularly, where impressions are often formed very quickly, appearance is a testimony to an individual's lifestyle and attitudes, the inference being that if someone looks professional in their appearance, then they will be professional in their work.

One study conducted in the US[13] examined the effects of cosmetic make-overs for women, which included change of hairstyle and colouring, on employment prospects and expected salary levels. The women were seen, in photographs, both before and after the make-over, by personnel interviewers in employment agencies. The interviewers were then asked to assess the employment prospects of the women and to estimate their expected salary level. Interestingly, it was found that if the woman was seen after the make-over, rather than before, then an average of 12% increase in salary level could be expected.

Clearly, the application of effective make-overs can have a marked effect on personal success, mood and well-being. These principles should apply equally for men, although currently there is no research data on the effects of make-overs for men. The implication is that whatever impression is conveyed in the style of a man's appearance, it must be reflected in his thinking, lifestyle, personality and work, as well.

The social and historical background

Having considered the psychology of cosmetics and fragrances in Western culture, it is necessary to put cosmetics and fragrances into perspective, by considering the historical background to the use of these products in other cultures. The term "adornments" has meant different things in different cultures, and at different times throughout history[14].

For example, in Japanese and Maori cultures, tattooing can be very elaborate and cover large areas of the face and body. The same is true in certain subcultures of our modern day Western society. The social meanings of particular forms of tattooing vary within their social contexts. They may be a means of displaying an aggressive or rebellious attitude, or they may reflect an attempt to ward off the "ageing" process, as in the case of the Maoris, for example. Today, in Western society, tattooing around the eyes is a beauty treatment that is sometimes carried out. One purpose of this is to render unnecessary the application of eyeliner on a daily basis; another is to camouflage the lack of eyelash growth.

In certain African cultures, wooden lip plugs denote the social status of the person; the larger the size of the plug, the higher the status of the person. Displaying neck adornments often involves stretching of the neck to elongate it and the number of displays indicate the wealth and social standing of the person. The larger the number of adornments worn, then the wealthier the individual, although some discomfort may need to be endured if such adornments are permanent fixtures to stretch the neck. Many facial and bodily adornments are used deliberately to display aggression. Perhaps one of the best known adornments, most closely linked with the use of cosmetics today, is the use of paint for ceremonial, ritual and religious occasions.

Anthropological research also reveals how cosmetic adornments and fragrances have been used by men in different cultures, during different periods in history. In certain cultures, men have been equally or sometimes more highly adorned than women. For example, among the people of Mount Hagen in New Guinea, who share in an elaborate network of exchange relationships, it is the men who express these relationships by decorating themselves lavishly on ceremonial occasions. The decoration also signifies the different roles of the sexes. The men, who are responsible for war preparations and exchange transactions, have more brilliant body decoration than the women, whose duties are very much lower in status being primarily domestic and agricultural. Historically, men have used adornments to accentuate physical strength, sexual attractiveness, wealth and status, an observation which is still valid in today's modern society.

In Western culture, attitudes to cosmetic adornments and use of fragranced products have varied at different times throughout history. In certain periods, such as the seventeenth century, it was socially highly acceptable for men in Europe to be very elaborately and fashionably dressed, to use powder for their wigs, paint beauty spots on their faces and to be lavish in their use of perfume. Louis XIV of France, who reigned in the seventeenth century, was devoted to splendour and luxury and is said to have controlled his nobles by pre-occupying them with the details and expense of high fashion, in all aspects of their attire. Thus, clothes became a national fixation and, whatever his profession, a man would create a particular look and consciously project whatever image he chose, to make a statement about himself. This is a good example of an occasion in history, when making high fashion a priority for men was actually sanctioned by those in authority.

In the nineteenth century, the impression that men gave in their fashions, posture and body language was basically rather effeminate, delicate and graceful. Moving on to the twentieth century, a rather different picture of how political, social and economic circumstances determine the emphasis on cosmetics, fragrance and fashion, emerges. This, of course, is true for women as well as men.

Throughout the periods of the two World Wars, the use of cosmetic or perfumed products by men, particularly those in the services, was regarded as effeminate and socially unacceptable, projecting completely the wrong image. There was something of a stigma associated with such things, at a time when men had to project the image of being strong, masculine and protective. During this period, man's conservatism prevailed very strongly, upholding the conviction that fashion and fragrance was of relevance for women only. This inclination was, of course, encouraged during the war years by economic constraints such as rationing, when men often gave their clothing coupons to their wives or girlfriends, their own appearance suffering and becoming very shabby, as a result.

After the war years, attitudes towards men's products changed, albeit rather slowly. For the conventional man in the mainstream of society, the conservative attitude towards fashion and the use of cosmetic or fragranced products changed very gradually, with only occasional use of after-shave products. During that period, men were still very much concerned with maintaining their sense of masculinity which, for a long time, had meant refraining from the use of such products. Only a slow growth occurred in the market for men's products during the 1970's and 1980's, partly because of the attitude and feminine image that was, for a long time, associated with use of cosmetics and fragrance.

It was not until the late 1980's, that the mainstream of society's men really accepted the idea that fashions for them were more acceptable and that more scope existed for self presentation. Men's fashion, cosmetics and fragrances then became more available across a wide spectrum of income, age, social background and occupation.

In conclusion, it is obvious how social, political and economic factors shaped attitudes and fashions, which, in turn, also influenced the trends of the cosmetic and fragrance markets.

Products for men

Today's men have managed to successfully develop their freedom of expression, by dressing how they wish and using personal care, hair care and fragrance products, without portraying a feminine image. As such, they have managed to retain the "macho" or "manly" look. This is being done more successfully now, than ever before in history[15]. Buying products has now become much easier for men, than it was in previous years. It is now socially acceptable for men to buy cosmetics and toiletry products, rather than leaving it to women to buy the products for them. Supermarket-style purchasing has made it easy for men to select the products they want, without fuss, in both large supermarkets and smaller self-service health and beauty stores. The placement of men's cosmetics and toiletries, adjacent to other men's products such as clothing, shoes and accessories, in large department stores, also makes purchase more psychologically comfortable. These developments, along with the overall image, packaging and presentation of cosmetics and fragrances, have significantly helped growth in this market. As long as the "macho" image is maintained, given that men's products have a much more socially acceptable image nowadays, then the growth in this market should continue.

The concept of the "New Man" that has emerged in the late 1980's and early 1990's, can only serve to strengthen the cosmetics market for men. With this concept, comes a new role for men to adopt, with the likely acceptance of many more fragranced products, provided the "macho" image is still portrayed. This concept also provides a framework for predicting future anticipated trends in the market. Men will feel more comfortable using personal care, hair care and fragranced products, as progress further into the 1990's takes place. Make-overs for men might eventually become commonplace and they may ultimately experience and enjoy a wider range of treatments for the skin, hair and body. Already the range of products available for men is wide but there is room for further growth as modern man becomes even more aware of, and concerned with, his appearance. This, in turn, will inevitably influence the next generation.

Men's products, and the directions for their future growth, are relatively less well explored than those for women. There are many interesting avenues for further research, such as identifying men's motivations for using colognes and other products, within different age groups and different market segments. The benefits, both psychologically and socially, can be specified and quantified and the use of products by men in different social and professional environments requires further study.

Children's products: tomorrow's market

In the 1990's, the children's market will be an important area of potential growth within the cosmetics and toiletries industry[15]. Children, from as young as five years old up to twelve years old, are the new generation of tomorrow's users. Even today, the market for children's cosmetic and toiletry products is growing rapidly. Children in modern day society, from about the age of eight years and upwards, are relatively well-informed on many aspects of fashion, hairstyles, and types of cosmetic product available. They are maturing much earlier and there is a great deal of interest in bath, hair and other fragranced products. There is also every indication that this interest will continue to grow.

The children's market of today and tomorrow is another interesting area for research and there are many products now available for different age groups, particularly of the "novelty" bath and hair care type. Other areas could also be developed for children, such as perfumes using simple, colourful, natural fragrances and products that are fun and enjoyable to use.

More research is needed to explore children's primary motivations for using such products, and the benefits they enjoy from so doing. The psychology associated with children's use of cosmetics and fragrance products can be explored further, examining the types of product they want to use, the influence of peer groups and the image that they are trying to project about themselves. A better understanding of the psychology of today's children, will provide an insight into the growth potential for the children's cosmetics and fragrances markets in the future.

References

1. Graham, J A and Jouhar, A J (1980) The effects of cosmetics on person perception. Int. J. Cos. Sci. 3 197-208.
2. Jouhar, A J, Louden, M, Graham, J A, and Benjamins, N, (1986). Psychological effects of fragrance. Soaps, Perfumery, Cosmetics, April 59, (4) 209-211.
3. Graham, J A, and Klingman, A M, (1984) Cosmetic Therapy for the elderly. J.Soc.Cos.Chem 35, 133-145.
4. Graham, J A, (1986) Relating Colour Cosmetics to Psychological Concepts. Unpublished Report.
5. Graham, J A, (1986) The image projected by fragrance. Unpublished Report, USA.
6. Graham, J A, and Furnham, A F, (1981) Sexual differences in attractiveness ratings of day/night cosmetic use. Cosmetic Technology 3, 36-42.
7. Graham, J A, (1982) Hair from the Inside Out. Silkience Seminar III. The Psychology of Hair. Audio Transcript, Burson-Marsteller: New York.
8. Graham, J A, (1988) The Psychology of Cosmetics and Cutaneous Aging. In Cutaneous Aging Kligman, A M, and Takase, Y, (eds) Tokyos University of Tokyo Press.
9. Graham, J A, and Kligman, A M, (1985) Physical attractiveness, cosmetic use and self perception in the elderly. Int. J.Cos.Sci 7, 85-97.
10. Graham, J A, and Wallace, L, (eds) (1990) The Complete Mind and Body Book: Total Bodycare New York: Simon and Schuster.

11. Graham, J A, (1990) The Psychology of Cosmetics. Paper presented at In-Cosmetics Scientific Symposium, NEC, Birmingham March 6–8.
12. Molloy, J T, (1977) The Woman's Dress for Success Book Follett, Chicago, Il.
13. Waters, J, (1985) Cosmetics and the Job Market. In Graham, J A, and Kligman, A M, (eds). The Psychology of Cosmetic Treatments. New York: Praeger.
14. Ebin, V, (1979) The Body Decorated. London: Thamer and Hudson.
15. Graham, J A, The Psychology of Fragrance. In Poucher's Perfumes, Cosmetics and Soaps Vol: 3, Ninth Edition, Part II.

SECTION 10

Glossary of Terms

GLOSSARY OF TERMS

A

Absolute - A mixture of natural materials, obtained by extraction of a *concrete* with alcohol. The extraction is carried out in order to separate the principle odoriferous chemicals from waxes and fats, producing a product with better solubility for use in fragrances.

Absorption - The penetration of a substance into the body of another.

Accuracy - The correctness of a reading. For example, if there is not a mistake, then a weight measurement of 10.1g is accurate. The analytical measurement of 10.125g is also accurate, the two readings differing only in their precision.

Accelerated Rancidity test - Any method of determining the relative storage properties of oils and fats, by speeding up the onset and progress of rancidity. This is usually achieved by increasing one or more of the factors contributing to rancidity. See *Schaal Oven Test* and *Active Oxygen Method.*

Acid - A substance containing hydrogen, which dissociates in water to produce one, or more, hydrogen ions.

Active Oxygen Method - An accelerated rancidity test, in which the sample is held at a high temperature (97.8°C) and air bubbled through at a specified rate. A peroxide value is determined at intervals, with the end point reported in hours required to reach a value of 100 meq/Kg.

Adsorption - The condensation of gases, liquids or dissolved substances on the surfaces of a solid.

Aerobic - An environment where micro-organisms which have the ability to grow in the presence of air or oxygen, are found.

Amino Acids - Carboxylic acids containing an amino group. They can link together to form polypeptides and proteins, and are therefore of fundamental importance to life, being involved in cell structure and metabolism. More than 80 amino acids occur in nature, all of which have the general formula, R-CH-NH_2-COOH. Twenty six occur in proteins and twenty are regularly involved in protein structure. Essential amino acids are those which an organism cannot produce for itself and must be obtained from dietary intake. Ten of these amino acids can be considered essential to man. A list of amino acids is given below.

TABLE 1

Name	*Formula*	*Molecular weight*
Glycine	NH_2CH_2COOH	75.1
Alanine	$CH_3CH(NH_2)COOH$	89.1
Phenylalanine*	$C_6H_5CH_2CH(NH_2)COOH$	165.2
Tyrosine	$HOC_6H_4CH_2CH(NH_2)COOH$	181.2
Valine*	$(CH_3)_2CHCH(NH_2)COOH$	117.1
Leucine*	$(CH_3)_2CHCH_2CH(NH_2)COOH$	131.2
Iso-leucine*	$CH_3CH_2CH(CH_3)CH(NH_2)COOH$	131.2
Serine	$HOCH_2CH(NH_2)COOH$	105.1
Threonine*	$CH_3CH(OH)CH(NH_2)COOH$	119.1
Cysteine	$HSCH_2CH(NH_2)COOH$	121.1
Cystine	$[HOOCCH(NH_2)CH_2S]_2$	240.3
Methionine*	$CH_3SCH_2CH_2CH(NH_2)COOH$	149.2
Asparagine	$NH_2COCH_2CH(NH_2)COOH$	132.1
Glutamine	$HOOCCH(NH_2)CH_2CH_2CONH_2$	146.1
Lysine*	$NH_2(CH_2)_4CH(NH_2)COOH$	146.2
Arginine*	$NH_2C(:NH)NH(CH_2)_3CH(NH_2)\ COOH$	174.2
Aspartic Acid	$HOOCCH_2CH(NH_2)COOH$	133.1
Glutamic Acid	$HOOCCH_2CH_2CH(NH_2)COOH$	147.1
Histidine*	$C_3H_3N_2CH_2CH(NH_2)COOH$	155.2
Tryptophan*	$C_6H_4NHCH:CCH_2CH(NH_2)COOH$	204.2
Proline	$HNCH_2CH_2CH_2CHCOOH$	115.1

* *Essential to man*

Anabolism - The building-up process by which new tissues are made from food.

Anaerobic - An environment where micro-organisms which have the ability to grow in the absence of oxygen or air, are found. Some anaerobic bacteria are killed by exposure to air or oxygen.

Anion - A negatively charged ion.

Anisidine Value - A measure of oxidative deterioration. Spectrophotographic measurements are used to estimate aldehydes, which are secondary oxidation products.

Anosmia - The lack of a sense of smell.

Antiseptic - A chemical agent capable of inhibiting or killing pathogens. Antiseptics are normally relatively harmless to the skin.

Aseptic - The state in which potentially harmful, resistant microbes, (in an industrial context), are absent, having been removed by disinfectants. Asepsis does not necessarily imply sterility.

Atom - The smallest part of an element which can enter into a chemical reaction.

Attributes - Used in statistical analysis and measurement of quality, this refers to the category variables of a product, for example, height being classified as tall, medium or short.

Audit - A systematic examination of the records of a business or quality management system, to ensure that execution of activities and generation of reports have been accurately made.

B

Bactericide - An agent able to kill at least some types of bacteria.

Bacteriostatic - Describes an agent able to inhibit the growth and reproduction of at least some types of bacteria.

Base - A substance which dissociates upon solution in water, to produce one, or more, hydroxyl ions.

Biofilm - A film of micro-organisms, usually embedded into pipe or tank surfaces, which adheres to surfaces submerged in, or subjected to, aquatic environments. Biofilms can frequently cause fouling and will interfere with sterilising procedures.

Bonded Goods - Imported goods on which duty is payable. They must be stored in a secure warehouse until duty is paid, or until they are exported.

Brix (°Brix) - The percent of sugar, normally sucrose, in a solution.

C

Carbohydrate - A group of organic compounds containing carbon, hydrogen and oxygen only, with the general formula $C_x(H_2O)_y$. Carbohydrates play an important role in the metabolism of many living beings. Energy is stored as starch, and cellulose is the principle structural material in plants.

Carboxylic acid - An organic compound that contains one or more carboxyl groups (-COOH).

Cation - A positively charged ion

Cell Membrane - The cell wall or membrane varies in structure and properties, depending upon whether the cell is *prokaryotic* or *eukaryotic*. The plasma membrane of the eukaryotic cell is about 9 nanometres thick and contains about equal amounts of lipids and proteins. The lipids are arranged in a bilayer and there is greater variety than that found in bacterial cells. The adhesive properties of the cell exterior are very important in cell-cell recognition and tissue organisation. The plasma membrane is selectively permeable and contains active-transport systems for sodium and potassium ions, glucose, amino acids, enzymes and some nutrients.

Cell Structure - The unit of life. All living organisms contain cells which are composed of discrete, membrane-bounded units, which usually consist of two types of *protoplasm*, that is the *nucleus* and the *cytoplasm*. Many micro-organisms consist of only one cell, whereas man is made up of many millions. A diagram of the basic structure of the eukaryotic cell is shown below:

FIGURE 1

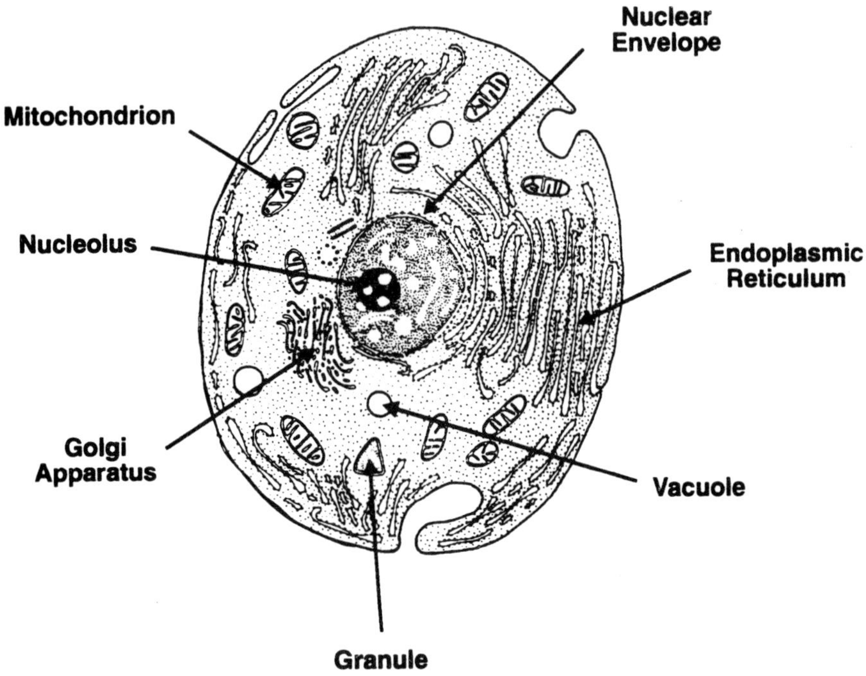

Centipoise - The standard cgs unit of viscosity, equal to 10^{-2} poise. Water at 20°C has a viscosity of 1.002 centipoise.

Cloud Point - The point at which an ethoxylated nonionic *surfactant*, when heated, just becomes turbid. Commonly expressed in degrees centigrade.

Cold Test - A test in which an oil is held in a bath at 0°C and the time required for the first appearance of cloudiness is noted. This value, referred to as the "cold test hours" value, gives an indication of the likelihood that the oil will crystallise during storage at low temperatures.

Colligative Property - A property numerically the same for a group of substances, independent of their chemical nature.

Colloid - A phase dispersed to such a degree that the surface forces become an important factor in determining its properties. Typically, the size of the dispersed particles is between about 10^{-7} and 10^{-9} mm and they can only be detected by an ultramicroscope.

Concrete - A perfumery raw material extracted from a plant, usually using a hydrocarbon solvent, and containing waxes and fatty acids, as well as the required aroma chemical.

Contact Angle - The angle that a liquid makes with other phases in contact with it, at equilibrium.

Contaminated - In the context of microbiology, a product that contains a high number of micro-organisms. Contamination does not necessarily imply the presence of pathogens.

Critical Path Analysis - A planning technique that details all the separate tasks within a project and the time they are estimated to take. The critical path is the order in which the tasks are organised, to ensure the fewest delays and lowest cost.

Customer - In the context of quality management systems, any individual or group that depends on an organisation for the provision of one or more products or services and that is in a position to reciprocate in some way. The customers of an organisation may include suppliers of goods, services and/or information, employees, shareholders and government departments, as well as the traditional customer.

Cytoplasm - The protoplasm of a cell, excluding the nucleus.

D

D-Value - Otherwise referred to as the D_{10}-Value or decimal reduction time. Defined as the time required at a given temperature to reduce the number of viable cells or spores of a given micro-organism to 10% of the initial number.

Density - The mass of a unit volume of substance, usually expressed in kilograms per cubic metre or grams per cubic centimetre. Density varies with temperature and the latter must be specified when quoting the former.

Dermis - The inner layer of the skin of a vertebrate.

Diffraction - In the context of the physics of light, the phenomena produced by the spreading of waves around and past obstacles which are comparable in size to their wavelength.

Disinfection - The removal or killing of micro-organisms likely to cause infection. In this context it refers to the treatment of surfaces and equipment by chemicals, ultra-violet radiation or steam, in order to free them from micro-organisms. Disinfection may lead to sterilisation.

Dynamic Viscosity - Referring to fluids, the ratio of the shear stress to the shear motion.

E

Eau de Cologne - An alcoholic solution containing about 2% - 3% of an extrait perfume formulation.

Eau de Toilette - An alcoholic solution containing about 4% - 6% of an extrait perfume formulation.

Eau de Parfum - An alcoholic solution containing about 10% of an extrait perfume formulation.

Elasticity - The property by virtue of which a body resists and recovers from a deformation caused by an external force.

Emulsifier - A *surfactant* used to disperse two immiscible or partially miscible liquid phases, one in the other, to create or stabilise an emulsion.

Emulsion - A dispersion of one or more immiscible liquid phases in another, the distribution being in the form of tiny droplets. Normally classified as either oil-in-water (O/W) or water-in-oil (W/O).

Endoplasmic Reticulum - The *cytoplasm* of cells consists of a matrix which contains ribosomes, fine fibrils and various structures surrounded by membranes, including the *golgi apparatus* and the endoplasmic reticulum. This reticulum is a continuous and variable system of flattened sacs and tubules, separated from the matrix by a membrane about 4 nanometres thick. Two types of endoplasmic reticulum exist. The first is heavily coated with ribosomal particles and is thought to be involved with protein synthesis. The second is smooth and not found in plant cells, occurring most often in cells secreting steroids. Both forms may be present in the same cell.

Enzyme - A large group of *proteins* produced within living cells and responsible for catalysing all chemical reactions upon which life depends. Enzymes are very active at minute quantities, typically having the ability to process between 10 and 1000 substrate molecules per second. It has been estimated that without the enzymes present in the gastric juices, it would take over 50 years to digest our food rather than just a few hours. All enzymes are *proteins*, with highly specific activity and they are sensitive to the conditions of their environment (temperature, *pH*, etc). Some enzymes contain a non-proteinaceous moiety, often a metal ion, and are known as conjugated enzymes. Certain enzymes require a co-factor for correct functioning. These are either a co-enzyme , a *protein* usually acting as a carrier of a chemical group, or an activator, usually a metal ion that acts by bringing the enzyme or enzyme/substrate complex into the active conformation.

Epidermis - The outermost layer of skin of a vertebrate.

Ergonomics - The scientific study of man in his working environment. Observations are used to make changes in equipment design, procedures and the environment, in order to improve safety, comfort and efficiency.

Essence - A term sometime used to describe a flavour formulation.

Essential Oil - A volatile oily substance derived from plants by a process of extraction.

Eukaryotic Cell - The cells of all higher animals and plants are *eukaryotic*, as are those of fungi, *protozoa* and most algae. Eukaryotic cells are between 1,000 and 10,000 times larger than *prokaryotic cells* and contain a *nucleus* surrounded by a membrane. They also contain internal membranes surrounding organelles, such as *mitochondria* and *golgi bodies.*

Extrait Perfume - The compounded concentrate of *essential oils* and synthetic raw materials, which can be subsequently diluted for use in an eau de toilette, etc.

F

Fat - An ester of glycerol and a fatty acid, which is not soluble in water. Natural fats are also mixtures of various glycerides.

FDA - The Food and Drug Administration, an American government body responsible for the regulation, monitoring and enforcement of the laws regarding cosmetics, toiletries, food and pharmaceuticals in the United States.

FEMA - The Flavour and Extract Manufacturers Association.

Fixative - A material which is added to a fragrance formulation, in order to slow the evaporation of the main aroma components.

Flash Point - The temperature at which the vapour from a substance ignites on contact with a naked flame. The figure is usually quoted in degrees centigrade, with a stated test method.

Franchise - An agreement allowing someone to sell a third-party product or service, subject to certain conditions, usually within a restricted geographical area.

Free Fatty Acids - The fatty acids that may be liberated from monoglycerides, diglycerides and triglycerides. These can be estimated by alkaline titration in ethanol.

Fruit - A ripened ovary containing seeds produced from a flower, following pollination and fertilisation.

Fungicide - An agent able to kill at least some types of fungi.

Fungistatic - An agent able to inhibit the growth and multiplication of at least some types of fungi.

Fungus - A colourless plant whose body (mycelium) consists of branching threads (hyphae). Reproduction occurs through the production of spores.

G

Golgi Apparatus - The golgi apparatus is a complex structure, consisting of flattened, single membrane vesicles, which are often stacked, some arising peripherally by a pinching process and some becoming vacuoles in which secretory products are concentrated. The golgi apparatus functions in the secretion of cell products, such as *proteins*, to the exterior. It also helps to form the plasma membrane and the membranes of *lysosomes*.

Gram's Stain - A stain used in bacteriology to differentiate between bacteria that take the stain, Gram positive, for example *Staphylococcus* sp., *Streptococcus* sp., and those that do not, the Gram negative, such as *Gonococcus* sp. and typhoid bacteria.

GRAS - Generally Recognised As Safe, an expression denoting approval of substances by FEMA and most often referring to flavour ingredients.

H

Head Space Analysis - Analysis, usually by gas chromatography, of the air present around a fragrance composition or an odour-emitting object, such as a fruit or flower. Head space analysis has proved invaluable in helping create a range of fragrances with odours similar to the living plant.

Hooke's Law - Within the elastic limit of a body, Hooke's Law states that the ratio of the *stress* applied to the *strain* produced, is a constant.

Hormone - A chemical substance found in small quantities in animals and plants, producing one or more effects on the growth and functioning of the body. Hormones vary widely in their structure, some being single *amino acids* such as adrenalin and thyroxine, which are derived from tyrosine. More complex hormones like testosterone and progesterone are steroids formed from cholesterol. Many are peptides, or small *protein* molecules, such as insulin. Hormones have an effect at very low concentrations (10^{-10} to 2×10^{-9} molar) and are involved in a wide range of biological processes. Control is usually exerted through modification of particular chemical reactions in responsive cells. The principle hormones, and their actions in man, are illustrated below.

FIGURE 2

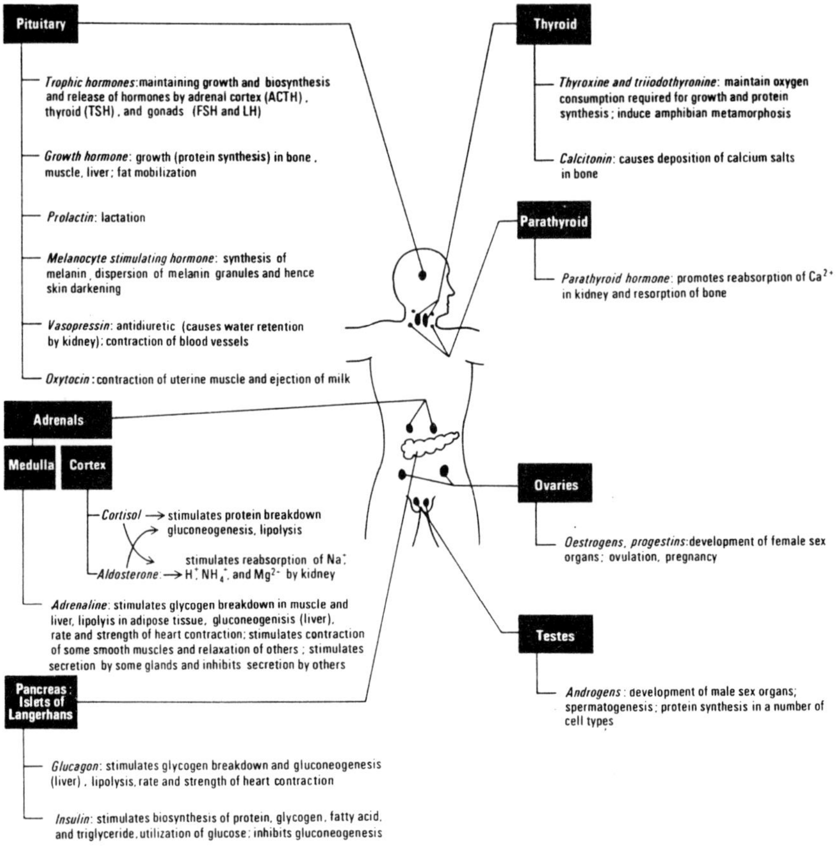

Humidity (Absolute) - The mass of water vapour present in unit volume of the atmosphere, usually expressed as grams per cubic metre.

Humidity (Relative) - The ratio of the quantity of water vapour present in the atmosphere to the quantity of water vapour which would saturate at the existing temperature.

Hydrolysis - A double decomposition reaction, involving the splitting of water into its ions and the formation of a weak *acid* or *base*, or both.

I

IFRA - The International Fragrance Association which provides information, recommendations and guidelines on the legislative, toxicological and dermatological aspects of perfumery.

IOFI - International Organisation of the Flavour Industry which provides information, recommendations and guidelines on the legislative, toxicological and dermatological aspects flavour production.

Iodine Value - An expression of the degree of unsaturation of an oil or *fat*. Iodine value is measured by determining the amount of iodine which will react with a sample of the oil or *fat*. Unsaturated oils have high iodine values and saturated or hydrogenated oils and *fats* have low iodine values.

Input Variable - In the context of quality management systems, any characteristic, feature or aspect of a *process* that contributes to the production of an output. Inputs are traditionally divided into six categories, people, materials, methods, machinery/ equipment, environment and measurement.

Irradiation - Normally refers to ionising radiation by gamma rays, beta rays and X-rays, for the sterilisation of raw materials and equipment. Ionising radiation may adversely affect the materials being sterilised. The levels of radiation are measured in Kilograys and the dose range for cosmetic materials is normally between 1 and 25 Kilograys. Ionising radiation functions by supplying energy which permits a great variety of chemical reactions to occur in the contaminating micro-organisms. In general, resistance to ionising radiation increases in the following order: multicellular organisms, gram negative bacteria, gram positive bacteria and fungi, bacterial *spores* and viruses. Sub-lethal doses may trigger mutagenicity and therefore irradiated raw materials must be sterile. Irradiation of certain foodstuffs has been allowed in the UK since 1990.

K

Key Input Variable - In the context of quality management systems, any *input variable* that has been assigned high priority.

Key Output Variable - In the context of quality management systems, any *output variable* that has been assigned high priority.

Kinematic Viscosity - A coefficient defined as the ratio of the *dynamic viscosity* of a fluid, to its *density*.

L

Lamellae - The thin films of liquid detergent solution that exist between individual gaseous (usually air) bubbles in a foam. When the liquid drains sufficiently from the lamellae, they become unstable causing the foam to collapse.

LD 50 - The lethal dose of a substance which, when administered to a group of experimental animals, causes the death of 50% of them. The figure is usually quoted in mg of substance per kg of body weight.

Lyophilisation - Often referred to as freeze-drying, a process whereby a product is frozen and placed under vacuum. The ice is vaporised and removed, leaving a dry product.

Lysosomes - Lysosomes are parts of a cell containing a great number of acidic materials that can destroy the whole cell, although this is normally prevented by the lysosome membrane, which breaks down during aging and after death. Lysosomes are thought to destroy unwanted products in the body, being triggered by deprivation of oxidation, and by some chemical stimuli. Lysosomes may offer some protection against viruses and bacteria and have been implicated in carcinogenesis after chemical or radioactive stimulation. They tend to concentrate any abnormal particles entering the cell, and generally offer protection to the cell until malfunction occurs.

M

Metabolism - The whole of the chemical processes, both anabolic and catabolic, occurring in cells of a plant or animal.

MIC - The Minimum Inhibitory Concentration. Referring to a preservative or antimicrobial agent and defined as the lowest concentration at which inhibition of specific micro-organisms occur, under a known set of conditions.

Mitochondria - Minute rod shaped or granular bodies, between 0.25 and 1.0 microns in diameter and up to 400 microns in length, found in the *cytoplasm* of most cells. Often termed the "powerhouse" of the cell, the mitochondria are responsible for the generation of energy and contain many cell enzymes, particularly those required by the citric acid cycle. There are large numbers of mitochondria (ca 500,000), in cells that require a lot of energy, for example muscle cells.

Mitosis - The normal division of the *nucleus*, involving no change in the number and constitution of the chromosomes. Mitosis is divided into four stages, prophase, metaphase, anaphase, and telophase.

Modulus of Elasticity - The stress required to produce unit strain, which may be a change of length (Young's modulus), a twist or shear (modulus of rigidity or modulus of torsion), or a change in volume (bulk modulus), expressed in units of dynes per square centimetre.

N

Nucleus - A dense area of the cell containing genetic information for control and procreation. Usually very important to the cell, although some cells do not possess a nucleus. Nuclei vary in size and shape, from species to species, and also from tissue to tissue within the same organism. The nucleus is only partly separated from the *cytoplasm* by its nuclear envelope, which contains pores.

O

Optical Rotation - The rotation of plane polarised light, due to the optical activity of a asymmetric carbon atom. This property is used in the quality assurance of perfumery materials, particularly to detect adulteration and purity of *essential oils*.

Organisation - Any group of people that undertakes responsibility for the performance of a task. The group may be large or small, formal or informal, permanent or temporary.

Osmosis - The diffusion of liquids, mostly water from solutions with a lower concentration, through a *semi-permeable membrane* into solutions with higher concentrations, until a equalisation of the concentrations occurs. The generation of osmotic pressure using sugar solutions can act as a preservative.

Output Variable - In the context of quality management systems, any characteristic, feature or aspect of a product or service supplied to a *customer*, the presence or absence of which will affect the customers perception of the value of the product or service supplied. Outputs may be specified or not and can be measurable or intangible. Outputs include such characteristics as dimensions, chemical condition, style or fashion, function, availability, pleasantness, identification, attendant documentation and timeliness. An output is produced by a *process*.

P

pH - A value expressed on a scale of 1-14, giving a measure of the acidity (between pH 0 and pH 7) or alkalinity (between pH 7 and pH 14). Strictly defined as the negative logarithm of the hydrogen ion concentration, $pH = -\log_{10}[H^+]$.

Parameter - A property of a population or a product, such as mean, variance, density, flash point, etc.

Pasteurisation - Often referred to as a form of "partial *sterilisation*", although this is technically inaccurate. Pasteurisation involves heating to a temperature sufficiently high enough to kill bacteria, but not bacterial spores.

Pathogen - Any micro-organism which, by direct contact with humans, causes disease. Even microbes which can normally be considered as harmless, can, in very high numbers, cause disease and therefore may be classified as pathogenic under these circumstances.

Peptone - A product of the partial digestion of *proteins*.

PIT - The phase inversion temperature. The temperature at which an emulsion's orientation is inverted (ie O/W inverts to W/O or W/O inverts to O/W), due to the balance of the hydrophilic and lipophilic character of the emulsifier system used.

Preservative - Any chemical used to kill or prevent the growth of micro-organisms which, by their growth, will spoil or contaminate a raw material or product. Preservatives are referred to as biocides in some industries.

Process - A combination of *input variables* necessary for the generation of an output. Also anything that does work.

Prokaryotic Cell - Prokaryotes are very small simple cells, having only a single cell membrane, which is usually surrounded by a rigid cell wall. They do not contain any internal membranes or organelles and contain only one chromosome, which consists of a single molecule of double helical DNA. Prokaryotes are believed to be the first cells to arise in biological evolution and include the eubacteria, the blue-green algae, the spirochetes, the rickettsiae, and the mycoplasma organisms.

Protozoa - Minute single-cell animals.

Protein - Complex organic compounds of high molecular weight (18,000-10,000,000). Protein molecules consist of many hundreds of *amino acid* molecules, joined by the peptide linkage to form polypeptide chains, which can be folded in a variety of ways. The sequence of the *amino acids* gives the protein its individual properties. Most proteins form colloidal solutions in water but some are insoluble. These include keratin, the protein responsible for the structure and properties of skin, hair and nails. Protein structure is defined in the following way:

1. The primary structure is the complete composition and sequence of the *amino acids* in each chain or segment.
2. The secondary structure is the precise folding or arrangement of the chains, for

example two chains may be wrapped together in a right-handed alpha-helix. This may involve the formation of inter-chain covalent disulphide bonds.

3. The tertiary structure describes the folding of the molecule upon itself, resulting in the formation of a globular *protein*. This may involve a great deal of internal hydrogen bonding.
4. The quarternary structure acknowledges the fact that there may be more than one segment to the whole protein. Some *enzymes*, for example, consist of several parts but are only active when all pieces are in the correct conformation.

Q

Quality - In the context of quality management systems, a *customers* perception of the total value of a product or service provided. Certain aspects of quality can be measured precisely, but total quality is always made up of disparate elements that are evaluated in combination by subjective judgment. Total quality includes aspects of cost, conformance to specification and service.

R

Refractive Index - A numerical expression of the ratio of the speed of light in a vacuum, compared to the speed of light in the test substance. For practical reasons, instruments usually compare the speed of light in the test substance to the speed of light in air. The refractive index value is characteristic for each substance and may be used as a check on material consistency for quality assurance purposes.

Resinoid - Similar to an oleoresin, typically a concentrated viscous extract of resins and *essential oils*, such as benzoin resin, and used as a *fixative* in perfumery.

Resistance - In the context of microbiology, a resistance to *preservatives* and *disinfectants*. An organism can be said to be resistant if it has developed a tolerance to a specific *preservative* or *disinfectant*. Gram negative bacteria are generally more resistant than gram positive bacteria and it is possible for different strains of the same organism to display a wide range of resistance to anti-microbials. The presence of certain *surfactants* can potentiate resistance, or it may be acquired by mutation. In the absence of anti-microbial agents, resistant organisms will grow slower than sensitive wild types. The practice of preservative/disinfectant rotation is strongly recommended, in order to minimise or avoid the development of microbial resistance.

Reverse Osmosis - The process of applying pressure to a dilute solution, forcing the water across a *semi-permeable membrane*, against the concentration gradient, thus concentrating further the more concentrated solution. Reverse osmosis is used to purify sea water and to concentrate dilute solutions of aqueous flavour or fragrance ingredients.

Ross-Miles Method - A standard test method, used in the cosmetic and toiletries industry, for assessing the foaming capability of *surfactants* or surfactant combinations.

S

Sampling Error - The difference between the estimated parameter from the sample and the true value for the population.

Saponification Value - Saponification is the process of hydrolysis of mono-, di- and trigycerides with alkali to form free fatty acids in the form of soaps. The saponification value is inversly related to the average molecular weight of the oil or *fat* and therefore provides an indication of the type of fatty acids present. The larger the fatty acid, the lower the saponification value.

Schaal Test - An accelerated method for the determination of the oxidative stability of an oil or *fat*. The sample is heated in an oven, in a covered glass container, and the time for a rancid odour to be detected is recorded. It is important to specify the sample size, oven temperature, etc., when reporting results.

Semi-Permeable Membrane - A membrane that allows the free passage of water but not certain materials in solution in the water.

Smoke Point - The temperature at which a sample of *fat* or oil produces a continuous thin stream of smoke, when heated under a defined set of conditions.

Soil - In the context of microbiology, any material found to cause hindrance to cleaning or *disinfection* procedures, for example product residues in vessels and piping.

Solubilisation - The micro-emulsification of small amounts of non-polar materials, by the micelles of a normally high HLB surfactant, to produce a clear dispersion.

Solubility - Referring to one liquid or solid in another, the mass of a substance contained in a solution, which is in equilibrium with an excess of the substance.

Specific Gravity - The ratio of the mass of a given volume of a sample, to the mass of the same volume of water at 4°C, or other specified temperature.

Spores - There are two main types of spore. The first is a specialised form of an organism, produced in response to stimulus of extreme duress, such as abnormal environmental conditions. This type of spore is dormant and very resistant to disinfectants. The second type of spore is that which is formed during the reproductive process, for example fungal spores. Spore germination occurs under favourable conditions for vegetative growth. Detergents and solvents break the dormancy of spores and activate germination, as will any physical process that damages the cell wall. Heating the spores to 50-60°C for 10-20 minutes, or subjecting them to cold temperatures will also induce germination.

Spray Drying - The process by which an *emulsion* of a material is sprayed, by one of various methods, into a chamber containing hot air at a temperature of up to 200°C. The rapid drying, and subsequent cooling, produces discreet capsules of product surrounded by a skin of maltodextrin or gum arabic, which extends the shelf life and enables the product to be used as a dry powder. Often used for flavours but can also be of use for cosmetic materials and fragrances.

Standard - In the context of quality management systems, a practical, documented description of the key inputs necessary for the production of a key output.

Standard Deviation (σ) - In statistics, the square root of the arithmetic mean of the squares of the deviations of the various items from the arithmetic mean of the whole. Also known as the root-mean-square error.

Sterilant - Any chemical agent which, under carefully controlled conditions, can sterilise objects, materials and environments. A sterilant may fail to sterilise if operational conditions such as concentration and temperature are not carefully controlled. Knowledge of the contaminant is also an important factor in selecting the correct sterilant.

Sterilisation - Any process by which objects, materials and environments may be rendered free from viable cells, spores and viruses. Sterilisation techniques include physical methods such as dry heat, moist heat, radiation, filtration and ultrasonics, or chemical methods such as the use of disinfectants and gaseous sterilants such as ethylene oxide. Some sterilisation procedures involve a combination of both physical and chemical techniques.

Strain - The deformation resulting from a *stress*, measured by the ratio of the change to the total value of the dimension in which the change occurred.

Stress - The force producing deformation in a body, measured by the force applied per unit area.

Sublimation - The direct transition of a solid substance to a gas, without passing through the liquid state.

Surfactant - Commonly used term for surface-active agent. A substance that, when present at low concentrations in a system, exhibits the property of adsorbing onto the surfaces or interfaces of the system and of altering, to a marked degree, the surface or interfacial free energies of those surfaces.

T

Taxonomy - The classification of micro-organisms into class, order, family, genus, species and strains. Colonial appearance on agar plates and examination of stains microscopically followed by biochemical tests, help the microbiologist to list and identify micro-organisms. Recent developments in biochemical test kits and microbiological profile software have accelerated the conventionally tedious task of microbiological taxonomy.

TEWL - Trans-epidermal Water Loss. A term used to describe the rate of loss of moisture from the skin, through the *epidermis*.

Thixotrope - A colloidal gel which will liquefy when pressure or physical shear is applied and then return to its original form when the stimulus is removed.

Threshold - In the context of perfumery, the concentration level at which it is just possible to perceive an odorous material.

Total Viable Count - Often designated as the TVC and defined as the total number of bacteria, fungi and yeasts detected in one gram or one millilitre of a raw material, or finished product. The results are expressed as a number of colony forming units (CFU) per gram or millilitre. Laboratory test methods of detection should be validated and adjusted to suit the nature and the type of material being tested. Some of the frequently used methods for determining the TVC in the cosmetics industry are the pour plate method, membrane filtration, the spread plate method and the most probable number (MPN) technique.

U

UV Radiation - Ultraviolet radiation. A part of the electromagnetic spectrum, occupying the wavelength values between 100nm and 400nm. Classically, ultraviolet radiation is divided into three types, UV-A (400nm - 320nm), UV-B (320nm - 280nm) and

UV-C (280nm - 100nm). UV-C does not reach the Earth's surface as it is filtered out by the upper atmosphere.

Unsaponifiable Matter - The substances that are found in oils and *fats* that do not react with alkali. Usually complex organic compounds soluble in the oil.

V

Vapour Pressure - The pressure exerted by a solid or liquid, in equilibrium with its own vapour. The vapour pressure of any substance will be a function of temperature.

Variables - In the context of quality management systems, these are continuous properties of a *process* or substance for example weight, height, size, colour etc.

Viscosity - Internal resistance to flow, or a degree of thickness of fluids and solutions, caused by intermolecular attraction and often related to the shape and size of the molecule. Viscosity often decreases with an increase in temperature.

Vitamins - A group of organic substances, occurring in various foods, that are essential for a normal healthy diet. Absence or shortage of vitamins frequently leads to deficiency diseases. The name vitamin is derived from the phrase "vital amines", which originated during the early days of research into their activities and properties. Before very much was known about vitamins, they were assigned letters as identification aids. Cosmetic claims are often made for the inclusion of vitamins in formulations. Vitamins can be divided into fat soluble and water soluble types. The fat soluble vitamins, principally vitamin A, vitamin E, vitamin D and vitamin K, can be stored in the body, whereas the water soluble vitamins, particularly vitamin B and vitamin C, need to included in the everyday diet. Water soluble vitamins are understood to act as coenzymes. The specific identification of the various vitamins is given below.

Fat Soluble

Vitamin A_4	Retinol, a terpinoid. It occurs in milk, butter, green vegetables and liver but can be synthesised in the body from carotene. Deficiency causes night blindness.
Vitamin B_4	Once thought to be a single agent but now known to be a complex of related materials. Occurs in wheat germ and yeast. Deficiency causes beriberi, which is rare in man.
Vitamin B_1	Thiamin
Vitamin B_2	Riboflavin
Vitamin B_6	Pyridoxine
Vitamin B_{12}	Cyanocobalamine
Vitamin B_c	Folic acid

Other members of the group include nicotinic acid, inositol, pantothenic acid, choline and biotin (also known as vitamin H).

Water Soluble

Vitamin C	Ascorbic acid. Occurs in fruit and vegetables and deficiency causes scurvy.
Vitamin D_4	Consists of several sterols, the most important of which is calciferol,

which is converted to the active vitamin by the action of *ultra-violet radiation* on the skin. The other principle source is fish liver oils. Deficiency leads to rickets in children.

Vitamin E — Tocopherol. Often used for its antioxidant and radical-scavenging properties. Very common in various vegetable oils, where it acts as a natural antioxidant preventing rancidity. Deficiency results in widespread effects in animals but is rare in man, although it is implicated in disorders of lipid metabolism and thought to play an important role in cell membrane stability.

Vitamin F_4 — Linoleic acid. No longer considered as a true vitamin.

Vitamin G_4 — Riboflavin.

Vitamin H_4 — Biotin. Occurs in chocolate, peanuts, eggs, beef, liver and yeast, as well as being produced by the intestinal fauna. Excretion by man usually exceeds the intake.

Vitamin K_4 — Consists of various napthoquinone compounds whose deficiency cause haemorrhaging. Deficiency is rare, as some is produced by the intestinal fauna.

Vitamin M — Folic acid.

W

Winterisation - The winterisation of an oil involves the removal of naturally occurring high melting-point solids. This is accomplished by holding the oil at a reduced temperature, whilst crystallisation occurs. The solids are then filtered out. Temperatures vary and should be stated on the product specification.

Z

Zeta Potential - The potential of a charged surface, at the plane of shear between the charged particle and the surrounding solution, as they are moved with respect to each other.

SECTION 11

Reference Section

BUFFER SYSTEMS
HLB VALUES
TABLES OF ATOMIC WEIGHTS
PHYSICAL CONSTANTS
ODOUR DESCRIPTIONS OF AN
ESSENTIAL OIL KEY COMPONENT
GREEK ALPHABET
STANDARD SIEVE SIZES
FRACTIONS & MULTIPLES
TABLE OF EQUIVALENT UNITS

REFERENCE SECTION

BUFFER SYSTEMS

CITRATE BUFFER

pH	*Citric Acid Monohydrate* g/l	*Sodium Citrate Dihydrate* g/l
2.5	64.4	7.8
3.0	57.4	17.6
3.5	47.6	31.4
4.0	40.6	41.2
4.5	30.8	54.9
5.0	19.6	70.6
5.5	9.8	84.3
6.0	4.2	92.1
6.5	1.8	95.6

SODIUM PHOSPHATE BUFFER

pH	*Sodium Phosphate* $Na_2HPO_4.7H_2O$	*Sodium Hydrogen Phosphate* $NaH_2PO_4.H_2O$
4.5	0.9	45.5
5.0	2.2	44.8
5.5	4.4	43.7
6.0	17.8	36.8
6.5	37.4	26.7
7.0	57.8	16.1
7.5	74.8	7.4
8.0	83.7	2.8
8.5	87.2	0.9

CITRATE/PHOSPHATE MIXED BUFFER

pH	*Citric Acid Monohydrate g/l*	*Sodium Phosphate Na_2HPO_4 g/l*
2.2	20.58	00.57
2.4	19.70	01.76
2.6	18.71	03.10
2.8	17.67	04.50
3.0	16.68	05.84
3.2	15.81	07.01
3.4	15.02	08.09
3.6	14.24	09.14
3.8	13.55	10.09
4.0	12.90	10.95
4.2	12.31	11.76
4.4	11.74	12.52
4.6	11.18	13.28
4.8	10.65	14.00
5.0	10.19	14.63
5.2	09.74	15.22
5.4	09.29	15.83
5.6	08.82	16.47
5.8	08.31	17.17
6.0	07.74	17.93
6.2	07.12	18.77
6.4	06.46	19.67
6.6	05.72	20.66
6.8	04.78	21.94
7.0	03.71	23.39
7.2	02.74	24.69
7.4	01.92	25.80
7.6	01.33	26.60
7.8	00.89	27.19
8.0	00.58	27.62

HLB VALUES

HLB NUMBERS FOR SURFACTANTS

Chemical designation and CAS registry number	*Type**	*HLB number*
Oleic acid (112-80-1)	N	1.0
Lanolin alcohols (61788-49-6)	N	1.0
Acetylated sucrose diester	N	1.0
Ethylene glycol distearate (627-83-8)	N	1.3
Acetylated monoglycerides	N	1.5
Sorbitan trioleate (26266-58-6)	N	1.8
Glycerol dioleate (25637-84-7)	N	1.8
Sorbitan tristearate (26658-19-5)	N	2.1
Ethylene glycol monostearate (111-60-4)	N	2.9
Sucrose distearate (27195-16-0)	N	3.0
Decaglycerol decaoleate (11094-60-3)	N	3.0
Propylene glycol monostearate (1323-39-3)	N	3.4
Glycerol Monoleate (25496-72-4)	N	3.4
Diglycerine sesquioleate	N	3.5
Sorbitan sesquioleate (8007-43-0)	N	3.7
Glycerol monostearate (31566-31-1)	N	3.8
Acetylated monoglycerides (stearate)	N	3.8
Decaglycerol octaoleate (66734-10-9)	N	4.0
Diethylene glycol monostearate (106-11-6)	N	4.3
Sorbitan monooleate (1333-68-2)	N	4.3
Propylene glycol monolaurate (10108-22-2)	N	4.5
High-molecular-weight fatty amine blend	C	4.5
POE (1.5) nonyl phenol (9016-45-9)	N	4.6
Sorbitan monostearate (1338-41-6)	N	4.7
POE (2) oleyl alcohol (25190-05-0)	N	4.9
POE (2) stearyl alcohol (9005-00-9)	N	4.9
POE sorbitol beeswax derivative	N	5.0
PEG 200 distearate (9005-08-7)	N	5.0
Calcium stearoxyl-2-lactylate (5793-94-2)	A	5.1
Glycerol monolaurate (27215-38-9)	N	5.2
POE (2) octyl alcohol (27252-75-1)	N	5.3
Sodium-O-stearyllactate (18200-72-1)	A	5.7
Decaglycerol tetraoleate	N	6.0
PEG 300 dilaurate (9005-02-1)	N	6.3
Sorbitan monopalmitate (26266-57-9)	N	6.7
N,N,-Dimethylstearamide (3886-90-6)	N	7.0
PEG 400 distearate (900508-7)	N	7.2
High-molecular-weight amine blend	C	7.5
POE (5) lanolin alcohol (61790-91-8)	N	7.7

HLB Numbers for Surfactants – Continued

Polyethylene glycol ether of linear alcohol	N	7.7
POE octylphenol (9002-93-1)	N	7.8
Soya lecithin (8020-84-6)	N	8.0
Diacetylated tartaric acid esters of monoglycerides	N	8.0
POE (4) stearic acid (monoester) (9004-99-3)	N	8.0
Sodium Stearoyllactylate (18200-72-1)	A	8.3
Sorbitan monolaurate (1338-43-8)	N	8.6
POE (4) nonylphenol (9016-45-9)	N	8.9
Calcium dodecyl benzene sulfonate (26264-06-2)	A	9.0
Isopropyl ester of lanolin fatty acids	N	9.0
POE (4) tridecyl alcohol (24938-91-8)	N	9.3
POE (4) lauryl alcohol (9002-92-0)	N	9.5
POP/POE condensate	N	9.5
POE (5) sorbitan monooleate (9005-65-6)	N	10.0
POE (40) sorbitol hexaoleate (9011-29-4)	N	10.2
PEG 400 dilaurate (9005-02-1)	N	10.4
POE (5) nonylphenol (9016-45-9)	N	10.5
POE (20) sorbitan tristearate (9005-71-4)	N	10.5
POP/POE condensate (9003-11-6)	N	10.6
POE (6) nonylphenol (9016-45-9)	N	10.9
Glycerol monostearate-self emulsifying (31566-31-1)	A	11.0
POE (20) lanolin (ether and ester)	N	11.0
POE (20) sorbitan trioleate (9005-70-3)	N	11.0
POE (8) stearic acid (monoester) (9004-99-3)	N	11.1
POE (50) sorbitol hexaoleate (9011-29-4)	N	11.4
POE (6) tridecyl alcohol (24938-91-8)	N	11.4
PEG 400 monostearate (9004-99-3)	N	11.7
Alkyl aryl sulfonate	A	11.7
Triethanolamine oleate soap (2717-15-9)	A	12.0
POE (8) nonylphenol (9016-45-9)	N	12.3
POE (10) stearyl alcohol (9005-00-9)	N	12.4
POE (8) tridecyl alcohol (24938-91-8)	N	12.7
POP/POE condensate	N	12.7
POE (8) lauric acid (monoester) (9004-81-3)	N	12.8
POE (10) cetyl alcohol (9004-95-9)	N	12.9
Acetylated POE (10) lanolin	N	13.0
POE (20) glycerol monostearate (53195-79-2)	N	13.1
PEG 400 monolaurate (9004-81-3)	N	13.1
POE (16) lanolin alcohol (61790-81-6)	N	13.2
POE (4) sorbitan monolaurate (9005-64-5)	N	13.3
POE (10) nonylphenol (9016-45-9)	N	13.3
POE (15) tall oil fatty acids (ester)	N	13.4
POE (10) octylphenol (9002-93-1)	N	13.6
PEG 600 monostearate (004-99-3)	N	13.6

HLB Numbers for Surfactants – Continued

POP/POE condensate	N	13.8
Tertiary amines: POE fatty amines	C	13.9
POE (24) cholesterol (27321-96-6)	N	14.0
POE (14) nonylphenol (9016-45-9)	N	14.4
POE (12) lauryl alcohol (9002-92-0)	N	14.5
POE (20) sorbitan monostearate (9005-67-8)	N	14.9
Sucrose monolaurate (25339-99-5)	N	15.0
POE (20) sorbitan monooleate (9005-65-6)	N	15.0
POE (16) lanolin alcohols (8051-96-5)	N	15.0
Acetylated POE (9) lanolin (68784-35-0)	N	15.0
POE (20) stearyl alcohol (9005-00-9)	N	15.3
POE (20) oleyl alcohol (25190-05-0)	N	15.3
PEG 1000 monooleate (9004-96-0)	N	15.4
POE (20) tallow amine (61790-82-7)	C	15.5
POE (20) sorbitan monopalmitate (9005-66-7)	N	15.6
POE (20) cetyl alcohol (9004-95-9)	N	15.7
POE (25) propylene glycol monostearate (37231-60-0)	N	16.0
POE (20) nonylphenol (9016-45-9)	N	16.0
PEG 1000 monolaurate (9004-81-3)	N	16.5
POP/POE condensate	N	16.8
POE (20) sorbitan monolaurate (9005-64-5)	N	16.9
POE (23) lauryl alcohol (9002-92-0)	N	16.9
POE (40) stearic acid (monester) (9004-99-3)	N	16.9
POE (50) lanolin (ether and ester) (61790-81-6)	N	17.0
POE (25) soyasterol (68648-64-6)	N	17.0
POE (30) nonylphenol (9016-45-9)	N	17.1
PEG 4000 distearate (9005-08-7)	N	17.3
POE (50) stearic acid (monoester) (9004-99-3)	N	17.9
Sodium Oleate (143-91-1)	N	18.0
POE (70) dinonylphenol (9014-93-1)	N	18.0
POE (20) castor oil (ether, ester) (61791-12-6)	N	18.1
POP/POE condensate	N	18.7
Potassium oleate (143-18-0)	A	20.0
N-cetyl-*N*-ethyl morpholinium ethyl sulfate (35%) (78-21-7)	C	30.0
Ammonium lauryl sulfate (2235-54-3)	A	31.0
Triethanolamine lauryl sulfate (139-96-8)	A	34.0
Sodium alkyl sulfate	A	40.0

POE = polyoxyethylene (25609-81-8)
PEG = poly(ethylene glycol) (25322-68-3)
POP = polyoxypropylene (34465-52-6)

*A = Anionic
*C = Cationic
*N = Nonionic

REQUIRED HLB NUMBERS FOR EMULSIFICATION OF OILS AND WAXES

Compound	*CAS registry*	*HLB Number*
O/W Emulsion		
Acetophenone	98-86-2	14
Beeswax	8012-89-3	9
Benzene	71-43-2	15
Benzonitrile	100-47-0	14
Bromobenzene	108-86-1	13
Butyl stearate	123-95-5	11
Carbon tetrachloride	56-23-5	16
Carnauba wax	8015-86-9	15
Castor oil	8001-79-4	14
Ceresine wax		8
Cetyl alcohol	36653-82-4	16
Chlorinated paraffin	8029-39-8	12-14
Chlorobenzene	108-90-7	13
Cocoa butter		6
Corn oil	8001-30-7	8
Cottonseed oil	8001-29-4	6
Cyclohexane	110-82-7	15
Decahydronaphthalene	91-17-8	15
Decyl acetate	112-30-1	11
Decyl alcohol	25339-17-7	15
Diethylaniline	91-66-7	14
Diisooctyl phthalate	27554-26-3	13
Diisopropylbenzene	25321-09-9	15
Dimer acid	61788-89-4	14
Dimethyl silicone	9016-00-6	9
Ethylaniline	103-69-5	13
Ethyl benzoate	93-89-0	13
Fenchone	1196-79-5	12
Hexadecyl alcohol	36653-82-4	11-12
Isodecyl alcohol	25339-17-7	14
Isopropyl myristate	110-27-0	12
Isopropyl palmitate	142-91-6	12
Isostearic acid	2724-58-5	15-16
Kerosene	8008-20-6	12
Lanolin, anhydrous	8006-54-1	12
Lard	61789-99-9	5
Lauramine	124-22-1	12

Lauric acid	143-07-7	16
Lauryl alcohol	112-53-8	14
Linoleic acid	60-33-3	16
Menhaden oil	8002-50-4	12
Methyl phenyl silicone	42557-10-8	7
Methyl silicone	9076-37-3	11
Mineral oil, aromatic	8012-95-1	12
Mineral oil, paraffinic	8012-95-1	10
Mineral spirits	8030-30-6	14
Mink oil	8023-74-3	9
Nitrobenzene	98-53-3	13
Nonylphenol	25154-52-3	14
o-Dichlorobenzene	95-50-1	13
Oleic acid	112-80-11	17
Oleyl alcohol	143-28-2	14
Palm oil		7
Paraffin wax	8002-74-2	10
Petrolatum	8009-03-8	7-8
Petroleum naphtha	8030-30-6	14
Pine oil	8002-09-3	16
Polyethylene wax	9002-88-4	15
PPG-15 Stearyl Ether	25231-24-4	7
Propylene tetramer	9003-97-0	14
Rapeseed oil	8002-13-9	7
Ricinoleic acid	141-22-0	16
Safflower		7
Soybean oil		6
Stearyl alcohol	112-92-5	15-16
Styrene	100-42-5	15
Tallow	61789-97-7	6
Toluene	108-88-3	15
Tridecyl alcohol	112-70-9	14
Trichlorotrifluoroethane	76-13-1	14
Tricresyl phosphate	1330-78-1	17
Xylene	1330-20-7	14
W/O Emulsion		
Gasoline		7
Kerosene	8008-20-6	6
Mineral oil		6
Stearyl alcohol	112-92-5	7

TABLE OF ATOMIC WEIGHTS

[A.W. values in brackets denote mass number of the most stable known isotope]

Element	*Symbol*	*At. No.*	*A.W.*
Actinium	Ac	89	[227.0278]
Aluminium	Al	13	26.9815
Americium	Am	95	[243.0614]
Antimony	Sb	51	121.75
Argon	Ar	18	39.948
Arsenic	As	33	74.9216
Astatine	At	85	[209.9871]
Barium	Ba	56	137.327
Berkelium	Bk	97	[247.0703]
Beryllium	Be	4	9.0122
Bismuth	Bi	83	208.9804
Boron	B	5	10.811
Bromine	Br	35	79.904
Cadmium	Cd	48	112.411
Caesium	Cs	55	132.9054
Calcium	Ca	20	40.078
Californium	Cf	98	[251.0796]
Carbon	C	6	12.011
Cerium	Ce	58	140.115
Chlorine	Cl	17	35.4527
Chromium	Cr	24	51.9961
Cobalt	Co	27	58.9332
Copper	Cu	29	63.546
Curium	Cm	96	[247.0703]
Dysprosium	Dy	66	162.50
Einsteinium	Es	99	[252.083]
Erbium	Er	68	167.26
Europium	Eu	63	151.965
Fermium	Fm	100	[257.0591]
Florine	F	9	18.9984
Francium	Fr	87	[223.0197]
Gadolinium	Gd	64	157.25
Gallium	Ga	31	69.723
Germanium	Ge	32	72.61
Gold	Au	79	196.9665
Hafnium	Hf	72	178.49
Helium	He	2	4.0026
Holmium	Ho	67	164.9303

Table of Atomic Weights – continued

Hydrogen	H	1	1.0079
Indium	In	49	114.82
Iodine	I	53	126.9044
Iridium	Ir	77	192.22
Iron	Fe	26	55.847
Krypton	Kr	36	83.80
Lanthanum	La	57	138.9055
Lawrencium	Lr	103	[262.11]
Lead	Pb	82	207.20
Lithium	Li	3	6.941
Lutetium	Lu	71	174.967
Magnesium	Mg	12	24.305
Manganese	Mn	25	54.9381
Mendelevium	Md	101	[258.10]
Mercury	Hg	80	200.59
Molybdenum	Mo	42	95.94
Neodymium	Nd	60	144.24
Neon	Ne	10	20.1797
Neptunium	Np	93	[237.0482]
Nickel	Ni	28	58.69
Niobium	Nb	41	92.9064
Nitrogen	N	7	14.0067
Nobelium	No	102	[259.1009]
Osmium	Os	76	190.2
Osygen	O	8	15.9994
Palladium	Pd	46	106.42
Phosphorus	P	15	30.9738
Platinum	Pt	78	195.08
Plutonium	Pu	94	[244.0642]
Polonium	Po	84	[208.9824]
Potassium	K	19	39.0983
Praseodymium	Pr	59	140.9077
Promethium	Pm	61	[144.9127]
Protactinium	Pa	91	[231.0359]
Radium	Ra	88	[226.0254]
Radon	Rn	86	[222.0176]
Rhenium	Re	75	186.207
Rhodium	Ph	45	102.9055
Rubidium	Rb	37	85.4678
Ruthenium	Ru	44	101.07
Samarium	Sm	62	150.36
Scandium	Sc	21	44.9559

Table of Atomic Weights – continued

Selenium	Se	34	78.96
Silicon	Si	14	28.0855
Silver	Ag	47	107.8682
Sodium	Na	11	22.9898
Strontium	Sr	38	87.62
Sulphur	S	16	32.066
Tantalum	Ta	73	180.9479
Technetium	Tc	43	[97.9072]
Tellurium	Te	52	127.60
Terbium	Tb	65	158.9253
Thallium	Tl	81	204.3833
Thorium	Th	90	232.0381
Thulium	Tm	69	168.9342
Tin	Sn	50	118.710
Titanium	Ti	22	47.88
Tungsten	W	74	183.85
Unnilquadium	Unq	104	[261.11]
Unnilpentium	Unp	105	[262.114]
Unnilhexium	Unh	106	[263.118]
Unnilseptium	Uns	107	[262.12]
Uranium	U	92	238.0289
Vanadium	V	23	50.9415
Xenon	Xe	54	131.29
Ytterbium	Yb	70	173.04
Yttrium	Y	39	88.9059
Zinc	Zn	30	65.39
Zirconium	Zr	40	91.224

ODOUR DESCRIPTIONS OF ESSENTIAL OIL KEY COMPONENTS

Component	*Odour description*
Aldehyde C9	Very powerful, floral, waxy
Aldehyde C10	Very powerful, orange peel odour
Alpha-guaiene	Mild, woody, peppery
Alpha-pinene	See Pinenes
Alpha-terpinene	Fresh, citrus type
Anethole	Sweet, aniseed
Apiol	Warm, herbaceous, Parsley-like
Benzaldehyde	Strong, bitter-Almond
Benzoic acid	Faint, balsamic
Benzyl acetate	Floral (Jasmin), fruity
Benzyl propionate	Fruity, floral (Jasmin)
Beta-pinene	See Pinenes
Bisabolene	Sweet, balsamic, spicy
Bornyl acetate	Camphoraceous, Pine-like
Cadinene	Herbal, woody, medicated
Camphene	Camphoraceous
Camphor	Fresh, warm, minty
Carvacrol	Tar-like, herbal, spicy
Caryophyllene	Woody, somewhat spicy, dry
Carvone	Herbal, spicy, floral
Cedrene	Woody (Cedar), camphoraceous
Cedrenol	Mild, balsamic, woody
Cedrol	Faint, woody (Cedar type)
Chamazulene	Odourless, dark-blue constituent of Chamomile oils
Chavicol	Powerful, tar-like, "disinfectant"-like
Cineole	Fresh, camphoraceous, Eucalyptus-like
Cinnamic aldehyde	Sweet, balsamic, Cinnamon-like
Cis-3-hexenyl acetate	Very powerful, grassy-green, fruity
Cis-jasmone	Powerful, spicy (Celery-like)
Citral	Powerful, citrus (Lemon)
Citronellal	Fresh, citrus, somewhat Rose-like
Citronellol	Fresh, rich, floral (Rose-like)
Delta-3-carene	Sweet, rather harsh, citrus (Lemon) type
Dihydropyrocurzerenone	Rich, sweet, resinous and incense-like
Esters of angelic acid	Generally herbal and fruity
Esters of benzoic acid	Generally warm, balsamic, medicated
Esters of cinnamic acid	Generally warm, balsamic, fruity

Odour descriptions of an essential oil key component – continued

Esters of citronellol	Generally Rose-like and fruity
Esters of geraniol	Generally Rose-like and fruity
Esters of linalool	Variations on the Lavender odour theme
Esters of nerol	Generally Rose-like and fruity
Esters of tiglic acid	Generally herbal and fruity
Estragole	Sweet, herbaceous and anisic
Eugenol	Warm, spicy (Clove-like)
Eugenyl acetate	Sweet, balsamic, spicy and fruity
Farnesol	Sweet, delicate, floral and green
Fenchone	Sweet, warm, camphoraceous
Gemma-terpinene	Warm, herbaceous, citrus
Geranoil	Sweet, Rose-like
Geranyl acetate	Sweet, floral (Rose), somewhat fruity
Indole	Naphthenic ("moth balls"). Faecal at 10% and Jasmin-like at 1% or less in an odourless solvent
Irones	These substances possess sweet, soft, violet-like odours which are very fatiguing to the nose
Isomenthone	Powerful, minty
Isopinocamphone	Camphoraceous, herbal, somewhat minty
Limonene	Fresh, weak citrus if pure
Linalool	Light, floral, woody, slightly spicy
Linalyl acetate	Fresh, light, herbal, somewhat fruity
Lindestrene	Rich, leathery, balsamic, incense-like
Longifolene	Warm, spicy, peppery
Menthol	Fresh, cooling, minty
Menthone	Fresh, minty, slightly woody
Menthofuran	Sharp, herbal, minty
Menthyl acetate	Mild, sweet, floral, minty
Methyl amyl ketone	Fruity, minty
Methyl anthranilate	Floral (Orange blossom), fruity, dry
Methyl benzoate	Sweet, floral, somewhat medicated
Methyl chavicol	(= Estragole, see above)
Methyl eugenol	Floral (Daffodil-like)
Methyl heptenone	Fruity-green, oily
Methyl jasmonate	Sweet, floral, herbaceous
Methyl para-cresol	Pungent, sweet, Wallflower-like on high dilution

Odour descriptions of an essential oil key component – continued

Methyl salicylate	Strong, medicated, fruity
Myrcene	Sweet, light, balsamic
Nerol	Sweet, floral, faintly seaweed-like
Nerolidol	Very delicate floral, woody and green notes
Neryl acetate	Sweet, floral (Rose-like) and fruity
Norpatchoulenol	Herbal, balsamic, Patchouli-like
Ocimene	Light, warm, herbaceous
Para-cymene	Fresh, citrusy, somewhat herbal
Patchoulol	Odourless major constituent of Patchouli Oil
Phellandrene	Fresh, citrus, woody, spicy
Phenylethyl alcohol	Mild, floral (Rose-like, Hyacinth-like)
Pinenes	Resinous, woody (Pine-like)
Pinocamphone	Camphoraceous, herbal
Pulegone	Herbal, resinous, minty
Pyrazines	Extremely powerful and penetrating, sharp green notes related to the odour of green peppers (in which pyrazines occur)
Sabinene	Warm, spicy, woody and herbaceous
Safrole	Sweet, warm, spicy-floral and woody
Santalenes	Mild and woody
Santalols	Fine, mild woody notes; very long lasting
Sclareol	Very delicate, Ambergris-like
Tagetone	Warm, herbal
Terpineol	Of the three terpineols, alpha-terpineol is the most common as a constituent of essential oils. It possesses a floral (Lilac-like) odour
Terpinenol-4	Mild, peppery
Terpinyl acetate	Fresh, herbal, somewhat fruity, slightly oily
Thujone	Powerful, herbaceous, camphoraceous, minty

THE GREEK ALPHABET

Letters		*Name*
Α	α	alpha
Β	β	beta
Γ	γ	gamma
Δ	δ	delta
Ε	ε	epsilon
Ζ	ζ	zeta
Η	η	eta
Θ	θ	theta
Ι	ι	iota
Κ	κ	kappa
Λ	λ	lambda
Μ	μ	mu
Ν	ν	nu
Ξ	ξ	xi
Ο	ο	omikron
Π	π	pi
Ρ	ρ	ro
Σ	σ	sigma
Τ	τ	tau
Υ	υ	upsilon
Φ	φ	phi
Χ	χ	khi
Ψ	ψ	psi
Ω	ω	omega

STANDARD SIEVE SIZE

BRITISH FINE MESH

(B.S.S. 410—193)

Sieve No.	*Nominal Aperture*		*Approx. Screen-ing Area %*
	in.	*mm.*	
300	.0021	0.053	41
240	.0026	0.066	38
200	.0030	0.076	36
170	.0035	0.089	35
150	.0041	0.104	37
120	.0049	0.124	35
100	.0060	0.152	36
85	.0070	0.178	35
72	.0083	0.211	36
60	.0099	0.251	35
52	.0116	0.295	37
44	.0139	0.353	38
36	.0166	0.422	36
30	.0197	0.500	35
25	.0236	0.599	35
22	.0275	0.699	36
18	.0336	0.853	36
16	.0395	1.003	40
14	.0474	1.204	44
12	.0553	1.405	44
10	.0660	1.676	44
8	.0810	2.057	42
7	.0949	2.411	44
6	.1107	2.812	44
5	.1320	3.353	44

I.M.M.

Sieve No.	*Nominal Aperture*		*Approx. Screen-ing Area %*
	in.	*mm.*	
200	.0025	0.063	25.0
150	.0033	0.084	24.5
120	.0042	0.107	25.4
100	.0050	0.127	25.0
90	.0055	0.139	24.5
80	.0062	0.157	24.6
70	.0071	0.180	24.7
60	.0083	0.211	24.8
50	.0100	0.254	25.0
40	.0125	0.347	25.0
30	.0166	0.421	24.8
20	.0250	0.635	25.0
16	.0312	0.792	24.9
12	.0416	1.056	24.9
10	.0500	1.270	25.0
8	.0620	1.574	24.6
5	.1000	2.540	25.0

TYLER

Sieve No.	Nominal Aperture		Approx. Screening Area
	in.	*mm.*	*%*
325	.0017	0.043	30.1
270	.0021	0.053	32.2
250	.0024	0.061	36.0
200	.0029	0.074	33.6
170	.0035	0.089	35.2
150	.0041	0.104	37.4
115	.0049	0.124	31.7
100	.0058	0.147	33.6
80	.0069	0.175	30.5
65	.0082	0.208	28.3
60	.0097	0.246	33.7
48	.0116	0.295	31.1
42	.0133	0.351	33.6
35	.0164	0.417	32.9
32	.0195	0.495	38.8
28	.0232	0.589	42.2
24	.0276	0.701	43.8
20	.0328	0.833	43.0
16	.0390	0.991	38.9
14	.0460	1.168	42.0
12	.0550	1.397	43.9
10	.0650	1.651	42.2
9	.0780	1.981	49.4
8	.0930	2.362	55.3
7	.1100	2.794	59.3
6	.1310	3.327	61.5
5	.1560	3.962	60.8
4	.1850	4.699	54.8

FRENCH *(A.F.N.O.R.)*

Sieve No.	Nominal Aperture	
	in.	*mm.*
17	.0015	.040
18	.0019	.050
19	.0024	.063
20	.0031	.080
21	.0039	.100
22	.0049	.125
23	.0063	.160
24	.0078	.200
25	.0098	.250
26	.0124	.315
27	.0157	.400
28	.0196	.500
29	.0248	0.63
30	.0315	0.80
31	.0393	1.00
32	.0492	1.25
33	.0630	1.60
34	.0787	2.00
35	.0984	2.50
36	.1240	3.15
37	.1570	4.00
38	.1970	5.00

U.S. *(and A.S.T.M.) sieve series*

Sieve No.	*A.S.T.M. designation microns*	*Nominal Aperture*		*Approx. Screening Area %*
		in.	*mm.*	
325	44	.0017	0.04	30.1
270	53	.0021	0.053	32.2
230	62	.0024	0.062	32.7
200	74	.0029	0.074	33.6
170	88	.0034	0.088	34.0
140	105	.0041	0.105	34.3
120	125	.0049	0.125	34.9
100	149	.0059	0.149	35.5
80	177	.0070	0.177	35.8
70	210	.0083	0.210	36.2
60	250	.0098	0.250	36.6
50	297	.0117	0.297	37.5
45	350	.0138	0.350	37.6
40	420	.0165	0.420	39.4
35	500	.0197	0.500	40.1
30	590	.0232	0.590	41.1
25	710	.0280	0.71	43.2
20	840	.0331	0.84	44.5
18	1000	.0394	1.00	45.6
16	1190	.0469	1.19	47.3
14	1410	.0555	1.41	48.7
12	1680	.0661	1.68	50.2
10	2000	.0787	2.00	52.5
8	2330	.0937	2.38	54.6
7	2830	.1110	2.83	57.0
6	3360	.1320	3.36	58.9
5	4000	.1570	4.00	61.0
4	4760	.1870	4.76	62.3

FRACTIONS AND MULTIPLES

multiple	*prefix*	*symbol*
10^{12}	tera	T
10^{9}	giga	G
10^{6}	mega	M
10^{3}	kilo	k
10^{2}	hecto	h
10	deca	da
10^{-1}	deci	d
10^{-2}	centi	c
10^{-3}	milli	m
10^{-6}	micro	μ
10^{-9}	nano	n
10^{-12}	pico	p
10^{-15}	femto	f
10^{-18}	atto	a

TABLE OF EQUIVALENT UNITS

Physical Quantity	*Unit*	*Equivalent*
length	Angstrom inch foot yard mile	10^{-10}m 0.0254 m 0.3048 m 0.9144 m 1.60934 km
area	square inch square foot square yard square mile	645.16 mm^2 0.092903 m^2 0.836127 m^2 2.58999 km^2
volume	cubic inch cubic foot U.K. gallon	1.63871 x $10^{-5}m^3$ 0.028316 8 m^3 0.004546 092 m^2
mass	pound	0.453592 37 kg
density	pound/cubic inch pound/cubic foot	2.76799 x 10^4kg m^{-3} 16.0185 kg m^{-3}
force	dyne poundal pound-force kilogramme-force	10^{-5} N 0.138255 N 4.44822 N 9.80665 N
pressure	atmosphere torr pound (f)/sq.in.	101.325 kNm^{-2} 133.322 Nm^{-2} 6894.76 Nm^{-2}
energy	erg calorie (I.T.) calorie (15°C) carloie (thermo-chemical) B.t.u. foot poundal foot pound (f)	10^{-7} J 4.1868 J 4.1855 J 4.184 J 1055.06 J 0.042 140 J 1.35582 J
temperature	degree Fahrenheit	$t(°F) = [{}^9/{}_5 T(°C)]+32$

INDICATORS — pH VALUES

Indicator	*pH Value*										
	1.0	*2.0*	*3.0*	*4.0*	*5.0*	*6.0*	*7.0*	*8.0*	*9.0*	*10.0*	*11.0*
Thymol Blue	Pink	Orange	Yellow								
Bromophenol Blue			Yellow	Green	Blue-violet						
Methyl Red				Pink	Orange	Yellow					
Bromocresol Purple					Yellow	Greenish	Purple				
Bromothymol Blue						Yellow	Green	Blue			
Phenol Red							Yellow	Orange	Red		
Phenolphthalein								← Colourless →		Pink	
Thymolphthalein									← Colourless →		Blue

Acid ←——— Neutral ———→ Alkaline

The indicators given above are used in 0.04 per cent solutions in water, with the exceptions of Methyl Red, which is 0.02 per cent in 50 per cent alcohol, Phenol Red, which is a 0.02 per cent aqueous solution, and Thymolphthalein, which is a 0.04 per cent solution in alcohol.

Advertisers' Buyer's Guide

Sub-section (a) Classified Index of Advertisers by Product Category
(b) Alphabetical Index of Advertisers with Addresses, Telephone and Facsimile Numbers

Sub-section (a)

CLASSIFIED INDEX OF ADVERTISERS BY PRODUCT CATEGORY

Absolutes

A & E Connock (perfumery & cosmetics) Ltd

Acetic acid/Acetates

A & E Connock (perfumery & cosmetics) Ltd
Rhône-Poulenc Chemicals

Acetone

Ellis & Everard

Alcohols

A & E Connock (perfumery & cosmetics) Ltd
Eggar & Co (Chemicals) Limited

Alginates

A & E Connock (perfumery & cosmetics) Ltd
Eggar & Co (Chemicals) Limited

Alkanolamides

A & E Connock (perfumery & cosmetics) Ltd
Henkel Kgaa, Division COSPHA
Lonza (UK) Ltd
S. Black (Import & Export) Ltd

Alkanolamines

A & E Connock (perfumery & cosmetics) Ltd
Angus Chemie Gmbh
K & K Greeff Ltd
Rhône-Poulenc Chemicals
S. Black (Import & Export) Ltd

Alkylaryl sulphonates

A & E Connock (perfumery & cosmetics) Ltd
Rhône-Poulenc Chemicals

Allantoin

A & E Connock (perfumery & cosmetics) Ltd
Chemie Linz
ISP Europe
K & K Greeff Ltd

Almond oil

A & E Connock (perfumery & cosmetics) Ltd
Alban Muller International
Croda Chemicals Limited
Ellis & Everard
Henry Lamotte - Bremen
S. Black (Import & Export) Ltd

Aloe vera

A & E Connock (perfumery & cosmetics) Ltd
Alban Muller International
Henry Lamotte - Bremen
S. Black (Import & Export) Ltd

Aluminium Chlorhydrate

Ellis & Everard
S. Black (Import & Export) Ltd

Aluminium chlorhydrate propylene glycol

S. Black (Import & Export) Ltd

Amine oxides

A & E Connock (perfumery & cosmetics) Ltd
Croda Chemicals Limited
Goldschmidt AG

Amino acids

A & E Connock (perfumery & cosmetics) Ltd
Croda Chemicals Limited
K & K Greeff Ltd

Amphoteric surfactants

A & E Connock (perfumery & cosmetics) Ltd
Croda Chemicals Limited
Ellis & Everard
Goldschmidt AG
Henkel Kgaa, Division COSPHA
Lonza (UK) Ltd
Rhône-Poulenc Chemicals
S. Black (Import & Export) Ltd

Anionic surfactants

Croda Chemicals Limited
Ellis & Everard
Henkel Kgaa, Division COSPHA
K & K Greeff Ltd
Rhône-Poulenc Chemicals
S. Black (Import & Export) Ltd

Anisoles

James Robinson Ltd

Anticaking agents

K & K Greeff Ltd
S. Black (Import & Export) Ltd

Antidandruff agents

Alban Muller International
Croda Chemicals Limited
K & K Greeff Ltd

Antifoaming agents

A & E Connock (perfumery & cosmetics) Ltd
K & K Greeff Ltd
Rhône-Poulenc Chemicals
S. Black (Import & Export) Ltd
Ellis & Everard

Antimicrobials

A & E Connock (perfumery & cosmetics) Ltd
Angus Chemie Gmbh
Ellis & Everard
ISP Europe
K & K Greeff Ltd
Lonza (UK) Ltd
Rhône-Poulenc Chemicals
S. Black (Import & Export) Ltd
Zeneca Biocides

Antioxidants

A & E Connock (perfumery & cosmetics) Ltd
Henkel Kgaa, Division COSPHA
K & K Greeff Ltd
Rhône-Poulenc Chemicals
S. Black (Import & Export) Ltd

Antistatic agents

A & E Connock (perfumery & cosmetics) Ltd
Goldschmidt AG

K & K Greeff Ltd
Lonza (UK) Ltd
Rhône-Poulenc Chemicals
S. Black (Import & Export) Ltd

Arnica extracts

A & E Connock (perfumery & cosmetics) Ltd
Alban Muller International
K & K Greeff Ltd
S. Black (Import & Export) Ltd

Avocado oil

A & E Connock (perfumery & cosmetics) Ltd
Alban Muller International
Croda Chemicals Limited
Ellis & Everard
Henry Lamotte - Bremen
S. Black (Import & Export) Ltd

Beeswax

A & E Connock (perfumery & cosmetics) Ltd
Alban Muller International
Eggar & Co (Chemicals) Limited
Henry Lamotte - Bremen
K & K Greeff Ltd
S. Black (Import & Export) Ltd

Bentonite

A & E Connock (perfumery & cosmetics) Ltd
S. Black (Import & Export) Ltd

Benzalkonium chloride

A & E Connock (perfumery & cosmetics) Ltd
Lonza (UK) Ltd
Rhône-Poulenc Chemicals

Betaines

A & E Connock (perfumery & cosmetics) Ltd
Croda Chemicals Limited
Goldschmidt AG
Henkel Kgaa, Division COSPHA
Rhône-Poulenc Chemicals
S. Black (Import & Export) Ltd

Binders

A & E Connock (perfumery & cosmetics) Ltd
Croda Chemicals Limited
Roquette
ISP Europe
S. Black (Import & Export) Ltd

Biological extracts

A & E Connock (perfumery & cosmetics) Ltd
Alban Muller International
Eggar & Co (Chemicals) Limited
K & K Greeff Ltd
S. Black (Import & Export) Ltd

Bismuth salts

S. Black (Import & Export) Ltd

Bleaching agents

Borax
Ellis & Everard

Buffers

Angus Chemie Gmbh

Calcium carbonate

Ellis & Everard

Calcium stearate

K & K Greeff Ltd
S. Black (Import & Export) Ltd

Calendula oil/extract

A & E Connock (perfumery & cosmetics) Ltd
Alban Muller International
Croda Chemicals Limited

Henry Lamotte - Bremen
K & K Greeff Ltd
S. Black (Import & Export) Ltd

Camomile extract

A & E Connock (perfumery & cosmetics) Ltd
Alban Muller International
K & K Greeff Ltd
S. Black (Import & Export) Ltd

Carotene

A & E Connock (perfumery & cosmetics) Ltd

Carrot oil

A & E Connock (perfumery & cosmetics) Ltd
Alban Muller International
Croda Chemicals Limited
S. Black (Import & Export) Ltd

Castor oil

A & E Connock (perfumery & cosmetics) Ltd
Alban Muller International
Eggar & Co (Chemicals) Limited

Cationic surfactants

A & E Connock (perfumery & cosmetics) Ltd
Croda Chemicals Limited
Ellis & Everard
Henkel Kgaa, Division COSPHA
K & K Greeff Ltd
Lonza (UK) Ltd
Rhône-Poulenc Chemicals
S. Black (Import & Export) Ltd

Cellulose, modified

A & E Connock (perfumery & cosmetics) Ltd
Croda Chemicals Limited
Ellis & Everard
S. Black (Import & Export) Ltd

Cetyl alcohol

A & E Connock (perfumery & cosmetics) Ltd
Croda Chemicals Limited
Eggar & Co (Chemicals) Limited
Ellis & Everard
Henkel Kgaa, Division COSPHA

Chitin/chitosan

A & E Connock (perfumery & cosmetics) Ltd
Allied Colloids
Ellis & Everard
S. Black (Import & Export) Ltd

Chlorhexidine salts

Rhône-Poulenc Chemicals

Chlorhydroxyaluminium allantoinate

K & K Greeff Ltd

Cholesterol

A & E Connock (perfumery & cosmetics) Ltd
Croda Chemicals Limited

Citric acid/Citrates

Croda Chemicals Limited
Ellis & Everard

Clays

Alban Muller International
Ellis & Everard
K & K Greeff Ltd
S. Black (Import & Export) Ltd

Coconut oil

A & E Connock (perfumery & cosmetics) Ltd
Givaudan-Roure SA
S. Black (Import & Export) Ltd
Alban Muller International

Collagen

A & E Connock (perfumery & cosmetics) Ltd
Croda Chemicals Limited
Henkel Kgaa, Division COSPHA
K & K Greeff Ltd
S. Black (Import & Export) Ltd

Colours

A & E Connock (perfumery & cosmetics) Ltd
Ellis & Everard
James Robinson Ltd
K & K Greeff Ltd
S. Black (Import & Export) Ltd

Concretes

A & E Connock (perfumery & cosmetics) Ltd

Conditioning agents (for skin, hair, etc)

A & E Connock (perfumery & cosmetics) Ltd
Allied Colloids
BASF Aktiengesellschaft
Croda Chemicals Limited
Ellis & Everard
Goldschmidt AG
Henkel Kgaa, Division COSPHA
ISP Europe
Lonza (UK) Ltd
MHP Shellac Gmbh
Rhône-Poulenc Chemicals
Roquette
S. Black (Import & Export) Ltd

Consultancy services

Angus Chemie Gmbh

Corrosion inhibitors

A & E Connock (perfumery & cosmetics) Ltd
Angus Chemie Gmbh
Chemie Linz
James Robinson Ltd
Lonza (UK) Ltd
Rhône-Poulenc Chemicals

Cumin seed oil

A & E Connock (perfumery & cosmetics) Ltd

Cysteine, cystine & derivatives

A & E Connock (perfumery & cosmetics) Ltd
Croda Chemicals Limited
K & K Greeff Ltd

Denaturants

Ellis & Everard

Deodorants

Henkel Kgaa, Division COSPHA

Detergent additives

A & E Connock (perfumery & cosmetics) Ltd
Allied Colloids
Croda Chemicals Limited
Henkel Kgaa, Division COSPHA
Rhône-Poulenc Chemicals
S. Black (Import & Export) Ltd

Dispersants

Angus Chemie Gmbh

Dyhydroxyaluminium allantoinate

K & K Greeff Ltd

Elastin

A & E Connock (perfumery & cosmetics) Ltd
Croda Chemicals Limited
Henkel Kgaa, Division COSPHA
K & K Greeff Ltd
S. Black (Import & Export) Ltd

Emulsifiers/Emulsifying waxes

A & E Connock (perfumery & cosmetics) Ltd
Alban Muller International
Angus Chemie Gmbh
BASF Aktiengesellschaft
Croda Chemicals Limited
Eggar & Co (Chemicals) Limited
Ellis & Everard
Givaudan-Roure SA
Goldschmidt AG
Henkel Kgaa, Division COSPHA
K & K Greeff Ltd
Lonza (UK) Ltd
Rhône-Poulenc Chemicals
S. Black (Import & Export) Ltd

Enzyme inhibitors

S. Black (Import & Export) Ltd

Essential oils

A & E Connock (perfumery & cosmetics) Ltd
Alban Muller International
Givaudan-Roure SA

Esters

A & E Connock (perfumery & cosmetics) Ltd
Chemie Linz
Croda Chemicals Limited
Goldschmidt AG
K & K Greeff Ltd
Lonza (UK) Ltd
Rhône-Poulenc Chemicals
S. Black (Import & Export) Ltd

Ethoxylates

A & E Connock (perfumery & cosmetics) Ltd
Croda Chemicals Limited
Goldschmidt AG
Henkel Kgaa, Division COSPHA
K & K Greeff Ltd
Lonza (UK) Ltd
Rhône-Poulenc Chemicals
S. Black (Import & Export) Ltd

Evening primrose

A & E Connock (perfumery & cosmetics) Ltd
Alban Muller International
Croda Chemicals Limited
Ellis & Everard
Henry Lamotte - Bremen
S. Black (Import & Export) Ltd

Fabric conditioners

A & E Connock (perfumery & cosmetics) Ltd
Goldschmidt AG
Lonza (UK) Ltd
S. Black (Import & Export) Ltd

Fatty amines

A & E Connock (perfumery & cosmetics) Ltd
Croda Chemicals Limited
Lonza (UK) Ltd
S. Black (Import & Export) Ltd

Fatty alcohols & derivatives

A & E Connock (perfumery & cosmetics) Ltd
Croda Chemicals Limited
Eggar & Co (Chemicals) Limited
Ellis & Everard
Henkel Kgaa, Division COSPHA
S. Black (Import & Export) Ltd

Film formers

A & E Connock (perfumery & cosmetics) Ltd
Allied Colloids
BASF Aktiengesellschaft
Croda Chemicals Limited
Ellis & Everard
ISP Europe
K & K Greeff Ltd

MHP Shellac Gmbh
S. Black (Import & Export) Ltd

Fixatives (for perfumes)

A & E Connock (perfumery & cosmetics) Ltd
Ellis & Everard
MHP Shellac Gmbh
Roquette

Flavours

A & E Connock (perfumery & cosmetics) Ltd
Belmay Ltd
Chemie Linz
International Flowers & Fragrances (GB) Ltd

Fluid extracts

A & E Connock (perfumery & cosmetics) Ltd
Alban Muller International
K & K Greeff Ltd

Fluid tinctures

A & E Connock (perfumery & cosmetics) Ltd
Alban Muller International
K & K Greeff Ltd

Fragrance compounds

A & E Connock (perfumery & cosmetics) Ltd
Allied Colloids
Chemie Linz
Eggar & Co (Chemicals) Limited
International Flowers & Fragrances (GB) Ltd

French chalk

Ellis & Everard

Fructose

A & E Connock (perfumery & cosmetics) Ltd

Fruit extracts

A & E Connock (perfumery & cosmetics) Ltd
Alban Muller International
Croda Chemicals Limited
K & K Greeff Ltd
S. Black (Import & Export) Ltd

Fungicides

A & E Connock (perfumery & cosmetics) Ltd
Angus Chemie Gmbh
Chemie Linz
Lonza (UK) Ltd
Rhône-Poulenc Chemicals
S. Black (Import & Export) Ltd

Gamma sterilization equipment

Angus Chemie Gmbh

Gelling agents

A & E Connock (perfumery & cosmetics) Ltd
Alban Muller International
Croda Chemicals Limited
Eggar & Co (Chemicals) Limited
ISP Europe
K & K Greeff Ltd
Rhône-Poulenc Chemicals
Roquette
S. Black (Import & Export) Ltd

Ginseng

A & E Connock (perfumery & cosmetics) Ltd
Alban Muller International
K & K Greeff Ltd
S. Black (Import & Export) Ltd

Glitter

S. Black (Import & Export) Ltd

Glossers

A & E Connock (perfumery & cosmetics) Ltd
Ellis & Everard

Glucose esters/ethoxylates

A & E Connock (perfumery & cosmetics) Ltd
Croda Chemicals Limited
Ellis & Everard
Rhône-Poulenc Chemicals
S. Black (Import & Export) Ltd

Glucose tyrosinate (sun tanning accelerator)

Alban Muller International
S. Black (Import & Export) Ltd

Glycerides

A & E Connock (perfumery & cosmetics) Ltd
Croda Chemicals Limited
Goldschmidt AG
Henkel Kgaa, Division COSPHA
K & K Greeff Ltd
S. Black (Import & Export) Ltd

Glycerin (Glycerol)

A & E Connock (perfumery & cosmetics) Ltd
Croda Chemicals Limited
Ellis & Everard
Henry Lamotte - Bremen
K & K Greeff Ltd

Glycol esters

A & E Connock (perfumery & cosmetics) Ltd
Croda Chemicals Limited
Ellis & Everard
Goldschmidt AG
K & K Greeff Ltd
Lonza (UK) Ltd
Rhône-Poulenc Chemicals
S. Black (Import & Export) Ltd

Glycols

Ellis & Everard
K & K Greeff Ltd

Glycyrrhetinic acid

Alban Muller International
K & K Greeff Ltd

Grapeseed oil

A & E Connock (perfumery & cosmetics) Ltd
Alban Muller International
Croda Chemicals Limited
Ellis & Everard
Henry Lamotte - Bremen
S. Black (Import & Export) Ltd

Guaiazulene

A & E Connock (perfumery & cosmetics) Ltd

Guar hydroxypropyl-triammonium chloride

Henkel Kgaa, Division COSPHA
Rhône-Poulenc Chemicals

Gums

A & E Connock (perfumery & cosmetics) Ltd
Eggar & Co (Chemicals) Limited
Rhône-Poulenc Chemicals
Ellis & Everard

Hair complexes

Alban Muller International
James Robinson Ltd

Hair relaxers

Chemie Linz

Hairspray resins

Chemie Linz

ISP Europe
MHP Shellac Gmbh
S. Black (Import & Export) Ltd

Henna

A & E Connock (perfumery & cosmetics) Ltd
Alban Muller International
James Robinson Ltd

Herbal extracts

A & E Connock (perfumery & cosmetics) Ltd
Alban Muller International
Croda Chemicals Limited
K & K Greeff Ltd
S. Black (Import & Export) Ltd

Honey

A & E Connock (perfumery & cosmetics) Ltd
Alban Muller International
Eggar & Co (Chemicals) Limited

Humectants

A & E Connock (perfumery & cosmetics) Ltd
Croda Chemicals Limited
Ellis & Everard
Goldschmidt AG
K & K Greeff Ltd
Roquette
S. Black (Import & Export) Ltd

Hyaluronic acid & derivatives

A & E Connock (perfumery & cosmetics) Ltd
Ellis & Everard
K & K Greeff Ltd
S. Black (Import & Export) Ltd

Hydrochloric acid

Croda Chemicals Limited
Ellis & Everard

Hydrogen peroxide

Ellis & Everard

Hydrotropes

Lonza (UK) Ltd
Rhône-Poulenc Chemicals

Imidazolines & derivatives

A & E Connock (perfumery & cosmetics) Ltd
Chemie Linz
Croda Chemicals Limited
Ellis & Everard
Henkel Kgaa, Division COSPHA
ISP Europe
K & K Greeff Ltd
Lonza (UK) Ltd
Rhône-Poulenc Chemicals

Insect repellants

A & E Connock (perfumery & cosmetics) Ltd

Intermediates

A & E Connock (perfumery & cosmetics) Ltd
Angus Chemie Gmbh
Chemie Linz
Lonza (UK) Ltd

Iodophors

Rhône-Poulenc Chemicals

Iridescent pigments

S. Black (Import & Export) Ltd
Ellis & Everard

Isopropyl esters

A & E Connock (perfumery & cosmetics) Ltd
Croda Chemicals Limited
Goldschmidt AG
Henkel Kgaa, Division COSPHA
K & K Greeff Ltd

Lonza (UK) Ltd
S. Black (Import & Export) Ltd

Jasmine oil

A & E Connock (perfumery & cosmetics) Ltd

Jojoba oil

A & E Connock (perfumery & cosmetics) Ltd
Alban Muller International
Goldschmidt AG
Henry Lamotte - Bremen
S. Black (Import & Export) Ltd

Kaolin

Ellis & Everard

Keratin

A & E Connock (perfumery & cosmetics) Ltd
Croda Chemicals Limited
K & K Greeff Ltd
S. Black (Import & Export) Ltd

Lactates

A & E Connock (perfumery & cosmetics) Ltd
Croda Chemicals Limited

Lactic acid

A & E Connock (perfumery & cosmetics) Ltd

Lanolin & derivatives

Croda Chemicals Limited
S. Black (Import & Export) Ltd
Henkel Kgaa, Division COSPHA
Ellis & Everard

Lanolin replacements

A & E Connock (perfumery & cosmetics) Ltd
Ellis & Everard
S. Black (Import & Export) Ltd

Lecithin

A & E Connock (perfumery & cosmetics) Ltd
Ellis & Everard
Rhône-Poulenc Chemicals

Liposoluble extracts

A & E Connock (perfumery & cosmetics) Ltd
Alban Muller International
K & K Greeff Ltd
S. Black (Import & Export) Ltd

Liposomes

A & E Connock (perfumery & cosmetics) Ltd
Alban Muller International
Henkel Kgaa, Division COSPHA
K & K Greeff Ltd
Rhône-Poulenc Chemicals

Lubricants

A & E Connock (perfumery & cosmetics) Ltd
Croda Chemicals Limited
Goldschmidt AG

Magnesium carbonates

Ellis & Everard

Magnesium stearates

K & K Greeff Ltd
S. Black (Import & Export) Ltd

Masking agents

A & E Connock (perfumery & cosmetics) Ltd
Roquette

Mercaptans

James Robinson Ltd

Metal powders

S. Black (Import & Export) Ltd

Metal stearates

A & E Connock (perfumery & cosmetics) Ltd
K & K Greeff Ltd
S. Black (Import & Export) Ltd

Methyl ethyl ketone

A & E Connock (perfumery & cosmetics) Ltd

Microcrystalline wax

Ellis & Everard
K & K Greeff Ltd
S. Black (Import & Export) Ltd

Milk extracts

A & E Connock (perfumery & cosmetics) Ltd
Croda Chemicals Limited
Ellis & Everard

Mink Oil

A & E Connock (perfumery & cosmetics) Ltd
Alban Muller International
Croda Chemicals Limited
S. Black (Import & Export) Ltd

Moisturisers

A & E Connock (perfumery & cosmetics) Ltd
Chemie Linz
Croda Chemicals Limited
Goldschmidt AG
K & K Greeff Ltd
S. Black (Import & Export) Ltd

Nail polish ingredients

A & E Connock (perfumery & cosmetics) Ltd
Ellis & Everard
MHP Shellac Gmbh
S. Black (Import & Export) Ltd

Neutralizing agents

Angus Chemie Gmbh

Nitric acid

Chemie Linz

Nonionic emulsifiers

A & E Connock (perfumery & cosmetics) Ltd
Croda Chemicals Limited
K & K Greeff Ltd
Rhône-Poulenc Chemicals
Lonza (UK) Ltd
S. Black (Import & Export) Ltd
Goldschmidt AG
Henkel Kgaa, Division COSPHA
Ellis & Everard

Nonionic surfactants

A & E Connock (perfumery & cosmetics) Ltd
Croda Chemicals Limited
Ellis & Everard
Henkel Kgaa, Division COSPHA
ISP Europe
K & K Greeff Ltd
Lonza (UK) Ltd
Rhône-Poulenc Chemicals
S. Black (Import & Export) Ltd

Nut extracts/oils

A & E Connock (perfumery & cosmetics) Ltd
Alban Muller International
Croda Chemicals Limited
Ellis & Everard
Henry Lamotte - Bremen
K & K Greeff Ltd
S. Black (Import & Export) Ltd

Odour absorbers

Roquette

Oils, non-vegetable

A & E Connock (perfumery & cosmetics) Ltd
Croda Chemicals Limited
Goldschmidt AG
Henkel Kgaa, Division COSPHA
K & K Greeff Ltd
S. Black (Import & Export) Ltd

Oils, vegetable

A & E Connock (perfumery & cosmetics) Ltd
Alban Muller International
Croda Chemicals Limited
Ellis & Everard
Goldschmidt AG
Henkel Kgaa, Division COSPHA
Henry Lamotte - Bremen
K & K Greeff Ltd
S. Black (Import & Export) Ltd

Oleoresins

A & E Connock (perfumery & cosmetics) Ltd
Alban Muller International
Eggar & Co (Chemicals) Limited
Henry Lamotte - Bremen

Opacifiers

A & E Connock (perfumery & cosmetics) Ltd
Croda Chemicals Limited
Ellis & Everard
Goldschmidt AG
ISP Europe
S. Black (Import & Export) Ltd

Orchid oil

A & E Connock (perfumery & cosmetics) Ltd
Alban Muller International
S. Black (Import & Export) Ltd

Panthenol

A & E Connock (perfumery & cosmetics) Ltd

Parabens

Croda Chemicals Limited
ISP Europe
S. Black (Import & Export) Ltd

Paraffin oils/waxes

Eggar & Co (Chemicals) Limited
K & K Greeff Ltd
S. Black (Import & Export) Ltd

Pearlescing agents

A & E Connock (perfumery & cosmetics) Ltd
Croda Chemicals Limited
Ellis & Everard
Goldschmidt AG
Henkel Kgaa, Division COSPHA
ISP Europe
K & K Greeff Ltd
S. Black (Import & Export) Ltd

Pearl pigments (natural & synthetic)

A & E Connock (perfumery & cosmetics) Ltd
Ellis & Everard
ISP Europe
S. Black (Import & Export) Ltd

Perfumery synthetics

A & E Connock (perfumery & cosmetics) Ltd
Givaudan-Roure SA
International Flowers & Fragrances (GB) Ltd

Petrolatum

Eggar & Co (Chemicals) Limited

Ellis & Everard
K & K Greeff Ltd

Placenta (non-human)

A & E Connock (perfumery & cosmetics) Ltd
S. Black (Import & Export) Ltd

Plant extracts

A & E Connock (perfumery & cosmetics) Ltd
Alban Muller International
Croda Chemicals Limited
Ellis & Everard
K & K Greeff Ltd
S. Black (Import & Export) Ltd

Plasticisers

S. Black (Import & Export) Ltd

Pollen

A & E Connock (perfumery & cosmetics) Ltd
Alban Muller International

Polyethylene glycols

BASF Aktiengesellschaft
Ellis & Everard
K & K Greeff Ltd
Lonza (UK) Ltd

Polymers

A & E Connock (perfumery & cosmetics) Ltd
Allied Colloids
BASF Aktiengesellschaft
Ellis & Everard
ISP Europe
K & K Greeff Ltd
S. Black (Import & Export) Ltd

Polyoxyethylene esters/ethers

A & E Connock (perfumery & cosmetics) Ltd
Croda Chemicals Limited
Goldschmidt AG
K & K Greeff Ltd
Lonza (UK) Ltd
Rhône-Poulenc Chemicals
S. Black (Import & Export) Ltd

Polysorbates

A & E Connock (perfumery & cosmetics) Ltd
Croda Chemicals Limited
Henkel Kgaa, Division COSPHA
Lonza (UK) Ltd
S. Black (Import & Export) Ltd

Preservatives

A & E Connock (perfumery & cosmetics) Ltd
Angus Chemie Gmbh
Chemie Linz
Croda Chemicals Limited
Ellis & Everard
Henkel Kgaa, Division COSPHA
ISP Europe
K & K Greeff Ltd
Lonza (UK) Ltd
Rhône-Poulenc Chemicals
S. Black (Import & Export) Ltd
Zeneca Biocides

Proteins

A & E Connock (perfumery & cosmetics) Ltd
Croda Chemicals Limited
K & K Greeff Ltd
S. Black (Import & Export) Ltd

PVP

ISP Europe

PVP iodine

ISP Europe

Quaternary polymers

A & E Connock (perfumery & cosmetics) Ltd
Allied Colloids
Ellis & Everard
Goldschmidt AG
Henkel Kgaa, Division COSPHA
ISP Europe
Roquette
S. Black (Import & Export) Ltd

Quillaia bark

K & K Greeff Ltd
Alban Muller International

Resinoids

A & E Connock (perfumery & cosmetics) Ltd

Resins

A & E Connock (perfumery & cosmetics) Ltd
Chemie Linz
MHP Shellac Gmbh

Rice oils

A & E Connock (perfumery & cosmetics) Ltd
Alban Muller International
Henry Lamotte - Bremen

Rice wax

A & E Connock (perfumery & cosmetics) Ltd
S. Black (Import & Export) Ltd

Ricinoleates

A & E Connock (perfumery & cosmetics) Ltd
Croda Chemicals Limited

Royal jelly

A & E Connock (perfumery & cosmetics) Ltd
Alban Muller International
Eggar & Co (Chemicals) Limited
S. Black (Import & Export) Ltd

Rubifacients

A & E Connock (perfumery & cosmetics) Ltd

Saccharides

A & E Connock (perfumery & cosmetics) Ltd

Saccharin

Ellis & Everard

Safflower oil

A & E Connock (perfumery & cosmetics) Ltd
Alban Muller International
Croda Chemicals Limited
Ellis & Everard
Henry Lamotte - Bremen
S. Black (Import & Export) Ltd

Salicylic acid

A & E Connock (perfumery & cosmetics) Ltd

Saponin

K & K Greeff Ltd
S. Black (Import & Export) Ltd

Sarcosinates

A & E Connock (perfumery & cosmetics) Ltd
Croda Chemicals Limite

Seaweed extracts

A & E Connock (perfumery & cosmetics) Ltd
Alban Muller International
Eggar & Co (Chemicals) Limited
Ellis & Everard

K & K Greeff Ltd
S. Black (Import & Export) Ltd

Sequestrants

Allied Colloids
Rhône-Poulenc Chemicals

Sesame oil

A & E Connock (perfumery & cosmetics) Ltd
Alban Muller International
Croda Chemicals Limited
Ellis & Everard
Henry Lamotte - Bremen
S. Black (Import & Export) Ltd

Shea butter

A & E Connock (perfumery & cosmetics) Ltd
Alban Muller International
Croda Chemicals Limited
Henkel Kgaa, Division COSPHA
Henry Lamotte - Bremen
S. Black (Import & Export) Ltd

Shellac

Eggar & Co (Chemicals) Limited
MHP Shellac Gmbh

Silicas

K & K Greeff Ltd

Silicones

A & E Connock (perfumery & cosmetics) Ltd
Ellis & Everard
Goldschmidt AG
Rhône-Poulenc Chemicals
S. Black (Import & Export) Ltd

Silk powder

A & E Connock (perfumery & cosmetics) Ltd
Croda Chemicals Limited
S. Black (Import & Export) Ltd

Sodium carbonate

Ellis & Everard

Sodium dehydroacetate

S. Black (Import & Export) Ltd

Sodium fluoride

Ellis & Everard

Sodium perborate

Ellis & Everard

Sodium tripolyphosphate

K & K Greeff Ltd

Solubilisers

A & E Connock (perfumery & cosmetics) Ltd
Angus Chemie Gmbh
Croda Chemicals Limited
Goldschmidt AG
Henkel Kgaa, Division COSPHA
K & K Greeff Ltd
S. Black (Import & Export) Ltd

Solvents

A & E Connock (perfumery & cosmetics) Ltd
BASF Aktiengesellschaft
Chemie Linz
Ellis & Everard
ISP Europe
K & K Greeff Ltd
Rhône-Poulenc Chemicals
Sorbitol
Croda Chemicals Limited
Roquette

Sorbitol esters

Henkel Kgaa, Division COSPHA
K & K Greeff Ltd
Rhône-Poulenc Chemicals
S. Black (Import & Export) Ltd
Spermaceti (synthetic)

A & E Connock (perfumery & cosmetics) Ltd
Croda Chemicals Limited
Eggar & Co (Chemicals) Limited
Henkel Kgaa, Division COSPHA
S. Black (Import & Export) Ltd

Spice oleoresins

A & E Connock (perfumery & cosmetics) Ltd
Eggar & Co (Chemicals) Limited

Spreading agents

A & E Connock (perfumery & cosmetics) Ltd
Croda Chemicals Limited
Goldschmidt AG
Henkel Kgaa, Division COSPHA
ISP Europe
K & K Greeff Ltd
S. Black (Import & Export) Ltd

Squalene (natural & synthetic)

A & E Connock (perfumery & cosmetics) Ltd
Henry Lamotte - Bremen
S. Black (Import & Export) Ltd

Stabilisers

A & E Connock (perfumery & cosmetics) Ltd
K & K Greeff Ltd

Starch

Roquette

Stearic acid

A & E Connock (perfumery & cosmetics) Ltd
Croda Chemicals Limited
Eggar & Co (Chemicals) Limited
K & K Greeff Ltd
S. Black (Import & Export) Ltd

Sucrose esters

A & E Connock (perfumery & cosmetics) Ltd
Croda Chemicals Limited
Rhône-Poulenc Chemicals
S. Black (Import & Export) Ltd

Sulphosuccinates

A & E Connock (perfumery & cosmetics) Ltd
Croda Chemicals Limited
Henkel Kgaa, Division COSPHA
Rhône-Poulenc Chemicals

Sunscreen agents

A & E Connock (perfumery & cosmetics) Ltd
Alban Muller International
Givaudan-Roure SA
ISP Europe
K & K Greeff Ltd
Rhône-Poulenc Chemicals
S. Black (Import & Export) Ltd

Surfactants (speciality)

A & E Connock (perfumery & cosmetics) Ltd
Alban Muller International
Angus Chemie Gmbh
Croda Chemicals Limited
Ellis & Everard
Goldschmidt AG
Henkel Kgaa, Division COSPHA
K & K Greeff Ltd
Lonza (UK) Ltd
Rhône-Poulenc Chemicals
S. Black (Import & Export) Ltd

Suspension aids

Angus Chemie Gmbh
S. Black (Import & Export) Ltd

Synthetic organic complexes

S. Black (Import & Export) Ltd

Synthetic perfume oils

A & E Connock (perfumery & cosmetics) Ltd

Talc (cosmetic)

Ellis & Everard
Luzenac Europe
S. Black (Import & Export) Ltd

Taurates

Croda Chemicals Limited
Rhône-Poulenc Chemicals
S. Black (Import & Export) Ltd

Tertiary amines

A & E Connock (perfumery & cosmetics) Ltd
James Robinson Ltd
Lonza (UK) Ltd

Thickening agents

A & E Connock (perfumery & cosmetics) Ltd
Alban Muller International
Allied Colloids
Ellis & Everard
Goldschmidt AG
Henkel Kgaa, Division COSPHA
ISP Europe
K & K Greeff Ltd
Rhône-Poulenc Chemicals
Roquette
S. Black (Import & Export) Ltd

Thioglycolates

Rhône-Poulenc Chemicals

Titanium dioxide

Ellis & Everard
K & K Greeff Ltd
S. Black (Import & Export) Ltd

Triglycerides

A & E Connock (perfumery & cosmetics) Ltd
Croda Chemicals Limited
Ellis & Everard
Goldschmidt AG
Henkel Kgaa, Division COSPHA
S. Black (Import & Export) Ltd

Urea

Ellis & Everard

Urocanic acid

K & K Greeff Ltd

Vanillin

Eggar & Co (Chemicals) Limited

Vegetable extracts

A & E Connock (perfumery & cosmetics) Ltd
Alban Muller International
Croda Chemicals Limited
Ellis & Everard
K & K Greeff Ltd
Roquette
S. Black (Import & Export) Ltd

Vitamin complexes (oil & water soluble)

A & E Connock (perfumery & cosmetics) Ltd
BASF Aktiengesellschaft
Henkel Kgaa, Division COSPHA
S. Black (Import & Export) Ltd

Water repellants

A & E Connock (perfumery & cosmetics) Ltd
K & K Greeff Ltd
Roquette
S. Black (Import & Export) Ltd

Waxes (natural & synthetic)

A & E Connock (perfumery & cosmetics) Ltd
Croda Chemicals Limited
Eggar & Co (Chemicals) Limited

Goldschmidt AG
Henry Lamotte - Bremen
K & K Greeff Ltd
MHP Shellac Gmbh
S. Black (Import & Export) Ltd

Wetting agents

A & E Connock (perfumery & cosmetics) Ltd
Croda Chemicals Limited
Henkel Kgaa, Division COSPHA
ISP Europe
K & K Greeff Ltd
Rhône-Poulenc Chemicals

Wheatgerm oil/extract

A & E Connock (perfumery & cosmetics) Ltd
Alban Muller International
Croda Chemicals Limited
Ellis & Everard
Henry Lamotte - Bremen
S. Black (Import & Export) Ltd

Witch hazel (Hamamelis)

A & E Connock (perfumery & cosmetics) Ltd
Alban Muller International
Croda Chemicals Limited
K & K Greeff Ltd

Zinc stearate

Ellis & Everard
K & K Greeff Ltd
S. Black (Import & Export) Ltd

Zinc phenolsulphonate

K & K Greeff Ltd
S. Black (Import & Export) Ltd

Zirconium compounds

S. Black (Import & Export) Ltd

Sub-section (b)

ALPHABETICAL INDEX OF ADVERTISERS WITH ADDRESSES, TELEPHONE AND FACSIMILE NUMBERS

A & E CONNOCK (Perfumery & Cosmetics) LTD, Alderholt Mill House, Fordingbridge, Hampshire, SP6 1PU, UK
Telephone: (0)425 653367 Facsimile: (0)425 656041 .... **Facing Page 1**

ALBAN MULLER INTERNATIONAL, 212 Rue De Rosny, Montreuil, F-93100, France
Telephone: (0)1 48 58 30 25 Facsimile: (0)1 48 58 03 71 .... **Facing Page vi**

ALLIED COLLOIDS, P O Box 38, Low Moor, Bradford, West Yorkshire, BD12 0JZ, UK
Telephone: (0)274 671267 Facsimile: (0)274 606499 .... **Facing Page viii**

ANGUS CHEMIE GmbH, 19, Moorgate Street, Rotherham, Yorkshire, S60 2DA, (0)409 377743
Telephone: (0)409 3377743 Facsimile: (0)409 370596 .... **Facing Page 244**

BASF Aktiengesellschaft, Marketing Kosmetik, Riech-und Aromastoffe, Ludwigshafen, D-6700, Germany
Telephone: (0)621 6043751 Facsimile: (0)621 6047806 .... **Facing Page 200**

BELMAY LTD, Turnells Mill Lane, Dennington Estate, Wellingborough, Northants, NN8 2RN, UK
Telephone: (0)933 440 343 Facsimile: (0)933 274414 .... **Page 316**

CHEMAPOL (U.K.) LTD, Cranford, Blackdown, Royal Leamington Spa, Warwickshire CV32 6RG, UK
Telephone: (0)926 450623 Facsimile: (0)926 881844 .... **Facing Page 21**

CHEMIE LINZ, St Peter Strasse 25, P O Box 246, Linz, A-4021, Austria
Telephone: (0)70 5916-0 Facsimile: (0)70 5916 2248 .... **Facing Page 32**

CRODA CHEMICALS LIMITED, Cowick Hall, Snaith, Goole, North Humberside, DN14 9AA, UK
Telephone: (0)405 860551 Facsimile: (0)405 860205 .... **Facing Page vii**

EGGAR & Co (Chemicals) LIMITED, High Street, Theale, Reading, Berkshire, RG7 5AR, UK
Telephone: (0)734 302379 Facsimile: (0)734 323224 .... **Facing Page 11**

ELLIS & EVERARD, Radford House, Radford Way, Billericay, Essex, CM12 0DE, UK
Telephone: (0)277 630063 Facsimile: (0)277 631356 .... **Facing Page 167**

GIVAUDAN-ROURE SA, 5, Chemin de la Parfumerie, Geneva, CH-1214, Switzerland
Telephone: (0)22 7809111 Facsimile: (0)22 7809150 .... **Facing Page 233**

GOLDSCHMIDT AG, Goldschmidtstr. 100, Essen 1, D-4300, Germany
Telephone: (0)201 173 01 Facsimile: (0)201 173 2838 .... **Facing Page 95**

HENKEL KGaA, DIVISION COSPHA, Henkelstr. 67, Düsseldorf, D-40191, Germany
Telephone: (0)211 797 0 Facsimile: (0)211 798 7696 .... **Facing Page 245**

HENRY LAMOTTE - BREMEN, Auf Dem Dreieck 3, P O Box 10 38 49, Bremen 1, D-2800, Germany
Telephone: (0)421 54706-0 Facsimile: (0)421 54706 99 .... **Facing Page 10**

INTERNATIONAL FLOWERS & FRAGRANCES (GB) LTD, Commonwealth House, Hammersmith International Centre, London, W6 8DN, UK
Telephone: (0)81 741 5771 Facsimile: (0)81 741 2566 .... **Facing Page 316**

ISP EUROPE, 40 Alan Turing Road, Surrey Research Park, Guildford, Surrey, GU2 5YF, (0)483 301757
Telephone: (0)483 302175 Facsimile: (0)483 302175 .... **Facing Page 201**

JAMES ROBINSON LTD, P O Box 33, Hillhouse Lane, Huddersfield, West Yorkshire, HD1 6BU, UK
Telephone: (0)484 43557 Facsimile: (0)484 435580 .... **Facing Page 120**

KAHL & CO, Otto-Hahn-Str. 2, D-2077 Trittau, Germany
Telephone: (0)4154 3011 Facsimile: (0)4154 81508 .... **Facing Page 232**

K & K GREEFF LTD, Suffolk House, George Street, Croydon, Surrey, CR9 3QL, UK
Telephone: (0)81 686 0544 Facsimile: (0)81 686 4792 .... **Facing Page 21**

LONZA (UK) LTD, Imperial House, Lypiatt Road, Cheltenham, Glos., GL50 2QJ, UK
Telephone: (0)242 513211 Facsimile: (0)242 222294 .... **Facing Section 9**

LUZENAC EUROPE, 2 Place E. Bouilleres, Toulouse Mirail, F-31100, UK
Telephone: (0)61 406333 Facsimile: (0)61 400623 .... **Facing Page 121**

MHP SHELLAC GMBH, Repsoldstr. 4, Hamburg 1, D-2000, Germany
Telephone: (0)40 2801126 Facsimile: (0)40 2801947 .... **Facing Page 232**

NORDION INTERNATIONAL, 447 March Road, P O Box 13500, Kanata, Ontario, K2K 1XB, Canada
Telephone: (0)613 592 2790 Facsimile: (0)613 592 5206 .... **Facing Page 235**

NOVOSPRAY SA, Place Cornavin 18, Geneva CH-1201 Switzerland
Telephone: (0)22 738 2211 Facsimile: (0)22 738 2138 .... **Facing Page 261**

PENTAPHARM LTD, CH-4002 Basel, Switzerland
Telephone: (0)61 312 9680 .... **Facing Page 260**

RHÔNE-POULENC CHEMICALS, Poleacre Lane, Woodley, Stockport, Cheshire, SK6 1PQ, UK
Telephone: (0)61 430 4391 Facsimile: (0)61 430 4364 .... **Facing Page 165**

ROQUETTE, 4 Rue Paton, Lille Cedex, F-59022, France
Telephone: (0)20 30 77 97 Facsimile: (0)20 30 96 00 .... **Facing Page 234**

S. BLACK (IMPORT & EXPORT) LTD, The Coronnade, High Street, Cheshunt, Herts, EN8 0DJ, UK
Telephone: (0)992 630751 Facsimile: (0)992 622838 .... **Facing Section 1**

ZENECA BIOCIDES, P O Box 42, Hexagon House, Blackley, Manchester, Lancs, M9 3DA, UK
Telephone: (0)61 740 1460 Facsimile: (0)61 721 4173 .... **Page 362**

Editorial Index

Editorial Index

A

"A not A" test, 469
Acceptable quality level (AQL), 408, 409
Acid value, 28
Acne, 200
Acrylic acid based thickeners, 17-18
Acrylonitrile-butadiene-styrene copolymer (ABS), 297, 304
Acyl sarcosinates, 13
Additives, 137, 225
Adsorption
  gas-solid interfaces, 76-8
  liquid-liquid interfaces, 85-7
  liquid-solid interfaces, 79-85
  measurements, 75
  phenomenon, 75
  preferential, 99-103
Advertising and brand awareness, 489
Aerosols, 263-85, 349-50
  actuators, 274-5
  classification, 265-7
  containers, 271-3
  corrosion, 283
  definition, 263
  filling, 281-3, 395
  flammability testing, 284
  history, 263-5
  mousses, 266-7
  packaging, 271-5
  perfumed, 280-1
  product testing, 283-5
  propellants, 267-71
  specific products, 276-81
  spray characteristics, 275-6, 285
  storage testing, 284
  three-phase, 266
  two-phase, 265
  valves, 273-4
  weight losses, 285
African cultures, 497
After-bath preparations, 242
"Airspray" system, 273
Alcohols, 323
Aldehydes, 323
Aliphatic perfume ingredients, 323
Alkanolamides, 13
Alkyl ether sulphate, 221-2
Alkyl sulphates, 221
Aluminium monobloc containers, 272
Aluminium tubes, 306
  quality assurance, 307
Aluminium/zirconium chlorhydrate, 195-6
Ambergris, 337
Aminoanthraquinones, 231
2-Amino-2-ethyl-1,3-propandiol (AMPD), 277
Aminomethyl propandiol (AMPD), 227
Aminomethyl propanol (AMP), 227
2-Amino-2-methyl-1-propanol (AMP), 277
2-Amino-2-methyl-1,3-propandiol (AMPD), 277
Ammonium lauryl sulphate, 241
Ammonium thioglycollate, 230
Amorphous silicon dioxide, 18
Amphomer 28-4910 resin, 278
Animal fats, 23
Animal testing, 49-50
  legislation, 49
Animal waxes, 24
Anionic emulsifiers, 110-11
Anosmia, 320-1
Anti-acne products, 200
Antil 141, 240
Antioxidants, 151
Antiperspirants, 195-7, 279-80, 349-50
  efficacy, 196-7

evaluation, 477-9
Anti-plaque agent, 256
Apocrine sweat glands, 169, 197
AQL, 418-19
Arrector pili muscle, 204
Artificial tanning products, 194
Axial flow, 380

B

Bacteria, 440
"Bag-in-can" containers, 273
Barium sulphate, 132
Batch coding, 427
Batch control, 405
Batch identification, 393-4
Bath cubes, 233-4
Bath gels, 241
Bath oils, 234-8, 350
based on silicones, 237
dispersible or blooming, 236
emulsifying, 236
floating, 234-6
foaming, 237-8
functional, 236
"special effect", 238
Bath products, 233-43, 350
Bath salts, 233-4
Bath tablets, 233-4
Beeswax, 148, 150, 177
Beeswax/borax emulsifier system, 111
Bentonite, 18-19
Benzenoids, 323
Bergamot oil, 331
Binder systems, 137
Bleaching products, 228-9
Blusher pigments, 137
Blushers, 143-4
Body odour, 198
Borax, 111, 233
Bottle design, 291-3
Brand markets, 487-9
Brand mix, elements of, 489
Brand strategy, 489
Braunaeur, Emmett and Toller (BET) isotherms, 78
British Standards, 421
2-Bromo-2-nitropropan-1,3-diol, 151
Brookfield viscometer, 7-8
Brownian motion, 97-8
BS 5750, 402, 420-31
field of application, 422
quality systems requirements, 422-31
scope, 422
BS 6000, 426
BS 6001, 408-10, 412, 420, 426
Bulk density, 146
Butylated hydroxytoluene (BHT), 151

C

Cake mascaras, 156-7
Calcium carbonate, 135, 185
Calcium thioglycollate, 185
Calculus, 246, 249
Calgon, 234
Candelilla wax, 149
Capillary rise method, 82-3
Capillary viscometers, 10
Capping operations, 393
Caprylic/capric triglyceride, 177
Carbomer resins, 17-18
Carnauba wax, 149
Carrageenan, 19
Carton-board, 289-90
Cartons, 394
Castor oil, 148
Castoreum, 337
Cedarwood oil, 333
Cellulase, 254
Cellulose, structure of, 14
Cellulose ethers, 13
applications, 14
Cellulose gum, 15
Cellulosic thickeners, 13-17
microbial attack, 15
Cerisine wax, 150
Cetyl alcohol, 176, 178, 185
Cetyl pyridinium chloride (CPC), 256
Chalone, 170
Charge shielding effect, 101-3
Children's products, 500
China clay, 135
Chi-squared test, 468, 469
Chlorfluorocarbons (CFCs), 263-5, 278
Chlorhexidene, 255
Chromium dioxides, 130-1
Circle of instability, 99
Cistus oil, 333-4
Citronella, 329
Citrus oils, 330-1
folding, 327
Civet, 337
Claim support information, 462
Cleansing products, 179
Climbazole, 181, 182
Clove oil, 332
Coaxial viscometers, 8-9

Cocoa butter, 152
Coconut diethanolamide, 13
Cohesive powders, 371-4
Cold creams, 96, 111
COLIPA, 56, 59
Collagen, 183
Collagen fibres, 170-1
Collagen fibrils, 171
Collapsible tubes, 306-8
Colloid systems, 67
 characteristic property, 67
 intermolecular and interfacial forces, 69-74
 lyophilic, 69-70
 lyophobic, 71, 73
 maintaining stability, 68-9
 preparation, 68
 quantification of repulsive/attractive forces, 72-4
 stabilisation, 67
Colloidal state, 67-9
Colognes, 348
Colony forming units (CFU), 441
Colour
 blindness, 123
 description of, 124
 hair, 210-11
 interference, 123-4
 lipstick products, 148
 measurements, 26
 mixing, 124-5
 tests, 146
 theory, 121-4
 vision, 122
 vision defects, 122-3
Colour Index (CI) system, 125
Colour Index Reference Manual, 125
Colourants
 EEC classification, 126-7
 FDA classification, 126
 for powder cosmetics, 137
 legislation, 125-7
 nomenclature, 126
 physical properties, 127-32
Coloured pearls, 133
Composite containers, 309
Compressed gas propellants, 270-1
Conception
 and perception, 486-7
 principle of, 485-6
Conditioners, 225
Conductivity measurements, 116
Cone and plate viscometers, 9-10
Constructive interference, 123
Consumer acceptance, 46
Consumer exposure, assessment of, 44
Consumer research, 480-92
 data collection, 490
 designing ad-hoc research programs, 490-1
 methodology, 489-92
 psychology, 480-7
 qualitative research, 491-2
 quantitative, 492
 reasons for, 489-90
Contact angle, 82
Contact dermatitis, 40
Contract review, 424
Corrective action, 430
Corrosion, aerosols, 283
Corrugated board cases, 312-13
Cortex, 206
Cosmetic Products (Safety) Regulations, 62
Cosmetics Ingredients Dictionary (CID), 36
Cosmetics, Toiletries and Perfumes Association (CTPA), 39
Couette viscometer, 8-9
Cream mascaras, 157-8
Creams, 349
 cold, 96, 111
Critical micelle concentration (CMC), 43, 91-3
Customer complaints, 46
Cuticle keratin, 205
Cyclomethicone, 178, 278

D

Dandruff, 180-2
 treatments, 181
Date coding, 393-4
DEB-100, 277
Decorative cosmetics, 121-64
Defect types, 408-9
Delivery, precautions against damage or deterioration, 430
Dental caries, 246-8
 prevention of, 254-5
Dental disease, 246-50
Dental hypersensitivity, 250, 256-7
Dental plaque, 246
 in vitro studies, 256
 in vivo studies, 256
 pH value, 247
Dentine, 245
Deodorants, 198-9, 280, 349-50
 efficacy testing, 198-9
 evaluation, 476
Depigmenting agents, 188-9
Depilatory products, 184-5
Dermal papilla, 203
Dermis, structure of, 170

Destructive interference, 123
Detergents in bath foams, 239
Diaphragm pumps, 388-9
Difference testing, 467
Diffey method, 192
Dihydroxy acetone, 194
Dilatent behaviour, 4
Dimethyl ether (DME), 269
Dimethyl polysiloxanes, 237
1,4-Dioxin, 38
Distillation methods, 326-7
Distilled water, 442
Disulphide links, 208-9
DLVO theory, 72
Document control, 424-5
Drop test, 146
Drop volume method, 83-4
Dry offset printing, 299
Dual standard test, 468
Duo/trio test, 469
Duplex film, 90
Dupre equation, 88, 89
D-value determinations, 457
Dyes
  natural, 128
  synthetic, 127
  uptake, 116
Dynamic viscosity, 1-2, 4
Dynamic viscosity:shear rate flow curves, 3
Dysosmia, 321

E

Eau de Cologne, 317-18
Eccrine sweat glands, 168-9, 194-5
EDTA, 225
EEC Cosmetics Directive, 36, 47, 52-60
  amendment procedure, 56-60
  amendments, 58-9
  Cosmetic Directive Annexes, 54-6
  Cosmetic Directive Articles, 52-4
  implementation of, 58-9
  sixth amendment, 59-60
  structure of, 52-6
EEC pre-packaging Directives, 60
EINECS, 60
Electrical charge, application of, 99
Electrical double layer, 73
Electrical impedance, 178
  measurements, 476
Electrical potential plot, 101
Electrolyte thickeners, 11-12
Emollient, 175
Emulsification, 95
Emulsifier systems, 109-12
Emulsions, 67, 90-1, 95-118
  addition of nonionic surfactants, 105
  ageing programmes, 118
  centrifugal force, 118
  charge shielding effect, 101-3
  classical systems, 95-7
  creaming/sedimentation, 117
  formation, 96-7
  freeze-thaw testing, 117-18
  heating and cooling, 385-7
  instability, 97-8
  interfacial film strengthening, 103
  low shear rate evaluation, 118
  miscellaneous properties, 116
  particle charge, 114
  particle size, 113-14
  particle type, 113
  phase separation, 117
  physical properties, 112-14
  polymer stabilisation, 103-5
  powder stabilisation, 103
  production of, 383-7
  rheological properties, 114-15
  sedimentation/creaming forces, 98-9
  stabilising, 90, 99-105
  stability, 116-17
  testing, 117-18
  system selection, 105-7
  temperature stability, 117
  visual assessment, 118
  see also Oil-in-water (O/W) emulsions; Water-in-oil (W/O) emulsions
Enfleurage, 328
Engineering function, 368
Epidermal cells, 203
Epidermis, structure, 169-70
Essential oils, 324
  availability, 334
  chemistry of, 334-5
  components of, 324
  definition, 324
  examples of, 361
  future, 336
  grass family, 329
  health and safety requirements, 335-6
  history of, 325-6
  price of, 334
  production and distribution, 326
  sales and distribution, 328
  specific types, 329-34
Esters, 323
Ethanol B, 277
European Inventory of Existing Chemical Substances (EINECS), 37
European Inventory of Existing Commercial Substances (EINECS), 49

Evaporimetry, 475
Expression, 328
Extrusion blow moulding, 305
Eye mascara products, 155-60
  evaluation of formulations, 159
  formulation and manufacture, 156-9
  packaging, 160
  quality assurance, 160
Eye-shadow pigments, 138
Eye-shadow powder pressing, 395-6
Eyes, irritant reaction, 46

F

Face powder pigments, 137
Factory layout, 366
Falling ball viscometers, 10
Fats, 21-4
  derivatives from, 25
  test methods, 26-32
Fatty acids, 21-3
  distribution, 30-2
Fatty oils, 21-2
Fibre board cases, 312
Fibroblasts, 171, 183
Filling areas, hygiene in, 443-8
Filling lines
  design of, 392
management and automation, 396
Filling techniques, 389-96
Finished product control, 406
  legislation, 407
Flammable liquids, filling of, 391
Flavour
  assessment, 462-4
  toothpastes, 253-4
Flexographic printing, 299
Flexometry, 475
Fluoride tooth protectant, 252
Fluorides, 254-5
Fluorocarbon propellants, 269-70
Foam baths, 238-41, 350
  detergents for, 239
  formulation, 239-41
  gel formulation, 241
  luxurious pearlescent, 241
  powders and tablets, 242
Foaming agents, toothpastes, 253
Foams, 67
Folding techniques, 327
Food and Drug Administration (FDA), 37
Fragrances, 146, 150, 240
  development of, 344
  fine, 348
  in cosmetic and toiletry products, 348
  manufacture of, 345
  psychological benefits, 493-4
  see also Perfumery
Frequency distribution, 413
Freundlich isotherms, 78, 80
Fumed silica, 18, 136
Fungi, 440-1

G

Gas-solid interfaces, adsorption, 76-8
Geranium oil, 329-30
Gibbs' adsorption isotherm, 85-7
Gingivitis, 248
Glass, in packaging, 290-3
Glass containers, 291-3
  closures for, 293
Glazing test, 147
Glycerine, 176
Glyceryl monostearate, 178
Godet, 395-6
Good Manufacturing Practice (GMP), 365-6, 444, 445, 447, 448
Graphical scales, 470
Guar gum, 19-20
Guar hydroxypropyltrimonium chloride, 19-20
Gum regression, 248
Gums, 11-20
  naturally occurring, 18

H

Hair
  body, 218
  bulk fibre properties, 214
  chemistry of, 207-10
  colour, 210-11
  cross-sectional shape, 202
  dimensions, 202
  ease of combing, 214-15
  effects of acid and alkali, 219-20
  effects of anionic and cationic materials, 220
  effects of heat, 219
  effects of reducing agents, 220
  effects of water, 219
  elastic properties, 212-13
  electrostatic behaviour, 211
  evironmental effects, 218-20
  fly-away, 215-16
  frictional behaviour, 211-12
  lustre, 211, 216-17
  manageability, 217-18
  morphology, 204-7
  physical properties, 210-18

physiology, 202-4
structure, 201-7
style retention, 215
tensile properties, 212-13
terminal, 201-2
Hair care, psychology of, 494-5
Hair care products, 225-32
evaluation, 471-4
salon testing, 472-3
Hair colourants, 230-2
Hair conditioners, 226
Hair dyes
permanent, 232
semi-permanent, 231-2
temporary, 231
Hair fibres, 204, 210, 211
Hair follicles, 168
physiology, 202-4
Hair growth, 184-7
Hair growth cycle, 201-2
Hair index, 202
Hair mousse formulations, 228
Hair reducing agents, 230
Hair spray resins, 277
Hair sprays, 226-7, 276-8, 349-50
Hair waving products, 229-30
Halitosis, 249-50
Hand creams, 177
Handling, precautions against damage or deterioration, 430
Heat exchangers, 389
Heat transfer printing, 302
Helmholtz double layer, 73
High shear mixing systems, 387-8
Historical background, 497-9
HLB (hydrophilic-lipophilic balance) value, 90-2, 107-12, 236
experimental determination, 108-9
Homogenisation, 387
Hookes Law, 213
Hot stamping, 302-3
Humectants, 176, 177
toothpastes, 252-3
Hydrated alumina, 131
Hydro-alcoholic emulsions, 242
Hydrocarbon propellants, 268-9, 276
Hydrogen bonding, 208
Hydroquinone, 189
Hydroxyapatite, 244
Hydroxyethylcellulose, 15
Hydroxyl value, 29-30
Hydroxypropyl cellulose, 15-16
Hydroxypropyl methylcellulose, 16
Hyperosmia, 321
Hyposmia, 320

I

Imaging making, 496-7
Immediate Pigment Darkening (IPD), 191
Immiscible liquids, 383-7
Impeller design and placement, 381-3
Infra-red radiation, 189
Infra-red spectroscopy, 178
Injection blow moulding, 305
Injection stretch blow moulding, 305
Inorganic pearls, 132
Insect waxes, 24
Inspection, 412, 427-9
Instrumental evaluation, 462
Intaglio printing, 301-2
Interfacial forces, 74-91
Iodine value, 29
Ionisation, 99
Iron blue, 131
Iron oxides, 130
Ishihara Colour Tests, 123
Isoelectric point, 70
Isopropyl myristate, 176, 178, 235
Isostearyl neopentanoate (ISNP), 137

J

Japan
culture, 497
legislation, 61
Jasmine, 330
Jasmine perfume, 346
Just-in-time (JIT), 405

K

Kaolin, 135
Keratin
amorphous, 208
chemistry of, 207-10
crystalline, 207-8
Keratinisation process, 170, 203
and dry skin, 173-5
Keratohyalin, 170
Ketones, 323
Kinematic viscosity, 2
Kraft point, 93

L

Labdenum oil, 333-4
Labelling, 393

Labiates, 331-2
Lactones, 323
Laminar flow, 376
Laminated tubes, 308
Langmuir film balance, 93
Langmuir isotherms, 77-8, 80
Lanolin, 176
Lanolin oil, 235
Lauric acid dialkanolamide, 240
Lavender oil, 331
Legislation, 52-63
  animal testing, 49
  colourants, 125-7
  control of ingredients, 36
  finished product control, 407
  Japan, 61
  new chemical substances, 49
  raw materials, 39
  reference sources, 62-3
  USA, 60-1
  see also Cosmetics Directive
Lemon oil, 331
Lemongrass, 329
Letterpress printing, 298
Lilac perfume, 347
Linear regression method, 457
Liophilic colloids, 69-70
Lipstick products, 147-55, 350
  colours, 148
  fatty materials, 149
  formulation acceptability, 154-5
  formulation and manufacture, 151-3
  formulations, 153-4
  oils and liquid additivies, 148-9
  pigments, 148
  raw materials, 147-51
Liquid bridges, 372
Liquid filling, 389
Liquid flow, 376
Liquid-liquid interfaces, 85-93
  adsorption, 85-7
Liquid-liquid mixing, 383-7
Liquid mixing equipment, 378-83
Liquid-solid interfaces
  adsorption, 79-85
  adsorption from solution, 80
Loose face powders, 139-40
Lotions/milks, 96, 349
Lovibond Tintometer, 26
Lyophobic colloids, 71

M

Macrofibrils, 206
Magnesium carbonate, 135
Malassezia ovale, 474
Manganese violet, 131
Manufacturing areas, hygiene in, 443-8
Manufacturing devices, 388-9
Manufacturing equipment
  cleansing of, 444
  hygiene, 443
Manufacturing operation, 370-1
Manufacturing process, 444
  see also Production process
Manufacturing techniques, 371-88
Maori culture, 497
Marketing, 367-8
  and brand role, 487-9
  see also under specific aspects
Mascaras. *See* Eye mascara products
Materia alba, 246
MED (Minimum Erythemal Dose), 191
Melanin, 187, 188, 191, 207, 228
Melting point test, 26-7
Men's products, 498-9
Mentha Arvensis, 254
Mentha Piperita, 254
Metal tubes, 306
  manufacture of, 306-8
Metallic soaps, 135
Methylcellulose, 16
Micas, 136, 142
Micelle formation, 91-3
Microbial contamination, 441
Microbial content, 441
Microbial quality, raw materials, 442-3
Microbial quality control, 448-9
Microbiological testing, 427, 454-7
Microbiology, 440-57
Micro-organisms, 440
Mineral oils, 23-4, 235
Mineral raw materials, specifications, 38-9
Mineral waxes, 24-5
Miscible liquids, 383
Moisture content, 27
Moisturising products, 177, 179, 475
  evaluation of, 178
  formulations, 176
Monolayers, 93-4
Muget perfume, 347
Musk, 337

N

Nail lacquers, 160-4
  assessment of formulations, 164
  formulations, 161
  manufacture, 163-4

plasticisers, 162-3
required properties, 160-1
secondary resins, 162
solvent systems, 161-2
Natural moisturising factor (NMF), 174-5
Natural oils, 453
Neroli oil, 331
Newtonian behaviour, 4, 10
Newtonian flow, 377
Nitroaminophenols, 231
Nitrophenylenediamines, 231
Non-conformity
product control, 429
review and disposition, 429
Nonionic emulsifiers, 109-10
Nonionic thickeners, 12-13
Normal distribution, 414
"Not Animal Tested" statement, 49-50
Numerical scales, 469

O

Odour, 319
assessment, 476
characteristics, 354
descriptions, 321, 323
detection, 319
fatigue, 320
sensitivity, 320
Offset lithographic printing, 299
Oflaction, 319-21
Oil-in-water (O/W) emulsions, 90, 95, 105-8, 110, 112, 114, 117, 176, 242, 266, 278
Oils, 21-4
derivatives from, 25
test methods, 26-32
Opacifiers, 224
Oral accumulations, 246
Organic pearls, 132
Organoleptic assessment, 427-8
Orifice viscometers, 10
Orris oil, 332
Ozokerite wax, 150

P

Packaging, 286-314, 394
aerosols, 271-5
choice of case for distribution, 313-14
components, 447-8
control system, 405
definition, 286
design and development, 286-7
eye mascara products, 160
functions, 286
general considerations, 286
precautions against damage or deterioration, 430
pressed face powders, 142
quality assurance of boards and cases, 314
secondary, 311-14
toothpastes, 259-60
use of glass, 290-3
use of paper, 287-90
use of plastics, 293-7
Packaging Directives, 60
Packers Code, 407, 408
Palmarosa oil, 329
Paper in packaging, 287-90
Paraffin wax, 150
Patchouli plant, 332
Pearlescent pigments, 139
PEG 7 glyceryl cocoate, 240
PEG 6000 distearate, 240
Pellicle, 246
Penetration tests, 146
Pentaerythritol tetraisostearate (PTIS), 137
Peppermint oil, 254, 332
Peptide bonds, 209
Perception
and conception, 486-7
principle of, 480-5
Perfumery, 136, 224, 317-61
absolutes, 337
animal derived products, 336-7
concretes, 337
creative, 341-5
definition, 318
early synthetic materials, 318
formulations, 345-7
functions, 318-19
history, 317-18
ingredients, 321-38
derived from plant materials, 324-36
miscellaneous, 336-8, 360
isolates, 337
psychological benefits, 493-4
quality control, 338-40
resinoids, 337
safety of compounds, 340-1
storage of compounds, 340
see also Fragrances
Periodontal disease, 248-9
prevention of, 255
Permanent wave lotions, 350
Permanent waving, 229-30
Peroxide value, 30
Personal hygiene, 447

Petitgrain oil, 331
pH value, 42-3, 185
  dental plaque, 247
  effect on hair, 219-20
  preservatives, 451-2
Phantosmia, 321
Phenol-formaldehyde, 294
Phenyl acetaldehyde, 341
Pigments
  inorganic, 130
  lake, 129
  organic, 128-9
  pearlescent, 132-4
  true, 130
Piroctone olamine, 181, 182
pKa value, 42
Planning process, 368-70
Planographic printing, 299-300
Plant materials, 38
  biological activities, 40
  contamination, 38
  perfumery ingredients derived from, 324-36
Plastic bottles, 303-6
  closures, 306
  decoration, 305
  design, 305
  manufacture of, 304-5
  quality assurance, 306
Plastic containers, 272, 308-9
Plastic tubes, 307-8
  filling and sealing of, 308
  manufacture of, 308
Plastics
  in packaging, 293-7
  in powder products, 136
Polydimethyl siloxanes, 237
Polyethylene, 294-6, 304
Polyethylene terephthalate (PET), 272, 297, 304
Polymers in emulsion stabilisation, 103-5
Polypropylene, 296
Polyquaternium-4, 228, 278
Polyquaternium-11, 228, 278
Polystyrene, 297, 304
Polyvalent soaps, 111-12
Polyvinyl chloride (PVC), 296-7, 304
Polyvinyl pyrrolidone (PVP), 277
Polyvinyl pyrrolidone/vinyl acetate copolymer (PVP/VA), 277
Polyvinylidene chloride, 297
Powdered cosmetic products, 134-47
  formulation, manufacture and filling, 139-46
  quality assurance tests on, 146-7
Preferential adsorption, 99-103
Preservatives, 137, 151, 224, 241, 449-54
  concentration effects, 452
  effect of natural oils, 453
  effect of product container, 454
  effect of solid particles, 453
  effect of temperature, 453
  factors affecting efficacy, 451-4
  microbiological challenge testing, 454-7
  micro-organisms present, 452-3
  partition coefficient, 452
  pH value, 451-2
  rapid screening methods, 457
  requirements of, 449-51
  toothpastes, 254
Pressed face powders, 141-3
  manufacture, 142-3
  packaging for, 142
Pressed powder blushers, 143-4
Pressed powder eye-shadows, 144-6
  manufacturing process, 145
Pressed powder rouges, 144
Prickle cells, 170
Printing processes, 297-303
Process capability, 412
Process control, 412, 414-15, 427
Process control charts, 416-20
  attributes, 418-20
Process specifications, 403
Product applicators, 447-8
Product assessment, action standard, 462
Product containers, 454
Product development
  brief for, 37
  new products, 461
Product dossier, 47
Product evaluation, 461-79
  objective, 462
  panelists for, 470-1
  purpose of, 461
  quality control, 462
  subjective, 462
  techniques, 471-9
  test methods, 465-70
  types of, 462-5
Product formulation
  assessing toxic properties, 41-3
  toxicological profile, 43
Product safety
  effects of interaction, 42
  evaluation programme, 35-6
  human volunteer studies, 45-6
  in vitro tests, 45
Product specifications, 403
Product stability, 435-9
  assessment, 439
  characteristics to be considered, 435-6
  test methods, 439

testing, 436-9
Production process, 365
and interdepartmental relationships, 366-9
see also Manufacturing process
Pseudoplastic behaviour, 4-5
with yield point, 5
Psychology, 493-501
benefits of cosmetics and fragrance, 493-4
consumer research, 480-7
Pumps, 388-9
Purchase orders, 426
Purchasing, 368
quality requirements, 425

Q

Quality assurance, 368, 401, 413
basic elements, 402-8
eye mascara products, 160
tests on powder cosmetics, 146-7
Quality audits, 430-1
Quality concept, 398-401
Quality control, 368, 400-1
microbial, 442, 448-9
perfumery, 338-40
product evaluation, 462
see also Statistical quality control
Quality function, 397-434
Quality manual, 423-4
Quality of design, 398-9
Quality of performance, 399-400
Quality policy statement, 422, 432
Quality procedures manual, 423-4
Quality records, 430
Quality systems, 397-434
documentation, 433
Quality training, 431
Quenching, 341

R

Radial flow, 378-9
Radio-tracer methods, 178, 258
Rating scales, 469
Raw materials
control system, 403-4
identification and traceability, 426-7
impurities or by-products, 38
legislation, 39
microbial quality, 442-3
microbiological status, 39
mineral-derived, 39, 40
natural, 38
"Not Animal Tested", 49-50
plant-derived, 39, 40
resourcing, 461
selection criteria, 37-9
specifications, 39, 403
storage conditions, 443
synthetic, 39
toxic properties of, 41-3
toxicological data, 40-1
Refractive index, 26
Relief printing processes, 298
Research and development, 368
Resinoids, 337
Reynolds number, 377
Rheological characteristics, measurement of, 6-10
Rheological flow curves, 2-5
Rheology, 1-10
Ribbon blender, 372
Risk assessment, 43
Rose oil, 330
Rose perfume, 345-6

S

Safety issues. *See* Legislation; Product safety
Sales function, 369
Salt linkages, 209
Sample size code letters, 410
Sampling plans, 408, 410-39
Sampling routine, 410-12
Sandalwood oil, 333
Sanitisation techniques, 445
Saponification equivalent, 28-9
Saponification value, 28-9
Saponified colour, 26
Scale of standards, 470
Scientific Committee on Cosmetology (SCC), 58
S-curve, 480
Searle viscometer, 8-9
Sebaceous glands, 168, 199-200, 204
Sebum, 168
Selenium disulphide, 181
Sensory assessment, 476, 477
Sensory evaluation
objective, 462, 463-5
subjective, 462, 464
Sequestering agents, 225, 233-4
Sesquiterpenes, 338
Setting lotions, 227-8
Shampoo products, 44, 220-5, 349
anti-dandruff, 181-2
anti-dandruff testing, 473-4
formulations, 220-1

Shaving products, 186-7
Shear rate, determination of, 388
Shear stress:shear rate curves, 3, 5
Shower gels, 243
Shrink-wrapping, 311-12, 394
Silicone glycols, 237
Silver pearls, 133
Skin
  abrasion, 171
  absorption, 43, 44
  ageing, 182-4, 189
  anti-ageing products, 183-4
  chemical damage, 172
  damage, 171-2
  dry, 172-8
  flaky, 179-82
  functions of, 167
  physiology, 168
  pigmentation, 187-8
  roughness, 175
  sensitisation, 172
  structure, 168
  sunlight effects, 189-90
Skin bronzers, 194
Skin care, psychology of, 495-6
Skin care products, 175, 176
  evaluation, 474-5
Skin care regimen, 178-9
Skin contact, 340-1
Skin cream, 44
Skin defects, cosmetic camouflaging of, 496
Skin irritancy, 44, 171-2
  surfactants, 43
  testing, 46
Skin lightening products, 188-9
Slip point test, 26
Smelling, mechanism of, 319-21
Soap
  carton wrapping, 310-11
  effects of pH, 43
  packaging, 309-11
  quality control of wrapping materials, 311
  shrink-wrapping, 311
  three-component wrapping, 310
  toilet, 348-9
Social background, 497-9
Sodium alkyl benzene sulphonate, 242
Sodium carboxymethylcellulose, 17
Sodium hexametaphosphate, 234
Sodium laureth sulphate, 240, 243
Sodium lauryl ether sulphate, 182
Sodium lauryl sulphate (SLS), 221
Sodium sesquicarbonate, 233
Sodium tripolyphosphate, 233
Solid board cases, 312
Solid/liquid mixing, 374-6
Sols, 67
Solvent-extraction, 327-8
Sorbitol, 176
Spearmint oil, 332
Specific gravity, 26
Specifications, 402-3
Spreading, 87-90
Spreading coefficients, 89
Squalane, 177
Stability testing. *See* Product stability
Standard deviation, 414
Staphylococcus aureus, 476
Starches, 136
Statistical quality control, 408-39
  techniques, 431
Stencil printing, 302
Stephan curve, 247
Stokes' law, 69, 98-9
Storage conditions, 437-8
  precautions against damage or deterioration, 430
  see also Product stability
Stratum corneum, 203
Strontium sulphide, 185
Styling mousses, 228, 278-9
Styrene-acrylonitrile copolymer (SAN), 142, 297, 304
Sub-contractors, assessment of, 425
Sugar-based nonionic emulsifiers, 109-10
Sulphosuccinates, 239
Sun protection factors (SPF), 190-2
Sun tanning, 192-3
Sunflower oil, 177
Sunlight effects on the skin, 189-90
Sunscreen products, 151, 190-3
Supplier certification, 405
Surface chemistry, 67-94
Surface pressure, 93-4
Surface profile method, 257-8
Surface tension, 74-5, 93
  measurement of, 82-5
Surfactants, 90, 91, 93, 94, 97, 239, 450
  amphoteric, 223, 243
  anionic, 222
  interactions, 43
  nonionic, 105, 222-3
  primary, 221
  secondary, 222
  skin irritancy, 43
Sweat glands, 168-9, 194-9
  apocrine, 169, 197
  eccrine, 168-9, 194-5
Sweating reduction (SR), 478-9

T

Talc, 134, 144, 242, 350
  packaging, 308-9
Talc containers, quality assurance, 309
Tally chart, 413
Tangential flow, 378
Tattooing, 497
Teeth, 244-50
  structure, 244-6
Terpenes, 335, 338
Terpenoids, 321-5, 355
Test methods, 427-9
  product evaluation, 465-70
Thermoplastics, 294
Thermoset resins, 293-4
Thickeners, 11-20, 224, 240
  functions, 11-18
  naturally occurring, 18
  selection of, 11
  synthetic, 11-18
  toothpastes, 253
Thioglycollic acid, 230
Thixotropy, 5-6
Tilting plate method, 82
Tin-plate containers, 271-2
Tissue products, 350-1
Titanium dioxide, 131
Titre determination, 27
Toilet soaps, 348-9
Tolerable negative error (TNE), 407
Toners, 129-30, 179
Toothpastes, 250-60
  abrasives, 251-2
  abrasivity measurement, 257-8
  efficacy measurement, 258-9
  flavour, 253-4
  foaming agents, 253
  formulations, 250-1
  functions of, 250
  humectants, 252-3
  packaging, 259-60
  preservatives, 254
  therapeutic agents, 254
  thickeners, 253
Total quality management (TQM), 401
Total-reflectance infra-red spectroscopy, 476
Total viable colony count (TVC), 441
Toxicological tests, in vitro, 45
Trade associations, 61-2
Trading Standards Office, 407
Transepidermal water loss (TEWL), 174, 475
Triangle test, 467-8
Triethanolamine, 42, 227
Triethanolamine lauryl sulphate, 241
Triglycerides, 21-3
Triisopropanolamine (TIPA), 277
Tri-lobe pumps, 388
Trisodium phosphate, 233
Tube filling, 394-5
Turbine mixer, 386
Turbulent flow, 377-8

U

Ultramarines, 131
Ultraviolet radiation, 189-93
Unsaponifiable matter, 30
Urea-formaldehyde, 294
USA, legislation, 60-1

V

Vacuum filling, 390-1
Vacuum vessels, 388
Van der Waals forces, 70, 74, 78, 98, 210, 373-4
Vanishing creams, 96
Vapour-phase tap (VPT), 274
Veegum, 20
Vegetable oils, 22-3, 148
Vegetable waxes, 24
Verbal scales, 469
Verification of purchased products, 426
Vessel shape and design, 380-1
Vetivert oil, 329
Viscometers, 7-10
Viscosity, 1-2
  measurement of, 6-7
Viscosity:shear rate curves, 5
Viscosity:time curve, 6
Vitamin A, 183, 189
Vitamin D, 189
Volatile organic compound (VOCs), 265, 278
Volumetric filling, 389-90

W

Water
  dispersibility, 116
  microbial quality, 442
  microbiological quality, 443
Water-in-oil (W/O) emulsions, 90, 95, 105-8, 112, 114, 117, 177
Waterproof mascaras, 158-9
Waxes, 21, 24-5, 149
  derivatives from, 25
  microcrystalline, 152

synthetic, 150
test methods, 26-32
Weights and Measures Act 1985, 407
Wetting, 80-2
White pigments, 131
Wilcoxon Signed Rank test, 479
Wilhemy plate method, 84-5

X

Xanthan gum, 20

Y

Yeasts, 441
Yield point, pseudoplastic behaviour with, 5
Ylang oil, 333
Young's equation, 82

Z

Zinc oxide, 131, 135
Zinc pyrithione (ZPT), 181, 182
Zinc stearate/pentaerythritol tetraisostearate, 144
Zirconium chlorhydrate, 195-6
Zwitterions, 70